Dynamics of Third-Order Rational Difference Equations with Open Problems and Conjectures

Advances in
Discrete Mathematics and Applications

Series Editors
Saber Elaydi and Gerry Ladas

Volume 1
Analysis and Modelling of Discrete Dynamical Systems
Edited by Daniel Benest and Claude Froeschlé

Volume 2
Stability and Stable Oscillations in Discrete Time Systems
Aristide Halanay and Vladimir Răsvan

Volume 3
Partial Difference Equations
Sui Sun Cheng

Volume 4
Periodicities in Nonlinear Difference Equations
E. A. Grove and G. Ladas

Volume 5
Dynamics of Third-Order Rational Difference Equations with Open Problems and Conjectures
Elias Camouzis and Gerasimos Ladas

Advances in

Discrete Mathematics and Applications

Volume 5

Dynamics of Third-Order Rational Difference Equations with Open Problems and Conjectures

Elias Camouzis

Gerasimos Ladas

CRC Press
Taylor & Francis Group
Boca Raton London New York

CRC Press is an imprint of the
Taylor & Francis Group, an informa business
A CHAPMAN & HALL BOOK

CRC Press
Taylor & Francis Group
6000 Broken Sound Parkway NW, Suite 300
Boca Raton, FL 33487-2742

First issued in paperback 2019

ISBN-13: 978-1-58488-765-2 (hbk)
ISBN-13: 978-0-367-38829-4 (pbk)

Library of Congress Cataloging-in-Publication Data

Camouzis, Elias.
Dynamics of third-order rational difference equations with open problems and conjectures / Elias Camouzis and G. Ladas.
p. cm. -- (Discrete mathematics and applications ; 48)
Includes bibliographical references and index.
ISBN 978-1-58488-765-2 (alk. paper)
1. Difference equations--Numerical solutions. I. Ladas, G. E. II. Title. III. Series.

QA431.C26 2007
518'.6--dc22 2007034508

Visit the Taylor & Francis Web site at
http://www.taylorandfrancis.com

and the CRC Press Web site at
http://www.crcpress.com

E. CAMOUZIS and G. LADAS

DYNAMICS OF THIRD-ORDER RATIONAL DIFFERENCE EQUATIONS with Open Problems and Conjectures

CRC PRESS
Boca Raton Ann Arbor London Tokyo

To

Lina and Mary

Contents

Preface

This book is about the global character of solutions of the third-order rational difference equation

$$x_{n+1} = \frac{\alpha + \beta x_n + \gamma x_{n-1} + \delta x_{n-2}}{A + Bx_n + Cx_{n-1} + Dx_{n-2}}, \quad n = 0, 1, ... \tag{1}$$

with nonnegative parameters $\alpha, \beta, \gamma, \delta, A, B, C, D$ and with arbitrary nonnegative initial conditions x_{-2}, x_{-1}, x_0 such that the denominator is always positive.

We are primarily concerned with the boundedness nature of solutions, the stability of the equilibrium points, the periodic character of the equation, and with convergence to periodic solutions including periodic trichotomies. However, our ultimate goal should be to extend and generalize the results of rational equations to equations

$$x_{n+1} = f(x_n, \ldots, x_{n-k}), \quad n = 0, 1, \ldots$$

of the most general pattern.

For Eq.(1) and for each of its 225 special cases, we present the known results and/or derive some new ones. We also pose a large number of thought-provoking open problems and conjectures on the boundedness character, the global stability, and the periodic behavior of solutions of various special cases of Eq.(1). The open problems are quite challenging and the conjectures are based on numerous computer observations and analytic investigations. We believe that research work on these open problems and conjectures is of paramount importance for the development of the basic theory of the global behavior of solutions of nonlinear difference equations of order greater than one.

The large number of open problems and conjectures in rational difference equations will be a great source of attraction for research investigators in this dynamic area where, at the beginning of the third millennium, we know so surprisingly little.

The methods and techniques that we develop to understand the dynamics of rational difference equations and the theory we obtain will be useful in analyzing the equations in any mathematical model that involves difference equations.

Chapter 1 contains some basic definitions and some general results needed throughout the monograph.

Chapter 2 deals with the special cases of Eq.(1) that have bounded solutions only and Chapter 3 deals with the remaining cases, where the equations have unbounded solutions in some range of their parameters.

Chapter 4 is about the seven nonlinear known periodic trichotomies of third-order rational difference equations.

Chapter 5 presents the known results on each of the 225 special cases of Eq.(1). This chapter is the reason we wrote this book. The four preceding chapters present general results needed in order to discuss the character of each equation and how it relates to the other special cases.

Appendix A at the end of the book presents at a glance the boundedness character of each of the 225 special cases of Eq.(1) and gives important results and references related to each special case.

Appendix B contains information on the boundedness character for all fourth-order rational difference equations. The large number of conjectures listed in Appendix B on the boundedness character of fourth-order rational difference equations will help give new directions for future investigations in this fascinating area.

Acknowledgments

This book is the outgrowth of lectures and seminars given at the University of Rhode Island during the last 10 years and is based on an extensive collaboration of the authors by e-mail during the last five academic years and in person during the last five beautiful summers in Greece. We are grateful to Professors E. A. Grove, J. Hoag, W. Kosmala, M. R. S. Kulenović, O. Merino, S. Schultz, and W. S. Sizer, who have participated in class discussions or seminar presentations at the University of Rhode Island and to former and present URI students A. M. Amleh, S. Basu, M. R. Bellavia, A. M. Brett, W. Briden, E. Chatterjee, C. A. Clark, R. C. DeVault, H. A. El-Metwalli, J. Feuer, C. Gibbons, E. Janowski, C. M. Kent, Y. Kostrov, Z. A. Kudlak, S. Kuruklis, L. C. McGrath, C. Overdeep, F. Palladino, M. Predescu, N. Prokup, E. P. Quinn, M. Radin, I. W. Rodrigues, C. T. Teixeira, S. Valicenti, P. Vlahos, and G. Tzanetopoulos for their contributions to research in this area and for their enthusiastic participation in the development of this subject. A particular debt of gratitude is due to Dr. Amal Amleh, who read with great care the entire manuscript and its many revisions and whose numerous corrections and criticisms have improved this book.

Introduction

In this book we are interested in the global character of solutions of the third-order rational difference equation

$$x_{n+1} = \frac{\alpha + \beta x_n + \gamma x_{n-1} + \delta x_{n-2}}{A + Bx_n + Cx_{n-1} + Dx_{n-2}}, \quad n = 0, 1, \ldots, \tag{0.0.1}$$

where the parameters $\alpha, \beta, \gamma, \delta, A, B, C, D$ are nonnegative real numbers and the initial conditions x_{-2}, x_{-1}, x_0 are arbitrary nonnegative real numbers such that the denominator is always positive.

We are primarily concerned with the boundedness nature of solutions, the stability of the equilibrium points, the periodic character of the equation, and with convergence to periodic solutions including periodic trichotomies.

If we allow one or more of the parameters in Eq.(0.0.1) to be zero, then we can see that Eq.(0.0.1) contains

$$(2^4 - 1)(2^4 - 1) = 225$$

special cases, each with positive parameters and positive or nonnegative initial conditions.

For Eq.(0.0.1) and for each of its 225 special cases, we present the known results and/or derive some new ones. For most of the equations we also pose some thought-provoking open problems and conjectures on the boundedness character, the global stability, and the periodic behavior of their solutions. The open problems we pose are quite challenging and the conjectures are thought provoking and based on numerous computer observations and analytic investigations. We believe that research work on these open problems and conjectures is of paramount importance for the development of the basic theory of the global behavior of solutions of nonlinear difference equations. The large number of interesting open problems and conjectures in rational difference equations will be a great source of attraction for future investigators in this dynamic area of research.

Out of the 225 special cases of Eq.(0.0.1), 39 cases are about equations that are linear or reducible to linear or Riccati difference equations, or equations reducible to Riccati. See Appendix A.

Another 28 equations were investigated in the Kulenovic/Ladas book [175], which deals with the second-order rational difference equation

$$x_{n+1} = \frac{\alpha + \beta x_n + \gamma x_{n-1}}{A + Bx_n + Cx_{n-1}}, \quad n = 0, 1, \ldots . \tag{0.0.2}$$

There remain 158 equations, each of which is a nonlinear third-order difference equation crying to be thoroughly investigated.

It is an amazing fact that Eq.(0.0.1) contains a large number of special cases whose dynamics have not been investigated yet.

According to David Hilbert "The art of doing mathematics consists in finding that special case which contains all the germs of generality" and according to Paul Halmos "The source of all good mathematics is the special case, the concrete example."

We strongly believe that the special cases of Eq.(0.0.1) contain a lot of the germs of generality of the theory of difference equations of order greater than one about which, at the beginning of the third millennium, we know so surprisingly little. We also believe that the mathematics behind the special cases of Eq.(0.0.1) is beautiful, surprising, and interesting.

The methods and techniques we develop to understand the dynamics of various special cases of rational difference equations and the theory that we obtain will also be useful in analyzing the equation in any mathematical model that involves difference equations.

1

Preliminaries

1.0 Introduction

In this chapter we state some definitions and some known results that will be useful in the subsequent chapters. For details, see [12], [13], [84], [95], [96], [130], [131], [147], [202], [211], and [213].

The results from Theorem 1.6.7 to the end of this chapter were recently obtained by the authors while working on various special cases of rational difference equations and provide useful generalizations and some unifications in some special cases.

1.1 Definitions of Stability

A *difference equation of order* $(k+1)$ is an equation of the form

$$x_{n+1} = F(x_n, x_{n-1}, \ldots, x_{n-k}), \quad n = 0, 1, \ldots \tag{1.1.1}$$

where F is a function that maps some set I^{k+1} into I. The set I is usually an interval of real numbers, or a union of intervals, or a discrete set such as the set of *integers* $\mathbf{Z} = \{\ldots, -1, 0, 1, \ldots\}$.

A *solution* of Eq.(1.1.1) is a sequence $\{x_n\}_{n=-k}^{\infty}$ that satisfies Eq.(1.1.1) for all $n \geq 0$.

A solution of Eq.(1.1.1) that is constant for all $n \geq -k$ is called an *equilibrium solution* of Eq.(1.1.1). If

$$x_n = \bar{x}, \quad \text{for all } n \geq -k$$

is an equilibrium solution of Eq.(1.1.1), then $\bar{x}$ is called an *equilibrium point*, or simply an *equilibrium* of Eq.(1.1.1).

DEFINITION 1.1 **(Stability)**

(i) *An equilibrium point* $\bar{x}$ *of Eq.(1.1.1) is called* locally stable *if, for every* $\varepsilon > 0$, *there exists* $\delta > 0$ *such that if* $\{x_n\}_{n=-k}^{\infty}$ *is a solution of Eq.(1.1.1) with*

$$|x_{-k} - \bar{x}| + |x_{1-k} - \bar{x}| + \cdots + |x_0 - \bar{x}| < \delta,$$

then

$$|x_n - \bar{x}| < \varepsilon, \qquad \text{for all} \qquad n \geq 0.$$

(ii) *An equilibrium point* $\bar{x}$ *of Eq.(1.1.1) is called* locally asymptotically stable *if,* $\bar{x}$ *is locally stable, and if in addition there exists* $\gamma > 0$ *such that if* $\{x_n\}_{n=-k}^{\infty}$ *is a solution of Eq.(1.1.1) with*

$$|x_{-k} - \bar{x}| + |x_{-k+1} - \bar{x}| + \cdots + |x_0 - \bar{x}| < \gamma,$$

then

$$\lim_{n\to\infty} x_n = \bar{x}.$$

(iii) *An equilibrium point* $\bar{x}$ *of Eq.(1.1.1) is called a* global attractor *if, for every solution* $\{x_n\}_{n=-k}^{\infty}$ *of Eq.(1.1.1), we have*

$$\lim_{n\to\infty} x_n = \bar{x}.$$

(iv) *An equilibrium point* $\bar{x}$ *of Eq.(1.1.1) is called* globally asymptotically stable *if* $\bar{x}$ *is locally stable, and* $\bar{x}$ *is also a global attractor of Eq.(1.1.1).*

(v) *An equilibrium point* $\bar{x}$ *of Eq.(1.1.1) is called* unstable *if* $\bar{x}$ *is not locally stable.*

1.2 Linearized Stability Analysis

Suppose that the function F is continuously differentiable in some open neighborhood of an equilibrium point $\bar{x}$. Let

$$q_i = \frac{\partial F}{\partial u_i}(\bar{x}, \bar{x}, \ldots, \bar{x}), \text{ for } i = 0, 1, \ldots, k$$

denote the partial derivative of $F(u_0, u_1, \ldots, u_k)$ with respect to u_i evaluated at the equilibrium point $\bar{x}$ of Eq.(1.1.1). Then the equation

$$y_{n+1} = q_0 y_n + q_1 y_{n-1} + \cdots + q_k y_{n-k}, \;\; n = 0, 1, \ldots \tag{1.2.1}$$

is called the *linearized equation of Eq.(1.1.1) about the equilibrium point* $\bar{x}$, and the equation

$$\lambda^{k+1} - q_0\lambda^k - \cdots - q_{k-1}\lambda - q_k = 0 \tag{1.2.2}$$

is called the *characteristic equation of Eq.(1.2.1) about* $\bar{x}$.

The following result, known as the *Linearized Stability Theorem*, is very useful in determining the local stability character of the equilibrium point $\bar{x}$ of Eq.(1.1.1). See [13], [95], [131], and [202].

Theorem 1.2.1 *(The Linearized Stability Theorem)*
Assume that the function F is a continuously differentiable function defined on some open neighborhood of an equilibrium point $\bar{x}$. *Then the following statements are true:*

1. *When all the roots of Eq.(1.2.2) have absolute value less than one, then the equilibrium point* $\bar{x}$ *of Eq.(1.1.1) is locally asymptotically stable.*

2. *If at least one root of Eq.(1.2.2) has absolute value greater than one, then the equilibrium point* $\bar{x}$ *of Eq.(1.1.1) is unstable.*

The equilibrium point $\bar{x}$ of Eq.(1.1.1) is called *hyperbolic* if no root of Eq.(1.2.2) has absolute value equal to one. If there exists a root of Eq.(1.2.2) with absolute value equal to one, then the equilibrium $\bar{x}$ is called *nonhyperbolic*.

An equilibrium point $\bar{x}$ of Eq.(1.1.1) is called a *saddle point* if it is hyperbolic and if there exists a root of Eq.(1.2.2) with absolute value less than one and another root of Eq.(1.2.2) with absolute value greater than one.

An equilibrium point $\bar{x}$ of Eq.(1.1.1) is called a *repeller* if all roots of Eq.(1.2.2) have absolute value greater than one.

A solution $\{x_n\}_{n=-k}^{\infty}$ of Eq.(1.1.1) is called *periodic with period p* if there exists an integer $p \geq 1$ such that

$$x_{n+p} = x_n, \qquad \text{for all} \qquad n \geq -k. \tag{1.2.3}$$

A solution is called periodic with *prime period p* if p is the smallest positive integer for which Eq.(1.2.3) holds.

The following three theorems state necessary and sufficient conditions for all the roots of a real polynomial of degree two, three, or four, respectively, to have modulus less than one. For every equation of order two, three, or four that we investigate in this book we have to use one of these three theorems to determine the local asymptotic stability of the equilibrium points of the equation.

Theorem 1.2.2 *Assume that a_1 and a_0 are real numbers. Then a necessary and sufficient condition for all roots of the equation*

$$\lambda^2 + a_1\lambda + a_0 = 0$$

to lie inside the unit disk is

$$|a_1| < 1 + a_0 < 2.$$

Theorem 1.2.3 *Assume that a_2, a_1, and a_0 are real numbers. Then a necessary and sufficient condition for all roots of the equation*

$$\lambda^3 + a_2\lambda^2 + a_1\lambda + a_0 = 0$$

to lie inside the unit disk is

$$|a_2 + a_0| < 1 + a_1, \qquad |a_2 - 3a_0| < 3 - a_1, \qquad \text{and} \qquad a_0^2 + a_1 - a_0 a_2 < 1.$$

Theorem 1.2.4 *Assume that a_3, a_2, a_1, and a_0 are real numbers. Then a necessary and sufficient condition for all roots of the equation*

$$\lambda^4 + a_3\lambda^3 + a_2\lambda^2 + a_1\lambda + a_0 = 0$$

to lie inside the unit disk is

$$|a_1 + a_3| < 1 + a_0 + a_2, \qquad |a_1 - a_3| < 2(1 - a_0), \qquad a_2 - 3a_0 < 3,$$

and

$$a_0 + a_2 + a_0^2 + a_1^2 + a_0^2 a_2 + a_0 a_3^2 < 1 + 2a_0 a_2 + a_1 a_3 + a_0 a_1 a_3 + a_0^3.$$

The following result is a sufficient condition for all roots of an equation of any order to lie inside the unit disk. See [74] or [157, p. 12].

Theorem 1.2.5 *Assume that $q_0, q_1, \ldots, q_k$ are real numbers such that*

$$|\, q_0 \,| + |\, q_1 \,| + \cdots + |\, q_k \,| < 1.$$

Then all roots of Eq.(1.2.2) lie inside the unit disk.

1.3 Semicycle Analysis

Let $\bar{x}$ be an equilibrium point of Eq.(1.1.1), and assume that $\{x_n\}_{n=-k}^{\infty}$ is a solution of the equation.

A *positive semicycle* of $\{x_n\}_{n=-k}^{\infty}$ is a "string" of terms $\{x_l, x_{l+1}, \ldots, x_m\}$, all greater than or equal to $\bar{x}$, with $l \geq -k$ and $m \leq \infty$ such that

$$\text{either } l = -k \text{ or } l > -k \text{ and } x_{l-1} < \bar{x}$$

and

$$\text{either } m = \infty \text{ or } m < \infty \text{ and } x_{m+1} < \bar{x}.$$

A *negative semicycle* of $\{x_n\}_{n=-k}^{\infty}$ is a "string" of terms $\{x_l, x_{l+1}, \ldots, x_m\}$, all less than $\bar{x}$, with $l \geq -k$ and $m \leq \infty$ such that

$$\text{either } l = -k \text{ or } l > -k \text{ and } x_{l-1} \geq \bar{x}$$

and

$$\text{either } m = \infty \text{ or } m < \infty \text{ and } x_{m+1} \geq \bar{x}.$$

A solution $\{x_n\}_{n=-k}^{\infty}$ of Eq.(1.1.1) is called *nonoscillatory* about $\bar{x}$, or simply *nonoscillatory*, if there exists $N \geq -k$ such that either

$$x_n \geq \bar{x}, \text{ for all } n \geq N$$

or

$$x_n < \bar{x}, \text{ for all } n \geq N.$$

Otherwise, the solution $\{x_n\}_{n=-k}^{\infty}$ is called *oscillatory* about $\bar{x}$, or simply *oscillatory*.

1.4 A Comparison Result

The following comparison result is a very useful tool in establishing bounds for solutions of nonlinear equations in terms of the solutions of equations with known behavior, for example, linear or Riccati.

Theorem 1.4.1 *Let I be an interval of real numbers, let k be a positive integer, and let*

$$F : I^{k+1} \to I$$

be a function increasing in all of its arguments. Assume that $\{x_n\}_{n=-k}^{\infty}$, $\{y_n\}_{n=-k}^{\infty}$, and $\{z_n\}_{n=-k}^{\infty}$ are sequences of real numbers such that

$$\begin{cases} x_{n+1} \leq F(x_n, \ldots, x_{n-k}), & n = 0, 1, \ldots \\ y_{n+1} = F(y_n, \ldots, y_{n-k}), & n = 0, 1, \ldots \\ z_{n+1} \geq F(z_n, \ldots, z_{n-k}), & n = 0, 1, \ldots \end{cases}$$

and

$$x_n \le y_n \le z_n, \ \ for\ \ all \ \ -k \le n \le 0.$$

Then

$$x_n \le y_n \le z_n, \ \ for\ \ all \ \ n > 0\,. \tag{1.4.1}$$

PROOF Clearly,

$$x_1 \le F(x_0, \ldots, x_{-k}) \le F(y_0, \ldots, y_{-k}) = y_1$$

and

$$y_1 = F(y_0, \ldots, y_{-k}) \le F(z_0, \ldots, z_{-k}) \le z_1\,.$$

Hence,

$$x_1 \le y_1 \le z_1$$

and (1.4.1) follows by induction. ∎

1.5 Full Limiting Sequences

The following result about full limiting sequences sometimes is useful in establishing that all solutions of a given difference equation converge to the equilibrium of the equation. See [101], [144], [145], and [208].

Theorem 1.5.1 *Consider the difference equation*

$$x_{n+1} = F(x_n, x_{n-1}, \ldots, x_{n-k}) \tag{1.5.1}$$

where $F \in C(J^{k+1}, J)$ for some interval J of real numbers and some non-negative integer k. Let $\{x_n\}_{n=-k}^{\infty}$ be a solution of Eq.(1.5.1). Set $I = \liminf_{n\to\infty} x_n$ and $S = \limsup_{n\to\infty} x_n$, and suppose that $I, S \in J$. Let $\mathcal{L}_0$ be a limit point of the solution $\{x_n\}_{n=-k}^{\infty}$. Then the following statements are true:

1. *There exists a solution $\{L_n\}_{n=-\infty}^{\infty}$ of Eq.(1.5.1), called a full limiting sequence of $\{x_n\}_{n=-k}^{\infty}$, such that $L_0 = \mathcal{L}_0$, and such that for every $N \in \{\ldots, -1, 0, 1, \ldots\}$, L_N is a limit point of $\{x_n\}_{n=-k}^{\infty}$. In particular,*

$$I \le L_N \le S, \qquad for\ all \qquad N \in \{\ldots, -1, 0, 1, \ldots\}.$$

2. *For every $i_0 \in \{\ldots, -1, 0, 1, \ldots\}$, there exists a subsequence $\{x_{r_i}\}_{i=0}^{\infty}$ of $\{x_n\}_{n=-k}^{\infty}$ such that*

$$L_N = \lim_{i\to\infty} x_{r_i+N}, \ \ for\ every\ \ N \ge i_0.$$

1.6 Convergence Theorems

The following convergence result will be useful in studying certain rational equations. See [101] and [103].

Theorem 1.6.1 *Let I be an interval of real numbers and let $F \in C(I^{k+1}, I)$. Assume that the following three conditions are satisfied:*

1. *F is increasing in each of its arguments.*

2. *$F(z_1, \ldots, z_{k+1})$ is strictly increasing in each of the arguments $z_{i_1}, z_{i_2}, \ldots, z_{i_l}$, where $1 \leq i_1 < i_2 < \ldots < i_l \leq k+1$, and the arguments $i_1, i_2, \ldots, i_l$ are relatively prime.*

3. *Every point c in I is an equilibrium point of Eq.(1.1.1).*

Then every solution of Eq.(1.1.1) has a finite limit.

The following convergence result is due to Hautus and Bolis. See [132] and Theorem 2.6.2 in [157, p. 53].

Theorem 1.6.2 *Let I be an open interval of real numbers, let $F \in C(I^{k+1}, I)$, and let $\bar{x} \in I$ be an equilibrium point of the Eq.(1.1.1). Assume that F satisfies the following two conditions:*

1. *F is increasing in each of its arguments.*

2. *F satisfies the **negative feedback** property:*

$$(u - \bar{x})[F(u, u, \ldots, u) - u] < 0, \qquad \text{for all} \qquad u \in I - \{\bar{x}\}.$$

Then the equilibrium point $\bar{x}$ is a global attractor of all solutions of Eq.(1.1.1).

The next two global attractivity results were motivated by second-order rational equations and have several applications.

Theorem 1.6.3 *[157, p. 27] Assume that the following conditions hold:*

(i) $f \in C[(0, \infty) \times (0, \infty), (0, \infty)]$.

(ii) $f(x, y)$ is decreasing in x and strictly decreasing in y.

(iii) $x f(x, x)$ is strictly increasing in x.

(iv) The equation

$$x_{n+1} = x_n f(x_n, x_{n-1}), \quad n = 0, 1, \ldots \tag{1.6.1}$$

has a unique positive equilibrium $\bar{x}$.

Then $\bar{x}$ is a global attractor of all positive solutions of Eq.(1.6.1).

Theorem 1.6.4 *[106] Assume that the following conditions hold:*

(i) $f \in C[[0,\infty) \times [0,\infty), (0,\infty)]$.

(ii) $f(x,y)$ *is decreasing in each argument.*

(iii) $xf(x,y)$ *is increasing in* x.

(iv) $f(x,y) < f(y,x) \Leftrightarrow x > y$.

(v) The equation

$$x_{n+1} = x_{n-1}f(x_{n-1}, x_n), \quad n = 0, 1, \ldots$$

has a unique positive equilibrium $\bar{x}$.

Then $\bar{x}$ is a global attractor of all positive solutions.

The following global attractivity result from [175] is very useful in establishing convergence results in many situations.

Theorem 1.6.5 *Let $[a,b]$ be a closed and bounded interval of real numbers and let $F \in C([a,b]^{k+1}, [a,b])$ satisfy the following conditions:*

1. *The function $F(z_1, \ldots, z_{k+1})$ is monotonic in each of its arguments.*
2. *For each $m, M \in [a,b]$ and for each $i \in \{1, \ldots, k+1\}$, we define*

$$M_i(m,M) = \begin{cases} M, & \text{if } F \text{ is increasing in } z_i \\ m, & \text{if } F \text{ is decreasing in } z_i \end{cases}$$

and

$$m_i(m,M) = M_i(M,m)$$

and assume that if (m,M) is a solution of the system:

$$\left.\begin{array}{l} M = F(M_1(m,M), \ldots, M_{k+1}(m,M)) \\ m = F(m_1(m,M), \ldots, m_{k+1}(m,M)) \end{array}\right\},$$

then $M = m$.

Then there exists exactly one equilibrium $\bar{x}$ of the equation

$$x_{n+1} = F(x_n, x_{n-1}, \ldots, x_{n-k}), \qquad n = 0, 1, \ldots \tag{1.6.2}$$

and every solution of Eq.(1.6.2) converges to $\bar{x}$.

The following period-two convergence result of Camouzis and Ladas was motivated by several period-two convergence results in rational equations. (See Chapters 4 and 5.) Thanks to this result, several open problems and conjectures posed in the Kulenovic and Ladas book have now been resolved and the character of solutions of many rational equations has now been clarified. See Theorems 4.2.2, 4.3.1, 5.74.2, 5.86.1, 5.109.1, 5.145.2.

Theorem 1.6.6 *[61] Let I be a set of real numbers and let*

$$F:\ I\times I\rightarrow I$$

be a function $F(u,v)$, which decreases in u and increases in v. Then for every solution $\{x_n\}_{n=-1}^{\infty}$ of the equation

$$x_{n+1}=F(x_n,x_{n-1}),\quad n=0,1,\ldots,$$

the subsequences $\{x_{2n}\}_{n=0}^{\infty}$ and $\{x_{2n+1}\}_{n=-1}^{\infty}$ of even and odd terms of the solution do exactly one of the following:

(i) They are both monotonically increasing.

(ii) They are both monotonically decreasing.

(iii) Eventually, one of them is monotonically increasing and the other is monotonically decreasing.

PROOF

Assume that (i) and (ii) are not true for a solution $\{x_n\}_{n=-1}^{\infty}$. Then for some N,

$$x_{2N+2}\geq x_{2N}\quad\text{and}\quad x_{2N+3}\leq x_{2N+1} \tag{1.6.3}$$

or

$$x_{2N+2}\leq x_{2N}\quad\text{and}\quad x_{2N+3}\geq x_{2N+1}. \tag{1.6.4}$$

Assume that (1.6.3) holds. The case where (1.6.4) holds is similar and will be omitted. Then

$$x_{2N+4}=F(x_{2N+3},x_{2N+2})\geq F(x_{2N+1},x_{2N})=x_{2N+2}$$

and

$$x_{2N+5}=F(x_{2N+4},x_{2N+3})\leq F(x_{2N+2},x_{2N+1})=x_{2N+3}$$

and the result follows by induction. ∎

The results in the remainder of this chapter were recently obtained by the authors while working on various special cases of rational difference equations and provide useful generalizations and some unifications in some special cases.

In order to simplify and unify several convergence results for the difference equation

$$x_n = f(x_{n-i_1}, \ldots, x_{n-i_k}), \quad n = 1, 2, \ldots, \tag{1.6.5}$$

where $k \geq 2$ and the function $f(z_1, \ldots, z_k)$ is monotonic in each of its arguments, we introduce some notation and state several hypotheses.

For every pair of numbers (m, M) and for each $j \in \{1, \ldots, k\}$, we define

$$M_j = M_j(m, M) = \begin{cases} M, & \text{if } f \textit{ is increasing in } z_j \\ m, & \text{if } f \textit{ is decreasing in } z_j \end{cases}$$

and

$$m_j = m_j(m, M) = M_j(M, m).$$

$(\mathbf{H_1})$ **:** $\mathbf{f \in C([0,\infty)^k, [0,\infty))}$ **and** $\mathbf{f(z_1, \ldots, z_k)}$ **is monotonic in each of its arguments.**

$(\mathbf{H_1^*})$ **:** $\mathbf{f \in C((0,\infty)^k, (0,\infty))}$ **and** $\mathbf{f(z_1, \ldots, z_k)}$ **is monotonic in each of its arguments.**

$(\mathbf{H_1'})$ **:** $\mathbf{f \in C([0,\infty)^k, [0,\infty))}$ **and** $\mathbf{f(z_1, \ldots, z_k)}$ **is strictly monotonic in each of its arguments.**

$(\mathbf{H_1''})$ **:** $\mathbf{f \in C((0,\infty)^k, (0,\infty))}$ **and** $\mathbf{f(z_1, \ldots, z_k)}$ **is strictly monotonic in each of its arguments.**

$(\mathbf{H_2})$ **: For each** $\mathbf{m \in [0,\infty)}$ **and** $\mathbf{M > m}$**, we assume that**

$$\mathbf{f(M_1, \ldots, M_k) \geq M} \tag{1.6.6}$$

implies

$$\mathbf{f(m_1, \ldots, m_k) > m.} \tag{1.6.7}$$

$(\mathbf{H_2'})$ **: For each** $\mathbf{m \in (0,\infty)}$ **and** $\mathbf{M > m}$**, we assume that**

$$\mathbf{f(M_1, \ldots, M_k) \geq M} \tag{1.6.8}$$

implies

$$\mathbf{f(m_1, \ldots, m_k) > m.} \tag{1.6.9}$$

$(\mathbf{H_3})$ **: For each** $\mathbf{m \in [0,\infty)}$ **and** $\mathbf{M > m}$**, we assume that**

either

$$\mathbf{(f(M_1, \ldots, M_k) - M)(f(m_1, \ldots, m_k) - m) > 0} \tag{1.6.10}$$

or

$$\mathbf{f(M_1, \ldots, M_k) - M = f(m_1, \ldots, m_k) - m = 0.} \tag{1.6.11}$$

$\mathbf{(H_3')}$: **For each** $\mathbf{m \in (0,\infty)}$ **and** $\mathbf{M > m}$**, we assume that**

either

$$\mathbf{(f(M_1,\dots,M_k) - M)(f(m_1,\dots,m_k) - m) > 0} \tag{1.6.12}$$

or

$$\mathbf{f(M_1,\dots,M_k) - M = f(m_1,\dots,m_k) - m = 0.} \tag{1.6.13}$$

We also define the following sets:

$$\mathbf{S = \{i_s \in \{i_1,\dots,i_k\} : f \text{ strictly increases in } x_{n-i_s}\} = \{i_{s_1},\dots,i_{s_r}\}}$$

and

$$\mathbf{J = \{i_j \in \{i_1,\dots,i_k\} : f \text{ strictly decreases in } x_{n-i_j}\} = \{i_{j_1},\dots,i_{j_t}\}.}$$

Clearly when H_1' or H_1'' holds,

$$S \bigcup J = \{i_1,\dots,i_k\}.$$

$\mathbf{(H_4)}$: **The set S consists of even indices only and the set J consists of odd indices only.**

$\mathbf{(H_5)}$: **Either the set S contains at least one odd index, or the set J contains at least one even index.**
$\mathbf{(H_6)}$: **The greatest common divisor of the indices in the union of the sets S and J is equal to 1.**

The next few theorems can be used to establish global attractivity and period-two convergence results in many special cases of rational equations including the following:

#20 − 22,	#24,	#27,	#29,	#31,
#54,	#58,	#63,	#66,	#77 − 78,
#83 − 84,	#89,	#91,	#96 − 97,	#101 − 106,
#108 − 110,	#112,	#118,	#123,	#128,
#134 − 136,	#146,	#149,	#165 − 166,	#171 − 172,
#178 − 179,	#184,	#189 − 191,	#196 − 197,	#202 − 203,
#205 − 207,	#209 − 211,	#213,	#217 − 223,	#225.

See Chapters 4 and 5.

Theorem 1.6.7 *The following statements are true:*
(a) Assume that (H_1) *and* (H_2) *hold for the function* $f(z_1,\dots,z_k)$ *of Eq.(1.6.5). Then every solution of Eq.(1.6.5) which is bounded from above converges to a finite limit.*

(a′) Assume that (H_1^*) *and* (H_2') *hold for the function* $f(z_1,\dots,z_k)$ *of Eq.(1.6.5). Then every solution of Eq.(1.6.5) which is bounded from above and from below by positive constants converges to a finite limit.*

PROOF (*a*) Let $\{x_n\}$ be a bounded solution of Eq.(1.6.5). Set

$$I = \liminf_{n\to\infty} x_n \quad \text{and} \quad S = \limsup_{n\to\infty} x_n$$

and assume, for the sake of contradiction, that

$$S > I.$$

Clearly, there exists a sequence of indices $\{n_m\}$ and positive numbers L_{-r}, for $r \in \{1, \dots, k\}$, such that

$$S = \lim_{m\to\infty} x_{n_m} \quad \text{and} \quad L_{-r} = \lim_{m\to\infty} x_{n_m - i_r}.$$

From Eq.(1.6.5) and the monotonic character of f we see that

$$S = f(L_{-1}, \dots, L_{-k}) \le f(M_1(I,S), \dots, M_k(I,S)). \tag{1.6.14}$$

Similarly, we see that

$$I \ge f(m_1(I,S), \dots, m_k(I,S)). \tag{1.6.15}$$

But from (1.6.14) and the Hypothesis (H_2) we see that

$$f(m_1(I,S), \dots, m_k(I,S)) - I > 0,$$

which contradicts (1.6.15). The proof is complete in this case.
(a') The proof in this case is similar to the proof in part (a) and will be omitted. ∎

Theorem 1.6.8 *Assume that for any of the following three equations of order three:*

$$x_{n+1} = f(x_n, x_{n-1}, x_{n-2}), \quad n = 0, 1, \dots \tag{1.6.16}$$

$$x_{n+1} = f(x_n, x_{n-2}), \quad n = 0, 1, \dots \tag{1.6.17}$$

or

$$x_{n+1} = f(x_{n-1}, x_{n-2}), \quad n = 0, 1, \dots \tag{1.6.18}$$

the hypotheses (H_1'') and (H_3') are satisfied for the arguments shown in the equation and, furthermore, assume that the function f is:

strictly increasing in x_n ***or*** x_{n-2}, ***or strictly decreasing in*** x_{n-1}.

Then every solution of this equation bounded from below and from above by positive constants converges to a finite limit.

PROOF Let $\{x_n\}$ be a solution bounded from above and from below by positive constants. Set

$$I = \liminf_{n\to\infty} x_n \quad \text{and} \quad S = \limsup_{n\to\infty} x_n.$$

Clearly, there exists a sequence of indices $\{n_i\}$ and positive numbers L_{-j}, for $j \in \{0, 1, \ldots\}$, such that

$$S = \lim_{i\to\infty} x_{n_i+1} \quad \text{and} \quad L_{-j} = \lim_{i\to\infty} x_{n_i-j}.$$

First we will consider Eq.(1.6.16) and give the proof when the function $f(z_1, z_2, z_3)$ is strictly increasing in z_3. The proof when the function $f(z_1, z_2, z_3)$ is strictly decreasing in z_2, or when the function $f(z_1, z_2, z_3)$ is strictly increasing in z_1, is similar and will be omitted.

Case 1: The function $f(z_1, z_2, z_3)$ is strictly increasing in each argument.

Actually in this case we can show that the Hypotheses of Theorem 1.6.1 are satisfied from which the result follows. However, we give the details of the proof for completeness and practice.

From Eq.(1.6.16) and the monotonic character of f we see that

$$S = f(L_0, L_{-1}, L_{-2}) \leq f(S, S, S).$$

Similarly, we find

$$I \geq f(I, I, I). \tag{1.6.19}$$

Now assume that

$$S < f(S, S, S).$$

Then from (1.6.12)

$$I < f(I, I, I),$$

which contradicts (1.6.19). Hence,

$$S = f(L_0, L_{-1}, L_{-2}) = f(S, S, S) > 0. \tag{1.6.20}$$

Then from (1.6.20) we find that

$$L_0 = L_{-1} = L_{-2} = S$$

otherwise and, because of the strict monotonicity of f in all of its arguments,

$$S = f(L_0, L_{-1}, L_{-2}) < f(S, S, S),$$

which is a contradiction.

Clearly, for an arbitrarily small positive number ϵ there exists N sufficiently large such that

$$S - \epsilon < x_{n_N}, x_{n_N-1}, x_{n_N-2} < S + \epsilon.$$

Also, clearly,

$$f(S+\epsilon, S+\epsilon, S+\epsilon) = S+\epsilon \text{ and } f(S-\epsilon, S-\epsilon, S-\epsilon) = S-\epsilon$$

because otherwise and in view of (1.6.12) with

$$m = S \text{ and } M = S+\epsilon$$

or with

$$m = S-\epsilon \text{ and } M = S$$

we would have

$$f(S, S, S) \neq S,$$

which contradicts (1.6.20). Hence,

$$S-\epsilon = f(S-\epsilon, S-\epsilon, S-\epsilon) < x_{n_N+1} = f(x_{n_N}, x_{n_N-1}, x_{n_N-2})$$

$$< f(S+\epsilon, S+\epsilon, S+\epsilon) = S+\epsilon$$

and by induction for all $k \geq 1$

$$S-\epsilon < x_{n_N+k} < S+\epsilon,$$

from which it follows that

$$\lim_{k\to\infty} x_{n_N+k} = S.$$

The proof is complete in this case.

Case 2: The function $f(z_1, z_2, z_3)$ is strictly increasing in z_3 and is strictly decreasing in one of the other two arguments.

From Eq.(1.6.16) and the monotonic character of f we see that

$$S = f(L_0, L_{-1}, L_{-2}) \leq f(M_1, M_2, S),$$

where

$$M_i = M_i(I, S).$$

Similarly, we find

$$I \geq f(m_1, m_2, I), \tag{1.6.21}$$

where

$$m_i = m_i(I, S).$$

Now assume that

$$S < f(M_1, M_2, S).$$

Then from (1.6.12)

$$I < f(m_1, m_2, I),$$

which contradicts (1.6.21). Hence,

$$S = f(L_0, L_{-1}, L_{-2}) = f(M_1, M_2, S) > 0, \tag{1.6.22}$$

which, in view of (1.6.13), implies that

$$I = f(m_1, m_2, I) > 0.$$

Then from (1.6.22) we find that

$$L_0 = M_1, \ \ L_{-1} = M_2, \ \ \textit{and} \ \ L_{-2} = S$$

otherwise and, because of the strict monotonicity of f in all of its arguments,

$$S = f(L_0, L_{-1}, L_{-2}) < f(M_1, M_2, S),$$

which is a contradiction. When

$$M_1 = S,$$

the function is strictly increasing in z_1 and so it must be strictly decreasing in z_2. Hence,

$$M_2 = I = L_{-1}.$$

From

$$M_1 = S = f(M_1, M_2, S) = f(S, I, S) = L_0 = f(L_{-1}, L_{-2}, L_{-3})$$

it follows that

$$L_{-1} = S.$$

Hence,

$$I = L_{-1} = S.$$

When

$$M_1 = I$$

clearly

$$m_1 = S$$

and from

$$M_1 = I = f(S, m_2, I) = f(L_{-1}, L_{-2}, L_{-3})$$

we obtain

$$m_2 = L_{-2} = S$$

and so clearly

$$M_2 = L_{-1} = I.$$

Hence,

$$I = L_{-1} = S.$$

The proof is complete in this case.

Next we will consider Eq.(1.6.17) and we give the proof when the function $f(z_1, z_3)$ is strictly increasing in z_1. The proof when the function is strictly decreasing in z_3 is similar and will be omitted.
We divide the proof into the following two cases:

Case 3: The function $f(z_1, z_3)$ is strictly increasing in each argument. In this case the proof is similar to the proof in case 1 and will be omitted.

Case 4: The function $f(z_1, z_3)$ is strictly increasing in z_1 and strictly decreasing in z_3.

From Eq.(1.6.17) and the monotonic character of f we see that

$$S = f(L_0, L_{-2}) \leq f(S, I).$$

Similarly, we find

$$I \geq f(I, S). \tag{1.6.23}$$

Now assume that

$$S < f(S, I).$$

Then from (1.6.12)

$$I < f(I, S),$$

which contradicts (1.6.23). Hence,

$$S = f(L_0, L_{-2}) = f(S, I) > 0, \tag{1.6.24}$$

which, in view of (1.6.13), implies that

$$I = f(I, S) > 0.$$

Then from (1.6.24) we find that

$$L_0 = S \text{ and } L_{-2} = I$$

otherwise and because of the strict monotonicity of f,

$$S = f(L_0, L_{-2}) < f(S, I),$$

which is a contradiction. Similarly, we find

$$L_{-1} = S \text{ and } L_{-3} = I$$

and

$$L_{-2} = S \text{ and } L_{-4} = I.$$

Hence,

$$I = L_{-2} = S.$$

The proof is complete in this case.

Finally, we consider Eq.(1.6.18) and we give the proof when the function $f(z_2, z_3)$ is strictly decreasing in z_2. The proof when the function is strictly increasing in z_3 is similar and will be omitted.
We divide the proof into the following two cases:

Case 5: The function $f(z_2, z_3)$ is strictly decreasing in z_2 and strictly increasing in z_3.

From Eq.(1.6.18) and the monotonic character of f we see that

$$S = f(L_{-1}, L_{-2}) \leq f(I, S).$$

Similarly, we find

$$I \geq f(S, I). \tag{1.6.25}$$

Now assume that

$$S < f(I, S).$$

Then from (1.6.12)

$$I < f(S, I),$$

which contradicts (1.6.25). Hence

$$S = f(L_{-1}, L_{-2}) = f(I, S) > 0 \tag{1.6.26}$$

which, in view of (1.6.13), implies that

$$I = f(S, I) > 0.$$

Then from (1.6.26) we find that

$$L_{-1} = I \text{ and } L_{-2} = S$$

otherwise and, because of the strict monotonicity of f in all of its arguments,

$$S = f(L_{-1}, L_{-2}) < f(I, S),$$

which is a contradiction. Similarly, we find

$$L_{-3} = S \text{ and } L_{-4} = I$$

and

$$L_{-4} = S \text{ and } L_{-5} = I.$$

Hence,

$$I = L_{-4} = S.$$

The proof is complete in this case.
Case 6: The function $f(z_2, z_3)$ is strictly decreasing in both arguments.

From Eq.(1.6.18) and the monotonic character of f we see that

$$S = f(L_{-1}, L_{-2}) \leq f(I, I).$$

Similarly, we find

$$I \geq f(S, S). \tag{1.6.27}$$

Now assume that

$$S < f(I, I).$$

Then from (1.6.12)

$$I < f(S, S),$$

which contradicts (1.6.27). Hence,

$$S = f(L_{-1}, L_{-2}) = f(I, I) > 0, \tag{1.6.28}$$

which, in view of (1.6.13), implies that

$$I = f(S, S) > 0.$$

Then from (1.6.28) we find that

$$L_{-1} = I \text{ and } L_{-2} = I$$

otherwise and because of the strict monotonicity of f,

$$S = f(L_{-1}, L_{-2}) < f(I, I),$$

which is a contradiction. Similarly, we find

$$L_{-3} = S \text{ and } L_{-4} = S$$

and

$$L_{-4} = S \text{ and } L_{-5} = S$$

and

$$L_{-5} = I \text{ and } L_{-6} = I.$$

Hence,

$$I = L_{-5} = S.$$

The proof is complete. ∎

Theorem 1.6.9 *Assume that for any of the three equations (1.6.16), (1.6.17), or (1.6.18) the hypotheses* (H_1'') *and* (H_3') *are satisfied for the arguments shown in the equation, and, furthermore, assume that the function* f *is:*

strictly decreasing in x_n, ***for Eqs.(1.6.16) and (1.6.17),***

and

$$\textit{strictly increasing in } x_{n-1} \textit{ for Eqs.(1.6.16) and (1.6.18)},$$

and

$$\textit{strictly decreasing in } x_{n-2}.$$

Then every solution of this equation bounded from above and from below by positive constants converges to a (not necessarily prime) period-two solution.

PROOF We will give the proof for Eq.(1.6.16) under the assumption that the function is strictly decreasing in x_n and x_{n-2} and strictly increasing in x_{n-1}. The proof for the other two equations is similar and will be omitted.

Let $\{x_n\}$ be a solution bounded from above and from below by positive constants. Set

$$I = \liminf_{n\to\infty} x_n \quad \text{and} \quad S = \limsup_{n\to\infty} x_n.$$

Clearly, there exists a sequence of indices $\{n_i\}$ and positive numbers $\{L_{-j}\}$, for $j \in \{0, 1, 2\}$, such that

$$S = \lim_{i\to\infty} x_{n_i+1} \quad \text{and} \quad L_{-j} = \lim_{i\to\infty} x_{n_i-j}.$$

From Eq.(1.6.16) and the monotonic character of f we see that

$$S = f(L_0, L_{-1}, L_{-2}) \leq f(I, S, I). \tag{1.6.29}$$

Similarly, we find

$$I \geq f(S, I, S). \tag{1.6.30}$$

Now assume that

$$S < f(I, S, I).$$

Then from (1.6.12) we see that

$$I < f(S, I, S),$$

which contradicts (1.6.30). Hence,

$$S = f(I, S, I),$$

which, in view of (1.6.13), implies that

$$I = f(S, I, S).$$

From (1.6.29) we find that

$$L_0 = L_{-2} = I \quad \text{and} \quad L_{-1} = S$$

otherwise and because of the strict monotonicity of f,

$$S < f(I, S, I),$$

which is a contradiction. At this point we claim that there exist arbitrarily small positive numbers, ϵ_1 and ϵ_2, such that

$$S - \epsilon_1 = f(I + \epsilon_2, S - \epsilon_1, I + \epsilon_2).$$

Assume, for the sake of contradiction and without loss of generality, that for all positive numbers ϵ_1 and ϵ_2 we have

$$S - \epsilon_1 < f(I + \epsilon_2, S - \epsilon_1, I + \epsilon_2).$$

By letting $\epsilon_1 \to 0$ we obtain

$$S \leq f(I + \epsilon_2, S, I + \epsilon_2),$$

from which it follows that

$$S \leq f(I + \epsilon_2, S, I + \epsilon_2) < f(I, S, I) = S$$

which is a contradiction and so our claim holds.

Let N be sufficiently large and such that

$$x_{n_N}, x_{n_N-2} < I + \epsilon_2 \quad \text{and} \quad x_{n_N-1} > S - \epsilon_1.$$

Then

$$x_{n_N+1} = f(x_{n_N}, x_{n_N-1}, x_{n_N-2}) > f(I + \epsilon_2, S - \epsilon_1, I + \epsilon_2) = S - \epsilon_1$$

and, similarly,

$$x_{n_N+2} < I + \epsilon_2.$$

Inductively, we find

$$x_{2j+1+n_N} > S - \epsilon_1 \quad \text{and} \quad x_{2j+n_N} < I + \epsilon_2,$$

from which the result follows. ∎

Theorem 1.6.10 *Assume that for any of the three equations (1.6.16), (1.6.17), or (1.6.18) the Hypotheses (H_1') and (H_3) are satisfied for the arguments shown in the equation, and, furthermore, assume that the function f is:*

strictly decreasing in x_n, ***for Eqs.(1.6.16) and (1.6.17),***

and

strictly increasing in $\mathbf{x_{n-1}}$ ***for Eqs.(1.6.16) and (1.6.18),***

and

strictly decreasing in $\mathbf{x_{n-2}}$.

Then every solution of this equation bounded from above converges to a (not necessarily prime) period-two solution.

The results in this section can be extended and generalized to higher order equations as follows:

Theorem 1.6.11 *The following statements are true:*
(a) Assume that (H_1'), (H_3), (H_4), *and* (H_6) *hold. Then every bounded solution of Eq.(1.6.5) converges to a (not necessarily prime) period-two solution.*

(a′) Assume that (H_1''), (H_3'), (H_4) *and* (H_6) *hold. Then every solution of Eq.(1.6.5) bounded from above and from below by positive constants converges to a (not necessarily prime) period-two solution.*

(b) Assume that (H_1'), (H_3), (H_5), *and* (H_6) *hold. Then every bounded solution of Eq.(1.6.5) converges to a finite limit.*

(b′) Assume that (H_1''), (H_3'), (H_5), *and* (H_6) *hold. Then every solution of Eq.(1.6.5) bounded from above and from below by positive constants converges to a finite limit.*

PROOF We will prove (a') and (b') together. Let $\{x_n\}$ be a solution bounded from above and from below by positive constants. Clearly, there exists a full limiting sequence $\{L_{-q}\}_{q=0}^{\infty}$ such that

$$L_0 = S.$$

We divide the proof into the following two cases:

Case 1: $S, J \neq \emptyset$ and all the indices $i_{s_1}, \ldots, i_{s_r}$ of the set S are even. The proof when $S = \emptyset$ is similar and will be omitted. The proof when $J = \emptyset$ follows from Theorem 1.6.1.

In this case when (H_1''), (H_3'), (H_4), and (H_6) hold, all the indices of the set J are odd and when (H_1''), (H_3'), (H_5), and (H_6) hold, there exists at least one index $j_1, \ldots, j_t$ of the set J that is odd and also at least one index that is even.

Clearly, from Eq.(1.6.5) and the monotonic character of f we obtain

$$L_0 = S = f(L_{-i_1}, \ldots, L_{-i_k}) \leq f(M_1(I,S), \ldots, M_k(I,S)).$$

Similarly, we find

$$I \geq f(m_1(I,S),\dots,m_k(I,S)).$$

Now assume that

$$S < f(M_1(I,S),\dots,M_k(I,S)).$$

Then from (1.6.12) we see that

$$I < f(m_1(I,S),\dots,m_k(I,S)),$$

which is a contradiction. Hence,

$$S = f(M_1(I,S),\dots,M_k(I,S)). \tag{1.6.31}$$

Similarly, we see that

$$I = f(m_1(I,S),\dots,m_k(I,S)). \tag{1.6.32}$$

Hence,

$$L_{-i_{s_1}} = \dots = L_{-i_{s_r}} = S \text{ and } L_{-i_{j_1}} = \dots = L_{-i_{j_t}} = I$$

otherwise and because of the strict monotonicity of f,

$$S < f(M_1(I,S),\dots,M_k(I,S)),$$

which is a contradiction. Similarly,

$$L_{-2i_{j_1}} = \dots = L_{-2i_{j_t}} = S$$

and also for every positive linear combination

$$T = \sum_{l=1}^{r} \phi_l i_{s_l} + \sum_{p=1}^{t} \psi_p i_{j_p}$$

we see that

$$L_{-T} \in \{I,S\}. \tag{1.6.33}$$

There exists n_0 large enough such that for each $n \geq n_0$ there exist integers $\{\phi_{l,n}\}_{l=1}^{r}$, $\{\psi_{p,n}\}_{p=1}^{t}$ such that

$$n = \sum_{l=1}^{r} \phi_{l,n} i_{s_l} + \sum_{p=1}^{t} \psi_{p,n} i_{j_p}.$$

From this and (1.6.33) it follows that for all $n \geq n_0$

$$L_{-n} \in \{I,S\}.$$

From Eqs.(1.6.5) and (1.6.31) we obtain

$$L_{-n} = f(L_{-n-i_1}, \ldots, L_{-n-i_k}) = f(M_1(I,S), \ldots, M_k(I,S)) \in \{I, S\}$$

or

$$L_{-n} = f(L_{-n-i_1}, \ldots, L_{-n-i_k}) = f(m_1(I,S), \ldots, m_k(I,S)) \in \{I, S\}.$$

From this it follows that for each $l \in \{1, \ldots, r\}$

$$L_{-n} = L_{-n-i_{s_l}} \text{for all } \; n \geq n_0$$

and for each $p \in \{1, \ldots, t\}$

$$L_{-n} = L_{-n-2i_{j_p}} \text{for all } \; n \geq n_0.$$

Therefore, the sequence $\{L_{-q}\}_{q=n_0}^{\infty}$ is periodic with periods

$$i_{s_1}, \ldots, i_{s_r}, 2i_{j_1}, \ldots, 2i_{j_t}.$$

But

$$gcd\{i_{s_1}, \ldots, i_{s_r}, 2i_{j_1}, \ldots, 2i_{j_t}\} = 2$$

and so the sequence $\{L_{-q}\}_{q=0}^{\infty}$ is periodic with period two. In fact, it has the following form,

$$\ldots, I, S, \ldots .$$

When (H_1''), (H_3'), (H_4), and (H_6) hold, assume without loss of generality that, for all $j \geq 0$,

$$L_{-2j} = I \quad \text{and} \quad L_{-2j-1} = S.$$

At this point we claim that there exist arbitrarily small positive numbers, ϵ_1 and ϵ_2, such that

$$S - \epsilon_1 = f(M_1(I,S) + X_1, \ldots, M_k + X_k)$$

and

$$I + \epsilon_2 = f(m_1(I,S) + x_1, \ldots, m_k + x_k)$$

where, for each $r \in \{1, \ldots, k\}$,

$$X_r = \epsilon_2$$

when f decreases in x_{n-i_r} and

$$X_r = -\epsilon_1$$

when f increases in x_{n-i_r} and

$$x_r = -\epsilon_1$$

when f decreases in x_{n-i_r} and

$$x_r = \epsilon_2$$

when f increases in x_{n-i_r}. Assume, for the sake of contradiction and without loss of generality, that for all positive numbers ϵ_1 and ϵ_2 we have

$$S - \epsilon_1 < f(M_1(I,S) + X_1, \ldots, M_k(I,S) + X_k).$$

By letting $\epsilon_1 \to 0$ we obtain

$$S \leq f(M_1(I,S) + Y_1, \ldots, M_k(I,S) + Y_k)$$

where, for each $r \in \{1, \ldots, k\}$,

$$Y_r = \epsilon_2$$

when f decreases in x_{n-i_r} and

$$Y_r = 0$$

when f increases in x_{n-i_r}. From this it follows that

$$S \leq f(M_1(I,S) + Y_1, \ldots, M_k + Y_k) < f(M_1(I,S), \ldots, M_k) = S,$$

which is a contradiction. Similarly, it follows that

$$I + \epsilon_2 = f(m_1(I,S) + x_1, \ldots, m_k + x_k)$$

and so our claim holds.

Let N be sufficiently large such that, for each $r \in \{1, \ldots, k\}$,

$$x_{n_N - i_r} < I + \epsilon_2$$

when the function f decreases in $x_{n_N - i_r}$ and

$$x_{n_N - i_r} > S - \epsilon_1$$

when the function f increases in $x_{n_N - i_r}$. Then

$$x_{n_N} = f(x_{n_N - i_1}, \ldots, x_{n_N - i_k}) > f(M_1(I,S) + x_1, \ldots, M_k(I,S) + x_k) = S - \epsilon_1$$

and, similarly,

$$x_{n_N + 1} < I + \epsilon_2.$$

Inductively, we find

$$x_{2j+n_N} > S - \epsilon_1 \quad \text{and} \quad x_{2j+1+n_N} < I + \epsilon_2,$$

from which the result follows.

When (H_1''), (H_3'), (H_5), and (H_6) hold, there exist two even indices $i_{s_1} \in S$ and $i_{j_{t_0}} \in J$ such that

$$L_{-i_{s_1}} = S$$

and

$$L_{-i_{j_{t_0}}} = I.$$

Due to the fact that $L_{-i_{s_1}}$ and $L_{-i_{j_{t_0}}}$ belong to the period-two solution $\{L_{-q}\}_{q=0}^{\infty}$, it follows that

$$I = S.$$

Case 2: The set I contains at least one index $i_{s_{r_0}}$ such that

$$i_{s_{r_0}} = 2R + 1.$$

In this case, clearly,

$$L_{-(2R+1)\omega} = S, \quad \omega = 0, 1 \ldots$$

and for each $i_{j_t} \in J$

$$L_{-(2\phi+1)j_t} = I, \quad \phi = 0, 1 \ldots$$

and

$$L_{-2\phi j_t} = S, \quad \phi = 0, 1 \ldots .$$

Hence, there exist positive integers ϕ and ω such that

$$L_{-(2R+1)\omega} = S = L_{-(2\phi+1)j_t} = I.$$

The proof is complete in this case.

The proofs of (a) and (b) are similar and will be omitted. ∎

2

Equations with Bounded Solutions

2.0 Introduction

Consider the third-order rational difference equation

$$x_{n+1} = \frac{\alpha + \beta x_n + \gamma x_{n-1} + \delta x_{n-2}}{A + Bx_n + Cx_{n-1} + Dx_{n-2}}, \quad n = 0, 1, \ldots \tag{2.0.1}$$

with nonnegative parameters $\alpha, \beta, \gamma, \delta, A, B, C, D$ and with arbitrary nonnegative initial conditions x_{-2}, x_{-1}, x_0, such that the denominator is always positive.

This equation contains 225 special cases of equations with positive parameters. It was conjectured in [69] that in 135 of these special cases, every solution of the equation is bounded and, in the remaining 90 cases, the equation has unbounded solutions in some range of their parameters and for some initial conditions.

For each of the 225 special cases of Eq.(2.0.1) we assign a number from 1 to 225. See Appendix A for the number assigned to each equation. See also [192].

In this chapter we present several theorems on the boundedness of every solution of several equations of the form of Eq.(2.0.1) and in particular we establish that in all 135 special cases of Eq.(2.0.1), every solution of the equation is bounded.

In Section 2.1 we present a large number of special cases of Eq.(2.0.1) where every solution of the equation is easily seen to be bounded. Actually, this section will account for the boundedness of every solution in 91 special cases of Eq.(2.0.1). The remaining sections will account for the remaining 44 additional special cases.

In Section 2.2 we present in detail the proof of the boundedness of solutions of the second-order rational difference equation

$$x_{n+1} = \frac{\alpha + \beta x_n + \gamma x_{n-1}}{A + B x_n + C x_{n-1}}, \quad n = 0, 1, \ldots \tag{2.0.2}$$

when

$$C > 0.$$

This section will account for the boundedness of every solution in eight additional special cases of Eq.(2.0.1).

When

$$C = 0 \text{ and } B = 0,$$

Eq.(2.0.2) reduces to a linear equation that has unbounded solutions in some range of its parameters, unless

$$\beta = \gamma = 0.$$

Finally, when

$$C = 0 \text{ and } B > 0 \tag{2.0.3}$$

that is for the equation

$$x_{n+1} = \frac{\alpha + \beta x_n + \gamma x_{n-1}}{A + B x_n}, \quad n = 0, 1, \ldots, \tag{2.0.4}$$

we will see in Chapters 3 and 4 that it has unbounded solutions if and only if

$$\gamma > \beta + A.$$

Equivalently, when $B > 0$, every solution of Eq.(2.0.4) is bounded if and only if

$$\gamma \leq \beta + A.$$

In Section 2.3 we establish the boundedness of 16 additional special cases of Eq.(2.0.1) by the **method of iteration**, that is, by observing that when we write x_{n+2} or x_{n+3} in terms of x_n, x_{n-1}, and x_{n-2}, every solution of the resulting equation is bounded.

In Section 2.4 we establish that every solution of the (normalized) special case

$$\#58: \quad x_{n+1} = \beta + \frac{x_{n-2}}{x_n}, \quad n = 0, 1, \ldots$$

with

$$\beta > 0$$

is bounded. By using similar ideas, we establish in Section 2.5 the boundedness of every solution of the equation

$$x_{n+1} = \frac{\alpha + \beta x_n + x_{n-2}}{A + x_n}, \quad n = 0, 1, \ldots \tag{2.0.5}$$

with

$$\alpha+\beta, \beta+A \in (0,\infty). \tag{2.0.6}$$

This confirms the boundedness of solutions in four additional special cases of Eq.(2.0.1).

In Section 2.6 we establish that every solution of the (normalized) special case

$$\#63: \ x_{n+1}=\gamma+\frac{x_{n-2}}{x_{n-1}}, \quad n=0,1,\ldots$$

with

$$\gamma>0$$

is bounded. By using similar ideas, in Section 2.7 we establish the boundedness of every solution of the equation

$$x_{n+1}=\frac{\alpha+\beta x_n+\gamma x_{n-1}+x_{n-2}}{A+x_{n-1}}, \quad n=0,1,\ldots \tag{2.0.7}$$

with

$$\gamma+A, \alpha+\beta+\gamma \in (0,\infty). \tag{2.0.8}$$

This confirms the boundedness of solutions in 10 additional special cases of Eq.(2.0.1).

In Section 2.8 we establish that every solution of the equation

$$x_{n+1}=\frac{\alpha+\beta x_n+\gamma x_{n-1}}{Cx_{n-1}+Dx_{n-2}}, \quad n=0,1,\ldots \tag{2.0.9}$$

with

$$\alpha \geq 0 \quad \text{and} \quad \beta, \gamma, C, D \in (0,\infty). \tag{2.0.10}$$

is bounded.

Finally, in Section 2.9 we establish that every solution of the equation

$$x_{n+1}=\frac{\alpha+\beta x_n+x_{n-2}}{Cx_{n-1}+x_{n-2}}, \quad n=0,1,\ldots \tag{2.0.11}$$

with

$$\alpha \geq 0 \ \text{and} \ \beta, C \in (0,\infty)$$

is bounded. This confirms the boundedness of solutions of the remaining two cases of Eq.(2.0.1).

2.1 Some Straightforward Cases

In this section we present several special cases of Eq.(2.0.1) where every solution of the equation is easily seen to be bounded. Actually, this section will account for the boundedness of every solution in 91 special cases of Eq.(2.0.1).

Clearly, in all four trivial linear cases of Eq.(2.0.1), every solution of the equation is bounded. They are the following special cases:

#1,	#6,	#11,	#16.

See Appendix A.

Next, in all 15 special cases of Eq.(2.0.1) that are either Riccati or of the Riccati type, every solution of the equation is bounded. They are the following special cases:

#2,	#3,	#4,	#17,	#18,	#19,	#23,	#30
#37,	#42,	#47,	#52,	#65,	#72,	#79.	

See Appendix A.

Actually, in each of the above 15 cases, either every solution of the equation is periodic or the solutions of the equation converge to an equilibrium point. See Section 5.65 on Riccati equations in Chapter 5.

One can see that in every special case of Eq.(2.0.1) where all of the terms in the numerator are also contained in the denominator, every solution of the equation is bounded. By this we mean that if the constant α is present in the numerator of this special case, so is the constant A in the denominator. If the coefficient β of x_n is present in the numerator, so is the coefficient B of x_n in the denominator, and so on. This idea establishes the boundedness of every solution in each of the following 52 additional special cases of Eq.(2.0.1):

#26,	#27,	#32,	#34,	#39,	#40,	#86,	#93,
#100,	#101,	#102,	#103,	#105,	#106,	#108,	#109,
#111,	#112,	#114,	#115,	#116,	#133,	#134,	#135,
#136,	#141,	#142,	#145,	#147,	#150,	#151,	#153,
#156,	#158,	#160,	#163,	#164,	#189,	#190,	#191,
#192,	#193,	#194,	#201,	#206,	#211,	#216,	#217,
#218,	#219,	#220,	#225.				

See Appendix A.

As an example of the preceding idea, every solution of the equation in the special case

$$\#217: \quad x_{n+1} = \frac{\alpha + \beta x_n + \gamma x_{n-1}}{A + Bx_n + Cx_{n-1} + Dx_{n-2}}, \quad n = 0, 1, \ldots$$

is bounded. Indeed, for $n \geq 0$,

$$x_{n+1} = \frac{\alpha + \beta x_n + \gamma x_{n-1}}{A + Bx_n + Cx_{n-1} + Dx_{n-2}} \leq \frac{\max\{\alpha, \beta, \gamma\}(1 + x_n + x_{n-1})}{\min\{A, B, C, D\}(1 + x_n + x_{n-1} + x_{n-2})}$$

$$\leq \frac{\max\{\alpha, \beta, \gamma\}}{\min\{A, B, C, D\}}.$$

A minor extension of the above idea establishes the boundedness of every solution of any rational equation of the form

$$x_{n+1} = \frac{\alpha + \beta x_n + \gamma x_{n-1} + \delta x_{n-2}}{Bx_n + Cx_{n-1} + Dx_{n-2}}, \quad n = 0, 1, \ldots \tag{2.1.1}$$

with $\alpha > 0$ under the condition that

$$\text{when } \beta > 0, \text{ then } B > 0,$$
$$\text{when } \gamma > 0, \text{ then } C > 0,$$

and

$$\text{when } \delta > 0, \text{ then } D > 0.$$

The above result for Eq.(2.1.1) establishes the boundedness of every solution in each of the following 20 special cases of Eq.(2.0.1):

#20,	#21,	#22,	#68,	#69,	#74,	#76,
#81,	#82,	#104,	#144,	#148,	#152,	# 168,
#175,	#182,	#204,	#208,	#212,	#224.	

See Appendix A.

In this section we accounted for the boundedness of every solution in 91 special cases of Eq.(2.0.1).

The following theorem unifies all the 91 special cases discussed in this section and extends the results to rational equations of any order k. It is amazing that for $k = 3$, this theorem presents with detailed proofs the boundedness of every solution in 91 special cases of Eq.(2.0.1). See [66].

Theorem 2.1.1 *Consider the* $(k+1)^{st}$*-order rational difference equation*

$$x_{n+1} = \frac{\alpha + \sum_{i=0}^{k} \beta_i x_{n-i}}{A + \sum_{i=0}^{k} B_i x_{n-i}}, \quad n = 0, 1, \ldots \tag{2.1.2}$$

with nonnegative parameters

$$\alpha,\ A,\ \beta_0, \ldots, \beta_k,\ B_0, \ldots, B_k$$

and with arbitrary nonnegative initial conditions $x_{-k}, \ldots, x_0$ *such that the denominator is always positive. Assume that for every* $i \in \{0, 1, \ldots, k\}$ *for which the parameter* β_i *in the numerator is positive, the corresponding parameter* B_i *in the denominator is also positive. Then every solution of Eq.(2.1.2) is bounded.*

PROOF Let us denote by I and I_0 the following subsets of $\{0, 1, \ldots, k\}$:

$$I = \{i \in \{0, 1, \ldots, k\} : \beta_i > 0 \text{ and } B_i > 0\}$$

and

$$I_0 = \{i \in \{0, 1, \ldots, k\} : \beta_i = 0 \text{ and } B_i > 0\}.$$

Then

$$I \cup I_0 \subset \{0, 1, \ldots, k\}$$

and Eq.(2.1.2) is equivalent to

$$x_{n+1} = \frac{\alpha + \sum_{i \in I} \beta_i x_{n-i}}{A + \sum_{i \in I} B_i x_{n-i} + \sum_{i \in I_0} B_i x_{n-i}}, \quad n = 0, 1, \ldots \tag{2.1.3}$$

with $\beta_i, B_i \in (0, \infty)$ for every $i \in I$ and with $B_i > 0$ for every $i \in I_0$. Of course, I or I_0, or both, may be empty sets.

First we show that when

$$A > 0 \text{ or } \alpha = 0,$$

every solution of Eq.(2.1.2) is bounded. Indeed, when $A > 0$,

$$x_{n+1} \leq \frac{\max_{i \in I}(\alpha, \beta_i)(1 + \sum_{i \in I} x_{n-i})}{\min_{i \in I}(A, B_i)(1 + \sum_{i \in I} x_{n-i})} = \frac{\max_{i \in I}(\alpha, \beta_i)}{\min_{i \in I}(A, B_i)}$$

and so every solution of Eq.(2.1.2) is bounded.

In the above inequality by $\max_{i \in I}(\alpha, \beta_i)$, we mean α if $I = \emptyset$ and the maximum of α and $\max_{i \in I} \beta_i$ otherwise. Similarly for the minimum. Also, if $I = \emptyset$, we define

$$\sum_{i \in I} x_{n-i} = 0.$$

Next assume that $\alpha = 0$. Then the set I must be nonempty and

$$x_{n+1} \leq \frac{\sum_{i\in I} \beta_i x_{n-i}}{\sum_{i\in I} B_i x_{n-i}} \leq \frac{(\max_{i\in I} \beta_i) \sum_{i\in I} x_{n-i}}{(\min_{i\in I} B_i) \sum_{i\in I} x_{n-i}} = \frac{\max_{i\in I} \beta_i}{\min_{i\in I} B_i}$$

and every solution is bounded.

In the remaining part of the proof we assume that

$$A = 0 \text{ and } \alpha > 0.$$

Now the proof depends on whether I or I_0 is empty.

Case 1: $I_0 = \emptyset$. Then, because $A = 0$, $I \neq \emptyset$ and

$$x_{n+1} = \frac{\alpha + \sum_{i\in I} \beta_i x_{n-i}}{\sum_{i\in I} B_i x_{n-i}} > \frac{\min_{i\in I} \beta_i}{\max_{i\in I} B_i}, \text{ for } n \geq 0.$$

So if we set

$$L = \frac{\min_{i\in I} \beta_i}{\max_{i\in I} B_i},$$

it follows that for $n \geq k$,

$$x_{n+1} \leq \frac{\alpha}{L \sum_{i\in I} B_i} + \frac{\max_{i\in I} \beta_i}{\min_{i\in I} B_i}$$

and every solution of Eq.(2.1.2) is bounded from below and from above. Actually in this case the equation is permanent.

Case 2: $I = \emptyset$. Then $I_0 \neq \emptyset$. In this case the Eq.(2.1.2) reduces to

$$x_{n+1} = \frac{\alpha}{\sum_{i\in I_0} B_i x_{n-i}}, \quad n = 0, 1, \ldots \tag{2.1.4}$$

with

$$\sum_{i\in I_0} B_i > 0.$$

We will prove that every solution of Eq.(2.1.4) is bounded. To this end, let $\{x_n\}$ be a solution of Eq.(2.1.4) and assume, without loss of generality, that the solution is positive for all $n \geq -k$. Let L,U be chosen in such a way that

$$x_{-k}, \ldots, x_0 \in (L, U)$$

and

$$LU = \frac{\alpha}{\sum_{i\in I_0} B_i}.$$

Then

$$L = \frac{\alpha}{U \sum_{i\in I_0} B_i} < x_1 = \frac{\alpha}{\sum_{i\in I_0} B_i x_{-i}} < \frac{\alpha}{L \sum_{i\in I_0} B_i} = U.$$

Hence,

$$x_1 \in (L, U)$$

and by induction

$$x_n \in (L, U) \ \text{ for } \ n \geq -k.$$

Case 3: Both I and I_0 are nonempty sets. In this case, as in case 2, we will assume, without loss of generality, that a solution $\{x_n\}$ is positive and show that there exists an interval (L, U) that contains the entire solution.

To see how the interval is found observe that

$$x_1 \in (L, U)$$

if and only if

$$L < \frac{\alpha + \sum_{i\in I} \beta_i x_{-i}}{\sum_{i\in I} B_i x_{-i} + \sum_{i\in I_0} B_i x_{-i}} < U$$

if and only if

$$\sum_{i\in I}(LB_i - \beta_i)x_{-i} + (L\sum_{i\in I_0} B_i x_{-i} - \alpha) < 0$$

and

$$\sum_{i\in I}(UB_i - \beta_i)x_{-i} + (U\sum_{i\in I_0} B_i x_{-i} - \alpha) > 0$$

if

$$L < \frac{\beta_i}{B_i} < U \ \text{ for all } \ i \in I$$

and

$$\frac{\alpha}{U} < \sum_{i\in I_0} B_i x_{-i} < \frac{\alpha}{L}.$$

But

$$L\sum_{i\in I_0} B_i < \sum_{i\in I_0} B_i x_{-i} < U\sum_{i\in I_0} B_i$$

and so it suffices to choose L and U such that

$$x_{-k}, \ldots, x_0 \in (L, U),$$

$$L < \min_{i\in I}(\frac{\beta_i}{B_i}, \frac{B_i}{\beta_i} \frac{\alpha}{\sum_{j\in I_0} B_j}),$$

and

$$LU = \frac{\alpha}{\sum_{j\in I_0} B_j}.$$

With the above choice of (L, U), it is now easy to show that

$$x_1 \in (L, U)$$

and then by induction

$$x_n \in (L, U), \quad \text{for } n \geq -k.$$

The proof is complete. ∎

One can see that Theorem 2.1.1 accounts for the boundedness of every solution in

$$1 + 2\sum_{i=1}^{k} \binom{k}{i} (2^{1+i} - 1) = 4(3^{k+1} - 2^k) - 1$$

special cases of Eq.(2.1.2). See [66].

2.2 The Second-Order Rational Equation

The second-order rational difference equation

$$x_{n+1} = \frac{\alpha + \beta x_n + \gamma x_{n-1}}{A + Bx_n + Cx_{n-1}}, \quad n = 0, 1, \ldots \tag{2.2.1}$$

was investigated in the book by Kulenovic and Ladas (see [175]). This equation contains 49 special cases and the boundedness character of most of them follows from the results in [175]. See also [39]. However, the boundedness character of the entire equation (2.2.1) including cases not discussed in [175], like #166, #168, and #201, was presented in [134]. Of the 49 special cases of Eq.(2.2.1), 35 cases have bounded solutions only and 14 special cases have unbounded solutions in some range of their parameters.

It is an amazing fact that when

$$C > 0,$$

every solution of Eq.(2.2.1) is bounded and when

$$C = 0 \text{ and } B > 0,$$

Eq.(2.2.1) has unbounded solutions if and only if

$$\gamma > \beta + A.$$

Finally, when

$$C = 0 \text{ and } B = 0$$

Eq.(2.2.1) reduces to a linear equation that has unbounded solutions in some range of its parameters unless

$$\beta = \gamma = 0,$$

in which case every solution of the equation is bounded.

The main result in this section is the following theorem, which establishes the boundedness of every solution in eight additional special cases of Eq.(2.0.1), namely:

#7,	#24,	#43,	#55,
#66,	#84,	#119,	#166.

See Appendix A.

The proof of the following theorem is a self-contained proof for all 28 special cases of Eq.(2.2.1) with $C > 0$, and in particular contains the proof of the boundedness of every solution of each of the above eight special cases.

Theorem 2.2.1 *Assume that $C > 0$. Then every solution of Eq.(2.2.1) is bounded.*

PROOF The proof is divided into the following eight cases:

Case 1:

$$A > 0 \text{ and } B > 0.$$

Here for $n \geq 0$,

$$x_{n+1} = \frac{\alpha + \beta x_n + \gamma x_{n-1}}{A + Bx_n + Cx_{n-1}} \leq \frac{\max\{\alpha, \beta, \gamma\}(1 + x_n + x_{n-1})}{\min\{A, B, C\}(1 + x_n + x_{n-1})}$$

$$= \frac{\max\{\alpha, \beta, \gamma\}}{\min\{A, B, C\}}$$

and so every solution is bounded.

Case 2:

$$A > 0 \text{ and } B = 0.$$

Here for $n \geq 1$,

$$x_{n+2} = \frac{\alpha + \beta x_{n+1} + \gamma x_n}{A + Cx_n} = \frac{\alpha + \gamma x_n}{A + Cx_n} + \frac{\beta}{A + Cx_n}\frac{\alpha + \beta x_n + \gamma x_{n-1}}{A + Cx_{n-1}}$$

$$= \frac{\alpha + \gamma x_n}{A + Cx_n} + \frac{\beta}{A + Cx_n}\frac{\alpha + \gamma x_{n-1}}{A + Cx_{n-1}} + \frac{\beta^2 x_n}{(A + Cx_n)(A + Cx_{n-1})}$$

$$\leq \frac{\max\{\alpha,\gamma\}}{\min\{A,C\}} + \frac{\beta}{A} \cdot \frac{\max\{\alpha,\gamma\}}{\min\{A,C\}} + \frac{\beta^2}{AC}$$

and so every solution is bounded.

Case 3:

$$A = 0 \text{ and } B > 0.$$

When $\alpha = 0$, every solution of Eq.(2.2.1) is bounded because for $n \geq 0$,

$$x_{n+1} = \frac{\beta x_n + \gamma x_{n-1}}{Bx_n + Cx_{n-1}} \leq \frac{\max\{\beta,\gamma\}}{\min\{B,C\}}.$$

On the other hand, when $\alpha > 0$ we consider the following four subcases:

Subcase 3(i):

$$A = 0, \;\; B > 0, \;\text{ and }\; \beta = \gamma = 0.$$

Let $\{x_n\}$ be a positive solution and choose $L, U \in (0, \infty)$ such that

$$x_{-1}, x_0 \in (L, U) \text{ and } LU = \frac{\alpha}{B + C}.$$

Then

$$L = \frac{\alpha}{(B+C)U} < x_1 = \frac{\alpha}{Bx_n + Cx_{n-1}} < \frac{\alpha}{(B+C)L} = U$$

and by induction,

$$x_n \in (L, U), \;\; \text{for all } \; n \geq 0.$$

Subcase 3(ii):

$$A = 0, \;\; B > 0, \;\text{ and }\; \beta, \gamma, C \in (0, \infty).$$

Here, clearly, for $n \geq 0$,

$$x_{n+1} = \frac{\alpha + \beta x_n + \gamma x_{n-1}}{Bx_n + Cx_{n-1}} > \frac{\beta x_n + \gamma x_{n-1}}{Bx_n + Cx_{n-1}} > \frac{\min\{\beta,\gamma\}}{\max\{B,C\}}.$$

Set

$$L = \frac{\min\{\beta,\gamma\}}{\max\{B,C\}}.$$

Then for $n \geq 2$,

$$x_{n+1} = \frac{\alpha}{Bx_n + Cx_{n-1}} + \frac{\beta x_n + \gamma x_{n-1}}{Bx_n + Cx_{n-1}}$$

$$\leq \frac{\alpha}{(B+1)L} + \frac{\max\{\beta,\gamma\}}{\min\{B,C\}}$$

and so every solution is bounded.

Subcase 3(iii):

$$A = 0, \quad B > 0, \quad \beta > 0, \quad \text{and} \quad \gamma = 0.$$

For simplicity we normalize the equation in the following form

$$x_{n+1} = \frac{\alpha + x_n}{x_n + Cx_{n-1}}, \quad n = 0, 1, \ldots .$$

Let $\{x_n\}$ be a positive solution and choose positive numbers L, U such that

$$x_{-1}, x_0 \in (L, U), \quad 0 < L < \min\{1, \frac{\alpha}{C}\}, \quad \text{and} \quad LU = \frac{\alpha}{C}. \tag{2.2.2}$$

Then

$$x_1 - U = \frac{\alpha + x_0}{x_0 + Cx_{-1}} - \frac{\alpha}{CL} = \frac{\alpha CL - \alpha Cx_{-1} + (CL - \alpha)x_0}{CL(x_0 + Cx_{-1})} < 0$$

and

$$x_1 - L = \frac{(\alpha - CLx_{-1}) + x_0(1 - L)}{x_0 + Cx_{-1}} > 0.$$

That is,

$$x_1 \in (L, U),$$

and the boundedness of $\{x_n\}$ follows by induction.

Subcase 3(iv):

$$A = 0, \quad B > 0, \quad \beta = 0, \quad \text{and} \quad \gamma > 0.$$

Here we can normalize the equation in the form

$$x_{n+1} = \frac{\alpha + x_{n-1}}{Bx_n + x_{n-1}}, \quad n = 0, 1, \ldots .$$

Let $\{x_n\}$ be a positive solution and as in the above case choose positive numbers L and U such that (2.2.2) holds with C replaced by B. Then, as in the above case, we can show that

$$x_1 \in (L, U)$$

and the proof follows by induction.

Subcase 4(i):

$$A = 0, \quad B = 0, \quad \text{and} \quad \gamma = 0.$$

Here the equation reduces by a change of variables to the well-known **Lyness's equation**

$$x_{n+1} = \frac{\alpha + x_n}{x_{n-1}}, \quad n = 0, 1, \dots .$$

Lyness's equation is gifted with the **invariant**

$$(a + x_n + x_{n-1})(1 + \frac{1}{x_{n-1}})(1 + \frac{1}{x_n}) = \text{constant}, \quad \text{for } n \geq 0.$$

It is now clear from this invariant that no subsequence of $\{x_n\}$ can converge to ∞ and so every solution of Lyness's equation is bounded.

Subcase 4(ii):

$$A = 0, \quad B = 0, \quad \text{and} \quad \gamma > 0.$$

A change of variables reduces the equation in this case to an equation of the form

$$y_{n+1} = \frac{\alpha + y_n}{\gamma + y_{n-1}}, \quad n = 0, 1, \dots$$

for which the boundedness was established in case 2. The proof is complete. ∎

2.3 Boundedness by Iteration

In this section we will confirm the boundedness of 16 additional special cases of Eq.(2.0.1), namely:

#25,	#60,	#67,	#91,	#107,	#124,	#143,	#155,
#159,	#173,	#188,	#200,	#203,	#207,	#215,	#223.

See Appendix A.

Theorem 2.3.1 *Assume that*

$$\beta, A, C, D \in (0, \infty) \quad \textit{and} \quad \alpha, \gamma, \delta \in [0, \infty).$$

Then every solution of the equation

$$x_{n+1} = \frac{\alpha + \beta x_n + \gamma x_{n-1} + \delta x_{n-2}}{A + C x_{n-1} + D x_{n-2}}, \quad n = 0, 1, \dots$$

is bounded.

PROOF The proof is a consequence of the fact that for $n \geq 1$ we have

$$x_{n+2} = \frac{\alpha + \gamma x_n + \delta x_{n-1}}{A + Cx_n + Dx_{n-1}} + \frac{\beta}{A + Cx_n + Dx_{n-1}} \cdot \frac{\alpha + \beta x_n + \gamma x_{n-1} + \delta x_{n-2}}{A + Cx_{n-1} + Dx_{n-2}}.$$

▮

Corollary 2.3.1 *Every solution of each of the following eight special cases is bounded:*

#107, #143, #155, #159, #203, #207, #215, #223.

See Appendix A.

Theorem 2.3.2 *Assume that*

$$\alpha, \beta \in [0, \infty) \quad \text{and} \quad \gamma, \delta, C, D \in (0, \infty).$$

Then every positive solution of the equation

$$x_{n+1} = \frac{\alpha + \beta x_n + \gamma x_{n-1} + \delta x_{n-2}}{Cx_{n-1} + Dx_{n-2}}, \quad n = 0, 1, \ldots$$

is bounded from above and from below by positive numbers.

PROOF We have

$$x_{n+1} \geq \frac{\gamma x_{n-1} + \delta x_{n-2}}{Cx_{n-1} + Dx_{n-2}} \geq \frac{\min\{\gamma, \delta\}}{\max\{C, D\}}$$

and so $\{x_n\}$ is bounded from below by the positive number

$$m = \frac{\min\{\gamma, \delta\}}{\max\{C, D\}}.$$

Furthermore, for $n \geq 1$,

$$x_{n+2} = \frac{\alpha + \beta x_{n+1} + \gamma x_n + \delta x_{n-1}}{Cx_n + Dx_{n-1}}$$

$$\leq \frac{\alpha}{(C+D)m} + \frac{\gamma x_n + \delta x_{n-1}}{Cx_n + Dx_{n-1}} + \frac{\beta}{Cx_n + Dx_{n-1}} \cdot \frac{\alpha + \beta x_n + \gamma x_{n-1} + \delta x_{n-2}}{Cx_{n-1} + Dx_{n-2}}$$

and so the solution is also bounded from above. ▮

Corollary 2.3.2 *Every solution of each of the two special cases, #188 and #200, is bounded.*

See Appendix A.

Theorem 2.3.3 *Assume that*

$$\alpha, \delta, A \in [0, \infty) \quad \text{and} \quad \beta, \delta + A \in (0, \infty).$$

Then every solution of

$$x_{n+1} = \frac{\alpha + \beta x_n + \delta x_{n-2}}{A + x_{n-2}}, \quad n = 0, 1, \ldots$$

is bounded.

PROOF Note that for $n \geq 2$,

$$x_{n+3} = \frac{\alpha + \beta x_{n+2} + \delta x_n}{A + x_n} = \frac{\alpha + \delta x_n}{A + x_n} + \frac{\beta}{A + x_n} \cdot \frac{\alpha + \beta x_{n+1} + \delta x_{n-1}}{A + x_{n-1}}$$

$$= \frac{\alpha + \delta x_n}{A + x_n} + \frac{\beta}{A + x_n}\left[\frac{\alpha + \delta x_{n-1}}{A + x_{n-1}} + \beta \cdot \frac{\alpha + \beta x_n + \delta x_{n-2}}{(A + x_{n-1})(A + x_{n-2})}\right]$$

$$= \frac{\alpha + \delta x_n}{A + x_n} + \frac{\beta}{A + x_n} \cdot \frac{\alpha + \delta x_{n-1}}{A + x_{n-1}} +$$

$$\frac{\beta^2}{(A + x_n)(A + x_{n-1})}\left[\frac{\alpha + \delta x_{n-2}}{A + x_{n-2}} + \frac{\beta x_n}{A + x_{n-2}}\right]$$

from which the boundedness of $\{x_n\}$ follows. ∎

Corollary 2.3.3 *Every solution of each of the following six special cases is bounded:*

#25, #60, #67, #91, #124, #173.

See Appendix A.

Theorems 2.3.1, 2.3.2, and 2.3.3 have straightforward extensions to fourth-order rational difference equations. See [66]. Actually, we can use these extensions to establish the boundedness of every solution in 104 special cases of the fourth-order rational difference equation

$$x_{n+1} = \frac{\alpha + \beta x_n + \gamma x_{n-1} + \delta x_{n-2} + \epsilon x_{n-3}}{A + Bx_n + Cx_{n-1} + Dx_{n-2} + Ex_{n-3}}, \quad n = 0, 1, \ldots . \tag{2.3.1}$$

When we deal with difference equations of order $k \geq 4$ it will be convenient to switch to a new notation for the special cases. In this notation the kth-order rational difference equation is written, for the sake of the new notation, in the form

$$x_n = \frac{\beta_0 + \sum_{i=1}^{k} \beta_i x_{n-i}}{B_0 + \sum_{i=1}^{k} B_i x_{n-i}}, \quad n = 0, 1, \ldots . \tag{2.3.2}$$

Now for a special case of Eq.(2.3.2), we define

$$v_i = \begin{cases} 2^{2i}, & \text{if } \beta_i > 0 \\ 0, & \text{if } \beta_i = 0 \end{cases} \quad \text{and} \quad V_i = \begin{cases} 2^{2i+1}, & \text{if } B_i > 0 \\ 0, & \text{if } B_i = 0. \end{cases}$$

Then the identifying number assigned to a special case of Eq.(2.3.2) is

$$\sum_{i=1}^{k} (v_i + V_i).$$

In this book, we will use the above numbering system for equations of order ≥ 4. For equations of order ≤ 3, we still use the numbers listed in Appendix A. See also [66].

In this notation 104 special cases of order 4 of Eq.(2.3.1) are the following:

#518 – 519	#694 – 695	#820 – 823
#530 – 531	#710 – 711	#902 – 903
#534 – 535	#722 – 723	#914 – 915
#538 – 539	#726 – 727	#918 – 919
#542 – 543	#730 – 731	#922 – 923
#550 – 551	#734 – 735	#926 – 927
#566 – 567	#742 – 743	#934 – 935
#646 – 647	#758 – 759	#950 – 951
#658 – 659	#772 – 775	#964 – 967
#662 – 663	#784 – 791	#976 – 983
#666 – 667	#794 – 797	#986 – 991
#670 – 671	#798 – 799	#998 – 999
#678 – 679	#806 – 807	#1012 – 1015.

See Appendix B.

For higher-order rational difference equations the following result extends Theorems 2.3.1, 2.3.2, and 2.3.3 and establishes by the **method of iteration** the boundedness of every solution of the following rational difference equation

$$x_n = \frac{\alpha + \sum_{i=1}^{k} \beta_i x_{n-i}}{A + \sum_{i=1}^{k} B_i x_{n-i}}, \quad n = 0, 1, \ldots \tag{2.3.3}$$

with nonnegative parameters and arbitrary nonnegative initial conditions such that the denominator is always positive. See [66].

Theorem 2.3.4 *Let I_β and I_B denote the following sets:*

$$I_\beta = \{j \in \{1, 2, \ldots, k\} : \beta_j > 0\}$$

and

$$I_B = \{j \in \{1, 2, \ldots, k\} : B_j > 0\}$$

and assume that either

$$A > 0$$

or

$$A = 0, \quad I_B \neq \emptyset, \quad \text{and} \quad I_B \subset I_\beta.$$

Furthermore, assume that for every infinite sequence

$$\{c_m\}_{m=1}^{\infty} \quad \text{with} \quad c_m \in I_\beta$$

there exist positive integers N_1 and N_2 such that

$$\left(\sum_{m=N_1}^{N_2} c_m\right) \in I_B.$$

Then every solution of Eq.(2.3.3) is bounded.

Theorem 2.3.4 establishes by **the method of iteration** the boundedness of every solution in 126 special cases of Eq.(2.3.3), with $k = 4$. These are the six second-order equations:

#24, #55, #66, #84, #119, #166,

the 16 third-order equations:

#25, #60, #67, #91, #107, #124, #143, #155
#159, #173, #188, #200, #203, #207, #215, #223,

and the 104 fourth-order equations listed previously.

Open Problem 2.3.1 *Assume that $k \geq 5$. How many special cases of Eq.(2.3.3) are predicted, by Theorem 2.3.4, that have bounded solutions only?*

Open Problem 2.3.2 *For $k \in \{3, 4\}$, determine all special cases of Eq.(2.3.3) with bounded solutions only. In particular, confirm or refute our conjecture that, when $k = 4$, Eq.(2.3.3) has 542 special cases with bounded solutions only.*

Conjecture 2.3.1 *For each equation listed in Appendices A and B with a* **B** *next to the equation, prove that local stability of an equilibrium point implies that the equilibrium point is a global attractor of all positive solutions.*

2.4 Boundedness of the Special Case #58

In this section we establish the boundedness of every solution of the special case #58 which for convenience we normalize in the form

$$\#58: \ x_{n+1} = \beta + \frac{x_{n-2}}{x_n}, \quad n = 0, 1, \ldots \tag{2.4.1}$$

with the parameter β positive and with arbitrary positive initial conditions x_{-2}, x_{-1}, x_0.

The following theorem is the main result in this section.

Theorem 2.4.1 *Every solution of Eq.(2.4.1) is bounded.*

PROOF First we make thc following useful general observations about the solutions of Eq.(2.4.1):

$$x_{n+1} > \beta, \quad \text{for } n \geq 0. \tag{2.4.2}$$

$$x_{n+1} < \beta + \frac{1}{\beta} x_{n-2}, \quad \text{for } n \geq 1. \tag{2.4.3}$$

$$\begin{aligned} x_{n+1} &< \beta + \frac{1}{\beta}\left(\beta + \frac{x_{n-5}}{x_{n-3}}\right) \\ &< \beta + 1 + \frac{1}{\beta^2} x_{n-5}, \quad \text{for } n \geq 4. \end{aligned} \tag{2.4.4}$$

$$x_{n_i+1} \to \infty \Rightarrow x_{n_i-2} \to \infty. \tag{2.4.5}$$

$$x_{n_i+1} \to \beta \Rightarrow x_{n_i} \to \infty. \tag{2.4.6}$$

Now assume for the sake of contradiction that Eq.(2.4.1) has an unbounded solution $\{x_n\}$. Then there exists a sequence of indices $\{n_i\}$ such that

$$x_{n_i+1} \to \infty \tag{2.4.7}$$

and, for every i,

$$x_{n_i+1} > x_j, \quad \text{for all } j < n_i + 1. \tag{2.4.8}$$

From (2.4.7) and (2.4.5) it follows that

$$x_{n_i-2} \to \infty, \quad x_{n_i-5} \to \infty, \quad \text{and } x_{n_i-8} \to \infty. \tag{2.4.9}$$

Next we claim that the subsequence $\{x_{n_i-4}\}$ is bounded. Otherwise, there would exist a subsequence of $\{n_i\}$, which we still denote by $\{n_i\}$, such that

$$x_{n_i-4} \to \infty, \quad x_{n_i-7} \to \infty, \quad \text{and } x_{n_i-10} \to \infty. \tag{2.4.10}$$

Note that, for every i,

$$x_{n_i-4} = \beta + \frac{x_{n_i-7}}{x_{n_i-5}}$$

and

$$x_{n_i-7} = \beta + \frac{x_{n_i-10}}{x_{n_i-8}}.$$

Hence, in view of (2.4.9) and (2.4.10), we have eventually

$$x_{n_i-7} > x_{n_i-5} \text{ and } x_{n_i-10} > x_{n_i-8} \tag{2.4.11}$$

and

$$\frac{x_{n_i-7}}{x_{n_i-5}} \to \infty \text{ and } \frac{x_{n_i-10}}{x_{n_i-8}} \to \infty. \tag{2.4.12}$$

Then, from (2.4.11) and (2.4.4), we see that eventually

$$\begin{aligned} x_{n_i+1} &< \beta + 1 + \frac{1}{\beta^2} x_{n_i-7} \\ &= \beta + 1 + \frac{1}{\beta^2}\left(\beta + \frac{x_{n_i-10}}{x_{n_i-8}}\right) \\ &= \beta + 1 + \frac{1}{\beta} + \frac{1}{\beta^2} \cdot \left(\frac{x_{n_i-10}}{x_{n_i-8}}\right). \end{aligned}$$

From (2.4.10), it follows that the right-hand side of the above inequality is eventually less than x_{n_i-10}, which contradicts (2.4.8) and establishes our claim that $\{x_{n_i-4}\}$ is bounded. From this and (2.4.9) we have

$$x_{n_i-1} = \beta + \frac{x_{n_i-4}}{x_{n_i-2}} \to \beta.$$

Also,

$$\liminf_{i\to\infty} x_{n_i-3} > \beta.$$

Otherwise, a subsequence of $\{x_{n_i-3}\}$ would converge to β and then from (2.4.6), $\{x_{n_i-4}\}$ would be unbounded, which is not true.

Hence, eventually,

$$x_{n_i} = \beta + \frac{x_{n_i-3}}{x_{n_i-1}} > \beta + 1$$

and so, for i sufficiently large,

$$x_{n_i+1} = \beta + \frac{x_{n_i-2}}{x_{n_i}} < \beta + \frac{x_{n_i-2}}{\beta+1} < x_{n_i-2},$$

which contradicts (2.4.8). The proof is complete. ∎

2.5 Boundedness of $x_{n+1} = \dfrac{\alpha + \beta x_n + x_{n-2}}{A + x_n}$

In this section we will confirm the boundedness of four additional special cases of Eq.(2.0.1), namely:

#77, #89, #122, #171.

See Appendix A.

Theorem 2.5.1 *Assume that*

$$\alpha + \beta > 0 \text{ and } \beta + A > 0. \tag{2.5.1}$$

Then every solution of Eq.(2.0.5) is bounded.

The proof of the theorem will be established through a series of lemmas. Throughout this section, unless otherwise stated, we will assume that (2.5.1) holds and also that

$$A < 1.$$

Lemma 2.5.1 *Every solution $\{x_n\}$ of Eq.(2.0.5) is eventually bounded from below by β.*

PROOF When $\beta = 0$, the result is trivial. Assume

$$\beta > 0.$$

Also assume, without loss of generality, that the initial conditions x_{-2}, x_{-1}, x_0 are positive and suppose for the sake of contradiction that there exists N sufficiently large such that

$$x_{N+1} = \frac{\alpha + \beta x_N + x_{N-2}}{A + x_N} \leq \beta.$$

Then

$$x_{N-2} \leq \beta A.$$

Similarly,

$$x_{N-5} \leq \beta A^2,$$

which inductively leads to a contradiction. ∎

In the sequel, whenever we refer to a solution $\{x_n\}$ of Eq.(2.0.5), unless otherwise stated, we will assume that, for $n \geq -2$,

$$x_n > \beta.$$

Lemma 2.5.2 *Every solution $\{x_n\}$ of Eq. (2.0.5) satisfies the following inequality, for $n \geq 0$ and $j \in \{-1, 0, \dots\}$,*

$$x_{n+1} < \alpha \sum_{s=1}^{j+2} \frac{1}{(A+\beta)^s} + \beta \sum_{s=0}^{j+1} \frac{1}{(A+\beta)^s} + \frac{x_{n-5-3j}}{(A+\beta)^{j+2}}. \tag{2.5.2}$$

PROOF The proof is by induction. Clearly, for $n \geq 0$,

$$x_{n+1} = \frac{\alpha + \beta x_n + x_{n-2}}{A + x_n} < \frac{\alpha}{A+\beta} + \beta + \frac{x_{n-2}}{A+\beta}$$

and so (2.5.2) holds, when $j = -1$. Assume that, for $n \geq 0$ and $j > -1$,

$$x_{n+1} < \alpha \sum_{s=1}^{j+2} \frac{1}{(A+\beta)^s} + \beta \sum_{s=0}^{j+1} \frac{1}{(A+\beta)^s} + \frac{x_{n-5-3j}}{(A+\beta)^{j+2}}.$$

Clearly, for $n \geq 0$ and $j > -1$,

$$x_{n-5-3j} = \frac{\alpha + \beta x_{n-6-3j} + x_{n-8-3j}}{A + x_{n-6-3j}} < \frac{\alpha}{A+\beta} + \beta + \frac{x_{n-5-3(j+1)}}{A+\beta}$$

and the result follows by combining the last two inequalities. ∎

Lemma 2.5.3 *Let $\{x_n\}$ be a solution of Eq.(2.0.5). Assume that $\{x_{n_i}\}$ is a subsequence of the solution $\{x_n\}$, which converges to β. Then the subsequence $\{x_{n_i-1}\}$ is unbounded.*

PROOF Suppose for the sake of contradiction that the subsequence $\{x_{n_i-1}\}$ is bounded. Rearranging Eq.(2.0.5) we have

$$Ax_{n_i} + x_{n_i-1}(x_{n_i} - \beta) = \alpha + x_{n_i-3}, \quad i = 0, 1, \dots \tag{2.5.3}$$

and so

$$A\beta = \alpha + \liminf_{i \to \infty} x_{n_i-3} \geq \alpha + \beta,$$

which implies that either

$$\beta > 0 \text{ and } 0 > (A-1)\beta \geq \alpha$$

or

$$\alpha = \beta = 0.$$

This is a contradiction and the proof is complete. ▮

Lemma 2.5.4 *Let $\{x_{n_i+1}\}$ be a subsequence of a solution $\{x_n\}$ such that*

$$\lim_{i\to\infty} x_{n_i+1} = \infty \tag{2.5.4}$$

with

$$\max\{x_n : -2 \leq n < n_i + 1\} < x_{n_i+1}, \quad i = 0, 1, \dots . \tag{2.5.5}$$

Then for $j \in \{0, 1, \dots\}$, the following hold:

$$\lim_{i\to\infty} x_{n_i-2-3j} = \infty \tag{2.5.6}$$

$$\limsup_{i\to\infty} x_{n_i-4-3j} < \infty \tag{2.5.7}$$

$$\lim_{i\to\infty} x_{n_i-1-3j} = \beta \tag{2.5.8}$$

$$\liminf_{i\to\infty} x_{n_i-3-3j} > \beta. \tag{2.5.9}$$

PROOF From (2.5.2), for $i \geq 0$ and $j \geq 0$,

$$x_{n_i+1} < \alpha \sum_{s=1}^{j+1} \frac{1}{(A+\beta)^s} + \beta \sum_{s=0}^{j} \frac{1}{(A+\beta)^s} + \frac{x_{n_i-2-3j}}{(A+\beta)^{j+1}}$$

and so, clearly, (2.5.6) is satisfied. To establish (2.5.7), suppose for the sake of contradiction that, for some $j \geq 0$,

$$x_{n_i-4-3j} \to \infty.$$

Then, clearly,

$$x_{n_i-7-3j},\ x_{n_i-10-3j} \to \infty. \tag{2.5.10}$$

Also from (2.5.6)

$$x_{n_i-5-3j},\ x_{n_i-8-3j} \to \infty. \tag{2.5.11}$$

For N sufficiently large, in view of (2.5.10) and (2.5.11), we see that, for $i \geq N$,

$$x_{n_i-4-3j} > \beta + 2 \text{ and } x_{n_i-5-3j} > \alpha - A(\beta + 2).$$

Now we claim that, for $i \geq N$,

$$x_{n_i-7-3j} > x_{n_i-5-3j}. \tag{2.5.12}$$

Suppose for the sake of contradiction that there exists $i_0 \geq N$ such that

$$x_{n_{i_0}-5-3j} \geq x_{n_{i_0}-7-3j}.$$

Then

$$x_{n_{i_0}-4-3j} = \frac{\alpha + \beta x_{n_{i_0}-5-3j} + x_{n_{i_0}-7-3j}}{A + x_{n_{i_0}-5-3j}} \leq \frac{\alpha + (\beta+1) x_{n_{i_0}-5-3j}}{A + x_{n_{i_0}-5-3j}} < \beta + 2,$$

which is a contradiction. The proof of (2.5.12) is complete. From (2.5.2), for $i \geq 0$ and $j \geq 0$, we have

$$x_{n_i+1} < \alpha \sum_{s=1}^{j+2} \frac{1}{(A+\beta)^s} + \beta \sum_{s=0}^{j+1} \frac{1}{(A+\beta)^s} + \frac{x_{n_i-5-3j}}{(A+\beta)^{j+2}}.$$

From this and (2.5.12) we obtain

$$x_{n_i+1} < \alpha \sum_{s=1}^{j+2} \frac{1}{(A+\beta)^s} + \beta \sum_{s=0}^{j+1} \frac{1}{(A+\beta)^s} + \frac{x_{n_i-7-3j}}{(A+\beta)^{j+2}}$$

and so, for $i \geq 0$ and $j \geq 0$,

$$x_{n_i+1} < \alpha \sum_{s=1}^{j+3} \frac{1}{(A+\beta)^s} + \beta \sum_{s=0}^{j+2} \frac{1}{(A+\beta)^s} + \frac{1}{(A+\beta)^{j+2}} \frac{x_{n_i-10-3j}}{x_{n_i-8-3j}}.$$

Then for N sufficiently large, in view of (2.5.10) and (2.5.11), we see that, for $j \geq 0$,

$$x_{n_N+1} < \max\{x_{n_N-10-3j}, x_{n_N-8-3j}\},$$

which contradicts (2.5.5). The proof of (2.5.7) is complete. To establish (2.5.8), note that, for $i \geq 0$ and $j \geq 0$,

$$x_{n_i-1-3j} = \frac{\alpha + \beta x_{n_i-2-3j}}{A + x_{n_i-2-3j}} + \frac{x_{n_i-4-3j}}{A + x_{n_i-2-3j}}.$$

By taking limits in the last equation as $i \to \infty$, in view of (2.5.6) and (2.5.7), we find that, for $j \geq 0$,

$$x_{n_i-1-3j} \to \beta$$

and so (2.5.8) is established. To establish (2.5.9), suppose for the sake of contradiction that, for some $j \geq 0$,

$$\liminf_{i\to\infty} x_{n_i-3-3j} = \beta.$$

From this and Lemma 2.5.3 we have

$$\limsup_{i\to\infty} x_{n_i-4-3j} = \infty$$

which contradicts (2.5.7). The proof is complete. ∎

PROOF The proof of the theorem is divided into the following eight cases:

Case 1: $\beta < A - 1$. Clearly, for $n \geq 0$,

$$x_{n+1} < \frac{\alpha}{A} + \beta + \frac{1}{A} x_{n-2}.$$

By using Theorem 1.4.1 it follows that

$$\limsup_{n\to\infty} x_n \leq \frac{\alpha + \beta A}{A - 1}.$$

Case 2: $\beta = A - 1$ and $\beta > 0$. Clearly, for $n \geq 0$,

$$x_{n+1} < \frac{\alpha}{\beta + 1} + \beta + \frac{1}{\beta + 1} x_{n-2}.$$

By using Theorem 1.4.1 it follows that

$$\limsup_{n\to\infty} x_n \leq \frac{\alpha + \beta(\beta + 1)}{\beta}.$$

Case 3: $\beta = 0$ and $A = 1$. Without loss of generality assume that the initial conditions x_{-2}, x_{-1}, x_0 are positive. Let $m > 0$ be such that

$$m < \min\{x_{-2}, x_{-1}, x_0, \frac{\alpha}{x_{-2}}, \frac{\alpha}{x_{-1}}, \frac{\alpha}{x_0}\}.$$

We claim that, for all $n \geq -2$,

$$m < x_n < \frac{\alpha}{m}. \tag{2.5.13}$$

Indeed,

$$m = \frac{\alpha + m}{1 + \frac{\alpha}{m}} < x_1 = \frac{\alpha + x_{-2}}{1 + x_0} < \frac{\alpha + \frac{\alpha}{m}}{1 + m} = \frac{\alpha}{m}$$

and the proof follows by induction.

Case 4: $\beta > A - 1 > 0$. Clearly, for $n \geq 0$,

$$x_{n+1} < \frac{\alpha}{A} + \beta + \frac{1}{A} x_{n-2}.$$

By using Theorem 1.4.1 it follows that

$$\limsup_{n\to\infty} x_n \leq \frac{\alpha + \beta A}{A - 1}.$$

Case 5: $\beta > 0$ and $A = 1$. Without loss of generality assume that the initial conditions x_{-2}, x_{-1}, x_0 are positive. Let $m > 0$ be such that

$$m < \min\{\beta, x_{-2}, x_{-1}, x_0\}.$$

We claim that, for all $n \geq -2$,

$$x_n > m. \tag{2.5.14}$$

Indeed,

$$x_1 - m = \frac{\alpha + (\beta - m)x_0 + (x_{-2} - m)}{1 + x_0} > 0$$

and the proof of (2.5.14) follows by induction. In view of (2.5.14) it follows that, for all $n \geq 0$,

$$x_{n+1} < \alpha + \beta + \frac{1}{1+m} x_{n-2}.$$

By using Theorem 1.4.1 it follows that

$$\limsup_{n\to\infty} x_n \leq \frac{(\alpha + \beta)(m+1)}{m}.$$

Case 6: $\beta > 1 - A > 0$. For $n \geq 0$, in view of Lemma 2.5.1 we have

$$x_{n+1} < \frac{\alpha}{A+\beta} + \beta + \frac{1}{A+\beta} x_{n-2}.$$

By using Theorem 1.4.1 it follows that

$$\limsup_{n\to\infty} x_n \leq \frac{\alpha + \beta(\beta + A)}{\beta + A - 1}.$$

Case 7: $1 - A > \beta \geq 0$. Suppose for the sake of contradiction that $\{x_{n_i+1}\}$ is a subsequence of the solution $\{x_n\}$ such that (2.5.4) and (2.5.5) hold. From (2.5.6) we have

$$\lim_{i\to\infty} x_{n_i-2} = \infty.$$

Without loss of generality, assume that, for $i \geq 0$,

$$x_{n_i-2} > \frac{\alpha + \beta\bar{x}}{A + \bar{x} - 1},$$

where

$$\bar{x} = \frac{\beta + 1 - A + \sqrt{(\beta + 1 - A)^2 + 4\alpha}}{2}$$

is the positive equilibrium of Eq.(2.5.6). Now we claim that, for $i \geq 0$,

$$x_{n_i} < \bar{x}. \tag{2.5.15}$$

Otherwise, for some i_0, $x_{n_{i_0}} \geq \bar{x}$. Then

$$x_{n_{i_0}+1} = \frac{\alpha + \beta x_{n_{i_0}} + x_{n_{i_0}-2}}{A + x_{n_{i_0}}} \leq \frac{\alpha + \beta \bar{x} + x_{n_{i_0}-2}}{A + \bar{x}} < x_{n_{i_0}-2},$$

which contradicts (2.5.5). Hence, (2.5.15) holds. Let $0 < \epsilon < 1 - A - \beta$ and let s be sufficiently large such that

$$\alpha \sum_{t=1}^{s+1} \frac{1}{(A + \beta + \epsilon)^t} > \bar{x}.$$

Then for N sufficiently large and $i \geq N$, and in view of (2.5.8) and (2.5.9), we see that, for $j \geq 0$,

$$x_{n_i-3-3j} > \beta + \epsilon \text{ and } x_{n_i-1-3j} < \beta + \epsilon.$$

Hence,

$$x_{n_i} = \frac{\alpha + \beta x_{n_i-1} + x_{n_i-3}}{A + x_{n_i-1}} > \frac{\alpha}{A + \beta + \epsilon} + \frac{x_{n_i-3}}{A + \beta + \epsilon} > \dots$$

$$> \alpha \sum_{t=1}^{s+1} \frac{1}{(A + \beta + \epsilon)^t} > \bar{x},$$

which contradicts (2.5.15).

Case 8: $\beta = 1 - A > 0$. The proof in this case is along the same lines as the proof in Case 7 and will be omitted. ∎

2.6 Boundedness of the Special Case #63

In this section we establish the boundedness of every solution of the special case #63, which for convenience we normalize in the form

$$\#63: \ x_{n+1} = \gamma + \frac{x_{n-2}}{x_{n-1}}, \quad n = 0, 1, \dots \tag{2.6.1}$$

with the parameter γ positive and with arbitrary nonnegative initial conditions x_{-2}, x_{-1}, x_0 such that the denominator is always positive.

The following theorem is the main result in this section.

Theorem 2.6.1 *Every solution of Eq.(2.6.1) is bounded.*

PROOF First we make the following useful general observations about the solutions of Eq.(2.6.1):

$$x_{n+1} > \gamma, \quad \text{for } n \geq 0. \tag{2.6.2}$$

$$x_{n+1} < \gamma + \frac{1}{\gamma} x_{n-2}, \quad \text{for } n \geq 2. \tag{2.6.3}$$

$$x_{n+1} < \gamma + \frac{1}{\gamma}\left(\gamma + \frac{x_{n-5}}{x_{n-4}}\right)$$

$$< \gamma + 1 + \frac{1}{\gamma^2} x_{n-5}, \quad \text{for } n \geq 5. \tag{2.6.4}$$

$$x_{n+1} \to \infty \Rightarrow x_{n-2} \to \infty. \tag{2.6.5}$$

$$x_{n+1} \to \gamma \Rightarrow x_{n-1} \to \infty. \tag{2.6.6}$$

Now assume for the sake of contradiction that Eq.(2.6.1) has an unbounded solution $\{x_n\}$. Then there exists a sequence $\{n_i\}$ such that

$$x_{n_i+1} \to \infty \tag{2.6.7}$$

and, for every i,

$$x_{n_i+1} > x_j, \quad \text{for all } j < n_i + 1. \tag{2.6.8}$$

From (2.6.7) and (2.6.5) it follows that

$$x_{n_i-2} \to \infty, \quad x_{n_i-5} \to \infty, \quad \text{and } x_{n_i-8} \to \infty. \tag{2.6.9}$$

Next we claim that the subsequence $\{x_{n_i-6}\}$ is bounded. Otherwise, there would exist a subsequence of $\{n_i\}$, which we still denote by $\{n_i\}$ such that

$$x_{n_i-6} \to \infty, \quad x_{n_i-9} \to \infty, \quad \text{and } x_{n_i-12} \to \infty. \tag{2.6.10}$$

From

$$x_{n_i-6} = \gamma + \frac{x_{n_i-9}}{x_{n_i-8}}, \quad i = 0, 1, \ldots$$

and

$$x_{n_i-9} = \gamma + \frac{x_{n_i-12}}{x_{n_i-11}}, \quad i = 0, 1, \ldots$$

and (2.6.10), we see that, eventually,

$$x_{n_i-9} > x_{n_i-8} \text{ and } x_{n_i-12} > x_{n_i-11}. \tag{2.6.11}$$

Then from (2.6.11) and (2.6.4), we see that, eventually,

$$x_{n_i+1} < \gamma + 1 + \frac{1}{\gamma^2} x_{n_i-5}$$

$$= \gamma + 1 + \frac{1}{\gamma^2}\left(\gamma + \frac{x_{n_i-8}}{x_{n_i-7}}\right)$$

$$< \gamma + 1 + \frac{1}{\gamma} + \frac{1}{\gamma^3} x_{n_i-8}$$

$$< \gamma + 1 + \frac{1}{\gamma} + \frac{1}{\gamma^3} x_{n_i-9}$$

$$= \gamma + 1 + \frac{1}{\gamma} + \frac{1}{\gamma^3}\left(\gamma + \frac{x_{n_i-12}}{x_{n_i-11}}\right).$$

From (2.6.9), (2.6.10), and (2.6.11), we can see that the right-hand side of the above inequality is eventually less than x_{n_i-12}, which contradicts (2.6.8) and establishes our claim that $\{x_{n_i-6}\}$ is bounded.

Therefore, from (2.6.9),

$$x_{n_i-3} = \gamma + \frac{x_{n_i-6}}{x_{n_i-5}} \to \gamma.$$

Also,

$$\liminf_{i\to\infty} x_{n_i-4} > \gamma.$$

Otherwise, a subsequence of $\{x_{n_i-4}\}$ would converge to γ and then from (2.6.6), $\{x_{n_i-6}\}$ would be unbounded, which is not true.

Hence, eventually,

$$x_{n_i-1} = \gamma + \frac{x_{n_i-4}}{x_{n_i-3}} > \gamma + 1$$

and so, for i sufficiently large,

$$x_{n_i+1} = \gamma + \frac{x_{n_i-2}}{x_{n_i-1}} < \gamma + \frac{x_{n_i-2}}{\gamma+1} < x_{n_i-2},$$

which contradicts (2.6.8) and completes the proof of the theorem. ∎

2.7 Boundedness of $x_{n+1} = \dfrac{\alpha + \beta x_n + \gamma x_{n-1} + x_{n-2}}{A + x_{n-1}}$

In this section we will confirm the boundedness of 10 additional special cases of Eq.(2.0.1), namely:

#78, #90, #96, #127, #131,
#139, #172, #178, #184, #196.

See Appendix A.

Theorem 2.7.1 *Assume that*

$$\gamma + A > 0 \text{ and } \alpha + \beta + \gamma > 0. \tag{2.7.1}$$

Then every solution of Eq.(2.0.7) is bounded.

The proof of the Theorem will be established through a series of lemmas. For the rest of this section, unless otherwise stated, we will assume that (2.7.1) holds and that

$$\beta + 1 > A\,.$$

Lemma 2.7.1 *Every solution $\{x_n\}$ of Eq.(2.0.7) is eventually bounded from below by γ.*

PROOF Suppose for the sake of contradiction that there exists N, sufficiently large, such that

$$x_{N+1} = \frac{\alpha + \beta x_N + \gamma x_{N-1} + x_{N-2}}{A + x_{N-1}} \leq \gamma.$$

Then clearly

$$\min\{x_N, x_{N-2}\} \leq \gamma \cdot \frac{A}{\beta + 1}.$$

Similarly

$$\min\{x_N, x_{N-1}, x_{N-2}, x_{N-3}, x_{N-5}\} \leq \gamma (\frac{A}{\beta + 1})^2.$$

Sufficient repetition of this argument leads to a contradiction. ∎

In the sequel, whenever we refer to a solution of Eq.(2.0.7), unless otherwise stated, we will assume that

$$x_n > \gamma, \quad n = -2, -1, \ldots .$$

Lemma 2.7.2 *Let $\{x_{n_i+1}\}$ be a subsequence of a solution $\{x_n\}$ such that*

$$\lim_{i\to\infty} x_{n_i+1} = \infty \tag{2.7.2}$$

with

$$\max\{x_n : -2 \le n < n_i + 1\} < x_{n_i+1}, \quad i = 0, 1, \ldots . \tag{2.7.3}$$

Also assume that

$$x_{n_i+1} \le P + \frac{Q + \sum_{t=0}^{r} a_t x_{n_i-t}}{R + x_{n_i-m}}, \quad i = 0, 1 \ldots, \tag{2.7.4}$$

where P, Q, R, m, and $\{a_r\}, r \in \{0, 1, \ldots\}$ are nonnegative real numbers. Then

$$\limsup_{i\to\infty} x_{n_i-m} < \infty. \tag{2.7.5}$$

PROOF Suppose for the sake of contradiction that $\{x_{n_i-m}\}$ is an unbounded sequence. Then for N sufficiently large,

$$x_{n_N+1} < \max\{\max_{0\le t\le r} x_{n_N-t}, x_{n_N-m}\}.$$

This contradicts (2.7.3). The proof is complete. ∎

Lemma 2.7.3 *Assume $\beta > 0$. Let $\{x_{n_i+1}\}$ be a subsequence of a solution $\{x_n\}$ that satisfies (2.7.2) and (2.7.3). Also assume that*

$$x_{n_i} \ge x_{n_i-2}, \quad i = 0, 1, \ldots . \tag{2.7.6}$$

Then

$$\limsup_{i\to\infty} x_{n_i-1} < \infty, \quad \limsup_{i\to\infty} x_{n_i-2} < \infty, \quad \limsup_{i\to\infty} x_{n_i-5} < \infty, \liminf_{i\to\infty} \frac{x_{n_i-4}}{x_{n_i-3}} > 0 \tag{2.7.7}$$

and

$$\lim_{i\to\infty} x_{n_i-3} = \infty, \quad \lim_{i\to\infty} x_{n_i-4} = \infty. \tag{2.7.8}$$

PROOF For $i \ge 0$,

$$x_{n_i+1} = \frac{\alpha + \beta x_{n_i} + \gamma x_{n_i-1} + x_{n_i-2}}{A + x_{n_i-1}} < \frac{\beta+1}{A+\gamma} x_{n_i} + \frac{\alpha}{A+\gamma} + \gamma$$

and so

$$\lim_{i\to\infty} x_{n_i} = \infty.$$

Clearly, for $i \geq 0$,

$$x_{n_i+1} = \frac{\alpha + \beta x_{n_i} + \gamma x_{n_i-1} + x_{n_i-2}}{A + x_{n_i-1}} < \frac{\alpha}{A+\gamma} + \gamma + \frac{(\beta+1)x_{n_i}}{A + x_{n_i-1}}.$$

From this and (2.7.5) it follows that

$$\limsup_{i\to\infty} x_{n_i-1} < \infty.$$

Furthermore,

$$x_{n_i} = \frac{\alpha + \beta x_{n_i-1} + \gamma x_{n_i-2} + x_{n_i-3}}{A + x_{n_i-2}}$$
$$< \frac{1}{A+\gamma} x_{n_i-3} + \gamma + \frac{\beta}{A+\gamma} x_{n_i-1} + \frac{\alpha}{A+\gamma}.$$

Hence,

$$\lim_{i\to\infty} x_{n_i-3} = \infty. \tag{2.7.9}$$

Furthermore, for $i \geq 0$,

$$x_{n_i+1} < \frac{\alpha}{A+\gamma} + \gamma + \frac{\beta+1}{A+\gamma} \frac{\alpha + \beta x_{n_i-1} + \gamma x_{n_i-2} + x_{n_i-3}}{A + x_{n_i-2}}.$$

From this and (2.7.5) it follows that

$$\limsup_{i\to\infty} x_{n_i-2} < \infty.$$

Also,

$$x_{n_i-2} > \beta \frac{x_{n_i-3}}{A + x_{n_i-4}} = \beta \frac{x_{n_i-3}}{x_{n_i-4}} \frac{1}{\frac{A}{x_{n_i-4}} + 1}.$$

From this, the fact that subsequence $\{x_{n_i-2}\}$ is bounded, and from (2.7.9) it follows that

$$\lim_{i\to\infty} x_{n_i-4} = \infty$$

and

$$\liminf_{i\to\infty} \frac{x_{n_i-4}}{x_{n_i-3}} > 0.$$

Also, for $i \geq 0$,

$$x_{n_i+1} < \frac{\alpha + \gamma(\beta+1)}{A+\gamma} + \frac{(\alpha+\beta)(\beta+1)}{(A+\gamma)^2} + \gamma$$
$$+ \frac{\beta+1}{(A+\gamma)^2} \frac{\alpha + \beta x_{n_i-4} + \gamma x_{n_i-5} + x_{n_i-6}}{A + x_{n_i-5}}.$$

From this and (2.7.5) it follows that

$$\limsup_{i\to\infty} x_{n_i-5} < \infty.$$

The proof of (2.7.7) and (2.7.8) is complete. ∎

The proof of the following identity is straightforward and will be omitted.

Lemma 2.7.4 *Every solution $\{x_n\}$ of Eq.(2.0.7) satisfies:*

$$x_{n+1} = \frac{\beta x_{n-3}(A+x_{n-3})(A+x_{n-4})}{\prod_{t=0}^{1}[A(A+x_{n-3-t})+\alpha+\beta x_{n-2-t}+\gamma x_{n-3-t}+x_{n-4-t}]}$$

$$+\frac{\alpha+\gamma x_{n-1}+x_{n-2}}{A+x_{n-1}}+\frac{\beta(\alpha+\beta x_{n-1}+\gamma x_{n-2})}{(A+x_{n-1})(A+x_{n-2})}. \quad (2.7.10)$$

Lemma 2.7.5 *Assume $\beta > 0$. Let $\{x_{n_i+1}\}$ be a subsequence of a solution $\{x_n\}$ that satisfies (2.7.2), (2.7.3), and (2.7.6). Then the following are true, with $\epsilon > 0$ and $i \geq 0$:*

$$A(A+x_{n_i-4})+\alpha+\beta x_{n_i-3}+\gamma x_{n_i-4}+x_{n_i-5} > (\beta+\epsilon)(A+x_{n_i-3}) \quad (2.7.11)$$

and

$$A(A+x_{n_i-3})+\alpha+\beta x_{n_i-2}+\gamma x_{n_i-3}+x_{n_i-4} > A+x_{n_i-4}. \quad (2.7.12)$$

PROOF The proof is a consequence of (2.7.7) and (2.7.8). ∎

Lemma 2.7.6 *Assume $\beta > 0$. Let $\{x_{n_i+1}\}$ be a subsequence of a solution $\{x_n\}$ such that (2.7.2) and (2.7.3) hold. Also assume that*

$$x_{n_i-2} > x_{n_i}, \quad i = 0, 1, \ldots . \quad (2.7.13)$$

Then

$$\limsup_{i\to\infty} x_{n_i-1}, \ \limsup_{i\to\infty} x_{n_i}, \ \limsup_{i\to\infty} \frac{x_{n_i-2}}{x_{n_i-3}}, \ \limsup_{i\to\infty} \frac{x_{n_i-3}}{x_{n_i-2}} < \infty \quad (2.7.14)$$

and

$$\lim_{i\to\infty} x_{n_i-2} = \lim_{i\to\infty} x_{n_i-3} = \infty. \quad (2.7.15)$$

PROOF For $i \geq 0$ and M sufficiently large,

$$x_{n_i+1} < \frac{\alpha + \gamma x_{n_i-1}}{A + x_{n_i-1}} + \frac{(\beta+1)x_{n_i-2}}{A + x_{n_i-1}} < M + \frac{(\beta+1)x_{n_i-2}}{A + x_{n_i-1}}.$$

From this and (2.7.5) it follows that

$$\limsup_{i\to\infty} x_{n_i-1} < \infty. \tag{2.7.16}$$

Also,

$$x_{n_i+1} < \frac{\alpha}{\gamma + A} + \gamma + \frac{\beta+1}{\gamma+A} x_{n_i-2}, \quad i = 0, 1, \ldots .$$

Hence,

$$\lim_{i\to\infty} x_{n_i-2} = \infty. \tag{2.7.17}$$

Furthermore,

$$x_{n_i-1} > \beta \frac{x_{n_i-2}}{A + x_{n_i-3}}, \quad i = 0, 1, \ldots$$

and so

$$\lim_{i\to\infty} x_{n_i-3} = \infty. \tag{2.7.18}$$

For $i \geq 0$,

$$x_{n_i+1} < \frac{\alpha}{A+\gamma} + \gamma + \frac{\beta+1}{A+\gamma} \frac{\alpha + \beta x_{n_i-3} + \gamma x_{n_i-4} + x_{n_i-5}}{A + x_{n_i-4}}.$$

From this and (2.7.5) it follows that

$$\limsup_{i\to\infty} x_{n_i-4} < \infty. \tag{2.7.19}$$

In view of (2.7.19),

$$\frac{x_{n_i-2}}{x_{n_i-3}} > \frac{\beta}{A + x_{n_i-4}} > m > 0$$

and so

$$\limsup_{i\to\infty} \frac{x_{n_i-3}}{x_{n_i-2}} < \infty. \tag{2.7.20}$$

Also,

$$x_{n_i} = \frac{\alpha}{A + x_{n_i-2}} + \frac{\beta x_{n_i-1}}{A + x_{n_i-2}} + \frac{\gamma x_{n_i-3}}{A + x_{n_i-2}}.$$

Thus, in view of (2.7.16), (2.7.17), and (2.7.20), we have

$$\limsup_{i\to\infty} x_{n_i} < \infty. \tag{2.7.21}$$

The proof of (2.7.14) and (2.7.15) is complete. ∎

Lemma 2.7.7 *Assume $\beta > 0$. Let $\{x_{n_i+1}\}$ be a subsequence of a solution $\{x_n\}$ such that (2.7.2), (2.7.3), and (2.7.13) hold. Then, for $s = 0, 1, \ldots,$ the following hold:*

$$\limsup_{i\to\infty} x_{n_i-4s-1},\ \limsup_{i\to\infty} x_{n_i-4s},\ \limsup_{i\to\infty} \frac{x_{n_i-4s-2}}{x_{n_i-4s-3}},\ \limsup_{i\to\infty} \frac{x_{n_i-4s-3}}{x_{n_i-4s-2}} < \infty \tag{2.7.22}$$

and

$$\lim_{i\to\infty} x_{n_i-4s-2} = \lim_{i\to\infty} x_{n_i-4s-3} = \infty. \tag{2.7.23}$$

PROOF The proof will be by induction. When $s = 0$, (2.7.22) and (2.7.23) follow from (2.7.14) and (2.7.15). Assume that, for $s = j$, (2.7.22) and (2.7.23) hold. In view of (2.7.22) and (2.7.23), there exists $M > 0$, such that, for $i \geq 0$,

$$x_{n_i-4j-1},\ x_{n_i-4j},\ \frac{x_{n_i-4j-1}}{x_{n_i-4j-2}},\ \frac{x_{n_i-4j-2}}{x_{n_i-4j-3}},\ \frac{x_{n_i-4j-3}}{x_{n_i-4j-2}} < M. \tag{2.7.24}$$

From Eq's.(2.0.7) and (2.7.24), it is easy to see that, for $i \geq 0$,

$$x_{n_i+1} < L + \frac{x_{n_i-3}}{\beta} < \ldots < L\sum_{t=0}^{j} \frac{1}{\beta^t} + \frac{x_{n_i-4j-3}}{\beta^{j+1}} \tag{2.7.25}$$

and

$$x_{n_i-4j-3} < Mx_{n_i-4j-2},$$

where

$$L = \frac{\alpha + \beta M}{A + \gamma} + \gamma + \frac{A}{\beta}.$$

Combining the last two inequalities we find

$$x_{n_i+1} < L\sum_{t=0}^{j} \frac{1}{\beta^t} + \frac{M}{\beta^{j+1}} \frac{\alpha + \beta x_{n_i-4j-3} + \gamma x_{n_i-4j-4} + x_{n_i-4j-5}}{A + x_{n_i-4j-4}}.$$

From this and (2.7.5) it follows that

$$\limsup_{i\to\infty} x_{n_i-4j-4} = \limsup_{i\to\infty} x_{n_i-4(j+1)} < \infty. \tag{2.7.26}$$

From (2.7.25) we have, for $i \geq 0$,

$$x_{n_i+1} < L\sum_{t=0}^{j} \frac{1}{\beta^t} + \frac{1}{\beta^{j+1}} \left(\frac{\alpha}{A + \gamma} + \gamma + \frac{\beta x_{n_i-4j-4} + x_{n_i-4j-6}}{A + \gamma} \right)$$

and so, clearly,

$$\lim_{i\to\infty} x_{n_i-4j-6} = \infty. \tag{2.7.27}$$

From (2.7.25) we have, for $i \geq 0$,

$$x_{n_i+1} < L\sum_{t=0}^{j}\frac{1}{\beta^t}+\frac{1}{\beta^{j+1}}\frac{\alpha+\beta x_{n_i-4j-4}+\gamma x_{n_i-4j-5}+x_{n_i-4j-6}}{A+x_{n_i-4j-5}}, \quad i=0,1,\ldots .$$

From this and (2.7.5) it follows that

$$\limsup_{i\to\infty} x_{n_i-4j-5} = \limsup_{i\to\infty} x_{n_i-4(j+1)-1} < \infty. \tag{2.7.28}$$

Finally, note that the following hold:

$$x_{n_i-4j-4} > \frac{x_{n_i-4j-7}}{A+x_{n_i-4j-6}}, \quad i=0,1,\ldots$$

and

$$x_{n_i-4j-5} > \frac{\beta x_{n_i-4j-6}}{A+x_{n_i-4j-7}}, \quad i=0,1,\ldots .$$

From these inequalities, in view of (2.7.26), (2.7.27), and (2.7.28) we have

$$\limsup_{i\to\infty}\frac{x_{n_i-4(j+1)-3}}{x_{n_i-4(j+1)-2}}, \quad \limsup_{i\to\infty}\frac{x_{n_i-4(j+1)-2}}{x_{n_i-4(j+1)-3}} < \infty$$

and

$$\lim_{i\to\infty} x_{n_i-4(j+1)-3} = \infty.$$

The proof of (2.7.22) and (2.7.23) is complete. ∎

Lemma 2.7.8 *Assume $\beta = 0$. Then every solution $\{x_n\}$ of Eq.(2.0.7) satisfies the following inequality for $j \geq -2$:*

$$x_{n+1} < \alpha\sum_{s=1}^{j+3}\frac{1}{(A+\gamma)^s}+\gamma\sum_{s=0}^{j+2}\frac{1}{(A+\gamma)^s}+\frac{x_{n-8-3j}}{(A+\gamma)^{j+3}}. \tag{2.7.29}$$

PROOF Clearly, for $n \geq 0$

$$x_{n+1} = \frac{\alpha+\gamma x_{n-1}+x_{n-2}}{A+x_{n-1}} < \frac{\alpha}{A+\gamma}+\gamma+\frac{x_{n-2}}{A+\gamma}.$$

From this it follows that (2.7.29) holds when $j=-2$. Assume that for $j>-2$

$$x_{n+1} < \alpha\sum_{s=1}^{j+2}\frac{1}{(A+\gamma)^s}+\gamma\sum_{s=0}^{j+1}\frac{1}{(A+\gamma)^s}+\frac{x_{n-5-3j}}{(A+\gamma)^{j+2}}, \quad n=0,1,\ldots .$$

Furthermore,

$$x_{n-5-3j} = \frac{\alpha+\gamma x_{n-7-3j}+x_{n-8-3j}}{A+x_{n-7-3j}} < \frac{\alpha}{A+\gamma}+\gamma+\frac{x_{n-5-3(j+1)}}{A+\gamma}, \quad n=0,1,\ldots .$$

The result follows by combining the last two inequalities. ∎

Lemma 2.7.9 *Assume* $\beta = 0$. *Let* $\{x_n\}$ *be a solution of Eq.(2.0.7). Assume that* $\{x_{n_i}\}$ *is a subsequence of* $\{x_n\}$ *that converges to* γ. *Then the subsequence* $\{x_{n_i-2}\}$ *is unbounded.*

PROOF Rearranging Eq.(2.0.7) we have

$$Ax_{n_i+1} + x_{n_i-1}(x_{n_i+1} - \gamma) = \alpha + x_{n_i-2}, \quad i = 0, 1, \dots . \tag{2.7.30}$$

Suppose for the sake of contradiction that the subsequence $\{x_{n_i-1}\}$ is bounded. Then from (2.7.30) we find that

$$A\gamma = \alpha + \liminf_{i\to\infty} x_{n_i-2} \geq \alpha + \gamma$$

and so either

$$\gamma > 0 \text{ and } 0 > \gamma(A-1) \geq \alpha$$

or

$$\alpha = \gamma = 0.$$

This is a contradiction and the proof is complete. ∎

Lemma 2.7.10 *Assume* $\beta = 0$. *Let* $\{x_{n_i+1}\}$ *be a subsequence of a solution* $\{x_n\}$ *that satisfies (2.7.2) and (2.7.3). Then, for* $j \geq 0$, *the following hold:*

$$\lim_{i\to\infty} x_{n_i-2-3j} = \infty \tag{2.7.31}$$

$$\lim_{i\to\infty} x_{n_i-3-3j} = \gamma \tag{2.7.32}$$

$$\liminf_{i\to\infty} x_{n_i-4-3j} > \gamma \tag{2.7.33}$$

and

$$\limsup_{i\to\infty} x_{n_i-6-3j} < \infty. \tag{2.7.34}$$

PROOF From (2.7.29) for $i \geq 0$

$$x_{n_i+1} < \alpha \sum_{s=1}^{j+1} \frac{1}{(A+\gamma)^s} + \gamma \sum_{s=0}^{j} \frac{1}{(A+\gamma)^s} + \frac{x_{n_i-2-3j}}{(A+\gamma)^{j+1}}, \quad j = 0, 1, \dots$$

and so, clearly, (2.7.31) is satisfied. To establish (2.7.34), suppose for the sake of contradiction that for some $j \geq 0$

$$\lim_{i\to\infty} x_{n_i-6-3j} = \infty.$$

Then, clearly, from (2.7.29)

$$\lim_{i\to\infty} x_{n_i-6-3j} = \lim_{i\to\infty} x_{n_i-9-3j} = \lim_{i\to\infty} x_{n_i-12-3j} = \infty. \tag{2.7.35}$$

Also from (2.7.31)

$$\lim_{i\to\infty} x_{n_i-5-3j} = \lim_{i\to\infty} x_{n_i-8-3j} = \lim_{i\to\infty} x_{n_i-11-3j} = \infty. \tag{2.7.36}$$

Without loss of generality, in view of (2.7.35) and (2.7.36), assume that

$$x_{n_i-6-3j} > \gamma + 2, \quad x_{n_i-8-3j} > \alpha - A(\gamma+2), \quad i = 0, 1, \dots . \tag{2.7.37}$$

We now claim that, for $i \geq 0$,

$$x_{n_i-9-3j} > x_{n_i-8-3j}. \tag{2.7.38}$$

Suppose that there exists $i_0 \geq 0$ such that

$$x_{n_{i_0}-8-3j} \geq x_{n_{i_0}-9-3j}.$$

From this and (2.7.37) we see that

$$x_{n_{i_0}-6-3j} = \frac{\alpha + \gamma x_{n_{i_0}-8-3j} + x_{n_{i_0}-9-3j}}{A + x_{n_{i_0}-8-3j}} \leq \frac{\alpha + (\gamma+1)x_{n_{i_0}-8-3j}}{A + x_{n_{i_0}-8-3j}} < \gamma + 2,$$

which contradicts (2.7.37). The proof of (2.7.38) is complete. From (2.7.29) for $i \geq 0$

$$x_{n_i+1} < \alpha \sum_{s=1}^{j+3} \frac{1}{(A+\gamma)^s} + \gamma \sum_{s=0}^{j+2} \frac{1}{(A+\gamma)^s} + \frac{x_{n_i-8-3j}}{(A+\gamma)^{j+3}}.$$

From this and (2.7.38) we have

$$x_{n_i+1} < \alpha \sum_{s=1}^{j+3} \frac{1}{(A+\gamma)^s} + \gamma \sum_{s=0}^{j+2} \frac{1}{(A+\gamma)^s} + \frac{x_{n_i-9-3j}}{(A+\gamma)^{j+3}}$$

and so for $i \geq 0$

$$x_{n_i+1} < \alpha \sum_{s=1}^{j+4} \frac{1}{(A+\gamma)^s} + \gamma \sum_{s=0}^{j+3} \frac{1}{(A+\gamma)^s} + \frac{x_{n_i-12-3j}}{(A+\gamma)^{j+3} x_{n_i-11-3j}}. \tag{2.7.39}$$

From this and (2.7.5) it follows that

$$\limsup_{i\to\infty} x_{n_i-11-3j} < \infty.$$

This is a contradiction. The proof of (2.7.34) is complete. For $i, j \geq 0$

$$x_{n_i-3-3j} = \frac{\alpha}{A + x_{n_i-5-3j}} + \frac{\gamma x_{n_i-5-3j}}{A + x_{n_i-5-3j}} + \frac{x_{n_i-6-3j}}{A + x_{n_i-5-3j}}.$$

By taking limits in the last equation as $i \to \infty$, in view of (2.7.34) and (2.7.36), we find that for $j \in \{0, 1, \ldots\}$

$$\lim_{i\to\infty} x_{n_i-3-3j} = \gamma.$$

Finally, from (2.7.34) and Lemma 2.7.9 we have for $j \in \{0, 1, \ldots\}$

$$\liminf_{i\to\infty} x_{n_i-4-3j} > \gamma.$$

The proof is complete. ∎

PROOF We divide the proof of Theorem 2.7.1 into the following nine cases:

Case 1: $\beta + 1 < A$. Clearly, for $n \geq 0$

$$x_{n+1} = \frac{\alpha + \beta x_n + \gamma x_{n-1} + x_{n-2}}{A + x_{n-1}} < \frac{\alpha}{A} + \gamma + \frac{\beta+1}{A}\max\{x_n, x_{n-2}\}.$$

By using Theorem 1.4.1 it follows that

$$\limsup_{n\to\infty} x_n \leq \frac{\alpha + \gamma A}{\beta + 1 - A}.$$

Case 2: $\beta + 1 = A,\ \ \gamma > 0,$ and $\alpha > 0$. Without loss of generality, assume that the initial conditions x_{-2}, x_{-1}, x_0 are positive. Let $m > 0$ be such that

$$m < \min\{\gamma, x_{-2}, x_{-1}, x_0\}.$$

We claim that

$$x_n > m, \quad \text{for } n = -2, -1, 0, \ldots . \tag{2.7.40}$$

Indeed,

$$x_1 - m = \frac{\alpha + \beta(x_0 - m) + (\gamma - m)x_{-1} + x_{-2} - m}{\beta + 1 + x_{-1}} > 0$$

and the proof of (2.7.40) follows by induction. Then

$$x_{n+1} = \frac{\alpha + \beta x_n + \gamma x_{n-1} + x_{n-2}}{\beta + 1 + x_{n-1}}$$

$$< \frac{\alpha}{\beta + 1 + m} + \gamma + \frac{\beta + 1}{\beta + 1 + m}\max\{x_n, x_{n-2}\}, \quad n = 0, 1, \ldots .$$

By using Theorem 1.4.1 we find that

$$\limsup_{n\to\infty} x_n \leq \frac{\alpha + \gamma(\beta + 1 + m)}{m}.$$

Case 3: $\beta+1=A,\ \ \gamma>0,\ $ and $\alpha=0$. Let $M>0$ be such that

$$M>\max\{\gamma,x_{-2},x_{-1},x_0\}.$$

We claim that

$$x_n<M,\ \ n=-2,-1,0,\dots\ . \tag{2.7.41}$$

Indeed,

$$x_1-M=\frac{\beta(x_0-M)+(\gamma-M)x_{-1}+x_{-2}-M}{\beta+1+x_{-1}}<0$$

and the proof follows by induction.

Case 4: $\beta+1=A,\ \ \alpha>0,\ $ and $\gamma=0$. Without loss of generality, assume that the initial conditions x_{-2}, x_{-1}, x_0 are positive. Let $m>0$ be such that

$$m<\min\{x_{-2},x_{-1},x_0,\frac{\alpha}{x_{-2}},\frac{\alpha}{x_{-1}},\frac{\alpha}{x_0}\}.$$

We claim that

$$m<x_n<\frac{\alpha}{m},\ \ n=-2,-1,0,\dots\ . \tag{2.7.42}$$

Indeed,

$$m=\frac{\alpha+\beta m+m}{\beta+1+\frac{\alpha}{m}}<x_1=\frac{\alpha+\beta x_0+x_{-2}}{\beta+1+x_{-1}}<\frac{\alpha+\beta\frac{\alpha}{m}+\frac{\alpha}{m}}{\beta+1+m}=\frac{\alpha}{m}$$

and the proof follows by induction.

Case 5: $\beta+1=A,\ \ \gamma=0,\ $ and $\alpha=0$. Clearly, for $n\geq 0$

$$x_{n+1}<\max\{x_n,x_{n-2}\}$$

and so

$$x_{n+1}<\max\{x_{-2},x_{-1},x_0\},\ \ n=0,1,\dots\ .$$

Case 6: $\beta+1>A\ $ and $\ \beta>0$. Suppose for the sake of contradiction that $\{x_{n_i+1}\}$ is a subsequence of a solution $\{x_n\}$ such that (2.7.2) and (2.7.3) hold.

We consider two subcases:

Subcase 6(i)**:** $x_{n_i}\geq x_{n_i-2}$, for $i\in\{0,1,\dots\}$. In view of (2.7.7) and (2.7.8), there exist $L,M>0$ such that

$$x_{n_i-2},x_{n_i-1}<M$$

and

$$\frac{\alpha+\gamma x_{n_i-1}+x_{n_i-2}}{A+x_{n_i-1}}+\frac{\beta(\alpha+\beta x_{n_i-1}+\gamma x_{n_i-2})}{(A+x_{n_i-1})(A+x_{n_i-2})}<L.$$

From (2.7.10), (2.7.11), (2.7.12), and the last inequality we have

$$x_{n_i+1} < L + \frac{\beta}{\beta+\epsilon} x_{n_i-3}$$

Then, for i sufficiently large,

$$x_{n_i+1} < x_{n_i-3},$$

which contradicts (2.7.3).

Subcase 6(*ii*)**:** In this case we assume that there exists a subsequence of $\{n_i\}$, which for the sake of simplicity we still denote by $\{n_i\}$ such that for $i \in \{0, 1, \ldots\}$

$$x_{n_i-2} > x_{n_i}.$$

Clearly, for $s \geq 0$

$$x_{n_i-4s} = \frac{\alpha}{A + x_{n_i-4s-2}} + \frac{\beta x_{n_i-4s-1}}{A + x_{n_i-4s-2}} + \frac{\gamma x_{n_i-4s-2}}{A + x_{n_i-4s-2}} + \frac{x_{n_i-4s-3}}{x_{n_i-4s-2}} \frac{1}{\frac{A}{x_{n_i-4s-2}} + 1} \tag{2.7.43}$$

and

$$\frac{x_{n_i-4s-2}}{x_{n_i-4s-3}} = \frac{\alpha}{x_{n_i-4s-3}(A + x_{n_i-4s-4})} + \frac{\beta}{A + x_{n_i-4s-4}} + \frac{\gamma x_{n_i-4s-4}}{x_{n_i-4s-3}(A + x_{n_i-4s-4})} + \frac{x_{n_i-4s-5}}{x_{n_i-4s-3}(A + x_{n_i-4s-4})}. \tag{2.7.44}$$

From (2.7.22) it follows that for $s \in \{0, 1, \ldots\}$ the subsequences $\{x_{n_i-4s}\}$ are bounded. Let l_{-4s} be accumulation points of the bounded subsequences $\{x_{n_i-4s}\}$. By taking limits in (2.7.43) and (2.7.44) as $i \to \infty$, in view of (2.7.22) and (2.7.23) we find

$$l_{-4s} = \gamma + \frac{l_{-4s-4}}{\beta}.$$

For s sufficiently large,

$$\gamma \sum_{t=0}^{s} \frac{1}{\beta^t} > l_0.$$

Then

$$l_0 = \gamma + \frac{l_{-4}}{\beta} = \ldots = \gamma \sum_{t=0}^{s} \frac{1}{\beta^t} + \frac{l_{-4s}}{\beta^{s+1}} > l_0,$$

which is a contradiction.

Case 7: $\gamma > 1 - A > 0$ and $\beta = 0$. Clearly, for $n \geq 0$,

$$x_{n+1} = \frac{\alpha + \gamma x_{n-1} + x_{n-2}}{A + x_{n-1}} < \frac{\alpha}{A+\gamma} + \gamma + \frac{1}{A+\gamma} x_{n-2}.$$

By using Theorem 1.4.1 it follows that

$$\limsup_{n \to \infty} x_n \leq \frac{\alpha + \gamma(\gamma + A)}{\gamma + 1 - A}.$$

Case 8: $1 - A > \gamma \geq 0$ and $\beta = 0$. Suppose for the sake of contradiction that $\{x_{n_i+1}\}$ is a subsequence of a solution $\{x_n\}$ such that (2.7.2) and (2.7.3) hold. From (2.7.31) we have

$$\lim_{i \to \infty} x_{n_i - 2} = \infty.$$

Without loss of generality, assume that, for $i \geq 0$,

$$x_{n_i-2} > \frac{\alpha + \gamma \bar{x}}{A + \bar{x} - 1}, \tag{2.7.45}$$

where

$$\bar{x} = \frac{\gamma + 1 - A + \sqrt{(\gamma + 1 - A)^2 + 4\alpha}}{2}$$

is the positive equilibrium of Eq.(2.0.7). We now claim that

$$x_{n_i - 1} < \bar{x}. \tag{2.7.46}$$

Otherwise, for some i_0,

$$x_{n_{i_0}-1} \geq \bar{x}.$$

Then

$$x_{n_{i_0}+1} = \frac{\alpha + \gamma x_{n_{i_0}-1} + x_{n_{i_0}-2}}{A + x_{n_{i_0}-1}} \leq \frac{\alpha + \gamma \bar{x} + x_{n_{i_0}-2}}{A + \bar{x}} < x_{n_{i_0}-2},$$

which contradicts (2.7.3). The proof of (2.7.46) is complete. Let $0 < \epsilon < 1 - A - \gamma$ and let s be sufficiently large such that

$$\alpha \sum_{t=1}^{s+1} \frac{1}{(A + \gamma + \epsilon)^{t+1}} > \bar{x}.$$

Then for N sufficiently large and $i \geq N$, in view of (2.7.32) and (2.7.33), we have, for $0 \leq j \leq s$,

$$x_{n_i-4-3j} > \gamma + \epsilon, \quad x_{n_i-3-3j} < \gamma + \epsilon.$$

Hence,

$$x_{n_i-1} = \frac{\alpha + \gamma x_{n_i-3} + x_{n_i-4}}{A + x_{n_i-3}} > \frac{\alpha}{A+\gamma+\epsilon} + \frac{x_{n_i-4}}{A+\gamma+\epsilon} > \dots$$

$$> \alpha \sum_{s=1}^{j+1} \frac{1}{(A+\gamma+\epsilon)^{s+1}}, > \bar{x}$$

which contradicts (2.7.46).

Case 9: $\gamma = 1 - A > 0$ and $\beta = 0$. The proof in this case is similar to the proof in Case 8 and will be omitted. ▮

2.8 Boundedness of $x_{n+1} = \dfrac{\alpha + \beta x_n + x_{n-1}}{x_{n-1} + D x_{n-2}}$

In this section we establish the boundedness of every solution of the equation in the title with

$$\alpha \geq 0 \text{ and } \beta, D \in (0, \infty).$$

This will confirm the boundedness of the following two special cases of Eq.(2.0.1), namely:

#88 and #170.

See Appendix A.

We present the proof for the case

$$\alpha = 0.$$

The case where α is positive is similar and will be omitted.

So for the remainder of this section we deal with the equation

$$x_{n+1} = \frac{\beta x_n + x_{n-1}}{x_{n-1} + D x_{n-2}}, \quad n = 0, 1, \dots \tag{2.8.1}$$

with positive parameters β and D and with arbitrary positive initial conditions x_{-2}, x_{-1}, x_0.

The main result in this section is the following.

Theorem 2.8.1 *Every solution of Eq.(2.8.1) is bounded.*

PROOF First we make the following useful observations:

$$x_{n+1} = \frac{x_{n-1}}{x_{n-1} + Dx_{n-2}} + \beta \cdot \frac{\beta x_{n-1} + x_{n-2}}{x_{n-1} + Dx_{n-2}} \cdot \frac{1}{x_{n-2} + Dx_{n-3}}, \quad n = 1, 2, \ldots \tag{2.8.2}$$

$$\frac{x_{n-1}}{x_{n-1} + Dx_{n-2}} \leq 1 \tag{2.8.3}$$

$$\frac{\min(\beta, 1)}{\max(D, 1)} \leq \frac{\beta x_{n-1} + x_{n-2}}{x_{n-1} + Dx_{n-2}} \leq \frac{\max(\beta, 1)}{\min(D, 1)} \tag{2.8.4}$$

$$x_{n+1} \to \infty \Rightarrow x_{n-2} + Dx_{n-3} \to 0 \Rightarrow x_{n-2} \to 0 \text{ and } x_{n-3} \to 0 \tag{2.8.5}$$

$$x_{n+1} \to 0 \Rightarrow x_{n-2} + Dx_{n-3} \to \infty\,. \tag{2.8.6}$$

Now assume for the sake of contradiction that Eq.(2.8.1) has a positive unbounded solution $\{x_n\}$. Then there exists a subsequence $\{x_{n_i}\}$ such that

$$x_{n_i+1} \to \infty \tag{2.8.7}$$

and for every i,

$$x_{n_i+1} > x_j, \quad \text{for all } j < n_i + 1. \tag{2.8.8}$$

Then

$$x_{n_i-2} \to 0 \text{ and } x_{n_i-3} \to 0 \tag{2.8.9}$$

and

$$x_{n_i-5} + Dx_{n_i-6} \to \infty \text{ and } x_{n_i-6} + Dx_{n_i-7} \to \infty. \tag{2.8.10}$$

Next we claim that

$$\liminf_{i\to\infty} x_{n_i-1} > 0. \tag{2.8.11}$$

Otherwise, there exists a subsequence of $\{n_i\}$, which for economy of notation we still denote by $\{n_i\}$, such that

$$x_{n_i-1} \to 0.$$

Then

$$x_{n_i-4} + Dx_{n_i-5} \to \infty.$$

Also from (2.8.2), there exists a positive constant M such that

$$x_{n_i+1} \leq 1 + M \cdot \frac{1}{x_{n_i-3}} = 1 + M \cdot \frac{x_{n_i-5} + Dx_{n_i-6}}{\beta x_{n_i-4} + x_{n_i-5}}$$

$$\leq 1 + M\left(1 + \frac{Dx_{n_i-6}}{\beta x_{n_i-4} + x_{n_i-5}}\right)$$

and eventually

$$x_{n_i+1} < x_{n_i-6},$$

which contradicts (2.8.8) and establishes our claim that (2.8.11) holds.

We have

$$x_{n_i-1} = \frac{\beta x_{n_i-2} + x_{n_i-3}}{x_{n_i-3} + D x_{n_i-4}} = \frac{\beta \cdot (\frac{x_{n_i-2}}{x_{n_i-3}}) + 1}{1 + D \cdot (\frac{x_{n_i-4}}{x_{n_i-3}})}$$

and

$$x_{n_i-1} = \frac{\beta x_{n_i-2} + x_{n_i-3}}{x_{n_i-3} + D x_{n_i-4}} = \frac{\beta + \frac{x_{n_i-3}}{x_{n_i-2}}}{\frac{x_{n_i-3}}{x_{n_i-2}} + D \cdot (\frac{x_{n_i-4}}{x_{n_i-2}})}$$

and so the following statements are true:

If the sequence $\{\frac{x_{n_i-2}}{x_{n_i-3}}\}$ is bounded, then

$$\liminf_{i\to\infty}(\frac{x_{n_i-3}}{x_{n_i-4}}) > 0$$

and if the sequence $\{\frac{x_{n_i-2}}{x_{n_i-3}}\}$ is unbounded, then

$$\liminf_{i\to\infty}(\frac{x_{n_i-2}}{x_{n_i-4}}) > 0.$$

Hence, there exists $\mu > 0$ such that, eventually,

$$\begin{array}{ll} \text{either} & x_{n_i-2} > \mu x_{n_i-4} \\ \text{or} & x_{n_i-3} > \mu x_{n_i-4}. \end{array}$$

Then from (2.8.2) and for some positive constant K,

$$\begin{aligned} x_{n_i+1} &\le 1 + \beta \cdot \frac{\max(\beta, 1)}{\min(D, 1)} \cdot \frac{1}{x_{n_i-2} + D x_{n_i-3}} \\ &< 1 + K \cdot \frac{1}{x_{n_i-4}} = 1 + K \cdot \frac{x_{n_i-6} + D x_{n_i-7}}{\beta x_{n_i-5} + x_{n_i-6}} \\ &\le 1 + K(1 + \frac{D x_{n_i-7}}{\beta x_{n_i-5} + x_{n_i-6}}). \end{aligned}$$

Therefore, eventually,

$$x_{n_i+1} < x_{n_i-7}$$

which contradicts (2.8.8) and completes the proof. ∎

2.9 Boundedness of $x_{n+1} = \frac{\alpha + \beta x_n + x_{n-2}}{C x_{n-1} + x_{n-2}}$

In this section we establish the boundedness of every solution of the following two special cases of Eq.(2.0.1), namely:

#94 and #176.

See [152] and Appendix A.

We present the proof for the case

$$\alpha = 0.$$

The case where α is positive is similar and will be omitted. So for the remainder of this section we deal with the equation

$$x_{n+1} = \frac{\beta x_n + x_{n-2}}{Cx_{n-1} + x_{n-2}}, \quad n = 0, 1, \ldots \tag{2.9.1}$$

with positive parameters β and C and with positive initial conditions x_{-2}, x_{-1}, x_0.

The main result in this section is the following.

Theorem 2.9.1 *Every solution of Eq.(2.9.1) is bounded from above and from below by positive constants.*

PROOF Note that for $n \geq 0$,

$$Cx_{n+2} + x_{n+1} = C \cdot \frac{\beta x_{n+1} + x_{n-1}}{Cx_n + x_{n-1}} + \frac{\beta x_n + x_{n-2}}{Cx_{n-1} + x_{n-2}}$$

$$> \frac{\beta x_n x_{n-1} + C^2 x_{n-1}^2 + x_{n-2}(Cx_n + x_{n-1})}{C^2 x_n x_{n-1} + Cx_{n-1}^2 + x_{n-2}(Cx_n + x_{n-1})} > \frac{\min\left(\beta, C^2, 1\right)}{\max\left(C^2, C, 1\right)} = K$$

Then for $n \geq 0$,

$$x_{n+4} = \frac{\beta x_{n+3} + x_{n+1}}{Cx_{n+2} + x_{n+1}} \leq \frac{\beta}{K} \cdot x_{n+3} + 1.$$

Now using the above two estimates we see that for $n \geq 0$,

$$x_{n+5} = \frac{\beta x_{n+4} + x_{n+2}}{Cx_{n+3} + x_{n+2}}$$

$$\leq \frac{\frac{\beta^2}{K} x_{n+3} + \beta + x_{n+2}}{Cx_{n+3} + x_{n+2}} \leq \frac{\beta}{K} + \frac{\max\left(\frac{\beta^2}{K}, 1\right)}{\min(C, 1)} = U$$

and so all solutions of Eq.(2.9.1) are eventually bounded from above by the positive constant U.

From

$$x_{n+7} = \frac{\beta x_{n+6} + x_{n+4}}{Cx_{n+5} + x_{n+4}} \geq \frac{\beta}{(C+1)U} \cdot x_{n+6}, \quad n = 0, 1, \ldots$$

and

$$x_{n+8} = \frac{\beta x_{n+7} + x_{n+5}}{Cx_{n+6} + x_{n+5}} \geq \frac{\beta \cdot \frac{\beta}{(C+1)U} \cdot x_{n+6} + x_{n+5}}{Cx_{n+6} + x_{n+5}}$$

$$\geq \frac{\min\left(\frac{\beta^2}{(C+1)U}, 1\right)}{\max(C, 1)} = L, \quad n = 0, 1, \ldots,$$

we see that that all solutions of Eq.(2.9.1) are eventually bounded from below by the positive constant L. The proof is complete. ∎

3

Existence of Unbounded Solutions

3.0 Introduction

Consider the third-order rational difference equation

$$x_{n+1} = \frac{\alpha + \beta x_n + \gamma x_{n-1} + \delta x_{n-2}}{A + Bx_n + Cx_{n-1} + Dx_{n-2}}, \quad n = 0, 1, ... \tag{3.0.1}$$

with nonnegative parameters $\alpha, \beta, \gamma, \delta, A, B, C, D$ and with arbitrary nonnegative initial conditions x_{-2}, x_{-1}, x_0, such that the denominator is always positive.

This equation contains 225 special cases of equations with positive parameters. It was conjectured in [69] that in 135 of these special cases, every solution of the equation is bounded and, in the remaining 90 cases, the equation has unbounded solutions in some range of their parameters and for some initial conditions.

In this chapter we present several theorems on the existence of unbounded solutions of some equations of the form of Eq.(3.0.1) and in particular we establish that in 85 special cases of Eq.(3.0.1) there exist unbounded solutions in some range of their parameters. The only five special cases of Eq.(3.0.1) where it has been conjectured that they have unbounded solutions but we are unable yet to confirm it are the following:

#28, #44, #56, #70, #120.

See Appendix A. These five special cases can be summarized in the following conjecture.

Conjecture 3.0.1 *Show that for each of the following five third-order rational difference equations, which are written in normalized form, there exist unbounded solutions in some region of its parameters and for some initial conditions:*

$$\#28: \quad x_{n+1} = \frac{x_n}{Cx_{n-1} + x_{n-2}}, \quad n = 0, 1, \ldots$$

$$\#44: \quad x_{n+1} = \frac{\alpha + x_n}{x_{n-2}}, \quad n = 0, 1, \ldots$$

$$\#56: \quad x_{n+1} = \frac{\beta x_n + x_{n-1}}{x_{n-2}}, \quad n = 0, 1, \ldots$$

$$\#70: \quad x_{n+1} = \frac{\alpha + x_n}{Cx_{n-1} + x_{n-2}}, \quad n = 0, 1, \ldots$$

$$\#120: x_{n+1} = \frac{\alpha + \beta x_n + x_{n-1}}{x_{n-2}}, n = 0, 1, \ldots .$$

The existence of unbounded solutions is obvious in each of the following 14 special cases of Eq.(3.0.1), which are linear but nontrivial:

#5,	#9,	#13,	#41,	#45,	#49,	#53,
#57,	#61,	#117,	#121,	#125,	#129,	#137,

and in each of the following five special cases, which can be transformed to linear equations:

#8,	#10,	#12,	#14,	#15.

See Appendix A.

The proof of the existence of unbounded solutions in the special cases #51, #59, and #123 is quite lengthy and, to economize in space, we refer the reader to the original source [47] for #51 and [150] for #59 and #123.

In the remaining six sections of this chapter we will establish the existence of unbounded solutions in 63 additional special cases of Eq.(3.0.1).

In Section 3.1 we will establish the existence of unbounded solutions of the equation

$$x_{n+1} = \frac{\alpha + \beta x_n + \gamma x_{n-1} + \delta x_{n-2}}{A + x_n}, \quad n = 0, 1, \ldots \tag{3.0.2}$$

with

$$\gamma, \alpha + \beta + \delta + A \in (0, \infty). \tag{3.0.3}$$

This confirms the existence of unbounded solution in 15 additional special cases of Eq.(3.0.1).

In Section 3.2 we will establish the existence of unbounded solutions of the equation

$$x_{n+1} = \frac{\alpha + \beta x_n + \gamma x_{n-1} + \delta x_{n-2}}{A + x_{n-2}}, \quad n = 0, 1, \ldots \tag{3.0.4}$$

with

$$\gamma, \alpha + \beta + \delta + A \in (0, \infty) \text{ and } \delta = A = 0 \text{ only if } \beta = 0. \tag{3.0.5}$$

This confirms the existence of unbounded solution in 13 additional special cases of Eq.(3.0.1).

In Section 3.3 we will establish the existence of unbounded solutions of the equation

$$x_{n+1} = \frac{\alpha + \beta x_n + \gamma x_{n-1} + \delta x_{n-2}}{A + Bx_n + x_{n-2}}, \quad n = 0, 1, \ldots \tag{3.0.6}$$

with

$$\gamma, B \in (0, \infty). \tag{3.0.7}$$

This confirms the existence of unbounded solution in 16 additional special cases of Eq.(3.0.1).

In Section 3.4 we will establish the existence of unbounded solutions of the equation

$$x_{n+1} = \frac{\alpha + \beta x_n + \gamma x_{n-1} + \delta x_{n-2}}{A + Bx_n + x_{n-1}}, \quad n = 0, 1, \ldots \tag{3.0.8}$$

with

$$\delta, B \in (0, \infty). \tag{3.0.9}$$

This confirms the existence of unbounded solution in 14 additional special cases of Eq.(3.0.1).

In Section 3.5 we will establish the existence of unbounded solutions of the equation

$$x_{n+1} = \frac{x_{n-2}}{A + Bx_n + Cx_{n-1}}, \quad n = 0, 1, \ldots \tag{3.0.10}$$

with

$$B + C \in (0, \infty). \tag{3.0.11}$$

This confirms the existence of unbounded solution in four additional special cases of Eq.(3.0.1).

Finally in Section 3.6 we will establish the existence of unbounded solutions of the special case

$$\#50: \quad x_{n+1} = \frac{\alpha + x_{n-2}}{x_n}, \quad n = 0, 1, \ldots .$$

3.1 Unbounded Solutions of $x_{n+1} = \dfrac{\alpha + \beta x_n + \gamma x_{n-1} + \delta x_{n-2}}{A + x_n}$

The first theorem in this section confirms the existence of unbounded solutions in 15 additional special cases of Eq.(3.0.1), namely:

#29,	#46,	#54,	#62,	#71,	#83,	#95,	#118,
#126,	#130,	#138,	#165,	#177,	#183,	#195.	

See Appendix A.

We will assume that (3.0.3) holds and we will establish that there exist solutions of Eq.(3.0.2) that are unbounded in some range of its parameters and for some initial conditions. Actually, we exhibit a huge set of initial conditions through which the subsequences of even and odd terms of the solutions converge, one of them to ∞ and the other to

$$\frac{\beta\gamma + \delta A}{\gamma - \delta}.$$

Furthermore, our proof here extends and unifies all previously known results on the existence of unbounded solutions for all special cases of Eq.(3.0.2). More precisely, we establish the following result.

Theorem 3.1.1 *Assume that (3.0.3) holds and*

$$\gamma > \beta + \delta + A.$$

Let k be any number such that

$$0 < k < \gamma - \beta - \delta - A.$$

Then every solution of Eq.(3.0.2) with initial conditions x_{-2}, x_{-1}, x_0 such that

$$x_{-2}, x_0 \in (0, \gamma - A) \;\; \text{and} \;\; x_{-1} > \frac{\alpha + \gamma(\gamma - A)}{k}$$

is unbounded and, more precisely,

$$\lim_{n\to\infty} x_{2n+1} = \infty \;\; \text{and} \;\; \lim_{n\to\infty} x_{2n} = \frac{\beta\gamma + \delta A}{\gamma - \delta}.$$

PROOF Observe that

$$x_1 - x_{-1} = \frac{\alpha + \beta x_0 + \gamma x_{-1} + \delta x_{-2}}{A + x_0} - x_{-1} = \frac{\alpha + \beta x_0 + (\gamma - A - x_0)\, x_{-1} + \delta x_{-2}}{A + x_0}$$

and so

$$x_1 > x_{-1}.$$

Also,

$$\begin{aligned} x_2 - (\beta + \delta + k) &\leq \frac{\alpha + \beta x_1 + \gamma x_0 + \delta x_{-1}}{x_1} - (\beta + \delta + k) \\ &= \frac{\alpha + \gamma x_0 + \delta\,(x_{-1} - x_1) - k x_1}{x_1} \\ &< \frac{\alpha + \gamma(\gamma - A) - \alpha - \gamma(\gamma - A)}{x_1} = 0. \end{aligned}$$

Therefore,

$$x_2 < \beta + \delta + k$$

and, furthermore,

$$x_3 = \frac{\alpha + \beta x_2 + \gamma x_1 + \delta x_0}{A + x_2} > \frac{\gamma}{\beta + \delta + A + k} x_1.$$

It follows by induction that for $n \geq 0$,

$$x_{2n} < \beta + \delta + k$$

and

$$x_{2n+1} > \frac{\gamma}{\beta + \delta + A + k} x_{2n-1}$$

and so, in particular,

$$\lim_{n \to \infty} x_{2n+1} = \infty. \tag{3.1.1}$$

Let S and I denote the following limits:

$$S = \limsup_{n \to \infty} x_{2n} \quad \text{and} \quad I = \liminf_{n \to \infty} x_{2n}$$

Note that from Eq.(3.0.2),

$$x_{2n+1} = \frac{\alpha + \beta x_{2n} + \gamma x_{2n-1} + \delta x_{2n-2}}{A + x_{2n}} > \gamma \frac{x_{2n-1}}{A + x_{2n}}$$

and so for $n \geq 0$,

$$\frac{x_{2n-1}}{x_{2n+1}} < \frac{A + x_{2n}}{\gamma}.$$

Let $\epsilon > 0$ be given. Then, clearly, in view of (3.1.1), there exists $N \geq 0$ such that

$$\frac{\alpha + \beta x_{2n+1} + \gamma x_{2n}}{A + x_{2n+1}} < \beta + \epsilon, \quad \text{for} \;\; n \geq N.$$

By using the preceding two inequalities, it follows from Eq.(3.0.2) that for $n \geq N$,

$$\begin{aligned} x_{2n+2} &= \frac{\alpha + \beta x_{2n+1} + \gamma x_{2n} + \delta x_{2n-1}}{A + x_{2n+1}} \\ &< (\beta + \epsilon) + \frac{\delta}{\gamma}(A + x_{2n}) \\ &= \left(\beta + \frac{\delta A}{\gamma} + \epsilon\right) + \frac{\delta}{\gamma} x_{2n}. \end{aligned}$$

By using Theorem 1.4.1 and by taking limit superiors we find

$$S \leq \frac{\beta\gamma + \delta A + \gamma\epsilon}{\gamma - \delta}$$

and so, clearly,

$$S \leq \frac{\beta\gamma + \delta A}{\gamma - \delta}.$$

When

$$\beta = A = 0,$$

we see that $S = 0$, that is, $\lim_{n\to\infty} x_{2n} = 0$ and the proof is complete. Next assume that

$$\beta + A > 0.$$

Clearly, there exists a sequence of indices $\{n_i\}$ and a number $L_0 \in [I, S]$ such that

$$\lim_{i\to\infty} x_{2n_i+2} = I \text{ and } \lim_{i\to\infty} x_{2n_i} = L_0.$$

From Eq.(3.0.2) we have

$$\frac{x_{2n+1}}{x_{2n-1}} = \frac{\alpha}{A + x_{2n}} \cdot \frac{1}{x_{2n-1}} + \frac{\beta x_{2n}}{A + x_{2n}} \cdot \frac{1}{x_{2n-1}} + \frac{\gamma}{A + x_{2n}} + \frac{\delta x_{2n-2}}{A + x_{2n}} \cdot \frac{1}{x_{2n-1}}$$

and

$$\frac{x_{2n+2}}{x_{2n}} = \frac{\alpha}{A + x_{2n+1}} \cdot \frac{1}{x_{2n}} + \frac{\beta x_{2n+1}}{A + x_{2n+1}} \cdot \frac{1}{x_{2n}} + \frac{\gamma}{A + x_{2n+1}} + \frac{\delta x_{2n-1}}{A + x_{2n+1}} \cdot \frac{1}{x_{2n}}.$$

By replacing n by n_i in the above two identities and then by taking limits as $i \to \infty$, we find

$$\lim_{i\to\infty} \left(\frac{x_{2n_i+1}}{x_{2n_i-1}}\right) = \frac{\gamma}{A + L_0}$$

and

$$\frac{I}{L_0} = \frac{\beta}{L_0} + \frac{\delta}{L_0} \cdot \frac{A + L_0}{\gamma}.$$

Therefore,

$$I = \beta + \frac{\delta}{\gamma}(A + L_0) \leq L_0 \tag{3.1.2}$$

and so

$$L_0 \geq \frac{\beta\gamma + \delta A}{\gamma - \delta} \geq S.$$

Hence,

$$L_0 = \frac{\beta\gamma + \delta A}{\gamma - \delta} = S \tag{3.1.3}$$

and so, from (3.1.2) and (3.1.3),

$$I = \beta + \frac{\delta}{\gamma}(A + L_0) = \frac{\beta\gamma + \delta A}{\gamma - \delta}.$$

The proof is complete. ∎

The following generalization of Theorem 3.1.1 establishes the existence of unbounded solutions in the following 39 special cases of Eq.(3.0.1):

#29,	#31,	#33,	#46,	#48,	#54,	#62
#64,	#71,	#73,	#75,	#83,	#87,	#95
#97,	#99,	#110,	#118,	#126,	#128,	#130
#138,	#146,	#154,	#162,	#165,	#169,	#177
#179,	#181,	#183,	#187,	#195,	#197,	#199
#202,	#210,	#214,	#222			

and in the following 44 special cases of fourth-order rational difference equations:

#280-287,	#344-351,	#400-403,	#408-415,	#464-479.

See Appendices A and B.

Theorem 3.1.2 *Assume that*

$$\beta_1 > 0 \text{ and } \sum_{j=0}^{m} B_{2j} > 0$$

and that either

$$B_0 > 0 \tag{3.1.4}$$

or

$$\beta_{2s} > 0 \Rightarrow B_{2s} > 0, \text{ for all } s \leq m. \tag{3.1.5}$$

Then the rational equation

$$x_{n+1} = \frac{\alpha + \sum_{i=0}^{k} \beta_i x_{n-i}}{A + \sum_{i=0}^{m} B_{2i} x_{n-2i}}, \quad n = 0, 1, \ldots \tag{3.1.6}$$

has unbounded solutions in some range of the parameters and for some initial conditions.

PROOF Assume that $k = 2t$ is an even number. The proof when k is odd is similar and will be omitted. Set

$$S = \{s \leq m : \ \beta_{2s} > 0\}.$$

We will establish that there exist unbounded solutions in the range of parameters where

$$\beta_1 > A + U \sum_{j=0}^{m} B_{2j}$$

with

$$U = \frac{\beta_0}{B_0} + \frac{\sum_{i=1}^{t} \beta_{2i}}{B_0}$$

when (3.1.4) holds and

$$U = \frac{\max_{i \in S} \beta_{2i}}{\min_{j \in S} B_{2j}} + \frac{\sum_{i=m+1}^{t} \beta_{2i}}{B_{2s}}, \quad \text{for some } s \in S,$$

when (3.1.5) holds and provided that $S \neq \emptyset$. Note that when $t \leq m$,

$$\sum_{i=t+1}^{m} \beta_{2i} = 0$$

and so in this case

$$U = \frac{\beta_0}{B_0} \quad \text{or} \quad U = \frac{\max_{i \in S} \beta_{2i}}{\min_{j \in S} B_{2j}}.$$

When $S = \emptyset$, we set

$$U = 0.$$

Choose a positive number ϵ such that

$$\beta_1 > A + (U + \epsilon) \sum_{j=0}^{m} B_{2j} \tag{3.1.7}$$

and let $\{x_n\}$ be any solution with the initial conditions chosen as follows:

$$x_{-1} > x_{-3} > \cdots > x_{1-2q} > \frac{\alpha + \sum_{i=1}^{t} \beta_{2i-1} x_{2-2i}}{\epsilon B_0}, \tag{3.1.8}$$

when (3.1.4) holds, and

$$x_{-1} > x_{-3} > \cdots > x_{1-2q} > \frac{\alpha + \sum_{i=1}^{t} \beta_{2i-1} x_{2-2i}}{\epsilon B_{2s}}, \tag{3.1.9}$$

when (3.1.5) holds, and

$$x_{-q}, x_{-q+2}, \ldots, x_0 < U + \epsilon, \tag{3.1.10}$$

where

$$q = \max(2m, 2t).$$

We claim that $\{x_n\}$ is an unbounded solution of Eq.(3.1.6). To this end, we first establish that for all $n \geq 0$,

$$x_{2n+1} > \frac{\alpha + \sum_{i=0}^{m} \beta_{2i} x_{n-2i}}{A + \sum_{j=0}^{m} B_{2j}} + x_{2n-1} \tag{3.1.11}$$

and

$$x_{2n+2} < U + \epsilon. \tag{3.1.12}$$

Indeed,

$$x_1 = \frac{\alpha + \sum_{i=0}^{k} \beta_i x_{-i}}{A + \sum_{i=0}^{m} B_{2i} x_{-2i}} \geq \frac{\alpha + \sum_{i=0}^{m} \beta_{2i} x_{-2i}}{A + \sum_{j=0}^{m} B_{2j} x_{-2j}} + \frac{\beta_1}{A + (U+\epsilon) \sum_{j=0}^{m} B_{2j}} x_{-1}$$

and, in view of (3.1.7),

$$x_1 > \frac{\alpha + \sum_{i=0}^{m} \beta_{2i} x_{-2i}}{A + \sum_{j=0}^{m} B_{2j}} + x_{-1}.$$

Also, when (3.1.4) holds,

$$x_2 = \frac{\beta_0 x_1}{A + \sum_{j=0}^{m} B_{2j} x_{1-2j}} + \frac{\sum_{i=1}^{t} \beta_{2i} x_{1-2i}}{A + \sum_{j=0}^{m} B_{2j} x_{1-2j}} + \frac{\alpha + \sum_{i=1}^{t} \beta_{2i-1} x_{2-2i}}{A + \sum_{j=0}^{m} B_{2j} x_{1-2j}}.$$

In view of (3.1.8),

$$x_2 \leq \frac{\beta_0}{B_0} + \frac{\sum_{i=1}^{t} \beta_{2i} \cdot x_1}{B_0 x_1} + \epsilon = U + \epsilon$$

and the proofs of (3.1.11) and (3.1.12) follow by induction. When (3.1.5) holds and $S \neq \emptyset$ for all $i \in \{m+1, \ldots, t\}$ and for all $j \in \{0, \ldots, m\}$, we have

$$1 - 2j \geq 1 - 2i \quad \text{and} \quad x_{1-2j} \geq x_{1-2i}.$$

Therefore, there exists $s \in S$ such that

$$x_{1-2s} \geq x_{1-2i}.$$

In view of (3.1.9)

$$\frac{\alpha+\sum_{i=1}^{t}\beta_{2i-1}x_{2-2i}}{A+\sum_{j=0}^{m}B_{2j}x_{1-2j}}<\epsilon$$

and so

$$x_2=\frac{\sum_{i=0}^{t}\beta_{2i}x_{1-2i}}{A+\sum_{j=0}^{m}B_{2j}x_{1-2j}}+\frac{\alpha+\sum_{i=1}^{t}\beta_{2i-1}x_{2-2i}}{A+\sum_{j=0}^{m}B_{2j}x_{1-2j}}$$

$$<\frac{\max_{i\in S}\beta_{2i}}{\min_{j\in S}B_{2j}}+\frac{\sum_{i=m+1}^{t}\beta_{2i}\cdot x_{1-2s}}{B_{2s}x_{1-2s}}+\epsilon=U+\epsilon.$$

When (3.1.5) holds and $S=\emptyset$, in view of (3.1.9),

$$x_2=\frac{\alpha+\sum_{i=1}^{t}\beta_{2i-1}x_{2-2i}}{A+\sum_{j=0}^{m}B_{2j}x_{1-2j}}<\epsilon=U+\epsilon$$

and the proofs of (3.1.11) and (3.1.12) follow by induction. Next assume that

$$\lim_{n\to\infty}x_{2n+1}=L_1<\infty.$$

Clearly, there exist nonnegative numbers

$$L_{-2t},\ldots,L_0\leq U+\epsilon$$

such that

$$L_1(A+\sum_{j=0}^{m}B_{2j}L_{-2j})=\alpha+\sum_{i=0}^{t}\beta_{2i}L_{-2i}+L_1\sum_{i=1}^{t}\beta_{2i-1}.$$

Hence,

$$L_1(A+(U+\epsilon)\sum_{j=0}^{m}B_{2j})>\alpha+\sum_{i=0}^{t}\beta_{2i}L_{-2i}+L_1(A+(U+\epsilon)\sum_{j=0}^{m}B_{2j}),$$

which is a contradiction. The proof is complete. ∎

Open Problem 3.1.1 *Assume that*

$$\gamma>\beta+\delta+A$$

and let k be a number such that

$$0<k<\gamma-\beta-\delta-A.$$

Let $\bar{x}$ denote the positive equilibrium point. Determine the character of solutions of Eq.(3.0.2) with

$$x_{-2},x_0\in[\gamma-A,\bar{x}]\ \text{ and }\ x_{-1}\in\left[\bar{x},\frac{\alpha+\gamma(\gamma-A)}{k}\right].$$

3.2 Unbounded Solutions of $x_{n+1} = \dfrac{\alpha + \beta x_n + \gamma x_{n-1} + \delta x_{n-2}}{A + x_{n-2}}$

In this section we will confirm the existence of unbounded solutions in 13 additional special cases of Eq.(3.0.1), namely:

#31,	#48,	#64,	#73,	#85,	#97,	#128,
#132,	#140,	#167,	#179,	#185,	#197.	

See Appendix A.

We will assume that (3.0.5) holds, and we will establish that there exist solutions of Eq.(3.0.4) that are unbounded in some range of its parameters and for some initial conditions. Actually, we exhibit a huge set of initial conditions through which the subsequences of even and odd terms of the solutions converge, one of them to ∞ and the other to

$$\frac{\delta - A + \sqrt{(\delta - A)^2 + 4\beta\gamma}}{2}.$$

More precisely, we establish the following result.

Theorem 3.2.1 *Assume that (3.0.5) holds and that*

$$\gamma > \beta + \delta + A.$$

Then Eq.(3.0.4) has unbounded solutions. In fact there exist initial conditions x_{-2}, x_{-1}, x_0 *such that*

$$\lim_{n\to\infty} x_{2n+1} = \infty \quad \textit{and} \quad \lim_{n\to\infty} x_{2n+2} = \frac{\delta - A + \sqrt{(\delta - A)^2 + 4\beta\gamma}}{2}. \tag{3.2.1}$$

PROOF We consider the following cases:

Case 1:

$$A > 0.$$

Let $\{x_n\}$ be a solution of Eq.(3.0.4) with initial conditions x_{-2}, x_{-1}, x_0 satisfying

$$\frac{\epsilon + (\beta + \delta + \epsilon)\gamma - A(\gamma - A)}{\gamma - A} < x_{-2}, x_0 < \gamma - A, \tag{3.2.2}$$

$$x_{-1} > \max\{\frac{\alpha + (\gamma - A)^2}{\epsilon}, \frac{(\beta + \delta + \epsilon)[\alpha + (\beta + \delta)(\gamma - A)]}{\epsilon}\} \tag{3.2.3}$$

where

$$0 < \epsilon < \min\{\frac{(\gamma(\gamma-\beta-\delta-A)}{\gamma+1}, \frac{A(\gamma-\beta-\delta-A)}{\gamma+2}\}. \tag{3.2.4}$$

We claim that

$$x_{2n+1} > \frac{\alpha+\beta x_{2n}+\delta x_{2n-2}}{A+x_{2n-2}} + x_{2n-1}, \;\; n = 0, 1, \ldots \tag{3.2.5}$$

and

$$\frac{\epsilon+(\beta+\delta+\epsilon)\gamma - A(\gamma-A)}{\gamma-A} < x_{2n+2} < \gamma - A, \;\; n = 0, 1, \ldots . \tag{3.2.6}$$

In view of (3.2.2),

$$x_1 = \frac{\alpha+\beta x_0+\delta x_{-2}}{A+x_{-2}} + \frac{\gamma x_{-1}}{A+x_{-2}} > \frac{\alpha+\beta x_0+\delta x_{-2}}{A+x_{-2}} + x_{-1} > x_{-1}. \tag{3.2.7}$$

We claim

$$(\beta+\delta+\epsilon)x_1 < (\gamma-A)x_{-1}. \tag{3.2.8}$$

Indeed, in view of (3.2.2),

$$\begin{aligned}&(\beta+\delta+\epsilon)x_1 - (\gamma-A)x_{-1}\\ &= \frac{\alpha(\beta+\delta+\epsilon)+\beta(\beta+\delta+\epsilon)x_0}{A+x_{-2}}\\ &\frac{+\gamma(\beta+\delta+\epsilon)x_{-1}+\delta(\beta+\delta+\epsilon)x_{-2}-(A+x_{-2})(\gamma-A)x_{-1}}{A+x_{-2}}\\ &< \frac{\alpha(\beta+\delta+\epsilon)+(\beta+\delta)(\beta+\delta+\epsilon)(\gamma-A)-\epsilon x_{-1}}{A+x_{-2}}.\end{aligned}$$

In view of (3.2.3)

$$\frac{\alpha(\beta+\delta+\epsilon)+(\beta+\delta)(\beta+\delta+\epsilon)(\gamma-A)-\epsilon x_{-1}}{A+x_{-2}} < 0.$$

In addition, in view of (3.2.3), (3.2.7), and (3.2.8),

$$\begin{aligned}x_2-\gamma+A &= \frac{\alpha+\beta x_1+\gamma x_0+\delta x_{-1}-A(\gamma-A)-(\gamma-A)x_{-1}}{A+x_{-1}}\\ &< \frac{\alpha+(\gamma-A)^2-\epsilon x_1}{A+x_{-1}} < 0.\end{aligned}$$

Furthermore, in view of (3.2.2), (3.2.3), and (3.2.4),

$$x_2 - \frac{\epsilon+(\beta+\delta+\epsilon)\gamma-A(\gamma-A)}{\gamma-A}$$

$$= \frac{\alpha(\gamma - A) + \beta(\gamma - A)x_1 + \gamma(\gamma - A)x_0}{(A + x_{-1})(\gamma - A)}$$

$$\frac{+\delta(\gamma - A)x_{-1} - A[\epsilon + (\beta + \delta + \epsilon)\gamma - A(\gamma - A)]}{(A + x_{-1})(\gamma - A)}$$

$$\frac{-[\epsilon + (\beta + \delta + \epsilon)\gamma - A(\gamma - A)]x_{-1}}{(A + x_{-1})(\gamma - A)}$$

$$> \frac{[A(\gamma - \beta - \delta - A) - \epsilon(1 + \gamma)]\,x_{-1} - [\epsilon + (\beta + \delta + \epsilon)\gamma - A(\gamma - A)]}{(A + x_{-1})(\gamma - A)}$$

$$> \frac{\epsilon x_{-1} - [\epsilon + (\beta + \delta + \epsilon)\gamma - A(\gamma - A)]}{(A + x_{-1})(\gamma - A)}$$

$$> \frac{(\gamma - A)^2 - [\epsilon + (\beta + \delta + \epsilon)\gamma - A(\gamma - A)]}{(A + x_{-1})(\gamma - A)} > 0.$$

The proofs of (3.2.5) and (3.2.6) follow inductively.

Let

$$I = \liminf_{n \to \infty} x_{2n}, \quad S = \limsup_{n \to \infty} x_{2n}, \quad \text{and} \quad L = \lim_{n \to \infty} x_{2n+1}.$$

In view of (3.2.5) and (3.2.6),

$$0 \leq I, \;\; S \leq \gamma - A, \;\; 0 < \gamma - A < L \leq \infty. \tag{3.2.9}$$

At this point we will show that

$$L = \infty. \tag{3.2.10}$$

We consider the following cases:

Subcase 1:

$$\alpha + \beta + \delta > 0 \;\text{ and }\; I > 0.$$

From this and in combination with (3.2.5) and (3.2.6) it follows that

$$x_{2n+1} > \frac{\alpha + (\beta + \delta)I}{\gamma} + x_{2n-1}, \;\; n = 0, 1, \ldots .$$

Hence,

$$L = \lim_{n \to \infty} x_{2n+1} = \infty.$$

The proof of (3.2.10) is complete in this case.

Subcase 2:

$$\alpha + \beta + \delta > 0 \;\text{ and }\; I = 0.$$

Suppose for the sake of contradiction that

$$L < \infty.$$

There exists a sequence of indices $\{n_i\}$ such that

$$\lim_{i \to \infty} x_{2n_i+2} = 0.$$

Furthermore, from Eq.(3.0.4) we have

$$x_{2n+2} \geq \frac{\alpha + \beta x_{2n+1} + \delta x_{2n-1}}{A + x_{2n-1}}, \quad n = 0, 1, \dots .$$

By replacing n by n_i in the above identity and then by taking limits, as $i \to \infty$, we find

$$0 \geq \frac{\alpha + (\beta + \delta)L}{A + L},$$

which implies that

$$L = 0.$$

This contradicts (3.2.9). The proof of (3.2.10) is complete in this case.

Subcase 3:

$$\alpha = \beta = \delta = 0.$$

Suppose for the sake of contradiction that

$$L < \infty.$$

By taking limits in Eq.(3.0.4), as $n \to \infty$, and in view of (3.2.9) we have

$$\gamma - A < L = \lim_{n \to \infty} x_{2n+1} = \lim_{n \to \infty} x_{2n+2} = \gamma - A,$$

which is a contradiction. The proof of (3.2.10) is complete.

Next we will establish (3.2.1) in this case ($A > 0$). Let $e > 0$. There exists $N = N(e)$ such that for all $n \geq N$

$$\frac{\gamma}{A + x_{2n-2}} < \frac{x_{2n+1}}{x_{2n-1}} < e + \frac{\gamma}{A + x_{2n-2}},$$

$$\delta - e < \frac{\alpha + \gamma x_{2n} + \delta x_{2n-1}}{A + x_{2n-1}} < \delta + e.$$

Then for $n \geq N$

$$\delta - e + \frac{\beta\gamma}{A + x_{2n-2}} < x_{2n+2} < \delta + e + \beta e + \frac{\beta\gamma}{A + x_{2n-2}} \quad n = 0, 1, \dots \tag{3.2.11}$$

and in view of (3.2.11) we find

$$S \leq \delta + e + \beta e + \frac{\beta\gamma}{A + I} \tag{3.2.12}$$

and

$$I \geq \delta - e + \frac{\beta\gamma}{A + S}. \tag{3.2.13}$$

Letting $e \to 0$, in (3.2.12) and (3.2.13), we get

$$A\delta + \delta S - AI \leq SI - \beta\gamma \leq A\delta + \delta I - AS$$

from which it follows that

$$(\delta + A)S \leq (\delta + A)I,$$

which implies

$$I = S.$$

Hence, by taking limits in (3.0.4), as $n \to \infty$, the proof of (3.2.1) follows. The proof is complete in this case.

Case 2:

$$A = 0 \text{ and } \delta > 0.$$

In this case the change of variables

$$y_n = x_n - \delta$$

reduces Eq.(3.0.4) into the equation

$$y_{n+1} = \frac{\alpha + \beta\delta + \gamma\delta + \beta y_n + \gamma y_{n-1}}{\delta + y_{n-2}}, \quad n = 0, 1, \ldots . \tag{3.2.14}$$

Since $\delta > 0$ we may apply the results of Case 1 to Eq.(3.2.14) and so the result follows. The proof is complete in this case.

Case 3:

$$A = \delta = 0.$$

In this case, in view of (3.0.5), it follows that

$$\beta = 0$$

and so Eq.(3.0.4) becomes

$$x_{n+1} = \frac{\alpha + \gamma x_{n-1}}{x_{n-2}}, \quad n = 0, 1, \ldots . \tag{3.2.15}$$

We choose initial conditions x_{-2}, x_{-1}, x_0 such that

$$0 < x_{-2}, x_0 < \gamma, \quad x_{-1} > \frac{\alpha + \gamma^2}{\gamma}.$$

From this it follows that

$$x_1 > \frac{\alpha}{\gamma} + x_{-1} \text{ and } x_2 < \gamma$$

and by induction

$$x_{2n+1} > \frac{\alpha}{\gamma} + x_{2n-1} \text{ and } x_{2n+2} < \gamma, \ n = 0, 1, \ldots .$$

Therefore,

$$\lim_{n\to\infty} x_{2n+1} = \infty.$$

Moreover, from Eq.(3.2.15) we have

$$x_{2n+2} = \frac{\alpha}{x_{2n-1}} + \frac{\gamma x_{2n}}{x_{2n-1}}, \ n = 0, 1, \ldots . \tag{3.2.16}$$

By taking limits in (3.2.16), as $n \to \infty$, it follows that

$$\lim_{n\to\infty} x_{2n+2} = 0.$$

The proof is complete. ∎

3.3 Unbounded Solutions of $x_{n+1} = \frac{\alpha + \beta x_n + \gamma x_{n-1} + \delta x_{n-2}}{A + Bx_n + x_{n-2}}$

In this section we will confirm the existence of unbounded solutions in 16 additional special cases of Eq.(3.0.1), namely:

#33,	#75,	#87,	#99,	#110,	#146,	#154,	#162
#169,	#181,	#187,	#199,	#202,	#210,	#214,	#222.

See Appendix A.

We will assume that (3.0.7) holds and we will establish that there exist solutions of Eq.(3.0.6) that are unbounded in some range of its parameters and for some initial conditions.

More precisely, we establish the following result.

Theorem 3.3.1 *Assume that (3.0.7) holds and that*

$$\gamma > \beta + \delta + A.$$

Then Eq.(3.0.6) has unbounded solutions.

PROOF We will consider two cases.

Case 1:

$$\beta > \delta B.$$

Let $\{x_n\}$ be a solution of Eq.(3.0.6) with initial conditions satisfying

$$\frac{\beta+\delta}{B+1} < x_{-2}, x_0 < \frac{\gamma - A}{B+1} \quad \text{and} \quad \frac{\alpha(B+1) - A(\beta+\delta+k) + \gamma(\gamma-A) + \beta\frac{L}{M}}{kB} < x_{-1}, \tag{3.3.1}$$

where

$$0 < k = \gamma - \beta - A - \delta, \ \ L = \max\{\alpha, \beta, \delta\}, \ \text{ and } \ M = \min\{A, B, 1\}\,.$$

We will show that for $n \geq 0$,

$$x_{2n+1} > \frac{\alpha + \beta x_{2n} + \delta x_{2n-2}}{A + Bx_{2n} + x_{2n-2}} + x_{2n-1} \ \text{ and } \ x_{2n} \in (\frac{\beta+\delta}{B+1}, \frac{\gamma - A}{B+1}). \tag{3.3.2}$$

We have

$$\frac{\alpha + \beta x_n + \delta x_{n-2}}{A + Bx_n + x_{n-2}} < \frac{L}{M}, \ n = 0, 1, \dots .$$

Also,

$$x_1 = \frac{\alpha + \beta x_0 + \delta x_{-2}}{A + Bx_0 + x_{-2}} + \frac{\gamma}{A + Bx_0 + x_{-2}} x_{-1},$$

from which it follows that

$$x_1 > \frac{\alpha + \beta x_0 + \delta x_{-2}}{A + Bx_0 + x_{-2}} + x_{-1}$$

and

$$x_1 < \frac{\alpha + \beta x_0 + \delta x_{-2}}{A + Bx_0 + x_{-2}} + \frac{\beta + k}{\beta} x_{-1}.$$

Also,

$$x_2 - \frac{\gamma - A}{B+1} = \frac{\alpha + \beta x_1 + \gamma x_0 + \delta x_{-1}}{A + Bx_1 + x_{-1}} - \frac{\beta + \delta + k}{B+1}$$

$$= \frac{\alpha(B+1) - A(\beta+\delta+k) + \gamma(B+1)x_0 + (\beta x_1 - (\beta+k)x_{-1}) + B\delta(x_{-1} - x_1) - Bkx_1}{(B+1)(A + Bx_1 + x_{-1})}$$

$$< \frac{\alpha(B+1) - A(\beta+\delta+k) + \gamma(\gamma-A) + \beta\frac{\alpha+\beta x_0+\delta x_{-2}}{A+Bx_0+x_{-2}} - Bkx_1}{(B+1)(A + Bx_1 + x_{-1})}$$

$$< \frac{\alpha(B+1) - A(\beta+\delta+k) + \gamma(\gamma-A) + \beta\frac{L}{M} - Bkx_1}{(B+1)(A + Bx_1 + x_{-1})} < 0.$$

Furthermore,

$$x_2 - \frac{\beta+\delta}{B+1} = \frac{\alpha + \beta x_1 + \gamma x_0 + \delta x_{-1}}{A + Bx_1 + x_{-1}} - \frac{\beta+\delta}{B+1}$$

$$= \frac{\alpha(B+1) - A(\beta+\delta) + \gamma(B+1)x_0 + (\beta - \delta B)(x_1 - x_{-1})}{(B+1)(A + Bx_1 + x_{-1})}$$

$$> \frac{\alpha(B+1) - A(\beta+\delta) + \gamma(\beta+1) + (\beta - \delta B)(x_1 - x_{-1})}{(B+1)(A + Bx_1 + x_{-1})} > 0.$$

The proof of (3.3.2) follows by induction.

Let

$$I = \liminf_{n\to\infty} x_{2n+2} \text{ and } L = \lim_{n\to\infty} x_{2n+1}.$$

We will show that

$$L = \infty. \tag{3.3.3}$$

We consider the following two subcases:

Subcase 1:

$$\beta + \delta > 0 \text{ or } \alpha > 0.$$

Then from (3.3.2) we have

$$x_{2n+1} > \frac{\alpha + (\beta+\delta)(I - \epsilon)}{A + \beta + \delta} + x_{2n-1}, \quad n = 0, 1, \ldots,$$

where ϵ is a positive number. From this it follows that

$$L = \lim_{n\to\infty} x_{2n+1} = \infty.$$

The proof of (3.3.3) is complete in this case.

Subcase 2:

$$\alpha = \beta = \delta = 0.$$

In this case Eq.(3.0.6) becomes

$$x_{n+1} = \frac{\gamma x_{n-1}}{A + Bx_n + x_{n-2}}, \quad n = 0, 1, \ldots .$$

In particular we have

$$x_{2n+2} = \frac{\gamma x_{2n}}{A + Bx_{2n+1} + x_{2n-1}}, \quad n = 0, 1, \ldots .$$

By taking limits in the last equation as $n \to \infty$ we find

$$\lim_{n\to\infty} x_{2n+2} = 0.$$

We choose a positive number m and a positive integer N such that

$$0 < m < \frac{\gamma - A}{A + B + 1}$$

and

$$x_{2n} < m, \quad n = N, N+1, \dots .$$

Hence,

$$x_{2n+1} = \frac{\gamma x_{2n-1}}{A + Bx_{2n} + x_{2n-2}} > \frac{\gamma x_{2n-1}}{A + (B+1)m} > (1+m)x_{2n-1}, \quad n = N, N+1, \dots .$$

From this it follows that

$$L = \lim_{n\to\infty} x_{2n+1} = \infty.$$

The proof of (3.3.3) is complete.

Case 2:

$$\delta B \geq \beta.$$

Let $\{x_n\}$ be a solution of Eq.(3.0.6) with initial conditions such that,

$$0 < x_{-2}, x_0 < \frac{\gamma - A}{B+1}, \quad \frac{\alpha(B+1) - A(\beta+\delta+k) + \gamma(\gamma - A)}{k(B+1)} < x_{-1}, \quad (3.3.4)$$

where $0 < k = \gamma - \beta - A - \delta$.

We will show that for $n \geq 0$,

$$x_{2n+1} > \frac{\alpha + \beta x_{2n} + \delta x_{2n-2}}{A + Bx_{2n} + x_{2n-2}} + x_{2n-1} \quad \text{and} \quad x_{2n+2} < \frac{\gamma - A}{B+1}. \qquad (3.3.5)$$

Indeed,

$$x_1 = \frac{\alpha + \beta x_0 + x_{-2}}{A + Bx_0 + x_{-2}} + \frac{\gamma}{A + Bx_0 + x_{-2}} x_{-1},$$

from which it follows that

$$x_1 > \frac{\alpha + \beta x_0 + \delta x_{-2}}{A + Bx_0 + x_{-2}} + x_{-1}.$$

Let $s = \delta B - \beta \geq 0$. Then

$$x_2 - \frac{\gamma - A}{B+1} = \frac{\alpha + \beta x_1 + \gamma x_0 + \delta x_{-1}}{A + Bx_1 + x_{-1}} - \frac{\beta + \delta + k}{B+1}$$

$$= \frac{\alpha(B+1) - A(\beta+\delta+k) + \gamma(B+1)x_0 - s(x_1 - x_{-1}) - k(1+B)x_{-1}}{(B+1)(A + Bx_1 + x_{-1})}$$

$$< \frac{\alpha(B+1) - A(\beta+\delta+k) + \gamma(\gamma - A) - k(1+B)x_{-1}}{(B+1)(A + Bx_1 + x_{-1})} < 0.$$

The proof of (3.3.5) follows inductively.

Let

$$I_1 = \liminf_{n\to\infty} x_{2n+2} \quad \text{and} \quad L_1 = \lim_{n\to\infty} x_{2n+1}.$$

In view of (3.3.5) it follows that

$$L_1 > 0. \tag{3.3.6}$$

We will show that

$$L_1 = \infty. \tag{3.3.7}$$

We consider the following three subcases:

Subcase 1:

$$\alpha + \beta + \delta > 0 \quad \text{and} \quad I_1 > 0.$$

Then from (3.3.5) we have

$$x_{2n+1} > \frac{\alpha + (\beta + \delta)(I - \epsilon)}{\gamma} + x_{2n-1}, \quad n = 0, 1, \ldots,$$

where ϵ is a positive number. From this it follows that

$$L_1 = \lim_{n\to\infty} x_{2n+1} = \infty.$$

The proof of (3.3.7) is complete in this case.

Subcase 2:

$$\alpha + \beta + \delta > 0 \text{ and } I_1 = 0.$$

Suppose for the sake of contradiction that

$$L_1 < \infty.$$

We choose a subsequence $\{x_{2n_i+2}\}$ of $\{x_{2n+2}\}$ such that

$$\lim_{i\to\infty} x_{2n_i+2} = 0.$$

Furthermore, from Eq.(3.0.6) we obtain

$$x_{2n+2} \geq \frac{\alpha + \beta x_{2n+1} + \delta x_{2n-1}}{A + B x_{2n+1} + x_{2n-1}}, \quad n = 0, 1, \ldots .$$

By replacing n by n_i in the preceding identity and then by taking limits, as $i \to \infty$, we find

$$0 \geq \frac{\alpha + (\beta + \delta)L}{A + (B + 1)L},$$

which implies that

$$L_1 = 0.$$

This contradicts (3.3.6). The proof of (3.3.7) is complete in this case.

Subcase 3:

$$\alpha = \beta = \delta = 0 .$$

In this case Eq.(3.0.6) becomes

$$x_{n+1} = \frac{\gamma x_{n-1}}{A + Bx_n + x_{n-2}}, \quad n = 0, 1, \dots .$$

In particular, we have

$$x_{2n+2} = \frac{\gamma x_{2n}}{A + Bx_{2n+1} + x_{2n-1}}, \quad n = 0, 1, \dots .$$

By taking limits in the last equation as $n \to \infty$ we find

$$\lim_{n\to\infty} x_{2n+2} = 0.$$

We now choose positive numbers m and N such that

$$0 < m < \frac{\gamma - A}{A + B + 1}$$

and

$$x_{2n} < m, \quad \text{for} \quad n = N, N + 1, \dots .$$

Hence,

$$x_{2n+1} = \frac{\gamma x_{2n-1}}{A + Bx_{2n} + x_{2n-2}} > \frac{\gamma x_{2n-1}}{A + (B + 1)m} > (1+m)x_{2n-1}, \quad n = N, N+1, \dots .$$

From this it follows that

$$L_1 = \lim_{n\to\infty} x_{2n+1} = \infty.$$

The proof is complete. ∎

3.4 Unbounded Solutions of $x_{n+1} = \dfrac{\alpha + \beta x_n + \gamma x_{n-1} + \delta x_{n-2}}{A + Bx_n + Cx_{n-1}}$

In this section we will confirm the existence of unbounded solutions in 14 additional special cases of Eq.(3.0.1), namely:

#80,	#92,	#98,	#149,	#157,	#161,	#174
#180,	#186,	#198,	#205,	#209,	#213,	#221.

See Appendix A.

In this section we will assume that

$$\delta > A + \gamma B + \frac{\beta}{B}. \tag{3.4.1}$$

and establish that there exist solutions of Eq.(3.0.8) that are unbounded.

More precisely, we establish the following result:

Theorem 3.4.1 *Assume that (3.0.9) and (3.4.1) hold. Then Eq.(3.0.8) has unbounded solutions.*

PROOF Choose positive numbers m and ϵ such that

$$m \in \left(0, \delta - A - \gamma B - \frac{\beta}{B}\right) \quad \text{and} \quad \epsilon \in \left(0, \frac{m}{1+B}\right).$$

Set

$$K = \frac{1}{\epsilon}\left[\alpha + \beta\left(\epsilon + \frac{\beta}{B}\right) + \delta\left(\epsilon + \gamma\right)\right]$$

and

$$L = \frac{1}{\epsilon B}\left[\alpha + \gamma\left(\epsilon + \gamma\right) + \delta\left(\epsilon + \frac{\beta}{B}\right)\right].$$

Let $\{x_n\}$ be a solution of Eq.(3.0.8) with initial conditions chosen as follows:

$$x_{-2} > \max\{K, L\}, \quad x_{-1} \in \left(0, \epsilon + \frac{\beta}{B}\right), \quad \text{and} \quad x_0 \in (0, \epsilon + \gamma).$$

Then we claim that

$$\lim_{n\to\infty} x_{3n+1} = \infty, \quad \lim_{n\to\infty} x_{3n+2} = \frac{\beta}{B}, \quad \text{and} \quad \lim_{n\to\infty} x_{3n+3} = \gamma.$$

Indeed,

$$x_1 = \frac{\alpha + \beta x_0 + \gamma x_{-1} + \delta x_{-2}}{A + B x_0 + x_{-1}} > \frac{\alpha + \beta x_0 + \gamma x_{-1}}{A + \gamma B + \frac{\beta}{B} + \epsilon(1+B)} + \frac{\delta x_{-2}}{A + \gamma B + \frac{\beta}{B} + \epsilon(1+B)}$$

$$> \frac{\alpha + \beta x_0 + \gamma x_{-1}}{\delta} + \frac{\delta}{A + \gamma B + \frac{\beta}{B} + m} x_{-2},$$

$$x_2 = \frac{\alpha + \beta x_1 + \gamma x_0 + \delta x_{-1}}{A + B x_1 + x_0} < \frac{\alpha + \gamma(\epsilon + \gamma) + \delta\left(\epsilon + \frac{\beta}{B}\right) + \beta x_1}{B x_1}$$

$$< \frac{\alpha + \gamma(\epsilon + \gamma) + \delta\left(\epsilon + \frac{\beta}{B}\right) + \beta L}{B L} = \epsilon + \frac{\beta}{B},$$

and

$$x_3 = \frac{\alpha + \beta x_2 + \gamma x_1 + \delta x_0}{A + Bx_2 + x_1} < \frac{\alpha + \beta\left(\epsilon + \frac{\beta}{B}\right) + \delta(\epsilon+\gamma) + \gamma x_1}{x_1}$$

$$< \frac{\alpha + \beta\left(\epsilon + \frac{\beta}{B}\right) + \delta(\epsilon+\gamma) + \gamma K}{K} = \epsilon + \gamma.$$

It follows by induction that for $n \geq 0$,

$$x_{3n+1} > \frac{\alpha + \beta x_{3n} + \gamma x_{3n-1}}{\delta} + \frac{\delta}{A + \gamma B + \frac{\beta}{B} + m} x_{3n-2},$$

$$x_{3n-2} < \epsilon + \frac{\beta}{B},$$

and

$$x_{3n+3} < \epsilon + \gamma.$$

Therefore,

$$\lim_{n\to\infty} x_{3n+1} = \infty,$$

$$x_{3n+2} = \frac{\frac{\alpha}{x_{3n+1}} + \beta + \gamma\frac{x_{3n}}{x_{3n+1}} + \delta\frac{x_{3n-1}}{x_{3n+1}}}{\frac{A}{x_{3n+1}} + B + \frac{x_{3n}}{x_{3n+1}}} \to \frac{\beta}{B} \quad \text{as} \quad n \to \infty,$$

and

$$x_{3n+3} = \frac{\frac{\alpha}{x_{3n+1}} + \beta\frac{x_{3n+2}}{x_{3n+1}} + \gamma + \delta\frac{x_{3n}}{x_{3n+1}}}{\frac{A}{x_{3n+1}} + B\frac{x_{3n+2}}{x_{3n+1}} + 1} \to \gamma \quad \text{as} \quad n \to \infty$$

and the proof is complete. ∎

Theorem 3.4.1 extends in a natural way to the $(k+1)^{\text{st}}$-order rational equation

$$x_{n+1} = \frac{\alpha + \sum_{i=0}^{k} \beta_i x_{n-i}}{A + \sum_{i=0}^{k-1} B_i x_{n-i}}, \quad n = 0, 1, \ldots \tag{3.4.2}$$

with

$$\beta_k, B_0, B_1, \ldots, B_{k-1} \in (0, \infty). \tag{3.4.3}$$

This result establishes the existence of unbounded solutions in the following 32 special cases of Eq.(3.4.2), with $k = 4$:

#424 – 431, #440 – 447, #488 – 495, #504 – 511 .

See Appendix B.

Theorem 3.4.2 *Assume that (3.4.3) holds and that*

$$\beta_k > A + \sum_{i=0}^{k-1}(B_i \cdot \frac{\beta_{k-1-i}}{B_{k-1-i}}).$$

Then Eq.(3.4.2) has unbounded solutions.

PROOF Let $\epsilon > 0$ be chosen such that

$$\beta_k > A + \sum_{i=0}^{k-1} B_i(\frac{\beta_{k-1-i}}{B_{k-1-i}} + \epsilon). \tag{3.4.4}$$

Let $\{x_n\}$ be any solution of Eq.(3.4.2) with initial conditions satisfying the following conditions:

$$x_{-k} > \frac{\alpha + \sum_{i=1}^{k} \beta_i(\frac{\beta_{k-i}}{B_{k-i}} + \epsilon)}{\epsilon B_0}, \tag{3.4.5}$$

$$x_{-k} > \frac{\beta_0 A + \beta_0 \sum_{i=1}^{k-1} B_i(\frac{\beta_{k-i}}{B_{k-i}} + \epsilon)}{\epsilon B_0^2}, \tag{3.4.6}$$

$$\cdots$$

$$x_{-k} > \frac{\alpha + \sum_{i=0}^{k-2} \beta_i(\frac{\beta_{k-2-i}}{B_{k-2-i}} + \epsilon) + \beta_k(\frac{\beta_{k-1}}{B_{k-1}} + \epsilon)}{\epsilon B_{k-1}}, \tag{3.4.7}$$

$$x_{-k} > \frac{A\beta_{k-1} + \beta_{k-1} \sum_{i=0}^{k-2} B_i(\frac{\beta_{k-i-2}}{B_{k-i-2}} + \epsilon)}{\epsilon B_{k-1}^2}, \tag{3.4.8}$$

$$\frac{\beta_0}{B_0} - \epsilon < x_{-k+1} < \frac{\beta_0}{B_0} + \epsilon \tag{3.4.9}$$

and

$$\cdots$$

$$\frac{\beta_{k-1}}{B_{k-1}} - \epsilon < x_0 < \frac{\beta_{k-1}}{B_{k-1}} + \epsilon. \tag{3.4.10}$$

Then we claim that

$$x_{(k+1)n+1} \to \infty \tag{3.4.11}$$

and

$$x_{(k+1)n+i} \to \frac{\beta_{i-2}}{B_{i-2}}, \quad i = 2, 3, \cdots, k+1. \tag{3.4.12}$$

To this end we first show that for all $n \geq 0$

$$x_{(k+1)n+1} > \frac{\alpha + \sum_{i=0}^{k-1} \beta_i (\frac{\beta_{k-1-i}}{B_{k-1-i}} - \epsilon)}{A + \sum_{i=0}^{k-1} B_i (\frac{\beta_{k-1-i}}{B_{k-1-i}} + \epsilon)} + \frac{\beta_k}{A + \sum_{i=0}^{k-1} B_i (\frac{\beta_{k-1-i}}{B_{k-1-i}} + \epsilon)} \cdot x_{(k+1)(n-1)+1} \tag{3.4.13}$$

and

$$\frac{\beta_{i-2}}{B_{i-2}} - \epsilon < x_{(k+1)n+i} < \frac{\beta_{i-2}}{B_{i-2}} + \epsilon, \quad i = 1, 2, \cdots k+1. \tag{3.4.14}$$

Indeed,

$$x_1 = \frac{\alpha + \sum_{i=0}^{k} \beta_i x_{-i}}{A + \sum_{i=0}^{k-1} B_i x_{-i}}$$

$$> \frac{\alpha + \sum_{i=0}^{k-1} \beta_i (\frac{\beta_{k-1-i}}{B_{k-1-i}} - \epsilon)}{A + \sum_{i=0}^{k-1} B_i (\frac{\beta_{k-1-i}}{B_{k-1-i}} + \epsilon)} + \frac{\beta_k}{A + \sum_{i=0}^{k-1} B_i (\frac{\beta_{k-1-i}}{B_{k-1-i}} + \epsilon)} \cdot x_{-k} > x_{-k}.$$

From the last inequality and (3.4.5), (3.4.6), (3.4.7), and (3.4.8), we have

$$\frac{\beta_0}{B_0} - \epsilon < x_2 < \frac{\beta_0}{B_0} + \epsilon$$

and

$$\cdots$$

$$\frac{\beta_{k-1}}{B_{k-1}} - \epsilon < x_{k+1} < \frac{\beta_{k+1}}{B_{k+1}} + \epsilon$$

and so (3.4.13) and (3.4.14) follow by induction. Clearly, (3.4.12) follows from (3.4.14). Also, (3.4.11) follows from (3.4.13) as long as

$$\alpha + \sum_{i=0}^{k-1} \beta_i \frac{\beta_{k-1-i}}{B_{k-1-i}} > 0.$$

On the other hand, when

$$\alpha + \sum_{i=0}^{k-1} \beta_i \frac{\beta_{k-1-i}}{B_{k-1-i}} = 0$$

if

$$x_{(k+1)n+1} \to U \in (0, \infty),$$

then

$$U = \frac{\beta_k U}{A + \sum_{i=0}^{k-1} B_i \frac{\beta_{k-1-i}}{B_{k-1-i}}},$$

which contradicts (3.4.4). The proof is complete.

∎

3.5 Unbounded Solutions of $x_{n+1} = \frac{x_{n-2}}{A + Bx_n + Cx_{n-1}}$

Theorem 3.5.1 *Assume that*

$$A, B, C \in [0, \infty) \text{ and } A + B, A + C, B + C > 0.$$

Then the rational equation

$$x_{n+1} = \frac{x_{n-2}}{A + Bx_n + Cx_{n-1}}, \quad n = 0, 1, ...$$

has unbounded solutions.

PROOF For the proof see Theorem 4.4.1 in Section 4.4. ∎

3.6 Unbounded Solutions in the Special Case #50

Theorem 3.6.1 *Assume that*

$$0 < \alpha < 1.$$

Then the rational equation

$$x_{n+1} = \frac{\alpha + x_{n-2}}{x_n}, \quad n = 0, 1, \ldots \tag{3.6.1}$$

has unbounded solutions.

Note that the following identities hold:

$$x_{n+4} - x_{n-1} = \frac{\alpha x_{n+3} - (\alpha - 1)x_{n+1} - x_{n-2}}{x_{n+3}}, \quad n = 0, 1, \ldots . \tag{3.6.2}$$

$$x_{n+6} - x_{n+1} = \frac{\alpha x_{n+4} - (\alpha - 1)x_{n+1} - x_{n-1}}{\alpha + x_{n+2}}, \quad n = 0, 1, \ldots . \tag{3.6.3}$$

Lemma 3.6.1 *Assume that* $0 < \alpha < 1$ *and let* $\{x_n\}_{n=-2}^{\infty}$ *be any solution of Eq.(3.6.1) for which there exists* $N \geq 3$ *such that*

$$x_N \geq x_n, \quad \text{for } n \geq -2. \tag{3.6.4}$$

Then

$$x_n = \bar{x}, \quad \text{for } n \geq -2.$$

PROOF From (3.6.4) with $n = N+5$, (3.6.3) implies that

$$\alpha x_{N+3} + (1-\alpha)x_N \leq x_{N-2}. \tag{3.6.5}$$

Furthermore, in view of (3.6.4), we have $x_N \geq x_{N+3}$ and so (3.6.5) implies that

$$x_{N+3} \leq x_{N-2}. \tag{3.6.6}$$

In view of (3.6.2), we have

$$x_{N+3} - x_{N-2} = \frac{x_{N+2} - x_{N-3}}{x_{N+2}} + \frac{(\alpha-1)(x_{N+2} - x_N)}{x_{N+2}}. \tag{3.6.7}$$

From (3.6.4), we have $x_N \geq x_{N+2}$. Therefore, in view of (3.6.6), (3.6.7) implies that

$$x_{N+2} \leq x_{N-3}. \tag{3.6.8}$$

In addition, from (3.6.4), we have

$$x_N \geq x_{N-5}, \tag{3.6.9}$$

and so

$$x_{N+3} = \frac{\alpha + x_N}{x_{N+2}} \geq \frac{\alpha + x_{N-5}}{x_{N-3}} = x_{N-2}. \tag{3.6.10}$$

From (3.6.6) and (3.6.10), we have $x_{N+3} = x_{N-2}$, and so (3.6.4) and (3.6.5) imply that

$$x_N = x_{N-2}.$$

In addition,

$$\alpha x_{N+3} + (1-\alpha)x_N = \alpha x_{N+3} + (1-\alpha)x_{N-2} = x_{N-2},$$

from which it follows with the use of (3.6.3) that $x_{N+5} = x_N$. It is also true that

$$\alpha x_N + (1-\alpha)x_{N-2} = x_N \geq x_{N-5}, \tag{3.6.11}$$

from which it follows with the use of (3.6.2) that $x_{N+1} \geq x_{N-4}$. Using Eq.(3.6.1) we get

$$x_{N+4} = \frac{\alpha + x_{N+1}}{x_{N+3}} \geq \frac{\alpha + x_{N-4}}{x_{N-2}} = x_{N-1}. \tag{3.6.12}$$

In view of (3.6.2) we have

$$x_{N+4} - x_{N-1} = \frac{x_{N+3} - x_{N-2}}{x_{N+3}} + \frac{(\alpha-1)(x_{N+3} - x_{N+1})}{x_{N+3}}. \tag{3.6.13}$$

Since $x_{N+3} = x_{N-2}$, (3.6.12) and (3.6.13) imply that

$$x_{N+3} \leq x_{N+1}. \tag{3.6.14}$$

From Eq.(3.6.1) with the use of (3.6.14) we get

$$x_N \leq x_{N-1},$$

from which it follows, with the use of (3.6.4), that

$$x_N = x_{N-1}.$$

Therefore, $x_{N+3} = x_N = x_{N-1} = x_{N-2}$. Using Eq.(3.6.1) we have

$$x_n = \bar{x}, \quad \text{for} \quad n \geq -2.$$

The proof is complete. ▮

Theorem 3.6.2 *Assume* $0 < \alpha < 1$. *Then every solution of Eq.(3.6.1) is either unbounded or converges to the equilibrium of Eq.(3.6.1).*

PROOF Let $\{x_n\}_{n=-2}^{\infty}$ be a solution of Eq.(3.6.1) bounded from above and from below. Set

$$S = \limsup_{n\to\infty} x_n \quad \text{and} \quad I = \liminf_{n\to\infty} x_n.$$

There exists subsequences $\{x_{n_i+k}\}_{i=1}^{\infty}$, $k = -2, -1, \ldots$, such that

$$\lim_{i\to\infty} x_{n_i+4} = l_4 = S \geq \lim_{i\to\infty} x_{n_i+k} = l_k. \tag{3.6.15}$$

In addition the sequence $\{l_k\}_{k=-2}^{\infty}$ satisfies Eq.(3.6.1). In view of (3.6.15) and Lemma 3.6.1, we have

$$l_k = \bar{x}, \quad \text{for} \quad k \geq -2,$$

and so

$$S = \bar{x}.$$

In addition, there exist subsequences: $\{x_{n_j+k}\}_{i=1}^{\infty}$, $k = -3, -2, \ldots$, of the solution $\{x_n\}_{n=-2}^{\infty}$ so that

$$\lim_{i\to\infty} x_{n_j} = I_0 = I \leq I_k = \lim_{i\to\infty} x_{n_j+k} \leq \bar{x}.$$

In addition, $\{I_k\}_{k=-3}^{\infty}$ is a solution of Eq.(3.6.1) and so

$$I_0 = \frac{\alpha + I_{-3}}{I_{-1}} \geq \frac{\alpha + I_0}{\bar{x}},$$

which implies

$$\bar{x} \geq \frac{\alpha + I_0}{I_0} \geq \bar{x}.$$

Hence, $I_0 = \bar{x}$. The proof is complete. ▮

Lemma 3.6.2 *Assume that* $0 < \alpha < 1$ *and let* $\{x_n\}_{n=-2}^{\infty}$ *be a nontrivial oscillatory solution of Eq.(3.6.1) bounded from above and from below. Then there exists* $-2 \leq N \leq 2$ *such that*

$$x_N = \sup\{x_n\}_{n=-2}^{\infty}. \tag{3.6.16}$$

PROOF If $N \geq 3$, in view of Lemma 3.6.1 we get a contradiction. On the other hand assume that

$$S = \sup\{x_n\}_{n=-2}^{\infty}$$

and

$$S > x_n, \quad \text{for} \quad n \geq -2.$$

Since $\{x_n\}_{n=-2}^{\infty}$ oscillates about $\bar{x}$, we have $S > \bar{x}$. Furthermore, there exists a subsequence $\{x_{n_i}\}_{i=1}^{\infty}$ of $\{x_n\}_{n=-2}^{\infty}$ so that

$$\lim_{i\to\infty} x_{n_i} = S > \bar{x},$$

which, in view of Theorem 3.6.2, is a contradiction. ∎

The proof of Theorem 3.6.1:

PROOF Assume that $0 < \alpha < 1$ and let $\{x_n\}_{n=-2}^{\infty}$ be any nontrivial oscillatory solution of Eq.(3.6.1) such that

$$\sup x_n \neq x_i, \quad \text{for} \quad i \in \{-2, -1, 0, 1, 2\}. \tag{3.6.17}$$

We will show that $\{x_n\}_{n=-2}^{\infty}$ is an unbounded solution of Eq.(3.6.1). Assume for the sake of contradiction that the solution $\{x_n\}_{n=-2}^{\infty}$ is bounded from above and below. This contradicts Lemma 3.6.2 and completes the proof. ∎

4

Periodic Trichotomies

4.0 Introduction

In this chapter we present all known periodic trichotomies of the third-order rational difference equation

$$x_{n+1} = \frac{\alpha + \beta x_n + \gamma x_{n-1} + \delta x_{n-2}}{A + Bx_n + Cx_{n-1} + Dx_{n-2}}, \quad n = 0, 1, \ldots \tag{4.0.1}$$

with nonnegative parameters $\alpha, \beta, \gamma, \delta, A, B, C, D$ and with arbitrary nonnegative initial conditions x_{-2}, x_{-1}, x_0, such that the denominator is always positive.

In Section 4.1 we present necessary and sufficient conditions for the existence of prime period-two solutions of Eq.(4.0.1). See also Section 5.9.

In Section 4.2 we present the period-two trichotomies known for the rational equations

$$x_{n+1} = \frac{\alpha + \beta x_n + \gamma x_{n-1}}{A + x_n}, \quad n = 0, 1, \ldots, \tag{4.0.2}$$

$$x_{n+1} = \frac{\alpha + \gamma x_{n-1} + \delta x_{n-2}}{A + x_{n-2}}, \quad n = 0, 1, \ldots, \tag{4.0.3}$$

and

$$x_{n+1} = \frac{\alpha + \gamma x_{n-1}}{A + Bx_n + x_{n-2}}, \quad n = 0, 1, \ldots . \tag{4.0.4}$$

Note that in addition to these three nonlinear period-two trichotomies, Eq.(4.0.1) contains a trivial period-two trichotomy for the linear equation

$$x_{n+1} = \frac{\gamma}{A} x_{n-1}, \quad n = 0, 1, \ldots .$$

In Section 4.3 we present a unified version of the three known nonlinear period-two trichotomies of Eq.(4.0.1).

In Section 4.4 we present the period-three trichotomy known for the rational equation

$$x_{n+1} = \frac{x_{n-2}}{A + Bx_n + Cx_{n-1}}, \quad n = 0, 1, \ldots . \tag{4.0.5}$$

In Section 4.5 we present the period-four trichotomy for the rational equation

$$x_{n+1} = \frac{\alpha + \beta x_n + x_{n-2}}{x_{n-1}}, \quad n = 0, 1, \dots . \tag{4.0.6}$$

In Section 4.6 we present the period-five trichotomy for the rational equation

$$x_{n+1} = \frac{\alpha + x_{n-2}}{x_n}, \quad n = 0, 1, \dots . \tag{4.0.7}$$

In Section 4.7 we present the period-six trichotomy that has been conjectured for the rational equation

$$x_{n+1} = \frac{\alpha + x_n}{Cx_{n-1} + x_{n-2}}, \quad n = 0, 1, \dots . \tag{4.0.8}$$

We offer the following conjecture for Eq.(4.0.1):

Conjecture 4.0.1 *No other periodic trichotomies are possible for any nonlinear special case of Eq.(4.0.1) without restricting the region of parameters of the special case.*

Open Problem 4.0.1 *For the $(k+1)^{st}$-order rational difference equation*

$$x_{n+1} = \frac{\alpha + \sum_{i=0}^{k} \beta_i x_{n-i}}{A + \sum_{i=0}^{k} B_i x_{n-i}}, \quad n = 0, 1, \dots$$

find all possible nonlinear periodic trichotomies that are not reduced to one of the seven known nonlinear trichotomies of Eq.(4.0.1).

4.1 Existence of Prime Period-Two Solutions

In this section we present necessary and sufficient conditions for the existence of prime period-two solutions of the rational equation

$$x_{n+1} = \frac{\alpha + \beta x_n + \gamma x_{n-1} + \delta x_{n-2}}{A + Bx_n + Cx_{n-1} + Dx_{n-2}}, \quad n = 0, 1, \dots \tag{4.1.1}$$

with nonnegative parameters α, β, γ, δ, A, B, C, D, and with arbitrary nonnegative initial conditions x_{-2}, x_{-1}, x_0 such that the denominator is always positive.

Let

$$\dots, \phi, \psi, \phi, \psi, \dots \tag{4.1.2}$$

be a prime period-two solution of Eq.(4.1.1). Then

$$\phi = \frac{\alpha + \beta\psi + \gamma\phi + \delta\psi}{A + B\psi + C\phi + D\psi} \quad \text{and} \quad \psi = \frac{\alpha + \beta\phi + \gamma\psi + \delta\phi}{A + B\phi + C\psi + D\phi}$$

and so

$$A\phi + (B + D)\phi\psi + C\phi^2 = \alpha + (\beta + \delta)\psi + \gamma\phi$$

and

$$A\psi + (B + D)\phi\psi + C\psi^2 = \alpha + (\beta + \delta)\phi + \gamma\psi.$$

By subtracting the above two inequalities and then by dividing the result by $\phi - \psi$ we obtain that

$$C(\phi + \psi) = \gamma - \beta - \delta - A \tag{4.1.3}$$

is a necessary condition for Eq.(4.1.1) to have a prime period-two solution.

It is easy to see now that when

$$C = 0,$$

the condition

$$\gamma = \beta + \delta + A$$

is necessary and sufficient for Eq.(4.1.1) to have a prime period-two solution. Actually, in this case, when $B + D > 0$, there is a "hyperbola" of prime period-two solutions of Eq.(4.1.1) of the form (4.1.2) given by

$$(B + D)\phi\psi = \alpha + (\beta + \delta)(\phi + \psi)$$

with

$$\phi, \psi \in [0, \infty) \quad \text{and} \quad \phi \neq \psi.$$

This observation is the key point behind the three known nonlinear period-two trichotomies of Eq.(4.1.1) with $C = 0$. Note that when

$$B = C = D = 0,$$

for a prime period-two solution to exist, it is necessary that

$$\alpha = \beta = \delta = 0.$$

In this case the equation reduces to the linear equation

$$x_{n+1} = \frac{\gamma}{A} x_{n-1}, \quad n = 0, 1, \ldots,$$

which is the only linear period-two trichotomy contained in Eq.(4.0.1).

On the other hand, when

$$C > 0,$$

it follows from (4.1.3) that

$$\gamma > \beta + \delta + A,$$

and, in particular, $\gamma > 0$ is a necessary condition for Eq.(4.1.1) to have a prime period-two solution. With the parameters C and γ positive, Eq.(4.1.1) can be written in the normalized form

$$x_{n+1} = \frac{\alpha + \beta x_n + x_{n-1} + \delta x_{n-2}}{A + B x_n + x_{n-1} + D x_{n-2}}, \quad n = 0, 1, \ldots. \tag{4.1.4}$$

When

$$\alpha = \beta = \delta = 0,$$

Eq.(4.1.4) is reduced to the equation

$$x_{n+1} = \frac{x_{n-1}}{A + B x_n + x_{n-1} + D x_{n-2}}, \quad n = 0, 1, \tag{4.1.5}$$

which is investigated in Section 5.135. The following statements are easily established for Eq.(4.1.5):

1. Eq.(4.1.5) has a unique prime period-two solution if and only if

$$0 < A < 1 \text{ and } B + D = 0 \text{ or } 0 \leq A < 1 \text{ and } 0 < B + D \neq 1. \tag{4.1.6}$$

2. Eq.(4.1.5) has infinitely many prime period-two solutions if and only if

$$0 \leq A < 1 \text{ and } B + D = 1. \tag{4.1.7}$$

3. Assume that (4.1.6) holds. Then the unique prime period-two solution of Eq.(4.1.5) is

$$\ldots, 0,\ 1 - A,\ 0,\ 1 - A,\ \ldots. \tag{4.1.8}$$

   (a) When $0 < A < 1$ and $B + D = 0$, the unique prime period-two solution (4.1.8) of Eq.(4.1.5) is unstable.

   (b) When

$$0 \leq A < 1 \text{ and } B + D > 1,$$

   the period-two solution (4.1.8) of Eq.(4.1.5) is locally asymptotically stable and is unstable when

$$0 < B + D < 1.$$

4. Assume that (4.1.7) holds. Then the prime period-two solutions of Eq.(4.1.5) are given by (4.1.2) with

$$\phi + \psi = 1 - A, \quad \phi, \psi \in [0, \infty), \quad \text{and } \phi \neq \frac{1 - A}{2}.$$

On the other hand, one can see that when

$$C > 0 \quad \text{and} \quad \alpha + \beta + \delta > 0,$$

Eq.(4.1.1) has a prime period-two solution (4.1.2) if and only if

$$\beta + \delta + A < 1$$

and

$$4\alpha < (1 - \beta - \delta - A)[(B + D)(1 - \beta - \delta - A) - (1 + 3\beta + 3\delta - A)].$$

Furthermore, when the above inequalities do hold, then Eq.(4.1.1) has the unique prime period-two solution (4.1.2) and the values ϕ and ψ are the two positive and distinct roots of the quadratic equation

$$t^2 - (1 - \beta - \delta - A)t + \frac{\alpha + (\beta + \delta)(1 - \beta - \delta - A)}{B + D - 1} = 0. \tag{4.1.9}$$

Open Problem 4.1.1 *From the discussion in this section one can see that the only special cases of Eq.(4.0.1) with a unique prime period-two solution are the following 49 equations:*

#30,	*# 32,*	*#34,*	*#74,*	*#76,*	*#86,*	*#88*
#98,	*#100,*	*#109,*	*#111,*	*#112,*	*#135,*	*#145*
#147,	*#148,*	*#153,*	*#155,*	*#156,*	*#161,*	*#163*
#164,	*#168,*	*#170,*	*#180,*	*#182,*	*#186,*	*#188*
#190,	*#192,*	*#194,*	*#198,*	*#200,*	*#201,*	*#203*
#204,	*#209,*	*#211,*	*#212,*	*#213,*	*#215,*	*#216*
#217,	*#219,*	*#220,*	*#221,*	*#223,*	*#224,*	*#225.*

For the local asymptotic stability of the following 12 special cases:

#30,	*#32,*	*#34,*	*#74,*	*#76,*	*#86,*
#100,	*#109,*	*#111,*	*#112,*	*#135,*	*#145,*

see [119], [140], and [175].

Determine the local asymptotic stability of the prime period-two solution of each of the remaining 37 equations.

Open Problem 4.1.2 *For the $(k+1)^{st}$-order rational difference equation*

$$x_{n+1} = \frac{\alpha + \sum_{i=0}^{k} \beta_i x_{n-i}}{A + \sum_{i=0}^{k} B_i x_{n-i}}, \quad n = 0, 1, \ldots$$

determine all special cases with a unique prime period-two solution and in each case determine the local asymptotic stability of the prime period-two solution.

4.2 Period-Two Trichotomies of Eq.(4.0.1)

In this section we present the statements of the three known nonlinear period-two trichotomies of third-order rational equations and present the proof of the first period-two trichotomy in a spirit that can be easily generalized to the other two period-two trichotomies. In the next section we unify the three trichotomies into a single statement and present the proof of the unified result.

The following **nonlinear, period-two trichotomy** result for Eq.(4.0.2) was first observed and established in [16] for the special case #54. See also [108], [112], [175].

Theorem 4.2.1 *[175] The following period-two trichotomy result is true for Eq.(4.0.2).*
(a) Assume that

$$\gamma < \beta + A.$$

Then every solution of Eq.(4.0.2) has a finite limit.

(b) Assume that

$$\gamma = \beta + A.$$

Then every solution of Eq.(4.0.2) converges to a (not necessarily prime) period-two solution of Eq.(4.0.2).

(c) Assume that

$$\gamma > \beta + A.$$

Then Eq.(4.0.2) has unbounded solutions for some initial conditions.

Before we present the proof of Theorem 4.2.1 we make some observations and establish a lemma. Our aim is to present the proof in a spirit useful for extensions and generalizations to other known trichotomies for rational equations. We define the function

$$f(u,v) = \frac{\alpha + \beta u + \gamma v}{A + u}$$

and observe that for $\gamma > 0$, $f(u,v)$ is strictly increasing in v. Also

$$\frac{\partial f}{\partial u} = \gamma \cdot \frac{\frac{\beta A - \alpha}{\gamma} - v}{(A+u)^2}$$

and so the monotonic character of $f(u,v)$ in the variable u depends on whether the solutions are bounded from below or from above by $\frac{\beta A - \alpha}{\gamma}$.

Also note that when

$$\beta A - \alpha \leq 0,$$

$f(u,v)$ is strictly decreasing in u and strictly increasing in v and Theorem 1.6.6 applies.

The following lemma determines the monotonic character of the function $f(u,v)$ on the γ-line.

Lemma 4.2.1 *The following statements are true eventually for every positive solution $\{x_n\}$ of Eq.(4.0.2) and for n sufficiently large:*

(*i*) $\gamma > A \quad \Rightarrow \quad x_n > \frac{\beta A-\alpha}{\gamma}.$

(*ii*) $0 < \gamma < A \quad$ *and* $\quad \beta A - \alpha > 0 \quad \Rightarrow \quad x_n \leq \frac{\beta A-\alpha}{\gamma}.$

PROOF (*i*) Otherwise, there exists an N, as large as we please, such that

$$x_{N+1} \leq \frac{\beta A - \alpha}{\gamma}.$$

That is,

$$\frac{\alpha + \beta x_N + \gamma x_{N-1}}{A + x_N} \leq \frac{\beta A - \alpha}{\gamma},$$

which implies

$$\alpha + \left(\beta - \frac{\beta A - \alpha}{\gamma}\right) x_N + \gamma x_{N-1} \leq A\left(\frac{\beta A - \alpha}{\gamma}\right)$$

and so

$$x_{N-1} \leq \frac{A}{\gamma} \cdot \left(\frac{\beta A - \alpha}{\gamma}\right).$$

This, by a similar argument, implies that

$$x_{N-3} \leq \left(\frac{A}{\gamma}\right)^2 \cdot \left(\frac{\beta A - \alpha}{\gamma}\right),$$

which eventually leads to a contradiction.

(*ii*) Otherwise, there exists an N, as large as we please, such that

$$x_{N+1} > \frac{\beta A - \alpha}{\gamma}.$$

That is,

$$\frac{\alpha + \beta x_N + \gamma x_{N-1}}{A + x_N} > \frac{\beta A - \alpha}{\gamma},$$

which implies

$$\alpha + \left(\beta - \frac{\beta A - \alpha}{\gamma}\right) x_N + \gamma x_{N-1} > A\left(\frac{\beta A - \alpha}{\gamma}\right). \qquad (4.2.1)$$

But from Eq.(4.0.2),

$$x_{n+1} \leq \frac{\alpha + \beta x_n}{A + x_n} + \frac{\gamma}{A} x_{n-1} \leq \frac{\max(\alpha, \beta)}{\min(A, 1)} + \frac{\gamma}{A} x_{n-1}.$$

Therefore, by the comparison principle, there exists $U > 0$ such that

$$x_n \leq U, \quad \text{for all} \quad n \geq 0.$$

Then from (4.2.1) we see that

$$x_{N-1} > \frac{A}{\gamma} \cdot M_0 - \frac{K}{\gamma},$$

where

$$M_0 = \frac{\beta A - \alpha}{\gamma} \quad \text{and} \quad K = \max\left(\alpha, \alpha + \beta - \frac{\beta A - \alpha}{\gamma}\right).$$

Set

$$M_{n+1} = \frac{A}{\gamma} \cdot M_n - \frac{K}{\gamma}, \quad n = 0, 1, \ldots . \qquad (4.2.2)$$

Then, clearly,

$$x_{N-3} > \frac{A}{\gamma} \cdot M_1 - \frac{K}{\gamma}$$

which eventually leads to a contradiction because $\{x_n\}$ is bounded and the solution of Eq.(4.2.2) with initial condition M_0 is unbounded. ∎

We are now ready to present the proof of Theorem 4.2.1.

PROOF The result is clearly true when $\gamma = 0$ and so in the sequel we assume that $\gamma > 0$. The proof of part (*a*) will be divided into four cases. Please note that the proof of part (*b*) of the Theorem is included in Case 1:

Case 1:

$$A < \gamma \leq \beta + A \qquad (4.2.3)$$

or

$$\gamma \leq A \quad \text{and} \quad \beta A - \alpha \leq 0. \qquad (4.2.4)$$

It follows from our discussion preceding Lemma 4.2.1 and from Lemma 4.2.1 that, in this case, the function $f(u, v)$ is strictly decreasing in u and strictly increasing in v. The proof will be complete in this case if we also show that every solution of the equation is also bounded. To this end we first claim that every solution $\{x_n\}$ is eventually bounded from below by $\gamma - A$. That is,

$$x_n > \gamma - A, \quad \text{eventually.} \tag{4.2.5}$$

This is clearly true when (4.2.4) holds. Now assume that (4.2.3) holds and suppose for the sake of contradiction that there exists N, as large as we please, such that

$$x_{N+1} \leq \gamma - A.$$

Then

$$\frac{\alpha + \beta x_N + \gamma x_{N-1}}{A + x_N} \leq \gamma - A,$$

which implies

$$\alpha + (\beta + A - \gamma) x_N + \gamma x_{N-1} + \leq A(\gamma - A)$$

and so

$$x_{N-1} \leq \frac{A}{\gamma}(\gamma - A).$$

Similarly, we find that

$$x_{N-3} \leq \left(\frac{A}{\gamma}\right)^2 (\gamma - A),$$

which eventually leads to a contradiction.

Hence, (4.2.5) is true and so there exist N sufficiently large and positive numbers L and U such that

$$\gamma - A < L \leq x_{N-1}, x_N \leq U = \frac{\alpha + \beta L}{L - (\gamma - A)}.$$

Now one can see that

$$L \leq \frac{\alpha + \beta U + \gamma L}{A + U} \leq x_{N+1} = \frac{\alpha + \beta x_N + \gamma x_{N-1}}{A + x_N} \leq \frac{\alpha + \beta L + \gamma U}{A + L} = U$$

and by induction

$$x_n \in [L, U], \quad \text{for all } n \geq N.$$

Case 2:

$$\gamma = A, \quad \alpha > 0, \quad \text{and } \beta A - \alpha > 0. \tag{4.2.6}$$

To establish the result in this case, it suffices to show that every solution is bounded and that the function $f(x_n, x_{n-1})$ is strictly decreasing in x_n. To this end, we first claim that, eventually,

$$x_n > \frac{\beta A - \alpha}{\gamma},$$

which will imply that $f(x_n, x_{n-1})$ decreases in x_n. Suppose for the sake of contradiction that there exists N, as large as we please, such that

$$x_{N+1} \leq \frac{\beta\gamma - \alpha}{\gamma}.$$

Then, clearly,

$$\frac{\alpha + \beta x_N + \gamma x_{N-1}}{\gamma + x_N} \leq \frac{\beta\gamma - \alpha}{\gamma},$$

which implies that

$$\alpha + \frac{\alpha}{\gamma} x_N + \gamma x_{N-1} \leq \beta\gamma - \alpha$$

and so

$$\min(x_N, x_{N-1}) < \left(\frac{\gamma^2}{\gamma^2 + \alpha}\right) \cdot \left(\frac{\beta\gamma - \alpha}{\gamma}\right).$$

Similarly,

$$\min(x_{N-1}, x_{N-2}, x_{N-3}) < \left(\frac{\gamma^2}{\gamma^2 + \alpha}\right)^2 \left(\frac{\beta\gamma - \alpha}{\gamma}\right),$$

which eventually leads to a contradiction.

Finally, note that in this case,

$$x_{n+1} \leq \frac{\alpha + \beta x_n}{\gamma + x_n} + \frac{\gamma}{\gamma + \left(\frac{\beta\gamma - \alpha}{\gamma}\right)} \cdot x_{n-1}$$

and so $\{x_n\}$ is bounded. The proof is complete in this case.

Case 3:

$$\gamma = A, \quad \alpha = 0, \quad \text{and } \beta A - \alpha > 0. \tag{4.2.7}$$

Here Eq.(4.0.2) reduces to

$$x_{n+1} = \frac{\beta x_n + \gamma x_{n-1}}{\gamma + x_n}, \quad n = 0, 1, \ldots . \tag{4.2.8}$$

Note that for $n \geq 0$,

$$x_{n+1} - \beta = \frac{\gamma}{\gamma + x_n}(x_{n-1} - \beta)$$

and

$$x_{n+1} - x_{n-1} = \frac{x_n}{\gamma + x_n}(\beta - x_{n-1}).$$

Hence,

$$x_{n-1} \leq x_n < \beta \text{ or } \beta < x_{n-1} \leq x_n,$$

from which it follows that

$$\lim_{n\to\infty} x_n = \beta.$$

The proof is complete in this case.

Case 4:

$$\gamma < A \text{ and } \beta A - \alpha > 0. \tag{4.2.9}$$

In this case

$$x_{n+1} \leq \frac{\alpha + \beta x_n}{A + x_n} + \frac{\gamma}{A} \cdot x_{n-1} \leq \frac{\max(\alpha, \beta)}{\min(A, 1)} + \frac{\gamma}{A} \cdot x_{n-1}$$

and so the solution is bounded. Also, by Lemma 4.2.1(ii) the function $f(x_n, x_{n-1})$ is strictly increasing in x_n and x_{n-1} and the result follows by Theorem 1.6.7.

Now we present the proof of part (c) on **the existence of unbounded solutions** when

$$\gamma > \beta + A.$$

To this end, let k be a number such that

$$0 < k < \gamma - \beta - A.$$

Then we claim that every solution of Eq.(4.0.2) with initial conditions x_{-1}, x_0 such that

$$x_0 \in (0, \gamma - A) \text{ and } x_{-1} > \frac{\alpha + \gamma(\gamma - A)}{k}$$

is unbounded and, more precisely,

$$\lim_{n\to\infty} x_{2n+1} = \infty \text{ and } \lim_{n\to\infty} x_{2n} = \beta.$$

Observe that

$$x_1 - x_{-1} = \frac{\alpha + \beta x_0 + \gamma x_{-1}}{A + x_0} - x_{-1} = \frac{\alpha + \beta x_0 + (\gamma - A - x_0)\, x_{-1}}{A + x_0}$$

and so

$$x_1 > x_{-1}.$$

Also,

$$x_2 - (\beta + k) \leq \frac{\alpha + \beta x_1 + \gamma x_0}{x_1} - (\beta + k)$$

$$= \frac{\alpha + \gamma x_0 - k x_1}{x_1} < \frac{\alpha + \gamma(\gamma - A) - \alpha - \gamma(\gamma - A)}{x_1} = 0.$$

Therefore,

$$x_2 < \beta + k.$$

Furthermore,

$$x_3 = \frac{\alpha + \beta x_2 + \gamma x_1}{A + x_2} > \frac{\gamma}{\beta + A + k} x_1.$$

It follows by induction that for $n \geq 1$,

$$x_{2n} < \beta + k$$

and

$$x_{2n+1} > \frac{\gamma}{\beta + A + k} x_{2n-1}$$

and so, in particular,

$$\lim_{n\to\infty} x_{2n+1} = \infty. \tag{4.2.10}$$

We also know from Lemma 4.2.1 that the function

$$f(x_n, x_{n-1}) = \frac{\alpha + \beta x_n + \gamma x_{n-1}}{A + x_n}$$

is strictly increasing in x_{n-1} and strictly decreasing in x_n and so by Theorem 1.6.6 the limit of the subsequence $\{x_{2n}\}$ exists and is a finite number. It now follows that

$$x_{2n+2} = \frac{\alpha + \beta x_{2n+1} + \gamma x_{2n}}{A + x_{2n+1}}$$

$$= \frac{\frac{\alpha+\gamma x_{2n}}{x_{2n+1}} + \beta}{\frac{A}{x_{2n+1}} + 1} \to \beta \text{ as } n \to \infty.$$

The proof is complete. ∎

Theorem 4.2.1 states that Eq.(4.0.2) has unbounded solutions when

$$\gamma > \beta + A. \tag{4.2.11}$$

Clearly, when (4.2.11) holds, by the stable manifold theorem, the equation has also bounded solutions because the equilibrium of the equation is a saddle point. We will now show that when (4.2.11) holds, every positive and bounded solution of Eq.(4.0.2) converges to the positive equilibrium.

Theorem 4.2.2 *Assume that*

$$\gamma > \beta + A.$$

Then every positive and bounded solution of Eq.(4.0.2) converges to the positive equilibrium

$$\bar{x} = \frac{\beta + \gamma - A + \sqrt{(\beta + \gamma - A)^2 + 4\alpha}}{2}.$$

PROOF Let $\{x_n\}$ be a positive and bounded solution of Eq.(4.0.2). We claim that, eventually,

$$x_n > \frac{\beta A - \alpha}{\gamma}. \tag{4.2.12}$$

Otherwise, for some N sufficiently large,

$$x_{N+1} = \frac{\alpha + \beta x_N + \gamma x_{N-1}}{A + x_N} \leq \frac{\beta A - \alpha}{\gamma},$$

which implies that

$$x_{N-1} < \frac{A}{\gamma}\frac{\beta A - \alpha}{\gamma},$$

which implies

$$x_{N-3} < (\frac{A}{\gamma})^2 \frac{\beta A - \alpha}{\gamma},$$

$$\ldots$$

which eventually leads to a contradiction. Therefore, (4.2.12) is true and so the function

$$F(u, v) = \frac{\alpha + \beta u + \gamma v}{A + u}$$

decreases in u and increases in v. Now by Theorem 1.6.6 it follows that the subsequences of the even and the odd terms converge to finite limits. Set

$$L_E = \lim_{n\to\infty} x_{2n} \text{ and } L_O = \lim_{n\to\infty} x_{2n+1}.$$

By taking limits in Eq.(4.0.2) we find

$$L_E = \frac{\alpha + \beta L_O + \gamma L_E}{A + L_O} \quad \text{and} \quad L_O = \frac{\alpha + \beta L_E + \gamma L_O}{A + L_E}.$$

Hence,

$$AL_E + L_O L_E = \alpha + \beta L_O + \gamma L_E$$
$$AL_O + L_E L_O = \alpha + \beta L_E + \gamma L_O$$

and by subtracting we find

$$(\gamma - \beta - A)(L_O - L_E) = 0.$$

In view of our assumption this is true if and only if

$$L_O = L_E.$$

It remains to show that

$$L_E = L_O > 0.$$

Otherwise,

$$L_E = L_O = 0,$$

that is

$$x_{2n} \searrow 0 \text{ and } x_{2n+1} \searrow 0.$$

However, this is impossible because, for n sufficiently large,

$$0 \geq x_{2n+1} - x_{2n-1} = \frac{\alpha + \beta x_{2n} + (\gamma - A - x_{2n})x_{2n-1}}{A + x_{2n}} > 0$$

and so the proof is complete.

∎

The following period-two trichotomy result for Eq.(4.0.3) was established in [70]. For the proof see Theorem 4.3.1.

Theorem 4.2.3 *Assume that*

$$\gamma + \delta + A > 0.$$

Then Eq.(4.0.3) has the following period-two trichotomy:
(a) Assume that

$$\gamma < \delta + A.$$

Then every solution of Eq.(4.0.3) has a finite limit.

(b) Assume that

$$\gamma = \delta + A.$$

Then every solution of Eq.(4.0.3) converges to a (not necessarily prime) period-two solution of Eq.(4.0.3).

(c) Assume that

$$\gamma > \delta + A.$$

Then Eq.(4.0.3) has unbounded solutions for some initial conditions.

The trichotomy results for Eqs.(4.0.2) and (4.0.3) were unified and extended in the higher order equation

$$x_{n+1} = \frac{\alpha + \gamma x_{n-(2l+1)} + \delta x_{n-2m}}{A + x_{n-2m}}, \quad n = 0, 1, \ldots \tag{4.2.13}$$

with nonnegative parameters and with arbitrary nonnegative initial conditions. See [129].

Let p be defined as follows:

$$p = \begin{cases} 2m+1, & \text{if } \gamma = \delta = A = 0 \\ l+1, & \text{if } \alpha = \delta = 0 \text{ and } A > 0 \\ \gcd(l+1, 2m+1), & \text{otherwise.} \end{cases}$$

Then the following trichotomy result is true for Eq.(4.2.13). A detailed proof is given in [124] and so it will be omitted here.

Theorem 4.2.4

1. *Assume that*
$$\gamma < \delta + A.$$
Then every solution of Eq.(4.2.13) converges to a finite limit.

2. *Assume that*
$$\gamma = \delta + A.$$
Then every solution of Eq.(4.2.13) converges to a (not necessarily prime) period-2p solution.

3. *Assume that*
$$\gamma > \delta + A.$$
Then Eq.(4.2.13) has unbounded solutions for some initial conditions.

Finally the following period-two trichotomy result was established in [72]. For the proof see Theorem 4.3.1.

Theorem 4.2.5 *Assume that*

$$\gamma + A + B > 0.$$

Then Eq.(4.0.4) has the following period-two trichotomy:
(a) Assume that
$$\gamma < A.$$
Then every solution of Eq.(4.0.4) has a finite limit.

(b) Assume that
$$\gamma = A.$$
Then every solution of Eq.(4.0.4) converges to a (not necessarily prime) period-two solution of Eq.(4.0.4).

(c) Assume that
$$\gamma > A.$$
Then Eq.(4.0.4) has unbounded solutions for some initial conditions.

Conjecture 4.2.1 *Show that the only possible nonlinear period-two trichotomies that Eq.(4.0.1) may have are those three presented in this section.*

4.3 Period-Two Trichotomy of $x_{n+1} = \dfrac{\alpha + \beta x_n + \gamma x_{n-1} + \delta x_{n-2}}{A + Bx_n + Dx_{n-2}}$

Consider the third-order rational difference equation

$$x_{n+1} = \frac{\alpha + \beta x_n + \gamma x_{n-1} + \delta x_{n-2}}{A + Bx_n + Dx_{n-2}}, \quad n = 0, 1, \ldots \tag{4.3.1}$$

with nonnegative parameters α, β, γ, δ, A, B, D and with arbitrary non-negative initial conditions x_{-2}, x_{-1}, x_0 such that the denominator is always positive.

The following result unifies Theorems 4.2.1, 4.2.2, and 4.2.3.

Theorem 4.3.1 *Assume that*

$$\beta+\gamma+\delta+A+B > 0, \quad \beta(\delta+D) = \delta(\beta+B) = 0, \quad \textit{and} \quad B+D > 0. \tag{4.3.2}$$

Then Eq.(4.3.1) has the following period-two trichotomy:
(a) Assume that

$$\gamma < \beta + \delta + A.$$

Then every solution of Eq.(4.3.1) has a finite limit.

(b) Assume that

$$\gamma = \beta + \delta + A.$$

Then every solution of Eq.(4.3.1) converges to a (not necessarily prime) period-two solution of Eq.(4.3.1).

(c) Assume that

$$\gamma > \beta + \delta + A.$$

Then Eq.(4.3.1) has unbounded solutions for some initial conditions.

PROOF We will give the proofs of statements (a) and (b). For the proof of statement (c), see the proofs of Theorems 3.1.1, 3.2.1, and 3.3.1. The result is clearly true when $\gamma = 0$ and so we assume that $\gamma > 0$. We divide the proof into four cases as follows:

Case 1:

$$A < \gamma \leq \beta + \delta + A \tag{4.3.3}$$

or

$$\gamma \leq A, \quad \beta A - B\alpha \leq 0, \quad \text{and} \quad \delta A - D\alpha \leq 0. \tag{4.3.4}$$

In this case we will show that the function

$$f(x_n, x_{n-1}, x_{n-2}) = \frac{\alpha + \beta x_n + \gamma x_{n-1} + \delta x_{n-2}}{A + Bx_n + Dx_{n-2}}$$

is eventually either strictly decreasing in x_n, or strictly decreasing in x_{n-2}, or strictly decreasing in both arguments x_n and x_{n-1}. Note that the above function is strictly decreasing in x_n if and only if

$$B\gamma x_{n-1} > \beta A - B\alpha \tag{4.3.5}$$

and also strictly decreasing in x_{n-2} if and only if

$$D\gamma x_{n-1} > \delta A - D\alpha. \tag{4.3.6}$$

This is obvious when (4.3.4) holds and so it remains to show that (4.3.3) implies (4.3.5) or (4.3.6). Suppose for the sake of contradiction that for some N, sufficiently large,

$$B\gamma x_{N+1} = \frac{B\gamma\alpha + B\gamma\beta x_N + B\gamma^2 x_{N-1} + B\gamma\delta x_{N-2}}{A + Bx_N + Dx_{N-2}} \leq \beta A - B\alpha$$

and

$$D\gamma x_{N+1} = \frac{D\gamma\alpha + D\gamma\beta x_N + D\gamma^2 x_{N-1} + D\gamma\delta x_{N-2}}{A + Bx_N + Dx_{N-2}} \leq \delta A - D\alpha.$$

Then, clearly,

$$B\gamma\alpha + (B\beta(\gamma - A) + B^2\alpha)x_N + B\gamma^2 x_{N-1} + (B\delta(\gamma - A) + BD\alpha)x_{N-2} \leq A(\beta A - B\alpha)$$

and

$$D\gamma\alpha + (\delta B(\gamma - A) + DB\alpha)x_N + D\gamma^2 x_{N-1} + (D\delta(\gamma - A) + D^2\alpha)x_{N-2} \leq A(\delta A - D\alpha),$$

from which it follows that either

$$B\gamma x_{N-1} < \frac{A}{\gamma} \cdot (\beta A - B\alpha)$$

or

$$D\gamma x_{N-1} < \frac{A}{\gamma} \cdot (\delta A - D\alpha).$$

Similarly, it follows that either

$$B\gamma x_{N-3} < (\frac{A}{\gamma})^2 \cdot (\beta A - B\alpha)$$

or

$$D\gamma x_{N-3} < (\frac{A}{\gamma})^2 \cdot (\delta A - D\alpha),$$

which eventually leads to a contradiction.

Next we will prove that there exist positive numbers L and U such that the solution $\{x_n\}$ lies eventually in the interval $[L, U]$. First we prove that the solution is eventually bounded from below by the positive constant $\frac{\gamma-A}{B+D}$. This is obvious when (4.3.4) holds. Now assume that (4.3.3) holds and suppose for the sake of contradiction that there exists N large enough such that

$$x_{N+1} = \frac{\alpha + \beta x_N + \gamma x_{N-1} + \delta x_{N-2}}{A + Bx_N + Dx_{N-2}} \leq \frac{\gamma - A}{B + D}.$$

Then, clearly,

$$\alpha + \frac{\beta B + \beta D - B(\gamma - A)}{B + D} x_N + \gamma x_{N-1} + \frac{\delta B + \delta D - D(\gamma - A)}{B + D} x_{N-2} \leq A \cdot \frac{\gamma - A}{B + D}.$$

Also,

$$\alpha + B \cdot \frac{\beta + \delta + A - \gamma}{B + D} x_N + \gamma x_{N-1} + D \cdot \frac{\beta + \delta + A - \gamma}{B + D} x_{N-2} \leq A \cdot \frac{\gamma - A}{B + D},$$

from which it follows that

$$x_{N-1} \leq \frac{A}{\gamma} \cdot \frac{\gamma - A}{B + D}.$$

Similarly,

$$x_{N-3} \leq (\frac{A}{\gamma})^2 \cdot \frac{\gamma - A}{B + D}$$

which eventually leads to a contradiction. Hence there exist N sufficiently large and positive numbers L,U with

$$L > \frac{\gamma - A}{B + D} \quad \text{and} \quad U = \frac{\alpha + (\beta + \delta)L}{(B + D)L - (\gamma - A)}$$

such that

$$x_{N-2}, x_{N-1}, x_N \in [L, U].$$

Then

$$L \leq \frac{\alpha + (\beta + \delta)U + \gamma L}{A + (B + D)U} \leq x_{N+1} = \frac{\alpha + \beta x_N + \gamma x_{N-1} + \delta x_{N-2}}{A + Bx_N + Dx_{N-2}}$$

$$\leq \frac{\alpha + (\beta + \delta)L + \gamma U}{A + (B + D)L} = U$$

and the result follows by induction. When

$$A < \gamma < \beta + \delta + A,$$

or

$$\gamma < A, \ \ \beta A - B\alpha \leq 0, \ \ \text{and} \ \ \delta A - D\alpha \leq 0,$$

or

$$\gamma = A, \ \ \beta A - B\alpha \leq 0, \ \ \delta A - D\alpha \leq 0, \ \ \text{and} \ \ \beta + \delta > 0,$$

the function

$$f(x_n, x_{n-1}, x_{n-2}) = \frac{\alpha + \beta x_n + \gamma x_{n-1} + \delta x_{n-2}}{A + Bx_n + Dx_{n-2}}$$

satisfies the hypotheses of Theorem 1.6.7 and so every solution of Eq.(4.3.1) converges to a finite limit in this case.

When

$$\gamma = \beta + \delta + A,$$

the function

$$f(x_n, x_{n-1}, x_{n-2}) = \frac{\alpha + \beta x_n + \gamma x_{n-1} + \delta x_{n-2}}{A + Bx_n + Dx_{n-2}}$$

satisfies the hypotheses of Theorem 1.6.9, or the hypotheses of Theorem 1.6.10, or the hypotheses of Theorem 1.6.6, and so every solution converges to a (not necessarily prime) period-two solution in this case.

Case 2:

$$\gamma < A \ \ \text{and} \ \ \beta A - B\alpha > 0 \tag{4.3.7}$$

or

$$\gamma < A \ \ \text{and} \ \ \delta A - D\alpha > 0. \tag{4.3.8}$$

In this case

$$x_{n+1} \leq \frac{\alpha + \beta x_n + \delta x_{n-2}}{A + Bx_n + Dx_{n-2}} + \frac{\gamma}{A} \cdot x_{n-1} \leq \frac{\max(\alpha, \beta, \delta)}{\min(A, B, D)} + \frac{\gamma}{A} \cdot x_{n-1}$$

and, by applying Theorem 1.4.1, we see that the solution is bounded from above.

We will give the proof in the case when (4.3.7) holds. The proof when (4.3.8) holds is similar and will be omitted. First note that

$$\beta A - B\alpha > 0 \Rightarrow \delta = D = 0$$

and so the function

$$f(x_n, x_{n-1}, x_{n-2}) = \frac{\alpha + \beta x_n + \gamma x_{n-1} + \delta x_{n-2}}{A + Bx_n + Dx_{n-2}} = \frac{\alpha + \beta x_n + \gamma x_{n-1}}{A + Bx_n}$$

strictly increases in all variables. It suffices to show that, eventually,

$$x_n \leq \frac{\beta A - B\alpha}{B\gamma}.$$

Suppose for the sake of contradiction that there exists N sufficiently large such that

$$x_{N+1} = \frac{\alpha + \beta x_N + \gamma x_{N-1}}{A + Bx_N} > \frac{\beta A - B\alpha}{B\gamma} = M_0.$$

Then, clearly,

$$\alpha + (\beta - \frac{\beta A - B\alpha}{\gamma})x_N + \gamma x_{N-1} > A \cdot \frac{\beta A - B\alpha}{B\gamma}.$$

Note that

$$\alpha + (\beta - \frac{\beta A - B\alpha}{\gamma})x_N < \max(\alpha, \alpha + (\beta - \frac{\beta A - B\alpha}{\gamma})U) = K$$

where U is an upper bound for the solution $\{x_n\}$. Then

$$x_{N-1} > \frac{A}{\gamma} \cdot \frac{\beta A - B\alpha}{B\gamma} - \frac{K}{\gamma} = \frac{A}{\gamma} \cdot M_0 - \frac{K}{\gamma} = M_1.$$

Similarly,

$$x_{N-3} > \frac{A}{\gamma} \cdot M_1 - \frac{K}{\gamma} = M_2,$$

which eventually leads to a contradiction. Hence, the function

$$f(x_n, x_{n-1}, x_{n-2}) = \frac{\alpha + \beta x_n + \gamma x_{n-1} + \delta x_{n-2}}{A + Bx_n + Dx_{n-2}}$$

satisfies the hypotheses of Theorem 1.6.7 and so every solution converges to a finite limit in this case.

Case 3:

$$\gamma = A, \ \alpha > 0, \ \text{and } \beta A - B\alpha > 0 \tag{4.3.9}$$

or

$$\gamma = A, \ \alpha > 0, \ \text{and } \delta A - D\alpha > 0. \tag{4.3.10}$$

We will give the proof when (4.3.9) holds. The proof when (4.3.10) holds is similar and will be omitted. First note that

$$\beta A - B\alpha > 0 \Rightarrow \delta = D = 0.$$

Hence, the function

$$f(x_n, x_{n-1}, x_{n-2}) = \frac{\alpha + \beta x_n + \gamma x_{n-1} + \delta x_{n-2}}{A + Bx_n + Dx_{n-2}} = \frac{\alpha + \beta x_n + \gamma x_{n-1}}{\gamma + Bx_n}$$

increases in x_{n-1}. We will show in this case that the solution $\{x_n\}$ is bounded from above and from below by positive constants and that f is eventually decreasing in x_n.

First we will show that the solution is eventually bounded from below by the positive constant $\frac{\beta A - B\alpha}{B\gamma}$. Suppose for the sake of contradiction that there exists N sufficiently large such that

$$x_{N+1} = \frac{\alpha + \beta x_N + \gamma x_{N-1}}{\gamma + B x_N} \leq \frac{\beta A - B\alpha}{B\gamma}.$$

Then, clearly,

$$\alpha + \frac{B\alpha}{\gamma} x_N + \gamma x_N \leq \frac{\beta A - B\alpha}{B\gamma},$$

from which it follows that

$$\min(x_N, x_{N-1}) < \frac{\gamma^2}{\gamma^2 + B\alpha} \cdot \frac{\beta A - B\alpha}{B\gamma}.$$

Similarly,

$$\min(x_{N-1}, x_{N-2}, x_{N-3}, x_{N-4}) < (\frac{\gamma^2}{\gamma^2 + B\alpha})^2 \cdot \frac{\beta A - B\alpha}{B\gamma},$$

which eventually leads to a contradiction. Hence, the function f increases in x_{n-1} and decreases in x_n. Also in this case,

$$x_{n+1} \leq \frac{\alpha + \beta x_n}{\gamma + B x_n} + \frac{\gamma}{\gamma + B \cdot \frac{\beta A - B\alpha}{B\gamma}} \cdot x_{n-1} \leq \frac{\max(\alpha, \beta)}{\min(\gamma, B)} + \frac{\gamma}{\gamma + B \cdot \frac{\beta A - B\alpha}{B\gamma}} \cdot x_{n-1}$$

and by applying Theorem 1.4.1, we see that the solution is bounded from above and so the function $f(x_n, x_{n-1}, x_{n-2})$ satisfies the hypotheses of Theorem 1.6.7. Hence, every solution converges to a finite limit in this case.

Case 4:

$$\gamma = A, \quad \alpha = 0, \quad \text{and} \quad \beta A - B\alpha > 0 \tag{4.3.11}$$

or

$$\gamma = A, \quad \alpha = 0, \quad \text{and} \quad \delta A - D\alpha > 0. \tag{4.3.12}$$

We will give the proof in the case when (4.3.11) holds. The proof when (4.3.12) holds is similar and will be omitted. Note that

$$\beta A - B\alpha > 0 \Rightarrow \delta = D = 0.$$

In this case Eq.(4.3.1) takes the normalized form

$$x_{n+1} = \frac{\beta x_n + \gamma x_{n-1}}{\gamma + x_n}. \tag{4.3.13}$$

Let $\{x_n\}$ be a positive solution of Eq.(4.3.13). Clearly, for all $n \geq 0$

$$x_{n+1} - \beta = \frac{\gamma(x_{n-1} - \beta)}{\gamma + x_n}$$

and

$$x_{n+1} - x_{n-1} = \frac{x_n(\beta - x_{n-1})}{\gamma + x_n},$$

from which it follows that

$$x_{2n} \to x \text{ and } x_{2n+1} \to y.$$

Then, clearly,

$$y = \frac{\beta x + \gamma y}{\gamma + x} \text{ and } x = \frac{\beta y + \gamma x}{\gamma + y},$$

from which it follows that

$$x = y = \beta.$$

The proof is complete.

∎

What is it that makes Eq.(4.3.1) possess a period-two trichotomy?

Could the period-two trichotomy of Eq.(4.3.1) be predicted from the linearized equation of Eq.(4.3.1) and its dominant characteristic root?

Open Problem 4.3.1 *Assume that (4.3.2) holds and that*

$$\gamma = \beta + \delta + A.$$

Determine the set of all initial conditions x_{-2}, x_{-1}, x_0 such that the solution $\{x_n\}$ of Eq.(4.3.1) converges to an equilibrium point of Eq.(4.3.1).

Open Problem 4.3.2 *Assume that (4.3.2) holds and that*

$$\gamma = \beta + \delta + A.$$

Let

$$\ldots, \phi, \psi, \ldots \tag{4.3.14}$$

be a prime period-two solution of Eq.(4.3.1). Determine the set of all initial conditions x_{-2}, x_{-1}, x_0 such that the solution $\{x_n\}$ of Eq.(4.3.1) converges to the prime period-two solution (4.3.14).

Open Problem 4.3.3 *Assume that (4.3.2) holds and that*

$$\gamma = \beta + \delta + A.$$

Let x_{-2}, x_{-1}, x_0 be given. Determine the period-two solution (4.3.14) to which the solution $\{x_n\}$ of Eq.(4.3.1) converges.

Open Problem 4.3.4 *Assume that (4.3.2) holds and that*

$$\gamma > \beta + \delta + A.$$

Show that every positive and bounded solution of Eq.(4.3.1) converges to the positive equilibrium.

4.4 Period-Three Trichotomy of $x_{n+1} = \dfrac{\delta x_{n-2}}{A + Bx_n + Cx_{n-1}}$

The only nonlinear period-three trichotomy result known for Eq.(4.0.1) is the following:

Theorem 4.4.1 *Assume that*

$$A, B, C \in [0, \infty) \text{ and } B + C > 0.$$

Then the solutions of the equation

$$x_{n+1} = \frac{x_{n-2}}{A + Bx_n + Cx_{n-1}}, \quad n = 0, 1, \ldots \tag{4.4.1}$$

have the following period-three trichotomy behavior.
(a) When

$$A > 1,$$

every solution of Eq.(4.4.1) converges to zero.
(b) When

$$A = 1,$$

every solution of Eq.(4.4.1) converges to a period-three solution of the form

$$\ldots, 0, 0, \phi, 0, 0, \phi, \ldots$$

with $\phi \geq 0$.
(c) When

$$0 \leq A < 1,$$

Eq.(4.4.1) has unbounded solutions for some initial conditions.

PROOF (*a*) The proof is a consequence of the inequality,

$$x_{n+1} \leq \frac{1}{A} x_{n-2}, \quad \text{for } n \geq 0$$

and the fact that Eq.(4.4.1) has no prime period-three solutions when $A > 1$.

(*b*) The proof is a consequence of the inequality

$$x_{n+1} \leq x_{n-2}, \quad \text{for } n \geq 0$$

and the fact that Eq.(4.4.1) has prime period-three solutions

$$\ldots, \phi, \psi, \omega, \ldots$$

if and only if $A = 1$, $\phi > 0$, and $\psi = \omega = 0$.

(*c*) When $A \in (0, 1)$, an unbounded solution of Eq.(4.4.1) is

$$0, 0, 1, 0, 0, \frac{1}{A}, 0, 0, \frac{1}{A^2}, \ldots .$$

When $A = 0$, $B > 0$, and $C > 0$, Eq.(4.4.1) can be written in the normalized form

$$x_{n+1} = \frac{x_{n-2}}{Bx_n + x_{n-1}}, \quad n = 0, 1, \ldots . \tag{4.4.2}$$

The solution of Eq.(4.4.2) with

$$x_{-2} = 0, \quad x_{-1} = x > 0, \quad \text{and } x_0 = y > 0$$

is explicitly given by (see [60])

$$\left.\begin{aligned} x_{3n+1} &= 0 \\ x_{3n+2} &= \frac{(Bx)^{F_n}}{By^{F_{n+1}}} \\ x_{3n+3} &= \frac{y^{F_{n+2}}}{(Bx)^{F_{n+1}}} \end{aligned}\right\} \quad \text{for } n = 0, 1, \ldots,$$

with the exponents F_n being the Fibonacci numbers with $F_0 = F_1 = 1$, and for certain values of x and y is unbounded.

Finally, when $A = B = 0$ or $A = C = 0$, Eq.(4.4.1) reduces to

$$x_{n+1} = \frac{x_{n-2}}{x_n} \quad \text{or} \quad x_{n+1} = \frac{x_{n-2}}{x_{n-1}}, \quad n = 0, 1, \ldots .$$

Both equations are reducible to linear third-order difference equations, that have a lot of unbounded solutions. The proof is complete. ∎

The following extension of Theorem 4.4.1 was presented in [146] for the rational equation

$$x_{n+1} = \frac{x_{n-k}}{A + \sum_{i=0}^{k-1} B_i x_{n-i}}, \quad n = 0, 1, \ldots \tag{4.4.3}$$

with nonnegative parameters and with arbitrary nonnegative initial conditions such that the denominator is always positive.

Theorem 4.4.2 *Assume $k \geq 2$. Then the following period-three trichotomy result holds for Eq.(4.4.3):*

(a) When

$$A > 1,$$

every solution of Eq.(4.4.3) converges to zero.

(b) When

$$A = 1,$$

every solution of Eq.(4.4.1) converges to a period-(k+1) solution of the form

$$\dots, 0, 0, \dots, \phi, 0, 0, \dots, \phi, \dots,$$

with $\phi \geq 0$.

(c) When

$$0 \leq A < 1,$$

Eq.(4.4.3) has unbounded solutions.

PROOF The proof is similar to the proof of Theorem 4.4.1 and is omitted. ∎

What is it that makes Eq.(4.4.3) possess a period-three trichotomy?

Could the period-three trichotomy of Eq.(4.4.3) be predicted from the linearized equation of Eq.(4.4.3) and its dominant characteristic root?

4.5 Period-Four Trichotomy of $x_{n+1} = \dfrac{\alpha + \beta x_n + \delta x_{n-2}}{C x_{n-1}}$

The only period-four trichotomy result known for Eq.(4.0.1) is the following conjecture. See [59] and [150].

Conjecture 4.5.1 *Assume that*

$$\alpha, \beta \in [0, \infty).$$

Then the following period-four trichotomy result is true for the rational equation

$$x_{n+1} = \frac{\alpha + \beta x_n + x_{n-2}}{x_{n-1}}, \quad n = 0, 1, \dots. \tag{4.5.1}$$

(a) Every solution of Eq.(4.5.1) converges to its positive equilibrium if and only if

$$\beta > 1.$$

(b) Every solution of Eq.(4.5.1) converges to a (not necessarily prime) period-four solution of Eq.(4.5.1) if and only if

$$\beta = 1.$$

(c) Eq.(4.5.1) has unbounded solutions if and only if

$$\beta < 1.$$

Part (a) of this conjecture has not been confirmed yet. Part (b) was confirmed in [59] and is based on the identity:

$$x_{n+3} - x_{n-1} = \frac{1}{x_{n+2}}(x_{n+2} - x_{n-2}), \quad n = 0, 1, \ldots .$$

Another proof of part (b) and a generalization of it is given in the following theorem. See [222].

Theorem 4.5.1 *Assume that*

$$\alpha \in [0, \infty) \text{ and } k \in \{1, 2, \ldots\}.$$

Then every solution of the equation

$$x_{n+1} = \frac{\alpha + x_n + x_{n-k}}{x_{n-k+1}}, \quad n = 0, 1, \ldots$$

converges to a periodic solution of period $2k$.

PROOF The proof is a consequence of the identity

$$x_{n+1} - x_{n-2k+1} = \frac{x_n - x_{n-2k}}{x_{n-k+1}}, \quad \text{for } n \geq 0.$$

∎

Finally, part (c) of this conjecture was confirmed in [150].

What is it that makes Eq.(4.5.1) possess a period-four trichotomy?

Could the period-four trichotomy of Eq.(4.5.1) be predicted from the linearized equation of Eq.(4.5.1) and its dominant characteristic root?

4.6 Period-Five Trichotomy of $x_{n+1} = \dfrac{\alpha + \delta x_{n-2}}{Bx_n}$

The only period-five trichotomy result known for Eq.(4.0.1) is the following. See [54], [59], and [151].

Theorem 4.6.1 *Assume $\alpha \geq 0$. Then the following statements are true for the difference equation*

$$x_{n+1} = \frac{\alpha + x_{n-2}}{x_n}, \quad n = 0, 1, \ldots . \tag{4.6.1}$$

(a) Assume that

$$\alpha > 1.$$

Then every solution of the Eq.(4.6.1) converges to its positive equilibrium point.

(b) Assume that

$$\alpha = 1.$$

Then every solution of Eq.(4.6.1) converges to a (not necessarily prime) period-five solution of Eq.(4.6.1).

(c) Assume that

$$\alpha < 1.$$

Then Eq.(4.6.1) has unbounded solutions.

Part (a) was established in [151]. See also [54]. Part (b) of the trichotomy was established in [59]. The proof is based on the identity

$$x_{n+4} - x_{n-1} = \frac{1}{x_{n+3}}(x_{n+3} - x_{n-2}), \quad n = 0, 1, \ldots .$$

For the complete proof of part (c) see Theorem 3.6.1 in Section 3.6.

What is it that makes Eq.(4.6.1) possess a period-five trichotomy?

Could the period-five trichotomy of Eq.(4.6.1) be predicted from the linearized equation of Eq.(4.6.1) and its dominant characteristic root?

4.7 Period-Six Trichotomy of $x_{n+1} = \dfrac{\alpha + \beta x_n}{Cx_{n-1} + x_{n-2}}$

The only period-six trichotomy result known for Eq.(4.0.1) is the following. See [59].

Conjecture 4.7.1 *Assume that*

$$\alpha, C \in [0, \infty).$$

Then the following period-six trichotomy result is true for the rational equation

$$x_{n+1} = \frac{\alpha + x_n}{Cx_{n-1} + x_{n-2}}, \quad n = 0, 1, \ldots . \tag{4.7.1}$$

(a) *Every solution of Eq.(4.7.1) converges to its positive equilibrium if and only if*

$$\alpha C^2 > 1.$$

(b) *Every solution of Eq.(4.7.1) converges to a (not necessarily prime) period-six solution of Eq.(4.7.1) if and only if*

$$\alpha C^2 = 1.$$

(c) *Eq.(4.7.1) has unbounded solutions if and only if*

$$\alpha C^2 < 1.$$

No part of this trichotomy has been confirmed yet.

What is it that makes Eq.(4.7.1) possess a period-six trichotomy?

Could the period-six trichotomy of Eq.(4.7.1) be predicted from the linearized equation of Eq.(4.7.1) and its dominant characteristic root?

5

Known Results for Each of the 225 Special Cases

5.0 Introduction

This chapter is the heart of this book. In this chapter we present the known results on each of the 225 special cases of the third-order rational difference equation

$$x_{n+1} = \frac{\alpha + \beta x_n + \gamma x_{n-1} + \delta x_{n-2}}{A + Bx_n + Cx_{n-1} + Dx_{n-2}}, \quad n = 0, 1, \ldots \tag{5.0.1}$$

with nonnegative parameters $\alpha, \beta, \gamma, \delta, A, B, C, D$ and with arbitrary nonnegative initial conditions x_{-2}, x_{-1}, x_0, such that the denominator is always positive.

In several special cases we also present some new results and pose some open problems and conjectures on the character of their solutions.

Whenever we can, we extend the results to the general $(k+1)^{st}$-order rational difference equation

$$x_{n+1} = \frac{\alpha + \sum_{i=0}^{k} \beta_i x_{n-i}}{A + \sum_{i=0}^{k} B_i x_{n-i}}, \quad n = 0, 1, \ldots . \tag{5.0.2}$$

The ultimate goal for the reader is to generalize each case in this chapter to the most general functional equation

$$x_{n+1} = f(x_n, \ldots, x_{n-k}), \quad n = 0, 1, \ldots .$$

A few of the 225 special cases we present in this chapter are trivial, linear, or reducible to linear. We only include them here for the sake of completeness and continuity of presentation.

Most nontrivial cases are written in normalized form by using a change of variables of the form $x_n = \lambda y_n$. This allows for two of the parameters in the equation to be assumed to be equal to 1. Of course in some equations there is some restriction, concerning which pair of parameters we assume equal to 1. For example, in Section 5.26 we cannot assume that both B and C can be taken equal to one.

5.1 Equation #1 : $x_{n+1} = \frac{\alpha}{A}$

The equation in this special case is trivial.

5.2 Equation #2 : $x_{n+1} = \frac{\alpha}{Bx_n}$

In this special case every nontrivial solution of the equation is periodic with period two.

It is interesting to note that periodicity may destroy the boundedness of solutions of the equation

$$x_{n+1} = \frac{\alpha}{x_n}, \quad n = 0, 1, \ldots \tag{5.2.1}$$

as the following example shows.

Example 5.2.1 *(see [62] Let*

$$\alpha_n = \begin{cases} \alpha_0, & \textit{if } n = 2k \\ \alpha_1, & \textit{if } n = 2k+1 \end{cases}, k = 0, 1, \ldots$$

with $\alpha_0,\ \alpha_1 \in (0, \infty)$. Then every solution of the equation,

$$x_{n+1} = \frac{\alpha_n}{x_n}, \quad n = 0, 1, \ldots \tag{5.2.2}$$

is ***unbounded*** *if and only if*

$$\alpha_0 \neq \alpha_1.$$

Indeed, one can see, by induction, that every solution of Eq.(5.2.2) is, for $n \geq 0$, given by

$$x_{2n+2} = (\frac{\alpha_1}{\alpha_0})^{n+1} \cdot x_0$$

and

$$x_{2n+3} = (\frac{\alpha_0}{\alpha_1})^{n+1} \cdot \frac{\alpha_0}{x_0}.$$

That is, in Eq.(5.2.1) **periodicity may destroy the boundedness of its solutions.**

Do you see a **pattern** in the periodic character of the equations

$$x_{n+1} = \frac{1}{x_n}, \quad n = 0, 1, \ldots$$

$$x_{n+1} = \frac{1}{x_n x_{n-1}}, \quad n = 0, 1, \ldots$$

$$\ldots$$

$$x_{n+1} = \frac{1}{x_n \cdots x_{n-k}}, \quad n = 0, 1, \ldots ?$$

Do you see a **pattern** in the periodic convergence character of the equations

$$x_{n+1} = \frac{1}{x_n} + \frac{1}{x_{n-2}}, \quad n = 0, 1, \ldots$$

$$x_{n+1} = \frac{1}{x_n x_{n-1}} + \frac{1}{x_{n-3} x_{n-4}}, \quad n = 0, 1, \ldots$$

$$\ldots$$

$$x_{n+1} = \frac{1}{\prod_{i=0}^{k} x_{n-i}} + \frac{1}{\prod_{j=k+2}^{2(k+1)} x_{n-j}}, \quad n = 0, 1, \ldots ?$$

What is going on? See [4], [32], [91], and [94].

Open Problem 5.2.1 *Let k be a nonnegative integer and let $\{\alpha_n\}$ be a nonnegative periodic sequence with prime period $p \geq 2$. Obtain necessary and sufficient conditions on p and*

$$\alpha_0, \ldots, \alpha_{p-1}$$

such that every positive solution of the equation

$$x_{n+1} = \frac{\alpha_n}{\prod_{i=0}^{k} x_{n-i}} + \frac{1}{\prod_{j=k+2}^{2(k+1)} x_{n-j}}, \quad n = 0, 1, \ldots \tag{5.2.3}$$

converges to a periodic solution with period $(k + 2)$.

Open Problem 5.2.2 *Let k be a nonnegative integer and let $\{\alpha_n\}$ be a convergent sequence with*

$$\lim_{n \to \infty} \alpha_n = \alpha > 0.$$

Investigate the global character of solutions of Eq.(5.2.3).

5.3 Equation #3 : $x_{n+1} = \dfrac{\alpha}{Cx_{n-1}}$

In this special case every nontrivial solution of the equation is periodic with period four. Clearly, periodicity may destroy the boundedness of solutions of the equation in the title. See [62].

5.4 Equation #4 : $x_{n+1} = \dfrac{\alpha}{Dx_{n-2}}$

In this special case every nontrivial solution of the equation is periodic with period six. Clearly, periodicity may destroy the boundedness of solutions of the equation in the title. See [62].

5.5 Equation #5 : $x_{n+1} = \dfrac{\beta}{A}x_n$

The equation in this special case is linear.

5.6 Equation #6 : $x_{n+1} = \dfrac{\beta}{B}$

The equation in this special case is trivial.

5.7 Equation #7 : $x_{n+1} = \dfrac{\beta x_n}{C x_{n-1}}$

The equation in this special case can be written in the normalized form

$$x_{n+1} = \frac{x_n}{x_{n-1}}, \quad n = 0, 1, \ldots$$

with arbitrary positive initial conditions. Every nontrivial solution of this equation is periodic with period six. If

$$x_{-1} = \phi \text{ and } x_0 = \psi,$$

the solution of the equation is the six-cycle:

$$\phi, \psi, \frac{\psi}{\phi}, \frac{1}{\phi}, \frac{1}{\psi}, \frac{\phi}{\psi}, \ldots .$$

What is it that makes every solution of a difference equation periodic with the same period?

Open Problem 5.7.1 *Assume $f \in C^1([0,\infty),[0,\infty))$ and let k be a given integer greater than one. Find necessary and sufficient conditions on f and k so that every positive solution of the difference equation*

$$x_{n+1} = \frac{f(x_n)}{x_{n-1}}, \quad n = 0, 1, \ldots$$

is periodic with period k.

For some work on this problem, see [1], [11], and [204].

It is interesting to note that periodicity may destroy the boundedness of solutions of the equation

$$x_{n+1} = \frac{\beta x_n}{x_{n-1}}, \quad n = 0, 1, \ldots \tag{5.7.1}$$

as the following example shows.

Example 5.7.1 *(see [62]) Let*

$$\beta_n = \begin{cases} 1, & \text{if } n = 6k + i \text{ with } i \in \{0,1,2,3,4\} \\ \beta, & \text{if } n = 6k + 5 \end{cases}, k = 0, 1, \ldots$$

with $\beta > 0$. *Then every solution of the equation*

$$x_{n+1} = \frac{\beta_n x_n}{x_{n-1}}, \quad n = 0, 1, \ldots \tag{5.7.2}$$

with initial conditions

$$x_{-1} = x_0 = 1 \tag{5.7.3}$$

is ***unbounded****, if and only if*

$$\beta \neq 1.$$

Indeed, one can see, by induction, that the solution of the IVP (5.7.2) and (5.7.3), for $n \geq 0$, is given by

$$\begin{aligned} x_{6n+5} &= 1 \\ x_{6n+6} &= \beta^{n+1} \\ x_{6n+7} &= \beta^{n+1} \\ x_{6n+8} &= 1 \\ x_{6n+9} &= \frac{1}{\beta^{n+1}} \\ x_{6n+10} &= \frac{1}{\beta^{n+1}}. \end{aligned}$$

Therefore, in Eq.(5.7.1), **periodicity may destroy the boundedness of its solutions.** For some results on the asymptotic behavior of the nonautonomous Eq.(5.7.2), see [10].

Open Problem 5.7.2 *Let* $\{\beta_n\}$ *be a positive periodic sequence with prime period* $p \geq 2$. *Obtain necessary and sufficient conditions on* p *and*

$$\beta_0, \ldots, \beta_{p-1}$$

such that every solution of Eq.(5.7.2) is bounded.

Remark 5.7.1 *It is interesting to note that every solution of the equation*

$$x_{n+1} = \frac{\beta x_n}{x_{n-1}} \cdots \frac{x_{n-2k}}{x_{n-(2k+1)}}, \quad n = 0, 1, \ldots$$

is periodic with period $(6 + 4k)$. *The case* $k = 0$ *is Eq.(5.7.1).*
Extend and generalize the open problem 5.7.2 using this equation.

5.8 Equation #8 : $x_{n+1} = \dfrac{\beta x_n}{D x_{n-2}}$

The change of variables, $x_n = e^{y_n}$, transforms Eq.(#8) to a linear equation.

5.9 Equation #9 : $x_{n+1} = \frac{\gamma}{A} x_{n-1}$

The equation in this special case is linear. This equation has a period-two trichotomy depending on whether

$$\gamma < A, \quad \gamma = A, \quad \text{or} \quad \gamma > A.$$

See Chapter 4 for the three nonlinear period-two trichotomies of Eq.(4.0.1).

This equation has infinitely many prime period-two solutions if and only if

$$\gamma = A$$

and no prime period-two solutions otherwise.

The only other second-order rational difference equations with infinitely many prime period-two solutions are, in normalized form, given by

$$x_{n+1} = \frac{\alpha + \beta x_n + \gamma x_{n-1}}{A + x_n}, \quad n = 0, 1, \ldots$$

and

$$x_{n+1} = \frac{x_{n-1}}{A + x_n + x_{n-1}}, \quad n = 0, 1, \ldots .$$

See Chapter 4 and Eq.(#109) in Section 5.109.

On the other hand, the only second-order rational difference equations with a "unique" prime period-two solution

$$\ldots, \phi, \psi, \ldots$$

are, in normalized form, given by

$$x_{n+1} = \frac{\alpha + \beta x_n + x_{n-1}}{A + B x_n + x_{n-1}}, \quad n = 0, 1, \ldots$$

with

$$\alpha+\beta > 0, \quad \beta+A < 1, \quad B > 1, \quad \text{and} \quad 4\alpha < (1-\beta-A)[B(1-\beta-A)-(1+3\beta-A)]$$

and

$$x_{n+1} = \frac{x_{n-1}}{A + B x_n + x_{n-1}}, \quad n = 0, 1, \ldots$$

with

$$A \in [0, 1), \quad B \neq 1, \quad \text{and} \quad A + B > 0.$$

Open Problem 5.9.1 (*a*) *Determine all special cases of Eq.(5.0.2) with infinitely many prime period-two solutions and investigate their global behavior.*

(*b*) *Determine all special cases of Eq.(5.0.2) with a "unique" prime period-two solution and investigate the local asymptotic stability of the period-two solution.*

5.10 Equation #10 : $x_{n+1} = \dfrac{\gamma x_{n-1}}{B x_n}$

The change of variables, $x_n = e^{y_n}$, transforms Eq.(#10) to a linear equation.

5.11 Equation #11 : $x_{n+1} = \dfrac{\gamma}{C}$

The equation in this special case is trivial.

5.12 Equation #12 : $x_{n+1} = \dfrac{\gamma x_{n-1}}{D x_{n-2}}$

The change of variables, $x_n = e^{y_n}$, transforms Eq.(#12) to a linear equation.

5.13 Equation #13 : $x_{n+1} = \dfrac{\delta}{A} x_{n-2}$

The equation in this special case is linear.

5.14 Equation #14 : $x_{n+1} = \frac{\delta x_{n-2}}{Bx_n}$

The change of variables, $x_n = e^{y_n}$, transforms Eq.(#14) to a linear equation.

5.15 Equation #15 : $x_{n+1} = \frac{\delta x_{n-2}}{Cx_{n-1}}$

The change of variables, $x_n = e^{y_n}$, transforms Eq.(#15) to a linear equation.

5.16 Equation #16 : $x_{n+1} = \frac{\delta}{D}$

The equation in this special case is trivial.

5.17 Equation #17 : $x_{n+1} = \frac{\alpha}{A + Bx_n}$

The following 11 special cases of Eq.(5.0.1)

#17,	#18,	#19,	#20,	#21,	#22,
#101,	#102,	#103,	#104,	#133	

are special cases of the more general $(k+1)^{st}$-order difference equation

$$x_{n+1} = \frac{1}{A + \sum_{i=0}^{k} B_i x_{n-i}}, \quad n = 0, 1, \ldots \tag{5.17.1}$$

with nonnegative parameters and with arbitrary nonnegative initial conditions $x_{-k}, \ldots, x_0$ such that the denominator is always positive.

Equation (5.17.1) was investigated in [91], [101], [103], [124], [157], and [208]. The following result establishes that when

$$A > 0,$$

the equilibrium of Eq.(5.17.1) is globally asymptotically stable.

Theorem 5.17.1 *Assume that*

$$A > 0.$$

Then the equilibrium $\bar{x}$ of Eq.(5.17.1) is globally asymptotically stable.

PROOF The characteristic equation of the linearized equation of Eq.(5.17.1) about its equilibrium $\bar{x}$ is

$$\lambda^{k+1} + \frac{\bar{x}}{A + \bar{x}\sum_{i=0}^{k} B_i} \cdot \sum_{i=0}^{k} B_{k-i}\lambda^i = 0. \tag{5.17.2}$$

It is now a consequence of Theorem 1.2.5 that all roots of Eq.(5.17.2) lie inside the unit disk and so $\bar{x}$ is locally asymptotically stable. It remains to show that $\bar{x}$ is a global attractor, that is, every solution of Eq.(5.17.1) converges to $\bar{x}$. To this end, first note that for $n \geq 1$,

$$x_n \leq \frac{1}{A}$$

and so also

$$x_n \geq \frac{1}{A + \frac{1}{A}\sum_{i=0}^{k} B_i}.$$

Hence, every solution of Eq.(5.17.1) is bounded from above and from below by positive numbers. Clearly, the function

$$f(z_0, \ldots, z_k) = \frac{1}{A + \sum_{i=0}^{k} B_i z_i}$$

satisfies the Hypotheses of Theorem 1.6.7 (a') from which the result follows. The proof is complete. ∎

The next result establishes the character of solutions of Eq.(5.17.1) when

$$A = 0.$$

In this case it is convenient to rewrite Eq.(5.17.1) in the form

$$x_n = \frac{1}{\sum_{i=1}^{k} B_i x_{n-l_i}}, \quad n = 0, 1, \ldots, \tag{5.17.3}$$

where

$$B_i > 0, \quad \text{for} \quad i = 1, \ldots, k$$

and

$$l_i \in \{1, 2, \ldots\}.$$

It is an amazing fact that the character of solutions of Eq.(5.17.3) **does not** depend on the size of the coefficients $B_1, \ldots, B_k$ but only on the parity of the delays $l_1, \ldots, l_k$ as the following theorem shows.

Theorem 5.17.2 *Let d_1 and d_2 be the greatest common divisors of the two sets of positive integers:*

$$\{l_1, l_2, \ldots, l_k\}$$

and

$$\{l_i + l_j : i, j \in \{1, \ldots, k\}\},$$

respectively. Then the following statements are true:

(a) The equilibrium $\bar{x}$ of Eq.(5.17.3) is globally asymptotically stable if and only if

$$d_1 = d_2.$$

(b) When

$$d_1 \neq d_2,$$

every solution of Eq.(5.17.3) converges to a (not necessarily prime) period-$(2 \cdot d_1)$ solution.

PROOF The proof is a consequence of Theorem 1.6.1. For the details see [124] or [101] and [103]. ∎

Open Problem 5.17.1 *(see [62]) Let $\{B_n\}$ be a periodic sequence of non-negative real numbers with prime period $k \geq 2$ and let l be a positive integer. Obtain necessary and sufficient conditions on k, l, and $B_0, \ldots, B_{k-1}$ such that every positive solution of the difference equation*

$$x_{n+1} = \frac{1}{B_n x_n + x_{n-l}}, \quad n = 0, 1, \ldots$$

is bounded. Extend and generalize.

Open Problem 5.17.2 *Let $\{B_n\}$ be a convergent sequence of positive real numbers and let l be a positive integer. Investigate the character of solutions of the difference equation*

$$x_{n+1} = \frac{1}{B_n x_n + x_{n-l}}, \quad n = 0, 1, \ldots.$$

Extend and generalize.

5.18 Equation #18 : $x_{n+1} = \frac{\alpha}{A + Cx_{n-1}}$

It is a consequence of Theorem 5.17.1 that the equilibrium of this equation is globally asymptotically stable.

5.19 Equation #19 : $x_{n+1} = \frac{\alpha}{A + Dx_{n-2}}$

It is a consequence of Theorem 5.17.1 that the equilibrium of this equation is globally asymptotically stable.

5.20 Equation #20 : $x_{n+1} = \frac{\alpha}{Bx_n + Cx_{n-1}}$

It is a consequence of Theorem 5.17.2 that the equilibrium of this equation is globally asymptotically stable.

Open Problem 5.20.1 *Let $\{B_n\}$ be a periodic sequence of nonnegative real numbers with prime period $k \geq 2$. Obtain necessary and sufficient conditions on k and $B_0, \ldots, B_{k-1}$ such that every positive solution of the difference equation*

$$x_{n+1} = \frac{1}{B_n x_n + x_{n-1}}, \quad n = 0, 1, \ldots$$

is bounded. Extend and generalize.

Open Problem 5.20.2 *Let $\{B_n\}$ be a convergent sequence of positive real numbers. Investigate the character of solutions of the difference equation*

$$x_{n+1} = \frac{1}{B_n x_n + x_{n-1}}, \quad n = 0, 1, \ldots .$$

Extend and generalize.

Open Problem 5.20.3 *Assume that B is a given real number. Determine the "good" set G of the equation*

$$x_{n+1} = \frac{1}{Bx_n + x_{n-1}}, \tag{5.20.1}$$

that is, the set of initial conditions

$$x_{-1}, x_0 \in \Re$$

such that the equation (5.20.1) is well defined for all $n \geq 0$. Determine the character of solutions of Eq.(5.20.1) for all initial conditions in the "good" set G.

5.21 Equation #21 : $x_{n+1} = \dfrac{\alpha}{Bx_n + Dx_{n-2}}$

This equation was investigated in [91] and is a special case of a more general equation investigated in Section 5.17. It follows from Theorem 5.17.2 that every solution of this equation converges to a period-two solution.

Open Problem 5.21.1 *Assume that*

$$B > 0$$

and that

$$\ldots, \phi, \psi, \phi, \psi, \ldots \tag{5.21.1}$$

is a given prime period-two solution of the equation

$$x_{n+1} = \frac{1}{Bx_n + x_{n-2}}, \quad n = 0, 1, \ldots . \tag{5.21.2}$$

Determine, explicitly in terms of B, ϕ, and ψ, the set of all nonnegative initial conditions x_{-2}, x_{-1}, x_0 such that the solution $\{x_n\}_{n=-2}^{\infty}$ converges to (5.21.1).

Open Problem 5.21.2 *Assume that*

$$B > 0$$

and that x_{-2}, x_{-1}, and x_0 are given positive numbers. Determine, explicitly in terms of B, x_{-2}, x_{-1}, and x_0, the values ϕ and ψ of the period-two solution (5.21.1) to which the solution $\{x_n\}_{n=-2}^{\infty}$ of Eq.(5.21.2) converges.

Open Problem 5.21.3 *Let k and l be given nonnegative integers. Determine the set of all initial conditions for which the solutions of the equation*

$$x_{n+1} = \frac{1}{x_{n-k}} + \frac{1}{x_{n-l}}, \quad n = 0, 1, \ldots$$

converge to $\sqrt{2}$.

Conjecture 5.21.1 *Show that the solution of the IVP*

$$x_{n+1} = \frac{1}{x_{n-2} + x_n},$$

$$x_{-2} = x_{-1} = x_0 = 1$$

converges to a prime period-two solution of the form (5.21.1).

Open Problem 5.21.4 *Assume that B is a given real number. Determine the "good" set G of the equation*

$$x_{n+1} = \frac{1}{Bx_n + x_{n-2}}, \quad n = 0, 1, \ldots, \tag{5.21.3}$$

that is, the set of all initial conditions

$$x_{-2}, x_{-1}, x_0 \in \Re$$

such that the equation (5.21.3) is well defined for all $n \geq 0$. Determine the character of solutions of Eq.(5.21.3) for all initial conditions in the "good" set G.

5.22 Equation #22 : $x_{n+1} = \dfrac{\alpha}{Cx_{n-1} + Dx_{n-2}}$

This equation is a special case of a more general equation investigated in Section 5.17. It follows from Theorem 5.17.2 that the equilibrium of the equation is globally asymptotically stable.

Open Problem 5.22.1 *Let $\{C_n\}$ be a periodic sequence of nonnegative real numbers with prime period $k \geq 2$. Obtain necessary and sufficient conditions on k and $C_0, \ldots, C_{k-1}$ such that every positive solution of the difference equation*

$$x_{n+1} = \frac{1}{C_n x_{n-1} + x_{n-2}}, \quad n = 0, 1, \ldots$$

is bounded. Extend and generalize.

Open Problem 5.22.2 *Let $\{C_n\}$ be a convergent sequence of positive real numbers. Investigate the character of solutions of the difference equation*

$$x_{n+1} = \frac{1}{C_n x_{n-1} + x_{n-2}}, \quad n = 0, 1, \dots .$$

Extend and generalize.

Open Problem 5.22.3 *Assume that C is a given real number. Determine the "good" set G of the equation*

$$x_{n+1} = \frac{1}{C x_{n-1} + x_{n-2}}, \tag{5.22.1}$$

that is, the set of initial conditions

$$x_{-2}, x_{-1}, x_0 \in \Re$$

such that the equation (5.22.1) is well defined for all $n \geq 0$. Determine the character of solutions of Eq.(5.22.1) for all initial conditions in the "good" set G.

5.23 Equation #23 : $x_{n+1} = \dfrac{\beta x_n}{A + B x_n}$

This equation or, more precisely, the difference equation

$$x_{n+1} = \frac{rKx_n}{K + (r-1)x_n}, \quad n = 0, 1, \dots \tag{5.23.1}$$

arises in application to **population dynamics** and it is known as the **Beverton -Holt equation**. See [83]. The parameter K is positive and is called the **"carrying capacity"** of the population and the parameter r is greater than 1 and is called the **"inherent growth rate"** of the population. As we will see in this section all solutions of Eq.(5.23.1), with $x_0 > 0$, approach the positive equilibrium K as $n \to \infty$. Actually, this equation is a special case of the so-called **Riccati difference equation**

$$x_{n+1} = \frac{\alpha + \beta x_n}{A + B x_n}, \quad n = 0, 1, \dots .$$

For the character of solutions of the Riccati equation with real parameters and real initial conditions see Section 5.65. The equation in the title can be written in the normalized form

$$x_{n+1} = \frac{x_n}{A + x_n}, \quad n = 0, 1, \dots \tag{5.23.2}$$

with $A > 0$ and $x_0 \geq 0$. Clearly,

$$x_{n+1} \leq \frac{1}{A} x_n, \quad \text{for } n \geq 0$$

and so when

$$A \geq 1,$$

every solution of Eq.(5.23.2) converges monotonically to the zero equilibrium of the equation.

On the other hand, when

$$A < 1,$$

we claim that every positive solution converges to the positive equilibrium

$$\bar{x} = 1 - A.$$

This result follows from Remark 5.65.1 (on the general Riccati equation) and the observation that if

$$\lim_{n \to \infty} x_n = 0,$$

then, eventually,

$$x_{n+1} = \frac{x_n}{A + x_n} \geq \frac{x_n}{A + 1 - A} = x_n,$$

which is a contradiction.

Another way to establish the character of solutions of Eq.(5.23.2) is to use a **stairstep diagram** or the following simple result.

Theorem 5.23.1 *Let I be a set of real numbers and let*

$$F : I \to I$$

be an increasing function. Then every solution of the difference equation

$$x_{n+1} = F(x_n), \quad n = 0, 1, \ldots$$

is increasing if and only if

$$x_1 \geq x_0$$

and is decreasing if and only if

$$x_1 \leq x_0.$$

PROOF The proof is a simple consequence of the monotonicity of the function F. ∎

We now present two general global asymptotic stability results that apply to several special cases of the $(k+1)^{st}$-order rational difference equation

$$x_{n+1} = \frac{\alpha + \sum_{i=0}^{k} \beta_i x_{n-i}}{A + \sum_{i=0}^{k} B_i x_{n-i}}, \quad n = 0, 1, \ldots \tag{5.23.3}$$

with $A > 0$, the remaining parameters nonnegative, with

$$\sum_{i=0}^{k} \beta_i \text{ and } \sum_{i=0}^{k} B_i \in (0, \infty),$$

and with arbitrary nonnegative initial conditions such that the denominator is always positive. For some general results on Eq.(5.23.3), see also [157].

The characteristic equation of the linearized equation of Eq.(5.23.3) about an equilibrium point $\bar{x}$ is

$$\lambda^{k+1} + \frac{1}{A + \bar{x} \cdot \sum_{i=0}^{k} B_i} \sum_{i=0}^{k} (B_i \bar{x} - \beta_i) \lambda^{k-i} = 0. \tag{5.23.4}$$

Zero is an equilibrium point of Eq.(5.23.3) if and only if

$$\alpha = 0 \text{ and } A > 0. \tag{5.23.5}$$

As we will see later, when (5.23.5) holds, the zero equilibrium of Eq.(5.23.3) is globally asymptotically stable when

$$A > \sum_{i=0}^{k} \beta_i \tag{5.23.6}$$

and unstable when

$$A < \sum_{i=0}^{k} \beta_i.$$

Eq.(5.23.3) has a positive equilibrium point if and only if

either

$$\alpha > 0 \tag{5.23.7}$$

or

$$\alpha = 0 \text{ and } A < \sum_{i=0}^{k} \beta_i. \tag{5.23.8}$$

When (5.23.7) holds, the equation has the unique equilibrium point

$$\bar{x} = \frac{\beta - A + \sqrt{(\beta - A)^2 + 4\alpha B}}{2B}, \tag{5.23.9}$$

where for simplicity we use the notation,

$$\beta = \sum_{i=0}^{k} \beta_i \text{ and } B = \sum_{i=0}^{k} B_i.$$

When (5.23.8) holds, Eq.(5.23.3) has the unique positive equilibrium point

$$\bar{x} = \frac{\beta - A}{B}.$$

Note that

$$\frac{1}{A + B\bar{x}} \sum_{i=0}^{k} |B_i\bar{x} - \beta_i| \leq \frac{1}{A + B\bar{x}} \cdot (B\bar{x} + \beta). \tag{5.23.10}$$

Therefore, by Theorem 1.2.5 and (5.23.10), the equilibrium of Eq.(5.23.3) is locally asymptotically stable when (5.23.6) holds.

Note that the condition (5.23.6) is at best a sufficient condition for the positive equilibrium of Eq.(5.23.3) to be locally asymptotically stable. **In every special case of Eq.(5.23.3) we should strive to determine the "entire" region of the local asymptotic stability of the positive equilibrium, when such equilibrium exists.**

Open Problem 5.23.1 *Assume that*

$$k \geq 4.$$

Obtain the region of the local asymptotic stability of the positive equilibrium of Eq.(5.23.3) (when a positive equilibrium exists) explicitly, in terms of the parameters of the equation.

For the values of $k \in \{1, 2, 3\}$, see Theorems 1.2.2, 1.2.3, and 1.2.4 in Chapter 1. The open problem 5.23.1 is asking for easily verifiable conditions in the spirit of Theorems 1.2.2, 1.2.3, and 1.2.4.

Theorem 5.23.2 *Assume that*

$$\beta = \sum_{i=0}^{k} \beta_i < A.$$

Then the following statements are true:

(i) If

$$\alpha = 0,$$

the zero equilibrium of Eq.(5.23.3) is globally asymptotically stable.

(ii) If

$$\alpha > 0,$$

the positive equilibrium of Eq.(5.23.3) is globally asymptotically stable.

PROOF As we saw in the discussion preceding the Theorem, the equilibrium is locally asymptotically stable when $\alpha \geq 0$. Furthermore, we have

$$x_{n+1} \leq \frac{\alpha}{A} + \frac{1}{A}\sum_{i=0}^{k} \beta_i x_{n-i},$$

which, together with Theorem 1.4.1, implies that the solution converges to zero in case (i) and also that the solution is bounded from above in case (ii). Now in case (ii) let

$$S = \limsup_{n\to\infty} x_n \quad \text{and} \quad I = \liminf_{n\to\infty} x_n.$$

Then, clearly,

$$S \leq \frac{\alpha + \beta S}{A + BI} \text{ and } I \geq \frac{\alpha + \beta I}{A + BS},$$

from which it follows that

$$\alpha + (\beta - A)I \leq BSI \leq \alpha + (\beta - A)S.$$

Hence,

$$S = I$$

and the proof is complete. ∎

In the very special case when

$$A = \sum_{i=0}^{k} \beta_i > 0 \text{ and } \alpha > 0,$$

the global character of solutions of Eq.(5.23.3) is completely described by the following result in [224]. In this case it is preferable to write the difference equation in the form

$$x_n = \frac{\alpha + \sum_{r=1}^{k} \beta_r x_{n-i_r}}{A + \sum_{t=1}^{m} B_j x_{n-j_t}}, \quad n = 1, 2, \ldots . \tag{5.23.11}$$

Also, by making a change of variables, if necessary, we may and do assume that the greatest common divisor of all "delays" in the numerator and denominator is 1, that is,

$$gcd\{i_1, \ldots, i_k, j_1, \ldots, j_m\} = 1.$$

Theorem 5.23.3 *Assume that*

$$\alpha, \beta_1, \ldots, \beta_k, B_1, \ldots, B_m \in (0, \infty) \text{ and } A = \sum_{i=1}^{k} \beta_i.$$

Then when the "delays" in the numerator

$$i_1, \ldots, i_k \text{ are all even}$$

and the "delays" in the denominator

$$j_1, \ldots, j_m \text{ are all odd,}$$

every solution of Eq.(5.23.11) converges to a period-two solution. In every other case of delays, every solution of Eq.(5.23.11) has a finite limit.

PROOF The proof is a straightforward application of Theorem 1.6.11 (a') and (b') and the fact that all solutions of Eq.(5.23.11) are bounded from above and from below by positive constants. ∎

Theorem 5.23.4 *Assume that*

$$\alpha = 0 \text{ and } \beta = \sum_{i=0}^{k} \beta_i = A. \tag{5.23.12}$$

and that one of the following three conditions is satisfied:

(a)

$$\beta_i B_i > 0 \text{ for some } i \in \{0, \ldots, k\}. \tag{5.23.13}$$

(b)

$$\beta_0 > 0. \tag{5.23.14}$$

(c)

$$B_0 > 0 \text{ and Eq.(5.23.3) has no period-two solutions.} \tag{5.23.15}$$

Then the zero equilibrium of Eq.(5.23.3) is globally asymptotically stable.

PROOF Observe that

$$x_{n+1} \leq \max_{0 \leq i \leq k} x_{n-i}.$$

From this it follows that the zero equilibrium is locally stable and also that every solution of the equation is bounded.

Let $\{x_n\}_{n=-k}^{\infty}$ be a positive solution of Eq.(5.23.3). It remains to show that when (5.23.12) holds and one of the three conditions (5.23.13), (5.23.14), or (5.23.15) is satisfied, then the zero equilibrium of Eq.(5.23.3) is a global attractor of all solutions. Set

$$S = \limsup_{n\to\infty} x_n \quad \text{and} \quad I = \liminf_{n\to\infty} x_n.$$

Then

$$S \le \frac{\beta S}{\beta + BI} \text{ and } I \ge \frac{\beta I}{\beta + BS},$$

from which it follows that

$$SI = 0\,.$$

Now, clearly, there exists a sequence of indices $\{n_i\}$ and positive numbers $\{L_{-r}\}_{r=0}^{k}$ such that

$$S = \lim_{i\to\infty} x_{n_i+1}$$

and

$$L_{-r} = \lim_{i\to\infty} x_{n_i-r}, \quad \text{for} \quad r = 0, \ldots, k.$$

Then from Eq.(5.23.3) we find

$$S = \frac{\sum_{i=0}^{k} \beta_i L_{-i}}{A + \sum_{i=0}^{k} B_i L_{-i}}. \tag{5.23.16}$$

When (5.23.13) is satisfied, it follows from (5.23.16) that

$$L_{-i} = S = I$$

for the i that (5.23.13) is satisfied. Otherwise,

$$L_{-i} < S \text{ or } L_{-i} > I$$

for the i that (5.23.13) is satisfied and so

$$SI < 0,$$

which is a contradiction.

When (5.23.14) is satisfied, it follows from (5.23.16) that

$$L_{-i} = S \text{ for all } i = 0, \ldots, k$$

and

$$L_{-i} = I \text{ for all } i \in \{0, \ldots, k\}, \text{ for which } B_i > 0.$$

Otherwise, there exists $i_0 \in \{0, \ldots, k\}$ such that

$$L_{-i_0} < S \text{ or } L_{-i_0} > I$$

and so

$$SI < 0,$$

which is a contradiction.

When (5.23.15) is satisfied, it follows from (5.23.16) that for all $m \geq 0$

$$L_{-2m} = I \text{ and } L_{1-2m} = S.$$

Otherwise, there exists $m_0 \geq 0$ such that

$$L_{-2m_0} > I \text{ or } L_{1-2m} < S$$

and so

$$SI < 0,$$

which is a contradiction. Hence,

$$\ldots, I, S, I, S, \ldots$$

is a period-two solution of Eq.(5.23.3), which is a contradiction. The proof is complete. ∎

Remark 5.23.1 *When*

$$\alpha = 0, \ \ \beta = A, \ \ \textit{and} \ \ B > 0,$$

the zero equilibrium is not always globally asymptotically stable. When the equation has periodic solutions, it is not. When $k = 2$, the cases where only the zero equilibrium exists and the equation has periodic solutions are the following:

#29,	*#35,*	*#36,*	*#109,*	*#110,*	*#113.*

In the special cases #29, #109, and #110 every solution of the equation converges to a (not necessarily prime) period-two solution, and in cases #35, #36, and #113 every solution converges to a (not necessarily prime) period-three solution. Furthermore, the zero equilibrium in all these special cases is only stable but not asymptotically stable.

Open Problem 5.23.2 *Assume that*

$$\alpha = 0 \ \ \textit{and} \ \ A = \sum_{i=0}^{k} \beta_i > 0.$$

Obtain necessary and sufficient conditions in terms of k, the delays in the equation, and $\beta_0, B_0, \ldots, \beta_k, B_k$ so that the zero equilibrium of Eq.(5.23.3) is globally asymptotically stable.

Conjecture 5.23.1 *Assume that*

$$\alpha = 0 \quad and \quad A = \sum_{i=0}^{k} \beta_i > 0$$

and that Eq.(5.23.3) has no periodic solutions with prime period $p \geq 2$. Show that the zero equilibrium of Eq.(5.23.3) is globally asymptotically stable.

5.24 Equation #24 : $x_{n+1} = \dfrac{\beta x_n}{A + C x_{n-1}}$

This equation, called **Pielou's equation**, was investigated in [186]. See also [154] and [157]. The more general equation

$$N_{t+1} = \frac{aN_t}{1 + bN_{t-k}},$$

where k is a nonnegative integer was proposed by Pielou in her books ([209, p. 22] and [210, p. 79]) as a discrete analogue of the delay logistic equation

$$N'(t) = rN(t)[1 - \frac{N(t-\tau)}{P}].$$

Eq.(#24) can be written in the normalized form

$$x_{n+1} = \frac{\beta x_n}{1 + x_{n-1}}, \quad n = 0, 1, \ldots \tag{5.24.1}$$

with positive parameter β and with arbitrary nonegative initial conditions x_{-1}, x_0.

Zero is always an equilibrium point of Eq.(5.24.1). The characteristic equation of the linearized equation of Eq.(5.24.1) about the zero equilibrium is

$$\lambda^2 - \beta\lambda = 0. \tag{5.24.2}$$

By Theorems 5.23.2 and 5.23.4 it follows that the zero equilibrium of Eq.(5.24.1) is globally asymptotically stable when

$$\beta \leq 1. \tag{5.24.3}$$

From (5.24.2) and Theorem 1.2.2 it follows that the zero equilibrium is unstable when

$$\beta > 1. \tag{5.24.4}$$

Furthermore, when (5.24.4) holds, Eq.(5.24.1) has also the unique positive equilibrium point

$$\bar{x} = \beta - 1.$$

The characteristic equation of the linearized equation of Eq.(5.24.1) about the positive equilibrium, $\bar{x} = \beta - 1$, is

$$\lambda^2 - \lambda + \frac{\beta - 1}{\beta} = 0.$$

From this and Theorem 1.2.2 it follows that $\bar{x} = \beta - 1$ is locally asymptotically stable when (5.24.4) holds.

In [186] it was shown that when (5.24.4) holds, every positive solution of Eq.(5.24.1) converges to the positive equilibrium, $\bar{x} = \beta - 1$.

5.24.1 The Autonomous Pielou's Equation

The main result in this section is the following new proof for the Pielou's equation 5.24.1, which in the next section will be adapted to the nonautonomous case.

Theorem 5.24.1 *Assume that*

$$\beta > 1.$$

Then every positive solution of Eq.(5.24.1) converges to the positive equilibrium

$$\bar{x} = \beta - 1.$$

PROOF Let $\{x_n\}$ be a positive solution of Eq.(5.24.1). Then for $n \geq 1$,

$$x_{n+1} = \frac{\beta x_{n-1}}{1 + x_{n-1}} \cdot \frac{\beta}{1 + x_{n-2}} \tag{5.24.5}$$

and so the solution is bounded from above by β^2. Next, we claim that the solution is also bounded from below by a positive constant. Otherwise, there exists a sequence of indices $\{n_i\}$ such that

$$x_{n_i+1} \to 0, \text{ and } x_{n_i+1} < x_j \text{ for all } j < n_i + 1. \tag{5.24.6}$$

Then from (5.24.1), the subsequences $\{x_{n_i}\}$ and $\{x_{n_i-1}\}$ converge to zero. Hence, eventually,

$$x_{n_i-1} < \beta - 1,$$

which implies that, eventually,

$$x_{n_i+1} = \frac{\beta x_{n_i}}{1 + x_{n_i-1}} > \frac{\beta x_{n_i}}{1 + (\beta - 1)} = x_{n_i}.$$

This contradicts (5.24.6) and establishes our claim that the solution is bounded from below by a positive constant.

Set

$$S = \limsup_{n\to\infty} x_n \quad \text{and} \quad I = \liminf_{n\to\infty} x_n.$$

Then it follows from (5.24.5) that

$$S \le \frac{\beta S}{1+S}\frac{\beta}{1+I} \quad \text{and} \quad I \ge \frac{\beta I}{1+I}\frac{\beta}{1+S}$$

which imply that

$$(1+S)(1+I) = \beta^2. \tag{5.24.7}$$

Clearly, there exists a sequence of indices $\{n_i\}$ and positive numbers $\{L_{-t}\}_{t=0}^{2}$ such that

$$x_{n_i+1} \to S$$

and for $t \in \{0, 1, 2\}$

$$x_{n_i-t} \to L_{-t}.$$

Thus, from (5.24.5) and (5.24.7) we see that

$$S = \frac{\beta^2 L_{-1}}{(1+L_{-1})(1+L_{-2})} = \frac{(1+S)(1+I)}{(1+L_{-1})(1+L_{-2})} \cdot L_{-1} \le S$$

and so

$$L_{-1} = S \ \text{and} \ L_{-2} = I.$$

Furthermore, from (5.24.1) and (5.24.7),

$$L_0 = \frac{\beta L_{-1}}{1+L_{-2}} = \frac{\beta S}{1+I} = \frac{S(1+S)}{\beta} \le S$$

and so $S = \beta - 1 = I$ and the proof is complete.

In addition to the proof given above, observe that the function

$$f(z_2, z_3) = \frac{\beta^2 z_2}{(1+z_2)(1+z_3)}$$

satisfies the Hypotheses of Theorem 1.6.9. Hence, every solution of Eq.(5.24.5) and consequently, every solution of Eq.(5.24.1) both converge to a (not necessarily prime) period-two solution. From this and the fact that Eq.(5.24.1) has no prime period-two solutions the result follows. The proof is complete. ∎

5.24.2 Periodically Forced Pielou's Equation

In this section we investigate the global character of solutions of the periodically forced Pielou's equation

$$x_{n+1} = \frac{\beta_n x_n}{1 + x_{n-1}}, \quad n = 0, 1, \ldots \tag{5.24.8}$$

and prove that when the sequence $\{\beta_n\}$ is periodic with prime period k, with positive values, and

$$\prod_{i=0}^{k-1} \beta_i > 1, \tag{5.24.9}$$

every positive solution converges to a periodic solution with prime period k.

Difference equations with periodic coefficients have been studied by several authors especially in connection with mathematical models in biology. See [62], [183], [75], [82], [83], [99], [100], [153], and [157].

In Section 5.24.1 we presented a new, simple, and elegant proof that, when (5.24.4) holds, every positive solution of Eq.(5.24.1) converges to the positive equilibrium $(\beta - 1)$. It is an amazing fact that the idea of our proof also extends to the periodically forced Eq.(5.24.8). This enables us in this section to establish that when the coefficient $\{\beta_n\}$ is periodic with period k, with positive values, that is:

$$\beta_n = \begin{cases} \beta_0, & \text{if } n = kj \\ \beta_1, & \text{if } n = kj + 1 \\ \quad \ldots & \\ \beta_{k-1}, & \text{if } n = kj + k - 1 \end{cases}, \quad j = 0, 1, \ldots$$

with

$$\beta_i \in (0, \infty), \quad i = 0, \ldots, k - 1,$$

and when (5.24.9) holds, every positive solution of Eq.(5.24.8) converges to a periodic solution with period k.

The special case where the sequence $\{\beta_n\}$ is periodic with period two was recently investigated in [183]. The method of that proof is different from our proof and does not seem to extend to higher periods.

The following theorem extends to the periodic case the result of the autonomous case when

$$\prod_{i=0}^{k-1} \beta_i \leq 1. \tag{5.24.10}$$

Its proof is simple and will be omitted.

Theorem 5.24.2 *Assume that (5.24.10) holds. Then every nonnegative solution of Eq.(5.24.8) converges to zero.*

It is easy to see that, as in the autonomous case, every positive solution of the periodically forced equation (5.24.8) is bounded from above and, furthermore, when (5.24.9) holds, every positive solution is also bounded from below by a positive constant.

Our goal now is to show that, when $\{\beta_n\}$ is a positive periodic sequence with prime period k and (5.24.9) holds, then every positive solution of Eq.(5.24.8) converges to a periodic solution with prime period k. To this end, let $\{y_n\}$ be an arbitrary, but fixed for the remaining part of this section, positive solution of Eq.(5.24.8).

For a fixed $k \in \{1, 2, \dots\}$ and for every integer i, we define the sequences $\{S_i\}$, $\{I_i\}$ as follows:

$$S_i = \limsup_{n\to\infty} y_{kn+i} \text{ and } I_i = \liminf_{n\to\infty} y_{kn+i}.$$

Clearly, for all integer values of j,

$$S_{j+k} = S_j \text{ and } I_{j+k} = I_j.$$

To make the proof very clear, we will first give the details for $k = 2$. The key idea now is to establish the following identities, which extend the Identity (5.24.7) of the autonomous case:

$$(1 + S_1)(1 + I_0) = (1 + S_0)(1 + I_1) = \beta_0\beta_1. \tag{5.24.11}$$

Lemma 5.24.1 *(5.24.11) holds.*

PROOF Clearly, for $n \geq 1$,

$$y_{2n+1} = \frac{\beta_0\beta_1 y_{2n-1}}{(1 + y_{2n-1})(1 + y_{2n-2})},$$

from which it follows that

$$(1 + S_1)(1 + I_0) \leq \beta_0\beta_1 \leq (1 + I_1)(1 + S_0).$$

Also, from

$$y_{2n+2} = \frac{\beta_1\beta_0 y_{2n}}{(1 + y_{2n})(1 + y_{2n-1})}$$

we obtain

$$(1 + S_0)(1 + I_1) \leq \beta_1\beta_0 \leq (1 + I_0)(1 + S_1),$$

from which (5.24.11) follows. ∎

Theorem 5.24.3 *Assume that $\{\beta_n\}$ is a positive periodic sequence with prime period two and that (5.24.9) holds. Then $\{y_n\}$ converges to a prime period-two solution.*

PROOF Clearly, there exist two sequences of indices, $\{n_i\}$ and $\{n_j\}$, and positive numbers, $\{U_{-t}\}_{t=0}^{3}$ and $\{L_{-t}\}_{t=0}^{3}$, such that

$$S_1 = \lim_{i\to\infty} y_{2n_i+1} \text{ and } I_1 = \lim_{j\to\infty} y_{2n_j+1}$$

and for each $t \in \{0, 1, 2, 3\}$,

$$U_{-t} = \lim_{i\to\infty} y_{2n_i-t} \text{ and } L_{-t} = \lim_{j\to\infty} y_{2n_j-t}.$$

Then,

$$S_1 = \frac{\beta_0\beta_1 U_{-1}}{(1+U_{-1})(1+U_{-2})}$$

from which it follows that

$$U_{-1} = S_1 \text{ and } U_{-2} = I_0$$

because otherwise

$$S_1 < \frac{\beta_0\beta_1 S_1}{(1+S_1)(1+I_0)},$$

which contradicts (5.24.11). Similarly,

$$U_{-3} = S_1.$$

Also,

$$U_0 = \frac{\beta_1 S_1}{1+I_0} = \frac{\beta_0\beta_1 I_0}{(1+I_0)(1+S_1)} = I_0$$

and so

$$\beta_1 S_1 = I_0(1+I_0).$$

Similarly,

$$\beta_1 I_1 = S_0(1+S_0).$$

Therefore,

$$I_0 = S_0 \text{ and } I_1 = S_1.$$

Hence, the two subsequences $\{y_{2n}\}$ and $\{y_{2n+1}\}$ converge to finite limits. Set

$$l_0 = \lim_{n\to\infty} y_{2n} \text{ and } l_1 = \lim_{n\to\infty} y_{2n+1}.$$

By taking limits in Eq.(5.24.8) we obtain

$$l_1 = \frac{\beta_0 l_0}{1+l_1} \text{ and } l_0 = \frac{\beta_1 l_1}{1+l_0}$$

and so, clearly, $\{y_n\}$ converges to the prime period-two solution of Eq.(5.24.8)

$$\ldots, l_0, l_1, \ldots .$$

The proof is complete. ∎

We now turn to the case where the period k is an arbitrary even number equal to $2p$. The key idea here is to observe that the following identities, which extend the Identities in (5.24.11), hold for the solution $\{y_n\}$:

$$\prod_{i=0}^{p-1}(1+S_{2i})(1+I_{2i+1}) = \prod_{i=0}^{p-1}(1+I_{2i})(1+S_{2i+1}) = \prod_{i=0}^{2p-1}\beta_i. \tag{5.24.12}$$

Theorem 5.24.4 *Assume that $\{\beta_n\}$ is a positive periodic sequence with prime period $k = 2p$ and that (5.24.9) holds. Then $\{y_n\}$ converges to a prime period-$2p$ solution.*

PROOF Clearly, there exist $2p$ sequences of indices,

$$\{n_{1,i}\}, \{n_{3,i}\}, \ldots, \{n_{2p-1,i}\}$$

and

$$\{n_{1,j}\}, \{n_{3,j}\}, \ldots, \{n_{2p-1,j}\},$$

and $2p$ sequences of positive numbers,

$$\{U_{1,-t}\}_{t=0}^{\infty}, \{U_{3,-t}\}_{t=-2}^{\infty}, \ldots, \{U_{2p-1,-t}\}_{t=2-2p}^{\infty},$$

$$\{L_{1,-t}\}_{t=0}^{\infty}, \{L_{3,-t}\}_{t=-2}^{\infty}, \ldots, \{L_{2p-1,-t}\}_{t=2-2p}^{\infty},$$

such that for $r \in \{1, \ldots, 2p-1\}$ and $t_r \in \{1-r, \ldots\}$

$$S_r = \lim_{i\to\infty} y_{(2p)\cdot n_{r,i}+r}, \quad I_r = \lim_{i\to\infty} y_{(2p)\cdot n_{r,j}+r},$$

$$U_{r,-t_r} = \lim_{i\to\infty} y_{(2p)\cdot n_{r,i}-t_r}, \text{ and } L_{r,-t} = \lim_{j\to\infty} y_{(2p)\cdot n_{r,j}-t_r}.$$

Then

$$S_1 = \frac{\beta_0\beta_{2p-1}U_{1,-1}}{(1+U_{1,-1})(1+U_{1,-2})}$$

$$S_3 = \frac{\beta_2\beta_1 U_{3,1}}{(1+U_{3,1})(1+U_{3,0})}$$

$$\ldots$$

$$S_{2p-1} = \frac{\beta_{2p-2}\beta_{2p-3}U_{2p-1,2p-3}}{(1+U_{2p-1,2p-3})(1+U_{2p-1,2p-4})},$$

from which it follows that

$$U_{1,-1} = S_{2p-1}, \quad U_{1,-2} = I_{2p-2}$$

$$U_{3,1} = S_1, \quad U_{3,0} = I_0$$

$$\dots$$

$$U_{2p-1,2p-3} = S_{2p-3}, \; U_{2p-1,2p-4} = I_{2p-4}$$

because otherwise

$$\prod_{i=0}^{p-1}(1+S_{2i-1})(1+I_{2i}) < \prod_{i=0}^{2p-1}\beta_i,$$

which contradicts (5.24.12). Similarly,

$$U_{1,-3} = S_{2p-3}, \; U_{1,-4} = I_{2p-4}$$

$$U_{3,-1} = S_{2p-1}, \; U_{3,-2} = I_{2p-2}$$

$$\dots$$

$$U_{2p-1,2p-5} = S_{2p-5}, \; U_{2p-1,2p-6} = I_{2p-6}$$

and, inductively,

$$U_{1,-(2j-1)} = S_{1,2p-(2j-1)} \text{ and } U_{1,-(2j)} = S_{2p-(2j)} \; j = 1, 2, \dots .$$

One can see, by iterating Eq.(5.24.8), that

$$S_1 = \frac{\beta_0\beta_{2p-1}\cdots\beta_2 U_{1,-(2p-2)}}{\prod_{i=1}^{2p-1}(1+U_{1,-i})},$$

from which it follows that

$$\beta_1 S_1 = \frac{\prod_{i=0}^{2p-1}\beta_i I_2}{\prod_{i=0}^{p-1}(1+S_{2i+1})\prod_{i=1}^{p-1}(1+I_{2i})}$$

and so

$$\beta_1 S_1 = I_2(1+I_0).$$

Similarly,

$$\beta_1 I_1 = S_2(1+S_0)$$

and so

$$S_0 = I_0, \; S_1 = I_1, \text{ and } S_2 = I_2$$

and, inductively,

$$S_i = I_i, \; i = 0, 1, \dots, 2p-1.$$

Hence, the $2p$ subsequences $\{y_{(2p)\cdot n+i}\}$ for $i \in \{0, \dots, 2p-1\}$ converge to finite limits. Set

$$l_i = \lim_{n\to\infty} y_{(2p)\cdot n+i}, \text{ for each } i \in \{0, \dots, 2p-1\}.$$

By taking limits in Eq.(5.24.8) we obtain

$$l_i = \frac{\beta_{i-1} l_{i-1}}{1 + l_{i-2}}, \quad \text{for each } i \in \{0, \ldots, 2p-1\}$$

and so, clearly, $\{y_n\}$ converges to the prime period-$2p$ solution

$$\ldots, l_0, \ldots, l_{2p-1}, \ldots .$$

The proof is complete. ∎

We now turn to the odd case $k = 2p + 1$. The key idea of the proof now is to establish the following identities:

$$S_i = \frac{\beta_{i-1}\beta_{i-2}S_{i-2}}{(1 + S_{i-2})(1 + I_{i-3})} \quad \text{and} \quad I_i = \frac{\beta_{i-1}\beta_{i-2}I_{i-2}}{(1 + I_{i-2})(1 + S_{i-3})}, \tag{5.24.13}$$

which are satisfied by the sequences $\{S_i\}$ and $\{I_i\}$.

Lemma 5.24.2 *(5.24.13) holds.*

PROOF We have

$$y_{(2p+1)\cdot n+1} = \frac{\beta_0\beta_{2p} y_{(2p+1)\cdot n-1}}{(1 + y_{(2p+1)\cdot n-1})(1 + y_{(2p+1)\cdot n-2})},$$

from which it follows that

$$S_1 \leq \frac{\beta_0\beta_{2p-1}S_{2p}}{(1 + S_{2p})(1 + I_{2p-1})} \quad \text{and} \quad I_1 \geq \frac{\beta_0\beta_{2p}I_{2p}}{(1 + I_{2p})(1 + S_{2p-1})}$$

or, equivalently,

$$\frac{S_1}{I_1} \cdot \frac{1 + S_{2p}}{1 + I_{2p}} \leq \frac{\beta_0\beta_{2p}S_{2p}}{I_1(1 + I_{2p})(1 + I_{2p-1})} \leq \frac{S_{2p}}{I_{2p}} \cdot \frac{1 + S_{2p-1}}{1 + I_{2p-1}}. \tag{5.24.14}$$

Similarly, we get

$$\frac{S_2}{I_2} \cdot \frac{1 + S_0}{1 + I_0} \leq \frac{\beta_1\beta_0 S_0}{I_2(1 + I_0)(1 + I_{2p})} \leq \frac{S_0}{I_0} \cdot \frac{1 + S_{2p}}{1 + I_{2p}} \tag{5.24.15}$$

$$\cdots$$

$$\frac{S_0}{I_0} \cdot \frac{1 + S_{2p-1}}{1 + I_{2p-1}} \leq \frac{\beta_{2p}\beta_{2p-1}S_{2p-1}}{I_0(1 + I_{2p-1})(1 + I_{2p-2})} \leq \frac{S_{2p-1}}{I_{2p-1}} \cdot \frac{1 + S_{2p-2}}{1 + I_{2p-2}}. \tag{5.24.16}$$

To complete the proof of (5.24.13), we need to establish that all the above inequalities reduce to equalities. To this end, it follows from (5.24.14) and (5.24.15) that

$$\frac{S_1}{I_1} \cdot \frac{S_2}{I_2} \cdot \frac{1 + S_0}{1 + I_0} \leq \frac{S_{2p}}{I_{2p}} \cdot \frac{S_0}{I_0} \cdot \frac{1 + S_{2p-1}}{1 + I_{2p-1}}. \tag{5.24.17}$$

Similarly,

$$\frac{S_2}{I_2}\cdot\frac{S_3}{I_3}\cdot\frac{1+S_1}{1+I_1}\leq\frac{S_0}{I_0}\cdot\frac{S_1}{I_1}\cdot\frac{1+S_{2p}}{1+I_{2p}}. \tag{5.24.18}$$

$$\cdots$$

$$\frac{S_0}{I_0}\cdot\frac{S_1}{I_1}\cdot\frac{1+S_{2p}}{1+I_{2p}}\leq\frac{S_{2p-1}}{I_{2p-1}}\cdot\frac{S_{2p}}{I_{2p}}\cdot\frac{1+S_{2p-2}}{1+I_{2p-2}}. \tag{5.24.19}$$

Hence,

$$\frac{S_0}{I_0}\cdot\frac{S_1}{I_1}\cdot\frac{1+S_{2p}}{1+I_{2p}}\leq\frac{S_{2p-1}}{I_{2p-1}}\cdot\frac{S_{2p}}{I_{2p}}\cdot\frac{1+S_{2p-2}}{1+I_{2p-2}}\leq\frac{S_{2p-3}}{I_{2p-3}}\cdot\frac{S_{2p-2}}{I_{2p-2}}\cdot\frac{1+S_{2p-4}}{1+I_{2p-4}}$$

$$\leq\ldots\leq\frac{S_1}{I_1}\cdot\frac{S_2}{I_2}\cdot\frac{1+S_0}{1+I_0}\leq\frac{S_{2p}}{I_{2p}}\cdot\frac{S_0}{I_0}\cdot\frac{1+S_{2p-1}}{1+I_{2p-1}}$$

$$\leq\frac{S_{2p-2}}{I_{2p-2}}\cdot\frac{S_{2p-1}}{I_{2p-1}}\cdot\frac{1+S_{2p-3}}{1+I_{2p-3}}$$

$$\leq\cdots\leq\frac{S_2}{I_2}\cdot\frac{S_3}{I_3}\cdot\frac{1+S_1}{1+I_1}\leq\frac{S_0}{I_0}\cdot\frac{S_1}{I_1}\cdot\frac{1+S_{2p}}{1+I_{2p}},$$

from which it follows that

$$\frac{S_0}{I_0}\cdot\frac{1+S_{2p-1}}{1+I_{2p-1}}=\frac{\beta_{2p}\beta_{2p-1}S_{2p-1}}{I_0(1+I_{2p-1})(1+I_{2p-2})}=\frac{S_{2p-1}}{I_{2p-1}}\cdot\frac{1+S_{2p-2}}{1+I_{2p-2}}$$

and so we establish equality in (5.24.16). The remaining cases are established in a similar fashion. ∎

Theorem 5.24.5 *Assume that $\{\beta_n\}$ is a positive periodic sequence of prime period $k=(2p+1)$ and that (5.24.9) holds. Then $\{y_n\}$ converges to a prime period-$(2p+1)$ solution.*

PROOF Clearly, there exist subsequences $\{y_{(2p+1)\cdot n_i+1}\}$ and $\{y_{(2p+1)\cdot n_i-t}\}_{t=0}^{\infty}$, and positive numbers $\{U_{-t}\}_{t=0}^{\infty}$ such that

$$S_1=\lim_{i\to\infty}y_{(2p+1)\cdot n_i+1}\ \text{ and }\ U_{-t}=\lim_{i\to\infty}y_{(2p+1)\cdot n_i-t},\ \text{ for }\ t\in\{0,1,\ldots\}.$$

Hence,

$$S_1=\frac{\beta_0\beta_{2p}U_{-1}}{(1+U_{-1})(1+U_{-2})},$$

from which it follows that

$$U_{-1}=S_{2p}\ \text{ and }\ U_{-2}=I_{2p-1}$$

because otherwise

$$S_1<\frac{\beta_0\beta_{2p}S_{2p}}{(1+S_{2p})(1+I_{2p-1})},$$

which contradicts (5.24.13). Similarly,

$$U_{-3} = S_{2p-2} \text{ and } U_{-4} = I_{2p-3}.$$

Also,

$$U_0 = \frac{\beta_{2p}\beta_{2p-1}U_{-2}}{(1+U_{-2})(1+U_{-3})} = \frac{\beta_{2p}\beta_{2p-1}I_{2p-1}}{(1+I_{2p-1})(1+S_{2p-2})} = I_0.$$

Hence,

$$S_1 = \frac{\beta_0 I_0}{1+S_{2p}}$$

or, equivalently,

$$\beta_0 I_0 = S_1(1+S_{2p}).$$

Similarly,

$$\beta_0 S_0 = I_1(1+I_{2p})$$

and so

$$I_0 = S_0, \quad I_1 = S_1, \quad \text{and} \quad I_{2p} = S_{2p}.$$

Inductively, it follows that

$$I_i = S_i, \quad i = 2, 3, \ldots, 2p-1.$$

Hence, the $2p+1$ subsequences $\{y_{(2p+1)\cdot n+i}\}$ for $i \in \{0, \ldots, 2p\}$ converge to finite limits. Set

$$l_i = \lim_{n\to\infty} y_{(2p+1)\cdot n+i}, \quad \text{for each } i \in \{0, \ldots, 2p\}.$$

By taking limits in Eq.(5.24.8) we obtain

$$l_i = \frac{\beta_{i-1}l_{i-1}}{1+l_{i-2}} \quad \text{for each } i \in \{0, \ldots, 2p\}.$$

and so, clearly, $\{y_n\}$ converges to the prime period-$(2p+1)$ solution

$$\ldots, l_0, \ldots, l_{2p}, \ldots .$$

The proof is complete. ∎

Open Problem 5.24.1 *Assume that $\{\beta_n\}$ is a convergent sequence of positive real numbers. Investigate the global character of solutions of the equation*

$$x_{n+1} = \frac{\beta_n x_n}{1+x_{n-1}}, \quad n = 0, 1, \ldots .$$

Open Problem 5.24.2 *Assume that β is a given real number. Determine the "good" set G of the equation*

$$x_{n+1} = \frac{\beta x_n}{1 + x_{n-1}}, \tag{5.24.20}$$

that is, the set of initial conditions

$$x_{-1}, x_0 \in \Re$$

such that the equation (5.24.20) is well defined for all $n \geq 0$. Determine the character of solutions of Eq.(5.24.20) for all initial conditions in the "good" set G.

5.25 Equation #25 : $x_{n+1} = \dfrac{\beta x_n}{A + Dx_{n-2}}$

This equation was investigated in [155] and [157]. See also Theorem 2.3.3, where we established that every solution of the equation in this special case is bounded. Eq.(#25) can be written in the normalized form

$$x_{n+1} = \frac{\beta x_n}{1 + x_{n-2}}, \quad n = 0, 1, \ldots \tag{5.25.1}$$

with positive parameter β and with arbitrary nonnegative initial conditions x_{-2}, x_{-1}, x_0.

Zero is always an equilibrium point of Eq.(5.25.1). From Theorems 5.23.2 and 5.23.4 it follows that the zero equilibrium of Eq.(5.25.1) is globally asymptotically stable when

$$\beta \leq 1 \tag{5.25.2}$$

and unstable when

$$\beta > 1\,. \tag{5.25.3}$$

Furthermore, when Eq.(5.25.3) holds, Eq.(5.25.1) has also the unique positive equilibrium point

$$\bar{x} = \beta - 1.$$

The characteristic equation of the linearized equation of Eq.(5.25.1) about the positive equilibrium, $\bar{x} = \beta - 1$, is

$$\lambda^3 - \lambda^2 + \frac{\beta - 1}{\beta} = 0.$$

From this and Theorem 1.2.3 it follows that $\bar{x} = \beta - 1$ is locally asymptotically stable when

$$1 < \beta < \frac{3+\sqrt{5}}{2} \tag{5.25.4}$$

and unstable when

$$\beta > \frac{3+\sqrt{5}}{2}.$$

When

$$\beta = \frac{3+\sqrt{5}}{2},$$

$\bar{x}$ is a nonhyperbolic equilibrium. In fact, in this case the three roots of the corresponding characteristic equation are:

$$\lambda_1 = \frac{1-\sqrt{5}}{2}, \quad \lambda_2 = \frac{1+\sqrt{5}-i\sqrt{10-2\sqrt{5}}}{4}, \quad \text{and } \lambda_3 = \frac{1+\sqrt{5}+i\sqrt{10-2\sqrt{5}}}{4}.$$

Conjecture 5.25.1 *Assume that (5.25.4) holds. Show that every positive solution of Eq.(5.25.1) converges to the positive equilibrium,* $\bar{x} = \beta - 1$.

Conjecture 5.25.2 *Assume that*

$$\beta > \frac{3+\sqrt{5}}{2}.$$

Show that Eq.(5.25.1) has solutions that do not converge to an equilibrium point or to a periodic solution.

Open Problem 5.25.1 *Assume that* $\{\beta_n\}$ *is a periodic sequence of positive real numbers with prime period* $k \geq 2$. *Investigate the global character of solutions of the equation*

$$x_{n+1} = \frac{\beta_n x_n}{1+x_{n-2}}, \quad n = 0, 1, \ldots . \tag{5.25.5}$$

Open Problem 5.25.2 *Assume that* $\{\beta_n\}$ *is a convergent sequence of positive real numbers. Investigate the global character of solutions of Eq.(5.25.5).*

Open Problem 5.25.3 *Assume that* β *is a given real number. Determine the "good" set G of the equation*

$$x_{n+1} = \frac{\beta x_n}{1+x_{n-2}}, \tag{5.25.6}$$

that is, the set of initial conditions

$$x_{-2}, x_{-1}, x_0 \in \Re$$

such that the equation (5.25.6) is well defined for all $n \geq 0$. *Determine the character of solutions of Eq.(5.25.6) for all initial conditions in the "good" set G.*

5.26 Equation #26 : $x_{n+1} = \dfrac{\beta x_n}{Bx_n + Cx_{n-1}}$

Eq.(#26) can be written in the normalized form

$$x_{n+1} = \frac{x_n}{Bx_n + x_{n-1}}, \quad n = 0, 1, \ldots \tag{5.26.1}$$

with positive parameter B and with arbitrary positive initial conditions x_{-1}, x_0.

Eq.(5.26.1) has the unique equilibrium point

$$\bar{x} = \frac{1}{B+1}.$$

The characteristic equation of the linearized equation of Eq.(5.26.1) about the equilibrium $\bar{x}$ is

$$\lambda^2 - \frac{1}{B+1}\lambda + \frac{1}{B+1} = 0.$$

It follows by Theorem 1.2.2 that $\bar{x}$ is locally asymptotically stable, as long as $B > 0$.

The change of variables

$$x_n = \frac{1}{B + y_n},$$

transforms Eq.(5.26.1) into the difference equation

$$y_{n+1} = \frac{B + y_n}{B + y_{n-1}}, \quad n = 0, 1, \ldots . \tag{5.26.2}$$

This equation, which is a special case of #66, was investigated in [158]. See also [157, p. 73] where it is shown that the equilibrium $\bar{y} = 1$ of Eq.(5.26.2) is globally asymptotically stable.

Here we give a new proof based on Theorem 1.6.7 that every solution of Eq.(5.26.2) converges to a finite limit.

Theorem 5.26.1 *Every solution of Eq.(5.26.2) converges to a finite limit.*

PROOF Let $\{y_n\}$ be a solution of Eq.(5.26.2). We will show that $\{y_n\}$ is bounded from above and from below by positive constants. Clearly, for $n \geq 1$,

$$y_{n+1} = \frac{B + y_n}{B + y_{n-1}} = \frac{B}{B + y_{n-1}} + \frac{1}{B + y_{n-2}} \leq 1 + \frac{1}{B}, \tag{5.26.3}$$

from which it follows that $\{y_n\}$ is bounded from above. In view of (5.26.3), we find that, for $n \geq 3$,

$$y_{n+1} \geq \frac{B}{B+1+\frac{1}{B}} + \frac{1}{B+1+\frac{1}{B}}$$

and so $\{y_n\}$ is also bounded from below.

We rewrite (5.26.2) in the following form:

$$B + y_{n+1} = B + \frac{B}{B+y_{n-1}} + \frac{1}{B+y_{n-2}}, \quad n = 1, 2, \ldots . \tag{5.26.4}$$

The change of variables

$$w_n = \frac{1}{B+y_n}$$

transforms Eq.(5.26.4) into the difference equation

$$w_{n+1} = \frac{1}{B + Bw_{n-1} + w_{n-2}}, \quad n = 1, 2, \ldots .$$

Clearly, the function

$$f(z_2, z_3) = \frac{1}{B + Bz_2 + z_3} \tag{5.26.5}$$

satisfies the Hypotheses of Theorem 1.6.7(a') from which the result follows. The proof is complete. ∎

Note that the conclusion of Theorem 5.26.1 also follows by employing Theorem 5.17.1 to (5.26.5).

Open Problem 5.26.1 *Let $\{\alpha_n\}$ be a periodic sequence with nonnegative values with period k. Investigate the global character of solutions of the equation*

$$x_{n+1} = \frac{\alpha_n + x_n}{\alpha_n + x_{n-1}}, \quad n = 0, 1, \ldots . \tag{5.26.6}$$

Open Problem 5.26.2 *Assume that $\{\alpha_n\}$ is a convergent sequence of positive real numbers. Investigate the global character of solutions of Eq.(5.26.6).*

Open Problem 5.26.3 *Let α be a real number. Investigate the "good" set G of the equation*

$$x_{n+1} = \frac{\alpha + x_n}{\alpha + x_{n-1}}, \tag{5.26.7}$$

with real initial conditions. That is, find all x_{-1}, $x_0 \in \Re$ such that the equation (5.26.7) is well defined for all $n \geq 0$. Investigate the character of solutions of Eq.(5.26.7) for all $x_{-1}, x_0 \in G$.

5.27 Equation #27 : $x_{n+1} = \dfrac{\beta x_n}{Bx_n + Dx_{n-2}}$

Eq.(#27) can be written in the normalized form

$$x_{n+1} = \frac{x_n}{Bx_n + x_{n-2}}, \quad n = 0, 1, \ldots \tag{5.27.1}$$

with positive parameter B and with arbitrary positive initial conditions x_{-2}, x_{-1}, x_0. The change of variables

$$x_n = \frac{1}{B + y_n}$$

transforms Eq.(5.27.1) into the difference equation

$$y_{n+1} = \frac{B + y_n}{B + y_{n-2}}, \quad n = 0, 1, \ldots . \tag{5.27.2}$$

By Theorem 2.3.3 it follows that every solution of Eq.(5.27.2) is bounded from above and clearly is also bounded from below by positive constants.

Eq.(5.27.2) has the unique equilibrium

$$\bar{y} = 1.$$

The characteristic equation of the linearized equation of Eq.(5.27.2) about the equilibrium $\bar{y}$ is

$$\lambda^3 - \frac{1}{B+1}\lambda^2 + \frac{1}{B+1} = 0.$$

From this and Theorem 1.2.3 it follows that the equilibrium $\bar{y}$ of Eq.(5.27.2) is locally asymptotically stable when

$$B > -1 + \sqrt{2} \tag{5.27.3}$$

and unstable when

$$B < -1 + \sqrt{2}.$$

Here we present a new proof about the global stability of the equilibrium $\bar{y}$ of Eq.(5.27.2) when

$$B \geq 1.$$

Theorem 5.27.1 *Assume that*

$$B \geq 1.$$

Then the equilibrium $\bar{y}$ of Eq.(5.27.2) is globally asymptotically stable.

PROOF Clearly, the equilibrium $\bar{y}$ of Eq.(5.27.2) is locally asymptotically stable. It suffices to show that the equilibrium of Eq.(5.27.2) is a global attractor of all solutions. When

$$B > 1,$$

the function

$$f(z_1, z_3) = \frac{B + z_1}{B + z_3}$$

satisfies the Hypotheses of Theorem 1.6.7 from which the result follows.

Also, when

$$B = 1,$$

the function

$$f(z_1, z_3) = \frac{1 + z_1}{1 + z_3}$$

satisfies the Hypotheses of Theorem 1.6.8 from which the result follows. ∎

Open Problem 5.27.1 *Let $\{\alpha_n\}$ be a periodic sequence of nonnegative real numbers with prime period $k \geq 2$. Investigate the global character of solutions of the equation*

$$x_{n+1} = \frac{\alpha_n + x_n}{\alpha_n + x_{n-2}}, \quad n = 0, 1, \dots . \tag{5.27.4}$$

Open Problem 5.27.2 *Assume that $\{\alpha_n\}$ is a convergent sequence of positive real numbers. Investigate the global character of solutions of Eq.(5.27.4).*

Open Problem 5.27.3 *Let α be a real number. Investigate the "good" set G of the equation*

$$x_{n+1} = \frac{\alpha + x_n}{\alpha + x_{n-2}}, \tag{5.27.5}$$

with real initial conditions. That is, find all x_{-2}, x_{-1}, $x_0 \in \Re$ such that the equation (5.27.5) is well defined for all $n \geq 0$. Investigate the character of solutions of Eq.(5.27.5) for all x_{-2}, x_{-1}, $x_0 \in G$.

Conjecture 5.27.1 *Assume that*

$$-1 + \sqrt{2} < B < 1.$$

Show that the equilibrium $\bar{y}$ of Eq.(5.27.2) is globally asymptotically stable.

Conjecture 5.27.2 *Assume that*

$$B < -1 + \sqrt{2}.$$

Show that Eq.(5.27.2) has solutions that do not converge to the equilibrium point $\bar{y}$ or to a periodic solution.

5.28 Equation #28 : $x_{n+1} = \dfrac{\beta x_n}{Cx_{n-1} + Dx_{n-2}}$

Eq.(#28) can be written in the normalized form

$$x_{n+1} = \frac{x_n}{Cx_{n-1} + x_{n-2}}, \quad n = 0, 1, \ldots \qquad (5.28.1)$$

with positive parameter C and with arbitrary positive initial conditions x_{-2}, x_{-1}, x_0.

The only equilibrium of Eq.(5.28.1) is

$$\bar{x} = \frac{1}{C+1}.$$

The characteristic equation of the linearized equation of Eq.(5.28.1) about the equilibrium $\bar{x}$ is

$$\lambda^3 - \lambda^2 + \frac{C}{C+1}\lambda + \frac{1}{C+1} = 0.$$

From this and Theorem 1.2.3 it follows that the positive equilibrium $\bar{x}$ is unstable for all positive values of the parameter C.

Conjecture 5.28.1 *Show that for all positive values of the parameter C, Eq.(5.28.1) possesses unbounded solutions.*

Conjecture 5.28.2 *Show that every bounded solution of Eq.(5.28.1) converges to the equilibrium $\bar{x}$.*

Open Problem 5.28.1 *Investigate the behavior of bounded solutions of Eq.(5.28.1).*

Open Problem 5.28.2 *Let $\{C_n\}$ be a periodic sequence of nonnegative real numbers with prime period $k \geq 2$. Investigate the global character of solutions of*

$$x_{n+1} = \frac{x_n}{C_n x_{n-1} + x_{n-2}}, \quad n = 0, 1, \ldots .$$

Open Problem 5.28.3 *Assume that C is a given real number. Determine the "good" set G of the equation*

$$x_{n+1} = \frac{x_n}{Cx_{n-1} + x_{n-2}}, \qquad (5.28.2)$$

that is, the set of initial conditions

$$x_{-2}, x_{-1}, x_0 \in \Re$$

such that the equation (5.28.2) is well defined for all $n \geq 0$. Determine the character of solutions of Eq.(5.28.2) for all initial conditions in the "good" set G.

5.29 Equation #29 : $x_{n+1} = \dfrac{\gamma x_{n-1}}{A + Bx_n}$

Eq.(#29) can be written in the normalized form

$$x_{n+1} = \frac{x_{n-1}}{A + x_n}, \quad n = 0, 1, \ldots \tag{5.29.1}$$

with positive parameter A and with arbitrary nonnegative initial conditions x_{-1}, x_0.

Eq.(5.29.1) possesses a period-two trichotomy depending on whether

$$A > 1, \quad A = 1, \quad \text{or} \quad A < 1.$$

This result is a special case of a more general period-two trichotomy result presented in Theorem 4.2.1.

The existence of solutions of Eq.(5.29.1) that converge to zero when

$$A = 1$$

and other similar results have been established, among other places, in [133], [143], [146], [148], [149], [226], [227], and [233].

When

$$A < 1,$$

it follows from Theorem 4.2.2 that every positive and bounded solution of Eq.(5.29.1) converges to the positive equilibrium, $\bar{x} = 1 - A$.

The following amazing result gives a set of initial conditions through which the solutions of Eq.(5.29.1), when $A = 1$, converge to a prime period-two solution.

Theorem 5.29.1 *Let $\{x_n\}_{n=-1}^{\infty}$ be a solution of*

$$x_{n+1} = \frac{x_{n-1}}{1 + x_n}, \quad n = 0, 1, \ldots \tag{5.29.2}$$

such that for some $N \geq 0$,

$$x_N \geq x_{N-1}.$$

Then

$$x_{2n+N-1} \downarrow 0 \text{ and } x_{2n+N} \to \text{to a positive limit.}$$

PROOF Note that if

$$x_N = x_{N-1}$$

then

$$x_{N+2} = \frac{x_N}{1 + x_{N+1}} = \frac{x_N}{1 + \frac{x_{N-1}}{1+x_N}} = \frac{x_N(1 + x_N)}{1 + 2x_N}$$

and

$$x_{N+1} = \frac{x_{N-1}}{1 + x_N} = \frac{x_N}{1 + x_N},$$

from which it follows that

$$x_{N+2} > x_{N+1}.$$

So without loss of generality we may assume that

$$x_N > x_{N-1}.$$

Now observe that for any n sufficiently large,

$$x_N - x_{N+2} = x_N - \frac{x_N}{1 + x_{N+1}} = \frac{x_N x_{N+1}}{1 + x_{N+1}}$$

$$= \frac{x_{N-1} - x_{N+1}}{x_{N+1} + 1} < \frac{x_{N-1} - x_{N+1}}{x_{2n+N+1} + 1}.$$

Similarly,

$$x_{N+2} - x_{N+4} = \frac{x_{N+1} - x_{N+3}}{x_{N+3} + 1} < \frac{x_{N+1} - x_{N+3}}{x_{2n+N+1} + 1}$$

$$\dots$$

$$x_{N+2n} - x_{N+2n+2} = \frac{x_{N+2n-1} - x_{N+2n+1}}{x_{N+2n+1} + 1}$$

and by summing up we find:

$$x_N - x_{2n+N+2} < \frac{x_{N-1} - x_{2n+N+1}}{x_{2n+N+1} + 1}$$

and so

$$x_{2n+N+2} > x_N + \frac{x_{2n+N+1} - x_{N-1}}{x_{2n+N+1} + 1}. \tag{5.29.3}$$

Also,

$$x_{N+1} - x_{N+3} = \frac{x_N - x_{N+2}}{x_{N+2} + 1}$$

$$x_{N+3} - x_{N+5} = \frac{x_{N+2} - x_{N+4}}{x_{N+4} + 1} > \frac{x_{N+2} - x_{N+4}}{x_{N+2} + 1}$$

$$\cdots$$

$$x_{2n+N+1} - x_{2n+N+3} = \frac{x_{2n+N} - x_{2n+N+2}}{x_{2n+N+2} + 1} > \frac{x_{2n+N} - x_{2n+N+2}}{x_{N+2} + 1}$$

and by summing up we find:

$$x_{N+1} - x_{2n+N+3} > \frac{x_N - x_{2n+N+2}}{x_{N+2} + 1}$$

and so

$$x_{2n+N+3} < x_{N+1} + \frac{x_{2n+N+2} - x_N}{x_{N+2} + 1}. \tag{5.29.4}$$

Now we claim that

$$1 + x_{N+2} < \frac{x_N}{x_{N+1}}.$$

This follows easily after we express all terms in terms of x_N and x_{N-1} and use the assumption that $x_N > x_{N-1}$.

Now assume for the sake of contradiction that

$$x_{2n+N+1} \to x \in (0, \infty).$$

Then, clearly,

$$x_{2n+N+2} \downarrow 0.$$

and (5.29.4) yields:

$$0 < x \leq x_{N+1} - \frac{x_N}{x_{N+2} + 1} < 0,$$

which is a contradiction. Hence,

$$x_{2n+N+1} \downarrow 0$$

and so from (5.29.3) we see that

$$\lim_{n \to \infty} x_{2n+N+2} \geq x_N - x_{N-1} > 0.$$

The proof is complete. ∎

The following amazing result is a corollary of Theorem 5.29.1.

Corollary 5.29.1 *A positive solution $\{x_n\}_{n=-1}^{\infty}$ of Eq.(5.29.2) converges to zero if and only if*

$$x_{n-1} > x_n, \quad \textit{for all } n \geq 0.$$

Open Problem 5.29.1 *Obtain "easily verifiable" conditions which determine the set of all positive initial conditions for which the solutions of Eq.(5.29.1) do exactly one of the following:*

(*i*) *converge to a prime period-two solution, when* $A = 1$

(*ii*) *converge to the positive equilibrium* $\bar{x}$, *when* $A < 1$

(*ii*) *are unbounded, when* $A < 1$

5.30 Equation #30 : $x_{n+1} = \dfrac{\gamma x_{n-1}}{A + C x_{n-1}}$

This is a Riccati-type equation. For the character of solutions of Riccati equations see Section 5.65.

5.31 Equation #31 : $x_{n+1} = \dfrac{\gamma x_{n-1}}{A + D x_{n-2}}$

This equation was investigated in [17] and [70]. Eq.(#31) possesses a period-two trichotomy depending on whether

$$\gamma < A, \quad \gamma = A, \quad \text{or} \quad \gamma > A.$$

This result is a special case of a more general period-two trichotomy result presented in Theorem 4.3.1.

Open Problem 5.31.1 *Assume that* A *is a given real number. Determine the "good" set* G *of the equation*

$$x_{n+1} = \frac{x_{n-1}}{A + x_{n-2}}, \tag{5.31.1}$$

that is, the set of initial conditions

$$x_{-2}, x_{-1}, x_0 \in \Re$$

such that the equation (5.31.1) is well defined for all $n \geq 0$. *Determine the character of solutions of Eq.(5.31.1) for all initial conditions in the "good" set* G.

Conjecture 5.31.1 *Assume that*

$$A < 1.$$

Show that every positive and bounded solution of the equation

$$x_{n+1} = \frac{x_{n-1}}{A + x_{n-2}}, \quad n = 0, 1, \ldots$$

converges to $1 - A$.

5.32 Equation #32 : $x_{n+1} = \dfrac{\gamma x_{n-1}}{Bx_n + Cx_{n-1}}$

Eq.(#32) can be written in the normalized form

$$x_{n+1} = \frac{x_{n-1}}{Bx_n + x_{n-1}}, \quad n = 0, 1, \ldots \tag{5.32.1}$$

with positive parameter B and with arbitrary nonnegative initial conditions x_{-1}, x_0 such that the denominator is always positive.

From Theorem 1.6.6 it follows that for all positive values of B, every solution of Eq.(5.32.1) converges to a (not necessarily prime) period-two solution.

The only equilibrium of Eq.(5.32.1) is

$$\bar{x} = \frac{1}{B+1}.$$

In addition, Eq.(5.32.1) has period-two solutions. When

$$B \neq 1,$$

Eq.(5.32.1) has the unique prime period-two solution

$$\ldots, 0, 1, 0, 1, \ldots \tag{5.32.2}$$

and when

$$B = 1$$

Eq.(5.32.1) has infinitely many prime period-two solutions of the form

$$\ldots, x, 1 - x, x, 1 - x, \ldots$$

with

$$x \in [0,1] \text{ and } x \neq \frac{1}{2}.$$

When

$$B < 1,$$

every positive solution of Eq.(5.32.1) converges to the positive equilibrium $\bar{x}$. This is because the change of variables, $x_n = \frac{1}{y_n}$, transforms Eq.(5.32.1) to the equation

$$y_{n+1} = \frac{y_n + By_{n-1}}{y_n}, \quad n = 0, 1, \ldots,$$

for which we know from Theorem 4.2.1 that every positive solution converges to the positive equilibrium.

When

$$B = 1,$$

every positive solution of Eq.(5.32.1) converges to a (not necessarily prime) period-two solution. This is because the change of variables, $x_n = \frac{1}{y_n}$, transforms Eq.(5.32.1) to the equation

$$y_{n+1} = \frac{y_n + y_{n-1}}{y_n}, \quad n = 0, 1, \ldots,$$

for which we know from Theorem 4.2.1 that every positive solution converges to a (not necessarily prime) period-two solution.

If a solution $\{x_n\}$ of Eq.(5.32.1) is not positive, then for all $n \geq 1$, we have

$$x_{2n} = 0 \text{ and } x_{2n-1} = 1 \text{ or } x_{2n} = 1 \text{ and } x_{2n-1} = 0$$

and so it is eventually equal with the period-two solution (5.32.2).

Open Problem 5.32.1 *Let $\{B_n\}$ be a periodic sequence of nonnegative real numbers with prime period $k \geq 2$. Determine the global character of solutions of the difference equation*

$$x_{n+1} = \frac{x_{n-1}}{B_n x_n + x_{n-1}}, \quad n = 0, 1, \ldots .$$

Extend and generalize.

Open Problem 5.32.2 *Let $\{B_n\}$ be a convergent sequence of positive real numbers. Investigate the character of solutions of the difference equation*

$$x_{n+1} = \frac{x_{n-1}}{B_n x_n + x_{n-1}}, \quad n = 0, 1, \ldots .$$

Extend and generalize.

Open Problem 5.32.3 *Assume that B is a given real number. Determine the "good" set G of the equation*

$$x_{n+1} = \frac{x_{n-1}}{Bx_n + x_{n-1}}, \tag{5.32.3}$$

that is, the set of all initial conditions

$$x_{-1}, x_0 \in \Re$$

such that the equation (5.32.3) is well defined for all $n \geq 0$. Determine the character of solutions of Eq.(5.32.3) for all initial conditions in the "good" set G.

5.33 Equation #33 : $x_{n+1} = \dfrac{\gamma x_{n-1}}{Bx_n + Dx_{n-2}}$

Eq.(#33) has unbounded solutions. This equation is part of a period-two trichotomy presented in Theorem 4.3.1.

Conjecture 5.33.1 *Show that every bounded solution of the rational equation*

$$x_{n+1} = \frac{\gamma x_{n-1}}{x_n + x_{n-2}}, \quad n = 0, 1, \ldots$$

converges to $\frac{\gamma}{2}$.

Open Problem 5.33.1 *Determine the set of all initial conditions x_{-2}, x_{-1}, x_0 so that every bounded solution of the rational equation*

$$x_{n+1} = \frac{\gamma x_{n-1}}{x_n + x_{n-2}}, \quad n = 0, 1, \ldots$$

converges to $\frac{\gamma}{2}$. Extend and generalize.

5.34 Equation #34 : $x_{n+1} = \dfrac{\gamma x_{n-1}}{Cx_{n-1} + Dx_{n-2}}$

Eq.(#34) can be written in the normalized form

$$x_{n+1} = \frac{x_{n-1}}{x_{n-1} + Dx_{n-2}}, \quad n = 0, 1, \ldots \tag{5.34.1}$$

with positive parameter D and with arbitrary nonnegative initial conditions x_{-2}, x_{-1}, x_0 such that the denominator is always positive.

The only equilibrium of Eq.(5.34.1) is

$$\bar{x} = \frac{1}{D+1}.$$

In addition, Eq.(5.34.1) has period-two solutions. When

$$D \neq 1,$$

Eq.(5.34.1) has the unique prime period-two solution

$$\ldots, 0, 1, 0, 1, \ldots \tag{5.34.2}$$

and when

$$D = 1,$$

Eq.(5.34.1) has infinitely many prime period-two solutions of the form

$$\ldots, x, 1-x, x, 1-x, \ldots$$

with

$$x \in [0,1] \text{ and } x \neq \frac{1}{2}.$$

When

$$D < 1,$$

every positive solution of Eq.(5.34.1) converges to the positive equilibrium $\bar{x}$. This is because the change of variables, $x_n = \frac{1}{y_n}$, transforms Eq.(5.34.1) to the equation

$$y_{n+1} = \frac{y_{n-2} + Dy_{n-1}}{y_{n-2}}, \quad n = 0, 1, \ldots,$$

for which we know from Theorem 4.3.1 that every positive solution converges to the positive equilibrium.

When

$$D = 1,$$

every positive solution of Eq.(5.34.1) converges to a (not necessarily prime) period-two solution. This is because the change of variables, $x_n = \frac{1}{y_n}$, transforms Eq.(5.34.1) to the equation

$$y_{n+1} = \frac{y_{n-2} + y_{n-1}}{y_{n-2}}, \quad n = 0, 1, \ldots,$$

for which we know from Theorem 4.3.1 that every positive solution converges to a (not necessarily prime) period-two solution.

If a solution $\{x_n\}$ of Eq.(5.34.1) is not positive, then for all $n \geq 1$, we have

$$x_{2n} = 0 \text{ and } x_{2n-1} = 1 \text{ or } x_{2n} = 1 \text{ and } x_{2n-1} = 0$$

and so it is eventually equal with the period-two solution (5.34.2).

Conjecture 5.34.1 *Assume that*

$$D > 1.$$

Show that every positive solution of Eq.(5.34.1) converges to a (not necessarily prime) period-two solution.

Open Problem 5.34.1 *Let $\{D_n\}$ be a periodic sequence of nonnegative real numbers with prime period $k \geq 2$. Determine the global character of solutions of the difference equation*

$$x_{n+1} = \frac{x_{n-1}}{x_{n-1} + D_n x_{n-2}}, \quad n = 0, 1, \ldots .$$

Extend and generalize.

Open Problem 5.34.2 *Let $\{D_n\}$ be a convergent sequence of positive real numbers. Investigate the character of solutions of the difference equation*

$$x_{n+1} = \frac{x_{n-1}}{x_{n-1} + D_n x_{n-2}}, \quad n = 0, 1, \ldots .$$

Extend and generalize.

Open Problem 5.34.3 *Assume that D is a given real number. Determine the "good" set G of the equation*

$$x_{n+1} = \frac{x_{n-1}}{x_{n-1} + D x_{n-2}}, \qquad (5.34.3)$$

that is, the set of initial conditions

$$x_{-2}, x_{-1}, x_0 \in \Re$$

such that the equation (5.34.3) is well defined for all $n \geq 0$. Determine the character of solutions of Eq.(5.34.3) for all initial conditions in the "good" set G.

5.35 Equation #35 : $x_{n+1} = \dfrac{\delta x_{n-2}}{A + Bx_n}$

Eq.(#35) possesses a period-three trichotomy depending on whether

$$\delta < A, \quad \delta = A, \quad \text{or} \quad \delta > A.$$

This result is a special case of a more general period-three trichotomy result presented in Theorem 4.4.1.

Open Problem 5.35.1 *(a) Determine all positive initial conditions x_{-2}, x_{-1}, and x_0 through which the solutions of the equation*

$$x_{n+1} = \frac{x_{n-2}}{A + x_n}, \quad n = 0, 1, \ldots \tag{5.35.1}$$

converge to zero.

(b) Determine all positive initial conditions x_{-2}, x_{-1}, and x_0 through which the solutions of Eq.(5.35.1) converge to the prime period-three solution

$$\ldots, 0, 0, 1, 0, 0, 1, \ldots .$$

(c) Determine the limit of solutions of Eq.(5.35.1) with initial conditions

$$x_{-2} = x_{-1} = x_0 = 1.$$

An unbounded solution of Eq.(5.35.1) when $A < 1$ is

$$0, 0, 1, 0, 0, \frac{1}{A}, 0, 0, \frac{1}{A^2}, \ldots .$$

Conjecture 5.35.1 *Assume $A < 1$. Show that Eq.(5.35.1) has positive unbounded solutions and that every positive and bounded solution converges to $1 - A$.*

5.36 Equation #36 : $x_{n+1} = \dfrac{\delta x_{n-2}}{A + Cx_{n-1}}$

Eq.(#36) possesses a period-three trichotomy depending on whether

$$\delta < A, \quad \delta = A, \quad \text{or} \quad \delta > A.$$

This result is a special case of a more general period-three trichotomy result presented in Theorem 4.4.1.

Open Problem 5.36.1 *Assume that $A < 1$. Determine the set of all initial conditions x_{-2}, x_{-1}, x_0 so that every positive and bounded solution of the rational equation*

$$x_{n+1} = \frac{x_{n-2}}{A + x_{n-1}}, \quad n = 0, 1, \ldots$$

converges to $1 - A$. Extend and generalize.

5.37 Equation #37 : $x_{n+1} = \dfrac{\delta x_{n-2}}{A + Dx_{n-2}}$

This is a Riccati-type equation. For the character of solutions of Riccati equations see Section 5.65.

5.38 Equation #38 : $x_{n+1} = \dfrac{\delta x_{n-2}}{Bx_n + Cx_{n-1}}$

Eq.(#38) has unbounded solutions. This equation is part of a period-three trichotomy presented in Theorem 4.4.1.

Open Problem 5.38.1 *Assume that B is a given real number. Determine the "good" set G of the equation*

$$x_{n+1} = \frac{x_{n-2}}{Bx_n + x_{n-1}}, \tag{5.38.1}$$

that is, the set of initial conditions

$$x_{-2}, x_{-1}, x_0 \in \Re$$

such that the equation (5.38.1) is well defined for all $n \geq 0$. Determine the character of solutions of Eq.(5.38.1) for all initial conditions in the "good" set G.

5.39 Equation #39 : $x_{n+1} = \dfrac{\delta x_{n-2}}{Bx_n + Dx_{n-2}}$

This equation was investigated in [60]. Eq.(#39) can be written in the normalized form

$$x_{n+1} = \frac{x_{n-2}}{Bx_n + x_{n-2}} \quad n = 0, 1, \ldots \tag{5.39.1}$$

with positive parameter B and with arbitrary nonnegative initial conditions x_{-2}, x_{-1}, x_0 such that the denominator is always positive.

The only equilibrium of Eq.(5.39.1) is

$$\bar{x} = \frac{1}{B+1}.$$

The characteristic equation of the linearized equation of Eq.(5.39.1) about the equilibrium $\bar{x}$ is

$$\lambda^3 + \frac{B}{B+1}\lambda^2 - \frac{B}{B+1} = 0.$$

From this and Theorem 1.2.3 it follows that the positive equilibrium $\bar{x}$ of Eq.(5.39.1) is locally asymptotically stable when

$$B < 1 + \sqrt{2} \tag{5.39.2}$$

and unstable when

$$B > 1 + \sqrt{2}.$$

When

$$B = 1 + \sqrt{2},$$

$\bar{x}$ is a nonhyperbolic equilibrium. In fact, the eigenvalues of the corresponding characteristic equation are:

$$\lambda_1 = \frac{\sqrt{2}}{2}, \quad \lambda_2 = -\frac{\sqrt{2}}{2} + i\frac{\sqrt{2}}{2}, \quad \text{and } \lambda_3 = -\frac{\sqrt{2}}{2} - i\frac{\sqrt{2}}{2}.$$

Note that λ_2 and λ_3 are eighth roots of unity .

In addition, Eq.(5.39.1) has period-three solutions. When

$$B < 1,$$

Eq.(5.39.1) has the unique prime period-three solution

$$\ldots, 0, 1, 1 - B, 0, 1, 1 - B, \ldots . \tag{5.39.3}$$

In Theorem 5.136.1 we will establish that when

$$B < 1,$$

every positive solution of Eq.(5.39.1) converges to its equilibrium point $\bar{x}$. See also [60].

When

$$B = 1,$$

every positive solution of Eq.(5.39.1) converges to its equilibrium point $\bar{x}$. This is because the change of variables, $x_n = \frac{1}{y_n}$, transforms Eq.(5.39.1) to the equation

$$y_{n+1} = 1 + \frac{y_{n-2}}{y_n}, \quad n = 0, 1, \ldots,$$

for which we know from Theorem 5.58.2 that every positive solution converges to the positive equilibrium.

If a solution of Eq.(5.39.1) is not positive it must be of the form

$$x_{3n-2} = 0 \text{ and } x_{3n-1}, x_{3n} > 0, \quad n = 0, 1, \ldots, \tag{5.39.4}$$

or

$$x_{3n-1} = 0 \text{ and } x_{3n-2}, x_{3n} > 0, \quad n = 0, 1, \ldots, \tag{5.39.5}$$

or

$$x_{3n} = 0 \text{ and } x_{3n-2}, x_{3n-1} > 0, \quad n = 0, 1, \ldots . \tag{5.39.6}$$

Assume that (5.39.4) holds. Then, for $n \geq 1$,

$$x_{3n-1} = \frac{x_{3n-4}}{Bx_{3n-2} + x_{3n-4}} = 1$$

and

$$x_{3n} = \frac{x_{3n-3}}{Bx_{3n-1} + x_{3n-3}} = \frac{x_{3n-3}}{B + x_{3n-3}}.$$

Hence, for $B < 1$ and $n \geq 1$,

$$x_{3n-2} = 0, \quad x_{3n-1} = 1, \quad \text{and } x_{3n} \to 1 - B$$

and, for $B \geq 1$ and $n \geq 1$,

$$x_{3n-2} = 0, \quad x_{3n-1} = 1, \quad \text{and } x_{3n} \to 0.$$

When (5.39.5) or (5.39.6) holds, the results are similar.

Remark 5.39.1 *Note that, when $B \geq 1$, every nonpositive solution of Eq.(5.39.1) converges to the three-cycle*

$$\ldots, 0, 1, 0, 0, 1, 0 \ldots,$$

which is not a solution of Eq.(5.39.1).

Conjecture 5.39.1 *Assume that (5.39.2) holds. Show that every positive solution of Eq.(5.39.1) converges to its equilibrium point $\bar{x}$.*

Conjecture 5.39.2 *Assume that*

$$B > 123.$$

Show that every positive solution of Eq.(5.39.1) converges to a periodic solution of period 19.

Conjecture 5.39.3 *Show that Eq.(5.39.1) has solutions that do not converge to the equilibrium point $\bar{x}$ or to a periodic solution.*

5.40 Equation #40 : $x_{n+1} = \dfrac{\delta x_{n-2}}{Cx_{n-1} + Dx_{n-2}}$

This equation was investigated in [60]. Eq.(#40) can be written in the normalized form

$$x_{n+1} = \frac{x_{n-2}}{Cx_{n-1} + x_{n-2}} \quad n = 0, 1, \ldots \tag{5.40.1}$$

with positive parameter C and with arbitrary nonnegative initial conditions x_{-2}, x_{-1}, x_0 such that the denominator is always positive.

The only equilibrium of Eq.(5.40.1) is

$$\bar{x} = \frac{1}{C+1}.$$

The characteristic equation of the linearized equation of Eq.(5.40.1) about the equilibrium $\bar{x}$ is

$$\lambda^3 + \frac{C}{C+1}\lambda - \frac{C}{C+1} = 0.$$

From this and Theorem 1.2.3 it follows that the equilibrium $\bar{x}$ of Eq.(5.40.1) is locally asymptotically stable when

$$C < \frac{1+\sqrt{5}}{2} \tag{5.40.2}$$

and unstable when

$$C > \frac{1+\sqrt{5}}{2}.$$

When

$$C = \frac{1+\sqrt{5}}{2},$$

$\bar{x}$ is a nonhyperbolic equilibrium. In fact, the eigenvalues of the corresponding characteristic equation are:

$$\lambda_1 = \frac{-1+\sqrt{5}}{2}, \quad \lambda_2 = \frac{1-\sqrt{5}-i\sqrt{10+2\sqrt{5}}}{4}, \quad \text{and} \quad \lambda_3 = \frac{1-\sqrt{5}+i\sqrt{10+2\sqrt{5}}}{4}.$$

In addition, Eq.(5.40.1) has period-three solutions. When

$$C < 1,$$

Eq.(5.40.1) has the unique prime period-three solution

$$\ldots, 0, 1-C, 1, 0, 1-C, 1, \ldots . \tag{5.40.3}$$

In Theorem 5.136.1 we will establish that when

$$C < 1,$$

every positive solution of Eq.(5.40.1) converges to its equilibrium point $\bar{x}$. See also [60].

When

$$C = 1$$

every positive solution of Eq.(5.40.1) converges to its equilibrium point $\bar{x}$. This is because the change of variables, $x_n = \frac{1}{y_n}$, transforms Eq.(5.40.1) to the equation

$$y_{n+1} = 1 + \frac{y_{n-2}}{y_{n-1}}, \quad n = 0, 1, \ldots,$$

for which we know from Theorem 5.63.2 that every positive solution converges to the positive equilibrium.

If a solution of Eq.(5.40.1) is not positive it must be of the form

$$x_{3n-2} = 0 \text{ and } x_{3n-1}, x_{3n} > 0, \quad n = 0, 1, \ldots, \tag{5.40.4}$$

or

$$x_{3n-1} = 0 \text{ and } x_{3n-2}, x_{3n} > 0, \quad n = 0, 1, \ldots, \tag{5.40.5}$$

or

$$x_{3n} = 0 \text{ and } x_{3n-2}, x_{3n-1} > 0, \quad n = 0, 1, \ldots . \tag{5.40.6}$$

Assume that (5.40.6) holds. Then, for $n \geq 0$,

$$x_{3n+2} = \frac{x_{3n-1}}{Cx_{3n} + x_{3n-1}} = 1$$

and

$$x_{3n+4} = \frac{x_{3n+1}}{Cx_{3n+2} + x_{3n+1}} = \frac{x_{3n+1}}{C + x_{3n+1}}.$$

Hence, for $C < 1$ and $n \geq 0$,

$$x_{3n+2} = 1, \quad x_{3n+3} = 0, \quad \text{and} \quad x_{3n+4} \to 1 - C$$

and, for $C \geq 1$ and $n \geq 0$,

$$x_{3n+2} = 1, \quad x_{3n+3} = 0, \quad \text{and} \quad x_{3n+4} \to 0.$$

When (5.40.4) or (5.40.5) holds, the results are similar.

Remark 5.40.1 *Note that, when $C \geq 1$, every nonpositive solution of Eq.(5.40.1) converges to the three-cycle*

$$\ldots, 1, 0, 0, 1, 0, 0 \ldots,$$

which is not a solution of Eq.(5.40.1).

Conjecture 5.40.1 *Assume that (5.40.2) holds. Show that every positive solution of Eq.(5.40.1) converges to its equilibrium point $\bar{x}$.*

Conjecture 5.40.2 *Assume that*

$$C > 8.$$

Show that every positive solution of Eq.(5.40.1) converges to a periodic solution of period 13.

Conjecture 5.40.3 *Show that Eq.(5.40.1) has solutions that do not converge to the equilibrium point $\bar{x}$ or to a periodic solution.*

5.41 Equation #41 : $x_{n+1} = \dfrac{\alpha + \beta x_n}{A}$

The equation in this special case is linear.

5.42 Equation #42 : $x_{n+1} = \dfrac{\alpha + \beta x_n}{Bx_n}$

This is a Riccati equation. By Remark 5.65.1 every solution of this equation converges to its equilibrium point. For the character of solutions of Riccati equations with real parameters and real initial conditions see Section 5.65.

Open Problem 5.42.1 *Let β be a given complex number. Determine the "good" set G of the equation*

$$x_{n+1} = \beta + \frac{1}{x_n}, \tag{5.42.1}$$

that is, the set of initial conditions x_0 in the complex plane such that Eq.(5.42.1) is well defined for all $n \geq 0$. Determine the character of solutions of Eq.(5.42.1) for all initial conditions $x_0 \in G$. Extend and generalize.

5.43 Equation #43 : $x_{n+1} = \dfrac{\alpha + \beta x_n}{Cx_{n-1}}$

This is the well-known **Lyness's equation**, which has been investigated by many authors. See [20], [21], [22], [116], [124], [157], [158], [174], [189], [193], [198], [199], [215], [216], and [237].

Eq.(#43) can be written in the normalized form

$$x_{n+1} = \frac{\alpha + x_n}{x_{n-1}}, \quad n = 0, 1, ... \tag{5.43.1}$$

with positive parameter α and with arbitrary positive initial conditions x_{-1}, x_0.

The only equilibrium of Eq.(5.43.1) is

$$\bar{x} = \frac{1 + \sqrt{1 + 4\alpha}}{2}.$$

The linearized equation of Eq.(5.43.1) about $\bar{x}$ is

$$y_{n+1} - \frac{2}{1 + \sqrt{1 + 4\alpha}} y_n + y_{n-1} = 0, \quad n = 0, 1, \ldots$$

with characteristic equation

$$\lambda^2 - \frac{2}{1+\sqrt{1+4\alpha}}\lambda + 1 = 0\,.$$

The characteristic roots are

$$\lambda_1 = \frac{1+i\sqrt{1+4\alpha+2\sqrt{1+4\alpha}}}{1+\sqrt{1+4\alpha}} \text{ and } \lambda_2 = \frac{1-i\sqrt{1+4\alpha+2\sqrt{1+4\alpha}}}{1+\sqrt{1+4\alpha}}.$$

For every $\alpha > 0$, it holds

$$|\lambda_1| = |\lambda_2| = 1$$

and so $\bar{x}$ is a nonhyperbolic equilibrium point.

In the special case where

$$\alpha = 1,$$

the characteristic roots of the corresponding characteristic equation are fifth roots of unity. In this special case Eq.(5.43.1) becomes

$$x_{n+1} = \frac{1+x_n}{x_{n-1}}, \quad n = 0, 1, \ldots\,. \tag{5.43.2}$$

Eq.(5.43.2) was discovered by Lyness in 1942 while working on a Number Theory problem. It is a fascinating fact that every solution of Eq.(5.43.2) is periodic with period five. In fact the solution of Eq.(5.43.2) with initial conditions x_{-1}, x_0 is the five-cycle

$$x_{-1}, x_0, \frac{1+x_0}{x_{-1}}, \frac{1+x_{-1}+x_0}{x_0 x_{-1}}, \frac{1+x_{-1}}{x_0}, x_{-1}, x_0, \ldots\,.$$

Eq.(5.43.1) possesses the invariant

$$I_n = (\alpha + x_{n-1} + x_n)(1+\frac{1}{x_{n-1}})(1+\frac{1}{x_n}) = (\alpha + x_{-1} + x_0)(1+\frac{1}{x_{-1}})(1+\frac{1}{x_0}).$$

From this it follows that every solution of Eq.(5.43.1) is bounded from above and from below by positive constants.

In [116] it was shown that no nontrivial solution of Eq.(5.43.1) has a limit.

Furthermore, in [193] it was shown using KAM theory that the positive equilibrium $\bar{x}$ of Eq.(5.43.1) is stable but not asymptotically stable. The same result was also established in [174] by using a Lyapunov function.

Conjecture 5.43.1 *Assume that*

$$\alpha \neq 1.$$

Show that Eq.(5.43.1) has solutions that do not converge to a periodic solution of prime period $p \geq 2$.

It is interesting to note that periodicity may destroy the boundedness of solutions of Lyness's equation as the following example shows.

Example 5.43.1 *(see [62]) Let*

$$\alpha_n = \begin{cases} 1, & \text{if } n = 5k+i \text{ with } i \in \{0,1,2,3\} \\ \alpha, & \text{if } n = 5k+4 \end{cases}, k = 0, 1, \ldots$$

with $\alpha \in (0, \infty)$. Then the solution of the difference equation

$$x_{n+1} = \frac{\alpha_n + x_n}{x_{n-1}}, \quad n = 0, 1, \ldots \tag{5.43.3}$$

with initial conditions

$$x_{-1} = x_0 = 1 \tag{5.43.4}$$

*is **unbounded**, if and only*

$$\alpha \neq 1.$$

Indeed, one can see, by induction, that the solution of the IVP (5.43.3) and (5.43.4), for $n \geq 0$, is given by

$$\begin{aligned} x_{5n+1} &= \left(\frac{\alpha+1}{2}\right)^n + 1 \\ x_{5n+2} &= 2 \cdot \left(\frac{2}{\alpha+1}\right)^n + 1 \\ x_{5n+3} &= 2 \cdot \left(\frac{2}{\alpha+1}\right)^n \\ x_{5n+4} &= 1 \\ x_{5n+5} &= \left(\frac{\alpha+1}{2}\right)^{n+1}. \end{aligned}$$

Thus, in Eq.(5.43.1), **periodicity may destroy the boundedness of its solutions.**

Open Problem 5.43.1 *Assume that $\{\alpha_n\}$ is a nonnegative periodic sequence with prime period $p \geq 2$.*

(*a*) *Obtain necessary and sufficient conditions on p such that every positive solution of Eq.(5.43.3) is bounded.*

(*b*) *Obtain conditions on p and $\alpha_0, \ldots, \alpha_{p-1}$ such that Eq.(5.43.3) has unbounded solutions in some region of the parameters and for some initial conditions.*

Open Problem 5.43.2 *Assume that $\{\alpha_n\}$ is a positive periodic sequence with prime period two. Determine the global asymptotic character and the periodic nature of solutions of Eq.(5.43.3). Extend and generalize.*

5.44 Equation #44 : $x_{n+1} = \dfrac{\alpha + \beta x_n}{D x_{n-2}}$

Eq.(#44) can be written in normalized form

$$x_{n+1} = \frac{\alpha + x_n}{x_{n-2}}, \quad n = 0, 1, \ldots \tag{5.44.1}$$

with positive parameter α and with arbitrary positive initial conditions x_{-2}, x_{-1}, x_0.

The only equilibrium of Eq.(5.44.1) is

$$\bar{x} = \frac{1 + \sqrt{1 + 4\alpha}}{2}.$$

The linearized equation of Eq.(5.44.1) about $\bar{x}$ is

$$y_{n+1} - \frac{2}{1 + \sqrt{1 + 4\alpha}} y_n + y_{n-2} = 0, \quad n = 0, 1, \ldots$$

with characteristic equation

$$\lambda^3 - \frac{2}{1 + \sqrt{1 + 4\alpha}} \lambda^2 + 1 = 0.$$

From this and Theorem 1.2.3 it follows that the equilibrium $\bar{x}$ is unstable.

Open Problem 5.44.1 *(i) Prove that for all positive values of the parameter α, Eq.(5.44.1) possesses unbounded solutions.*

(ii) Investigate the global behavior of bounded solutions of Eq.(5.44.1).

5.45 Equation #45 : $x_{n+1} = \dfrac{\alpha + \gamma x_{n-1}}{A}$

The equation in this special case is linear.

5.46 Equation #46 : $x_{n+1} = \dfrac{\alpha + \gamma x_{n-1}}{B x_n}$

Eq.(#46) has unbounded solutions. This equation is part of a period-two trichotomy presented in Theorem 4.2.1. See also Section 3.1. Eq.(#46) can be written in the normalized form

$$x_{n+1} = \frac{\alpha + x_{n-1}}{x_n}, \quad n = 0, 1, \ldots \tag{5.46.1}$$

with positive parameter α and with arbitrary positive initial conditions x_{-1}, x_0. By Theorem 4.3.1 this equation has unbounded solutions and by Theorem 4.2.2 every bounded solution converges to the equilibrium. Also, Theorem 1.6.6 applies and describes the monotonic character of solutions of Eq.(5.46.1).

Open Problem 5.46.1 *Obtain easily "verifiable" conditions that determine the set of all positive initial conditions for which the solutions of Eq.(5.46.1) do exactly one of the following:*

(i) converge to the equilibrium

(ii) are unbounded

Open Problem 5.46.2 *Let* $k \in (0,1)$. *Then by Theorem 3.1.1 every solution of Eq.(5.46.1) with*

$$x_0 \in (0,1) \text{ and } x_{-1} > \frac{\alpha + 1}{k}$$

is such that

$$\lim_{n\to\infty} x_{2n+1} = \infty \text{ and } \lim_{n\to\infty} x_{2n} = 0.$$

Investigate the global character of solutions of Eq.(5.46.1) with $x_0 \in [1, \bar{x}]$ *and* $x_{-1} \in \left[\bar{x}, \frac{\alpha+1}{k}\right]$ *where* $\bar{x}$ *denotes the equilibrium of Eq.(5.46.1).*

Open Problem 5.46.3 *Could the global character of solutions of Eq.(5.46.1) be predicted from the characteristic roots of the linearized equation*

$$\lambda^2 + \lambda - \frac{1}{\bar{x}} = 0$$

about the positive equilibrium $\bar{x}$*? Extend and generalize.*

Open Problem 5.46.4 *Assume that* α *is a given real number. Determine the "good" set* G *of the equation*

$$x_{n+1} = \frac{\alpha + x_{n-1}}{x_n}, \tag{5.46.2}$$

that is, the set of initial conditions

$$x_{-1}, x_0 \in \Re$$

such that the equation (5.46.2) is well-defined for all $n \geq 0$. *Determine the character of solutions of Eq.(5.46.2) for all initial conditions in the "good" set* G.

5.47 Equation #47 : $x_{n+1} = \frac{\alpha + \gamma x_{n-1}}{C x_{n-1}}$

This is a Riccati-type equation. For the character of solutions of Riccati equations see Section 5.65.

5.48 Equation #48 : $x_{n+1} = \frac{\alpha + \gamma x_{n-1}}{D x_{n-2}}$

Eq.(#48) has unbounded solutions. This equation is part of a period-two trichotomy presented in Theorem 4.3.1.

Conjecture 5.48.1 *Show that every bounded solution of Eq.(#48) converges to the equilibrium.*

Open Problem 5.48.1 *Assume that* α *is a given real number. Determine the "good" set* G *of the equation*

$$x_{n+1} = \frac{\alpha + x_{n-1}}{x_{n-2}}, \tag{5.48.1}$$

that is, the set of initial conditions

$$x_{-2}, x_{-1}, x_0 \in \Re$$

such that the equation (5.48.1) is well defined for all $n \geq 0$. Determine the character of solutions of Eq.(5.48.1) for all initial conditions in the "good" set G.

5.49 Equation #49 : $x_{n+1} = \dfrac{\alpha + \delta x_{n-2}}{A}$

The equation in this special case is linear.

5.50 Equation #50 : $x_{n+1} = \dfrac{\alpha + \delta x_{n-2}}{B x_n}$

This equation was investigated in [54], [59], and [151]. Eq.(#50) can be written in the normalized form

$$x_{n+1} = \frac{\alpha + x_{n-2}}{x_n}, \quad n = 0, 1, \ldots \tag{5.50.1}$$

with positive parameter α and with arbitrary positive initial conditions, x_{-2}, x_{-1}, x_0.

Eq.(5.50.1) possesses a period-five trichotomy. For details see Section 4.6.

Open Problem 5.50.1 *Let $\{x_n\}$ be the solution of the equation*

$$x_{n+1} = \frac{1 + x_{n-2}}{x_n}, \quad n = 0, 1, \ldots \tag{5.50.2}$$

with initial conditions

$$x_{-2} = x_{-1} = x_0 = 1.$$

Determine the period-five solution of Eq.(5.50.2) to which $\{x_n\}$ converges.

Open Problem 5.50.2 *Let $\{\phi_n\}_{n=-2}^{\infty}$ be a given five-cycle of Eq.(5.50.2). Determine the set of all initial conditions x_{-2}, x_{-1}, x_0 such that the solution $\{x_n\}_{n=-2}^{\infty}$ of Eq.(5.50.2) converges to $\{\phi_n\}_{n=-2}^{\infty}$.*

Open Problem 5.50.3 *Determine the set of all initial conditions x_{-2}, x_{-1}, x_0 for which the solutions of Eq.(5.50.1), with $\alpha < 1$, are bounded.*

5.51 Equation #51 : $x_{n+1} = \dfrac{\alpha + \delta x_{n-2}}{C x_{n-1}}$

Eq.(#51) can be written in the normalized form

$$x_{n+1} = \frac{\alpha + x_{n-2}}{x_{n-1}}, \quad n = 0, 1, \ldots \tag{5.51.1}$$

with positive parameter α and with arbitrary positive initial conditions, x_{-2}, x_{-1}, x_0. It was shown in [47] that Eq.(5.51.1) has unbounded solutions.

Open Problem 5.51.1 *Determine the set of all initial conditions x_{-2}, x_{-1}, x_0 for which the solutions of Eq.(5.51.1) are bounded.*

Conjecture 5.51.1 *Show that the solution of the equation*

$$x_{n+1} = \frac{1 + x_{n-2}}{x_{n-1}}, \quad n = 0, 1, \ldots$$

with

$$x_{-2} = x_{-1} = x_0 = 1$$

is unbounded.

5.52 Equation #52 : $x_{n+1} = \dfrac{\alpha + \delta x_{n-2}}{D x_{n-2}}$

This is a Riccati-type equation. For the character of solutions of Riccati equations see Section 5.65.

Open Problem 5.52.1 *Assume that α is a given real number. Determine the "good" set G of the equation*

$$x_{n+1} = \frac{\alpha + x_{n-2}}{x_{n-2}}, \tag{5.52.1}$$

that is, the set of initial conditions

$$x_{-2}, x_{-1}, x_0 \in \Re$$

such that the equation (5.52.1) is well defined for all $n \geq 0$. Determine the character of solutions of Eq.(5.52.1) for all initial conditions in the "good" set G.

5.53 Equation #53 : $x_{n+1} = \dfrac{\beta x_n + \gamma x_{n-1}}{A}$

The equation in this special case is linear.

5.54 Equation #54 : $x_{n+1} = \dfrac{\beta x_n + \gamma x_{n-1}}{B x_n}$

This equation can be written in the normalized form

$$x_{n+1} = \beta + \frac{x_{n-1}}{x_n}, \quad n = 0, 1, \ldots \tag{5.54.1}$$

with positive parameter β and with arbitrary positive initial conditions x_{-1}, x_0.

Eq.(5.54.1) was investigated in [16] where it was shown that it possesses a period-two trichotomy depending on whether

$$\beta < 1, \quad \beta = 1, \quad \text{or} \quad \beta > 1.$$

This was the very first period-two trichotomy result discovered for rational equations.

This result is a special case of a period-two trichotomy result presented in Theorem 4.2.1.

When

$$\beta < 1,$$

it follows from Theorem 4.2.2 that every bounded solution of Eq.(5.54.1) converges to the equilibrium, $\bar{x} = \beta + 1$.

For some work on the forbidden set of Eq.(5.54.1), see [53] and [55].

Open Problem 5.54.1 *Assume that $\beta \in (0, 1)$ and let $k \in (0, 1 - \beta)$. Then by Theorem 3.1.1 every solution of Eq.(5.54.1) with*

$$x_0 \in (0, 1) \ \text{ and } \ x_{-1} > \frac{1}{k}$$

is such that

$$\lim_{n\to\infty} x_{2n+1} = \infty \ \text{ and } \ \lim_{n\to\infty} x_{2n} = \beta.$$

Investigate the global character of solutions of Eq.(5.54.1) with

$$x_0 \in [1, 1+\beta] \ \text{ and } \ x_{-1} \in \left[1 + \beta, \frac{1}{k}\right].$$

5.55 Equation #55 : $x_{n+1} = \dfrac{\beta x_n + \gamma x_{n-1}}{C x_{n-1}}$

Eq.(#55) can be written in the normalized form

$$x_{n+1} = \gamma + \frac{x_n}{x_{n-1}}, \quad n = 0, 1, \ldots \tag{5.55.1}$$

with positive parameter γ and with positive initial conditions x_{-1}, x_0.

The change of variables

$$x_n = y_n + \gamma,$$

transforms Eq.(5.55.1) into the difference equation

$$y_{n+1} = \frac{\gamma + y_n}{\gamma + y_{n-1}}, \quad n = 0, 1, \ldots . \tag{5.55.2}$$

This equation, which is a special case of #66, was investigated in [158]. In Theorem 5.26.1 we established that the equilibrium of Eq.(5.55.2), $\bar{y} = 1$, is globally asymptotically stable. For another proof of the global asymptotic stability of the equilibrium of Eq.(5.55.2), see [157, p. 73].

Open Problem 5.55.1 *Assume that γ is a given real number. Determine the "good" set G of the equation*

$$x_{n+1} = \gamma + \frac{x_n}{x_{n-1}}, \tag{5.55.3}$$

that is, the set of initial conditions

$$x_{-1}, x_0 \in \Re$$

such that the equation (5.55.3) is well defined for all $n \geq 0$. Determine the character of solutions of Eq.(5.55.3) for all initial conditions in the "good" set G. Extend and generalize.

5.56 Equation #56 : $x_{n+1} = \dfrac{\beta x_n + \gamma x_{n-1}}{D x_{n-2}}$

This equation is a special case of a more general equation that will be investigated in Section 5.120.

Conjecture 5.56.1 *Show that Eq.(#56) has bounded solutions that do not converge to the equilibrium point or to a periodic solution.*

Open Problem 5.56.1 *Determine the "good" set G of the equation*

$$x_{n+1} = \frac{x_n + x_{n-1}}{x_{n-2}}, \tag{5.56.1}$$

that is, the set of initial conditions

$$x_{-2}, x_{-1}, x_0 \in \Re$$

such that the equation (5.56.1) is well defined for all $n \geq 0$. Determine the character of solutions of Eq.(5.56.1) for all initial conditions in the "good" set G. Extend and generalize.

Open Problem 5.56.2 *Let $\{\beta_n\}$ be a nonnegative periodic sequence with prime period $k \geq 2$. Determine the global character of solutions of the difference equation*

$$x_{n+1} = \frac{\beta_n x_n + x_{n-1}}{x_{n-2}}, \quad n = 0, 1, \ldots .$$

Extend and generalize.

5.57 Equation #57 : $x_{n+1} = \dfrac{\beta x_n + \delta x_{n-2}}{A}$

The equation in this special case is linear.

5.58 Equation #58 : $x_{n+1} = \dfrac{\beta x_n + \delta x_{n-2}}{B x_n}$

This equation was investigated in [49] and [87]. See also Section 2.4 where we established that every solution of the equation is bounded. This equation can be written in the normalized form

$$x_{n+1} = \beta + \frac{x_{n-2}}{x_n}, \quad n = 0, 1, \ldots \tag{5.58.1}$$

with positive parameter β and with arbitrary positive initial conditions x_{-2}, x_{-1}, x_0.

The only equilibrium of Eq.(5.58.1) is

$$\bar{x} = \beta + 1.$$

The characteristic equation of the linearized equation of Eq.(5.58.1) about the equilibrium $\bar{x}$ is

$$\lambda^3 + \frac{1}{\beta+1}\lambda^2 - \frac{1}{\beta+1} = 0. \tag{5.58.2}$$

From this and Theorem 1.2.3 it follows that the equilibrium $\bar{x}$ of Eq.(5.58.2) is locally asymptotically stable when

$$\beta > -1 + \sqrt{2} \tag{5.58.3}$$

and unstable when

$$\beta < -1 + \sqrt{2}. \tag{5.58.4}$$

When

$$\beta = -1 + \sqrt{2},$$

two of the characteristic roots of Eq.(5.58.2) are eighth roots of unity and the third root lies within the interval $(0, 1)$.

For equation (5.58.1) and for any equation of the form

$$x_{n+1} = f(x_n, x_{n-2}), \quad n = 0, 1, \ldots \tag{5.58.5}$$

with a unique equilibrium point $\bar{x}$ and with the function $f(u, v)$ decreasing in the first argument u and increasing in the second argument v, the following result holds.

Lemma 5.58.1 *Assume Eq.(5.58.5) has a unique equilibrium point $\bar{x}$ and that $f(u,v)$ decreases in u and increases in v. Then for any solution $\{x_n\}_{n=-2}^{\infty}$ of Eq.(5.58.5) one of the following three statements is true:*

(i) $x_n \geq \bar{x}$, *for* $n \geq -2$.

(ii) $x_n < \bar{x}$, *for* $n \geq -2$.

(iii) There exists an $N \geq -2$ such that

either

$$x_n \geq \bar{x}, \text{ for } -2 \leq n \leq N$$

or

$$x_n < \bar{x}, \text{ for } -2 \leq n \leq N$$

and for $n > N$ the solution is strictly oscillatory about $\bar{x}$ with semicycles of length one or two. Furthermore, after the first semicycle, every semicycle of length two is followed by a semicycle of length one.

PROOF Assume that neither (*i*) nor (*ii*) holds. Then there exists some $N \geq -2$ such that
either

$$x_n \geq \bar{x}, \text{ for } -2 \leq n \leq N \text{ and } x_{N+1} < \bar{x} \tag{5.58.6}$$

or

$$x_n < \bar{x}, \text{ for } -2 \leq n \leq N \text{ and } x_{N+1} \geq \bar{x}. \tag{5.58.7}$$

We will assume that (5.58.6) holds. The case where (5.58.7) holds is similar and will be omitted. Now it suffices to show that if

$$x_{N+2} < \bar{x},$$

then

$$x_{N+3} > \bar{x}.$$

Indeed, this is true because

$$x_{N+3} = f(x_{N+2}, x_N) > f(\bar{x}, \bar{x}) = \bar{x}. \tag{5.58.8}$$

To show that a semicycle of length two is followed by a semicycle of length one, assume that for some $N \geq 0$,

$$x_N \geq \bar{x}, \quad x_{N+1} < \bar{x} \text{ and } x_{N+2} < \bar{x}.$$

The other case is similar and will be omitted. Then by (5.58.6),

$$x_{N+3} > \bar{x}$$

and

$$x_{N+4} = f(x_{N+3}, x_{N+1}) < f(\bar{x}, \bar{x}) = \bar{x}$$

and the proof is complete. ∎

The following additional properties can be established for the solutions of Eq.(5.58.1). See [87].

Lemma 5.58.2 *(a) No solution of Eq.(5.58.1) has semicycles that are all eventually of length one.*

(b) After the second semicycle, the maximum term in a positive semi-cycle of length two is always less than or equal to the last term in the previous positive semi-cycle.

(c) After the second semicycle, the minimum term in a negative semi-cycle of length two is always greater than or equal to the last term in the previous negative semicycle.

PROOF Assume without loss of generality that there exists some $N \geq 0$ such that

$$x_{2n-1} < \beta + 1 \leq x_{2n}, \quad \text{for } n \geq N.$$

Then

$$x_{2n+2} = \beta + \frac{x_{2n-1}}{x_{2n+1}} \geq \beta + 1$$

and so

$$x_{2n-1} \geq x_{2n+1} \geq \beta.$$

Therefore, the subsequence of the odd terms decreases to a positive limit. Also,

$$x_{2n+3} = \beta + \frac{x_{2n}}{x_{2n+2}} < \beta + 1$$

and so

$$\beta + 1 \leq x_{2n} < x_{2n+2}. \tag{5.58.9}$$

Therefore, the subsequence of the even terms is strictly increasing and

$$x_{2n+2} = \beta + \frac{x_{2n-1}}{x_{2n+1}} \to \beta + 1,$$

which contradicts (5.58.9).
(b) Assume that for some $N \geq 0$,

$$x_N \geq \beta + 1 \text{ and } x_{N+1} \geq \beta + 1.$$

Then, clearly,

$$x_{N-1} < \beta + 1 < x_{N-2}$$

because a semicycle of length two must be preceded by a semicycle of length one. Also,

$$\beta + 1 \leq x_{N+1} = \beta + \frac{x_{N-2}}{x_N}$$

and so

$$x_N \leq x_{N-2}.$$

Furthermore,

$$x_{N+1} = \beta + \frac{x_{N-2}}{x_N} < \beta + \frac{x_{N-2}}{\beta + 1} \leq x_{N-2}$$

and the proof is complete.
(c) The proof is similar to the proof in (b) and will be omitted. ∎

The following theorem establishes the existence of nonoscillatory solutions of Eq.(5.58.1). See [226].

Theorem 5.58.1 *Eq.(5.58.1) has infinitely many nonoscillatory solutions, which decrease to the equilibrium,* $\bar{x} = \beta + 1$.

PROOF For all $n \geq 0$, we define

$$A_n = \{(x_{-2}, x_{-1}, x_0) \in (\beta, \infty)^3 : \ \beta + 1 \leq x_{n+1} \leq x_n \leq x_{n-1} \leq x_{n-2}\}$$

We claim that for all $n \geq 0$

$$\emptyset \neq A_{n+1} \subseteq A_n.$$

It suffices to show that

$$x_{n+1} \leq x_n \leq x_{n-1} \leq x_{n-2}$$

when

$$x_{n+2} \leq x_{n+1} \leq x_n \leq x_{n-1}.$$

Indeed,

$$x_{n-1} = (x_{n+2} - \beta)x_{n+1} \leq (x_{n+1} - \beta)x_n = x_{n-2}.$$

Also,

$$A_n \neq \emptyset$$

because, for all $n \geq 0$,

$$(\beta + 1, \beta + 1, \beta + 1) \in A_n.$$

Set

$$F(x, y, z) = (y, z, \beta + \frac{x}{z})$$

with $x, y, z \in [\beta + 1, \infty)$. Clearly, the function F is continuous and one to one. We claim that

$$A_n = F(A_{n+1}).$$

Let $(y_{-2}, y_{-1}, y_0) \in F(A_{n+1})$. Then

$$(y_{-2}, y_{-1}, y_0) = F(x_{-2}, x_{-1}, x_0) = (x_{-1}, x_0, x_1)$$

and

$$F^{(n)}(y_{-2}, y_{-1}, y_0) = (y_{n-2}, y_{n-1}, y_n) = F^{(n+1)}(x_{-2}, x_{-1}, x_0) = (x_{n-1}, x_n, x_{n+1})$$

and

$$F^{(n+1)}(y_{-2}, y_{-1}, y_0) = (y_{n-1}, y_n, y_{n+1})$$
$$= F^{(n+2)}(x_{-2}, x_{-1}, x_0) = (x_n, x_{n+1}, x_{n+2}) \, .$$

Hence,

$$\beta + 1 \leq y_{n+1} \leq y_n \leq y_{n-1} \leq y_{n-2},$$

which implies that $(y_{-2}, y_{-1}, y_0) \in A_n$.

On the other hand, assume that $(x_{-2}, x_{-1}, x_0) \in A_n$. Set

$$y_{-2} = (x_0 - \beta)x_{-1}, \quad y_{-1} = x_{-2}, \ \text{ and } \ y_0 = x_{-1}.$$

Then

$$(x_{-2}, x_{-1}, x_0) = (y_{-1}, y_0, \beta + \frac{y_{-2}}{y_0}) = F(y_{-2}, y_{-1}, y_0) = (y_{-1}, y_0, y_1).$$

Then

$$(y_{n-1}, y_n, y_{n+1}) = F^{(n+1)}(y_{-2}, y_{-1}, y_0) = F^{(n)}(x_{-2}, x_{-1}, x_0) = (x_{n-2}, x_{n-1}, x_n)$$

and

$$(y_n, y_{n+1}, y_{n+2}) = F^{(n+2)}(y_{-2}, y_{-1}, y_0) = F^{(n+1)}(x_{-2}, x_{-1}, x_0) = (x_{n-1}, x_n, x_{n+1}) .$$

Hence,

$$\beta + 1 \leq y_{n+2} \leq y_{n+1} \leq y_n \leq y_{n-1} .$$

From this it follows that $(y_{-2}, y_{-1}, y_0) \in A_{n+1}$. Also $(x_{-2}, x_{-1}, x_0) = F(y_{-2}, y_{-1}, y_0) \in F(A_{n+1})$. The proof of our claim is complete. Since F is invertible it also holds

$$A_{n+1} = F^{-1}(A_n), \text{ for all } n \geq 0.$$

Set

$$\Omega = \bigcap_{n=0}^{\infty} A_n.$$

Then

$$\Omega = \bigcap_{n=0}^{\infty} F^{(-n)}(A_0).$$

Note that A_0 is a nonempty, closed, connected, and unbounded subset of R^3. Also, A_1 is a nonempty and closed subset of R^3 and since $F^{-1}(A_0) = A_1$, it follows that A_1 is connected and unbounded. Inductively, it follows that each one of the $F^{(-n)}(A_0) = A_n$'s is a nonempty, closed, connected, and unbounded subset of R^3. Furthermore, the family $\{A_n\}_{n=0}^{\infty}$ satisfies the finite intersection property because

$$\bigcap_{k=0}^{n} A_k = A_n \neq \emptyset.$$

Then, clearly, Ω is a nonempty, closed, connected, and unbounded subset of R^3. By choosing the initial conditions x_{-2}, x_{-1}, x_0 in Ω, the solution $\{x_n\}$ that is generated satisfies for all $n \geq 0$,

$$\beta + 1 \leq x_{n+1} \leq x_n$$

and so converges to $\beta + 1$. The proof is complete. ∎

Next we would like to show that when

$$\beta > 1,$$

5.59 Equation #59 : $x_{n+1} = \dfrac{\beta x_n + \delta x_{n-2}}{Cx_{n-1}}$

This equation is part of a more general equation for which we conjecture that has a period-four trichotomy. See Section 5.123.

Open Problem 5.59.1 *Assume that β is a given real number. Determine the "good" set G of the equation*

$$x_{n+1} = \frac{\beta x_n + x_{n-2}}{x_{n-1}}, \tag{5.59.1}$$

that is, the set of initial conditions

$$x_{-2}, x_{-1}, x_0 \in \Re$$

such that the equation (5.59.1) is well defined for all $n \geq 0$. Determine the character of solutions of Eq.(5.59.1) for all initial conditions in the "good" set G.

Conjecture 5.59.1 *Assume that*

$$\beta > 1.$$

Show that every solution of Eq.(5.59.1) converges to the equilibrium.

Conjecture 5.59.2 *Assume that*

$$\beta < 1.$$

Show that every bounded solution of Eq.(5.59.1) converges to the equilibrium.

5.60 Equation #60 : $x_{n+1} = \dfrac{\beta x_n + \delta x_{n-2}}{Dx_{n-2}}$

Eq.(#60) can be written in the normalized form

$$x_{n+1} = \delta + \frac{x_n}{x_{n-2}}, \quad n = 0, 1, \ldots \tag{5.60.1}$$

with positive parameter δ and with positive initial conditions x_{-2}, x_{-1}, x_0.

The change of variables

$$x_n = y_n + \delta$$

transforms Eq.(5.60.1) into the difference equation

$$y_{n+1} = \frac{\delta + y_n}{\delta + y_{n-2}}, \quad n = 0, 1, \ldots . \tag{5.60.2}$$

For this equation, which is a special case of #67, see Section 5.67.

Open Problem 5.60.1 *Assume that δ is a given real number. Determine the "good" set G of the equation*

$$x_{n+1} = \delta + \frac{x_n}{x_{n-2}}, \tag{5.60.3}$$

that is, the set of initial conditions

$$x_{-2}, x_{-1}, x_0 \in \Re$$

such that the equation (5.60.3) is well defined for all $n \geq 0$. Determine the character of solutions of Eq.(5.60.3) for all initial conditions in the "good" set G. Extend and generalize.

Conjecture 5.60.1 *Assume that*

$$\delta > -1 + \sqrt{2}.$$

Show that every solution of Eq.(5.60.1) has a finite limit.

Conjecture 5.60.2 *Assume that*

$$\delta < -1 + \sqrt{2}.$$

Show that Eq.(5.60.1) has solutions that do not converge to the equilibrium point or to a periodic solution.

Open Problem 5.60.2 *Let $\{\delta_n\}$ be a periodic sequence of nonnegative real numbers with prime period $k \geq 2$. Determine the global character of solutions of the difference equation*

$$x_{n+1} = \delta_n + \frac{x_n}{x_{n-2}}, \quad n = 0, 1, \ldots .$$

Extend and generalize.

5.61 Equation #61 : $x_{n+1} = \dfrac{\gamma x_{n-1} + \delta x_{n-2}}{A}$

The equation in this special case is linear.

5.62 Equation #62 : $x_{n+1} = \frac{\gamma x_{n-1} + \delta x_{n-2}}{B x_n}$

This equation was investigated in [76]. See also [67]. Eq.(#62) can be written in the normalized form

$$x_{n+1} = \frac{\gamma x_{n-1} + x_{n-2}}{x_n}, \quad n = 0, 1, \ldots \tag{5.62.1}$$

with positive parameter γ and with arbitrary positive initial conditions x_{-2}, x_{-1}, x_0.

It follows from the work in [76] (see also Theorem 3.1.1), that when

$$\gamma > 1,$$

Eq.(5.62.1) possesses unbounded solutions. It was also shown in [76] that when

$$\gamma = 1,$$

the subsequences

$$\{x_{2n}\}_{n=0}^{\infty} \text{ and } \{x_{2n+1}\}_{n=-1}^{\infty}$$

of every solution are eventually monotonic and one of them may be unbounded.

The only equilibrium of Eq.(5.62.1) is

$$\bar{x} = \gamma + 1.$$

The characteristic equation of the linearized equation of Eq.(5.62.1) about the equilibrium $\bar{x}$ is

$$\lambda^3 + \lambda^2 - \frac{\gamma}{\gamma+1}\lambda - \frac{1}{\gamma+1} = 0. \tag{5.62.2}$$

From this and Theorem 1.2.3 it follows that the equilibrium $\bar{x}$ of Eq.(5.62.1) is locally asymptotically stable when

$$\frac{\sqrt{3}-1}{2} < \gamma < 1 \tag{5.62.3}$$

and unstable when

$$\gamma < \frac{\sqrt{3}-1}{2}.$$

When

$$\gamma = \frac{\sqrt{3}-1}{2},$$

one solution of Eq.(5.62.2) lies within the interval $(-1, 0)$ and the other two solutions are 12th roots of unity.

Conjecture 5.62.1 *Assume that*

$$\gamma > 1.$$

Show that every bounded solution of Eq.(5.62.1) converges to the equilibrium $\bar{x}$.

Open Problem 5.62.1 *Assume that*

$$\gamma > 1.$$

(a) Determine the set of all initial conditions through which the solutions of Eq.(5.62.1) converge to the equilibrium $\bar{x}$.

(b) Determine the set of all initial conditions through which the solutions of Eq.(5.62.1) are unbounded.

Open Problem 5.62.2 *Assume that*

$$\gamma = 1.$$

(a) Determine the set of all initial conditions through which the solutions of Eq.(5.62.1) converge to the equilibrium $\bar{x}$.

(b) Determine the set of all initial conditions through which the solutions of Eq.(5.62.1) converge to a prime period-two solution.

(c) Determine the set of all initial conditions through which the solutions of Eq.(5.62.1) are unbounded.

Conjecture 5.62.2 *Assume that (5.62.3) holds. Show that the equilibrium* $\bar{x}$ *of Eq.(5.62.1) is globally asymptotically stable.*

Conjecture 5.62.3 *Assume that*

$$\gamma < \frac{\sqrt{3}-1}{2}.$$

Show that Eq.(5.62.1) has bounded solutions that do not converge to the equilibrium point $\bar{x}$ *or to a periodic solution.*

5.63 Equation #63 : $x_{n+1} = \dfrac{\gamma x_{n-1} + \delta x_{n-2}}{C x_{n-1}}$

This equation was investigated in [49]. See also Section 2.6 where we established that every solution of the equation is bounded. This equation can be written in the normalized form

$$x_{n+1} = \gamma + \frac{x_{n-2}}{x_{n-1}}, \quad n = 0, 1, ... \tag{5.63.1}$$

with positive parameter γ and with arbitrary positive initial conditions x_{-2}, x_{-1}, x_0.

The only equilibrium of Eq.(5.63.1) is

$$\bar{x} = \gamma + 1 \,.$$

The characteristic equation of the linearized equation of Eq.(5.63.1) about the equilibrium $\bar{x}$ is

$$\lambda^3 + \frac{1}{\gamma+1}\lambda - \frac{1}{\gamma+1} = 0. \tag{5.63.2}$$

From this and Theorem 1.2.3 it follows that the equilibrium $\bar{x}$ is locally asymptotically stable when

$$\gamma > \frac{-1+\sqrt{5}}{2} \tag{5.63.3}$$

and unstable when

$$\gamma < \frac{-1+\sqrt{5}}{2}. \tag{5.63.4}$$

For equation (5.63.1) and for any equation of the form

$$x_{n+1} = f(x_{n-1}, x_{n-2}), \quad n = 0, 1, \ldots \tag{5.63.5}$$

with a unique equilibrium point $\bar{x}$ and with the function $f(u, v)$ decreasing in the first argument u and increasing in the second argument v, the following result holds.

Lemma 5.63.1 *Assume Eq.(5.63.5) has a unique equilibrium point $\bar{x}$ and that $f(u, v)$ decreases in u and increases in v. Then for any solution $\{x_n\}_{n=-2}^{\infty}$ of Eq.(5.63.5) one of the following three statements is true:*

(i) $x_n \geq \bar{x}$, *for* $n \geq -2$.

(ii) $x_n < \bar{x}$, *for* $n \geq -2$.

(iii) There exists an $N \geq -2$ *such that*

either

$$x_n \geq \bar{x}, \; for \; -2 \leq n \leq N$$

or

$$x_n < \bar{x}, \; for \; -2 \leq n \leq N$$

and where for $n > N$, the solution is strictly oscillatory about $\bar{x}$ with semicycles of length one or two. Furthermore, no solution of Eq.(5.63.5) has semicycles that are all eventually of length one.

PROOF Assume that neither (*i*) nor (*ii*) holds. Then there exists some $N \geq -2$ such that
either

$$x_n \geq \bar{x}, \text{ for } -2 \leq n \leq N \text{ and } x_{N+1} < \bar{x} \tag{5.63.6}$$

or

$$x_n < \bar{x}, \text{ for } -2 \leq n \leq N \text{ and } x_{N+1} \geq \bar{x}. \tag{5.63.7}$$

We will assume that (5.63.6) holds. The case where (5.63.7) holds is similar and will be omitted. Now it suffices to show that if

$$x_{N+2} < \bar{x},$$

then

$$x_{N+3} > \bar{x}.$$

Indeed, this is true because

$$x_{N+3} = f(x_{N+1}, x_N) > f(\bar{x}, \bar{x}) = \bar{x}. \tag{5.63.8}$$

To show that no solution of Eq.(5.63.5) has semicycles that are all eventually of length one, assume that for some $N \geq 0$,

$$x_N > \bar{x}, \ x_{N+1} < \bar{x} \text{ and } x_{N+2} > \bar{x}.$$

The other case is similar and will be omitted. Then

$$x_{N+3} = f(x_{N+1}, x_N) > f(\bar{x}, \bar{x}) = \bar{x}$$

and the proof is complete. ∎

The following additional properties can be established for the solutions of Eq.(5.63.1).

Lemma 5.63.2 *(a) When the maximum in a positive semicycle of length two is in the first term, then the negative semicycle that follows has length one, and when the maximum in a positive semicycle of length two is in the second term, then the negative semicycle that follows has length two.*

(b) When the minimum in a negative semicycle of length two is in the first term, then the positive semicycle that follows has length one, and when the minimum in a negative semicycle of length two is in the second term, then the positive semicycle that follows has length two.

PROOF (a) Assume that for some $N \geq 0$,

$$x_{N-2} \geq x_{N-1} \geq \gamma + 1.$$

The other case is similar and will be omitted. Clearly,

$$x_N < \gamma + 1$$

because a semicycle can have at most two terms. Also

$$x_{N+1} = \gamma + \frac{x_{N-2}}{x_{N-1}} \geq \gamma + 1.$$

and the proof is complete.
(b) The proof is similar to the proof in (a) and will be omitted. ∎

The following theorem establishes the existence of nonoscillatory solutions of Eq.(5.63.1).

Theorem 5.63.1 *Eq.(5.63.1) has infinitely many nonoscillatory solutions that decrease to equilibrium* $\bar{x} = \gamma + 1$.

PROOF For all $n \geq 0$, we define

$$A_n = \{(x_{-2}, x_{-1}, x_0) \in \Re^3 : \ \gamma + 1 \leq x_{n+1} \leq x_n \leq x_{n-1} \leq x_{n-2}\}.$$

We claim that for all $n \geq 0$

$$\emptyset \neq A_{n+1} \subseteq A_n.$$

It suffices to show that

$$x_{n+1} \leq x_n \leq x_{n-1} \leq x_{n-2}$$

when

$$x_{n+2} \leq x_{n+1} \leq x_n \leq x_{n-1}.$$

Indeed,

$$x_{n-1} = (x_{n+2} - \gamma)x_n \leq (x_{n+1} - \gamma)x_{n-1} = x_{n-2}.$$

Also,

$$A_n \neq \emptyset$$

because, for all $n \geq 0$,

$$(\gamma + 1, \gamma + 1, \gamma + 1) \in A_n.$$

Set

$$F(x, y, z) = (y, z, \gamma + \frac{x}{y})$$

with $x, y, z \in [\gamma+1, \infty)$. Clearly, the function F is continuous and one to one. We claim that

$$A_n = F(A_{n+1}).$$

Let $(y_{-2}, y_{-1}, y_0) \in F(A_{n+1})$. Then

$$(y_{-2}, y_{-1}, y_0) = F(x_{-2}, x_{-1}, x_0) = (x_{-1}, x_0, x_1),$$

and

$$F^{(n)}(y_{-2}, y_{-1}, y_0) = (y_{n-2}, y_{n-1}, y_n) = F^{(n+1)}(x_{-2}, x_{-1}, x_0) = (x_{n-1}, x_n, x_{n+1}),$$

and

$$F^{(n+1)}(y_{-2}, y_{-1}, y_0) = (y_{n-1}, y_n, y_{n+1}) = F^{(n+2)}(x_{-2}, x_{-1}, x_0) = (x_n, x_{n+1}, x_{n+2}).$$

Hence,

$$\gamma + 1 \leq y_{n+1} \leq y_n \leq y_{n-1} \leq y_{n-2},$$

which implies that $(y_{-2}, y_{-1}, y_0) \in A_n$.

On the other hand, assume that $(x_{-2}, x_{-1}, x_0) \in A_n$. Set

$$y_{-2} = (x_0 - \gamma)x_{-2}, \quad y_{-1} = x_{-2}, \quad \text{and} \quad y_0 = x_{-1}.$$

Then

$$(x_{-2}, x_{-1}, x_0) = (y_{-1}, y_0, \gamma + \frac{y_{-2}}{y_{-1}}) = F(y_{-2}, y_{-1}, y_0) = (y_{-1}, y_0, y_1),$$

and

$$(y_{n-1}, y_n, y_{n+1}) = F^{(n+1)}(y_{-2}, y_{-1}, y_0) = F^{(n)}(x_{-2}, x_{-1}, x_0) = (x_{n-2}, x_{n-1}, x_n),$$

and

$$(y_n, y_{n+1}, y_{n+2}) = F^{(n+2)}(y_{-2}, y_{-1}, y_0) = F^{(n+1)}(x_{-2}, x_{-1}, x_0) = (x_{n-1}, x_n, x_{n+1}).$$

Hence,

$$\gamma + 1 \leq y_{n+2} \leq y_{n+1} \leq y_n \leq y_{n-1}.$$

From this it follows that $(y_{-2}, y_{-1}, y_0) \in A_{n+1}$. Also

$$(x_{-2}, x_{-1}, x_0) = F(y_{-2}, y_{-1}, y_0) \in F(A_{n+1}).$$

The proof of our claim is complete. Since F is invertible it also holds

$$A_{n+1} = F^{-1}(A_n), \quad \text{for all} \quad n \geq 0.$$

Set

$$\Omega = \bigcap_{n=0}^{\infty} A_n.$$

Then

$$\Omega = \bigcap_{n=0}^{\infty} F^{(-n)}(A_0).$$

Note that A_0 is a nonempty, closed, connected, and unbounded subset of R^3. Also, A_1 is a nonempty and closed subset of R^3 and since $F^{-1}(A_0) = A_1$, it follows that A_1 is connected and unbounded. Inductively, it follows that each one of the $F^{(-n)}(A_0) = A_n$'s is a non-empty, closed, connected, and unbounded subset of R^3. Furthermore, the family $\{A_n\}_{n=0}^{\infty}$ satisfies the finite intersection property because

$$\bigcap_{k=0}^{n} A_k = A_n \neq \emptyset.$$

Then, clearly, Ω is a non-empty, closed, connected, and unbounded subset of R^3. By choosing the initial conditions x_{-2}, x_{-1}, x_0 in Ω, the solution $\{x_n\}$ that is generated satisfies for all $n \geq 0$,

$$\gamma + 1 \leq x_{n+1} \leq x_n$$

and so converges to $\gamma + 1$. The proof is complete. ∎

Next we would like to show that when

$$\gamma > 1$$

and $\{x_n\}_{n=-2}^{\infty}$, is a solution of Eq.(5.63.1) there exists an interval $[L, U]$, with $0 < L < U$, which contains the entire solution $\{x_n\}$ except possibly the first three terms x_{-2}, x_{-1}, x_0. We want to show the details of how the interval $[L, U]$ is found. Clearly,

$$x_n > \gamma, \quad \text{for } n \geq 1.$$

Now choose positive numbers L and U such that

$$x_1, x_2, x_3 \in [L, U].$$

We also want

$$L \leq x_4 \leq U.$$

Note that

$$\gamma + \frac{L}{U} \leq x_4 = \gamma + \frac{x_1}{x_2} \leq \gamma + \frac{U}{L}.$$

We need to choose L, U such that

$$L \leq \gamma + \frac{L}{U} \text{ and } \gamma + \frac{U}{L} \leq U.$$

Is it possible? The answer is yes. Just choose

$$L = \gamma + \frac{L}{U} \text{ and } L \in (\gamma, \gamma + 1).$$

Indeed, in this case

$$U = \frac{L}{L-\gamma}$$

and

$$L = \gamma + \frac{L}{U} \leq x_4 \leq \gamma + \frac{U}{L} = \gamma + \frac{1}{L-\gamma} \leq \frac{L}{L-\gamma}.$$

By using induction it follows that the interval $[L, U]$ contains the entire solution $\{x_n\}$ except possibly the fisrt three terms x_{-2}, x_{-1}, x_0. By employing Theorem 1.6.5 and the earlier local stability result of the equilibrium $\bar{x}$, one can easily see that when

$$\gamma > 1,$$

the equilibrium, $\bar{x} = \gamma + 1$, of Eq.(5.63.1) is globally asymptotically stable.

We can extend this global stability result to $\gamma = 1$ as the following theorem shows.

Theorem 5.63.2 *Assume that*

$$\gamma \geq 1.$$

Then the equilibrium, $\bar{x} = \gamma + 1$, *of Eq.(5.63.1) is globally asymptotically stable.*

PROOF The function

$$f(z_2, z_3) = \gamma + \frac{z_3}{z_2}$$

satisfies the Hypotheses of Theorem 1.6.7, when $\gamma > 1$, and the Hypotheses of Theorem 1.6.8, when $\gamma = 1$, and so the result follows. ∎

Conjecture 5.63.1 *Assume that*

$$\frac{-1+\sqrt{5}}{2} < \gamma < 1.$$

Show that every solution of Eq.(5.63.1) converges to the equilibrium $\bar{x}$.

Conjecture 5.63.2 *Assume that*

$$\gamma < \frac{1}{8}.$$

Show that every solution of Eq.(5.63.1) converges to a (not necessarily prime) periodic solution of period 13.

Conjecture 5.63.3 *Show that Eq.(5.63.1) has solutions that do not converge to the equilibrium point $\bar{x}$ or to a periodic solution.*

Conjecture 5.63.4 *Assume that $\gamma > 0$ and that m is a positive integer. Show that every solution of the equation*

$$x_{n+1} = \gamma + \frac{x_{n-2m}}{x_{n-1}}, \quad n = 0, 1, \ldots \tag{5.63.9}$$

is bounded.

Open Problem 5.63.1 *Assume that $\gamma > 0$ and that m and l are nonnegative integers. Investigate the global stability of solutions of the equation*

$$x_{n+1} = \gamma + \frac{x_{n-2m}}{x_{n-2l-1}}, \quad n = 0, 1, \ldots .$$

Open Problem 5.63.2 *Assume that $\{\gamma_n\}$ is a positive periodic sequence with prime period $k \geq 2$. Investigate the character of solutions of the difference equation*

$$x_{n+1} = \gamma_n + \frac{x_{n-2}}{x_{n-1}}, \quad n = 0, 1, \ldots .$$

Open Problem 5.63.3 *Assume that γ is a real number.*

(a) *Determine the set G of all initial conditions $(x_{-2}, x_{-1}, x_0) \in \Re^3$ such that the equation*

$$x_{n+1} = \gamma + \frac{x_{n-2}}{x_{n-1}}$$

is well defined for all $n \geq 0$.

(b) *Determine the character of solutions of the equation (5.63.1) for all $(x_{-2},\, x_{-1},\, x_0) \in G$.*

5.64 Equation #64 : $x_{n+1} = \dfrac{\gamma x_{n-1} + \delta x_{n-2}}{D x_{n-2}}$

Eq.(#64) possesses a period-two trichotomy depending on whether

$$\gamma < \delta, \quad \gamma = \delta, \quad \gamma > \delta.$$

This result is a special case of a more general period-two trichotomy result presented in Theorem 4.3.1.

Open Problem 5.64.1 *Assume that δ is a real number.*

(a) Determine the "good" set G of all initial conditions $(x_{-2}, x_{-1}, x_0) \in \Re^3$ such that the equation

$$x_{n+1} = \delta + \frac{x_{n-1}}{x_{n-2}} \tag{5.64.1}$$

is well defined for all $n \geq 0$.

(b) Determine the character of solutions of the equation (5.64.1) for all $(x_{-2}, x_{-1}, x_0) \in G$.

Open Problem 5.64.2 *Assume that $\delta \in (0,1)$ and let $k \in (0, 1-\delta)$. Investigate the global character of solutions of*

$$x_{n+1} = \delta + \frac{x_{n-1}}{x_{n-2}}, \quad n = 0, 1, \ldots \tag{5.64.2}$$

with

$$x_{-2}, x_0 \in [1, 1+\delta] \text{ and } x_{-1} \in \left[1+\delta, \frac{1}{k}\right].$$

Conjecture 5.64.1 *Assume that*

$$\delta < 1.$$

Show that every bounded solution of Eq.(5.64.2) converges to the equilibrium.

5.65 Equation #65 : $x_{n+1} = \dfrac{\alpha + \beta x_n}{A + Bx_n}$

The equation in the title is the well-known **Riccati difference equation.** It is one of the very few nonlinear difference equations that can be solved explicitly. See any book on difference equations about it, for example, [12], [95], [147], or [175].

In this section we will present the character of solutions of the equation

$$x_{n+1} = \frac{\alpha + \beta x_n}{A + Bx_n}, \quad n = 0, 1, \ldots \tag{5.65.1}$$

not only for the positive values of the parameters and for nonnegative initial conditions, as we do in every other section, but also for any real values of the parameters α, β, A, B and for any real initial condition x_0. We wish we could do this for every equation in this book. Unfortunately, it is a problem of great difficulty to determine the set of initial conditions for which a solution of the equation is well defined for all $n \geq 0$. For Eq.(5.65.1) this was done in [126].

To avoid degeneracies we will assume that

$$B \neq 0 \quad \text{and} \quad \alpha B - \beta A \neq 0\,.$$

The following result for equation (5.65.1) has a straightforward proof.

Theorem 5.65.1 *Every solution of Eq.(5.65.1) is periodic with period two if and only if*

$$\beta + A = 0\,. \tag{5.65.2}$$

When (5.65.2) holds, Eq.(5.65.1) becomes

$$x_{n+1} = \frac{\alpha - Ax_n}{A + Bx_n}, \quad n = 0, 1, \ldots \tag{5.65.3}$$

and every solution of Eq.(5.65.3) with

$$x_n \neq \frac{-A}{B}$$

is the two-cycle

$$x_0, \frac{\alpha - Ax_0}{A + Bx_0}, x_0, \frac{\alpha - Ax_0}{A + Bx_0}, \ldots\,.$$

In the remaining part of this section we will assume, without further mention, that

$$B \neq 0, \ \ \alpha B - \beta A \neq 0, \ \text{ and } \ \beta + A \neq 0\,.$$

Then the change of variables

$$x_n = \frac{\beta + A}{B} \cdot w_n - \frac{A}{B}, \quad n = 0, 1, \ldots \tag{5.65.4}$$

transforms Eq.(5.65.1) to the equation

$$w_{n+1} = 1 - \frac{R}{w_n}, \quad n = 0, 1, \ldots, \tag{5.65.5}$$

where

$$R = \frac{\beta A - \alpha B}{(\beta + A)^2}. \tag{5.65.6}$$

It is an amazing fact that when Eq.(5.65.1) has no period-two solutions, the change of variables (5.65.4) reduces it to Eq.(5.65.5), which depends on a single parameter. The parameter R, which is called the **Riccati number**, of Eq.(5.65.1) is the nonzero number given by (5.65.6).

Now the change of variables

$$w_n = \frac{y_{n+1}}{y_n}, \quad n = 0, 1, \ldots \tag{5.65.7}$$

with

$$y_0 = 1 \quad \text{and} \quad y_1 = w_0 \tag{5.65.8}$$

transforms Eq.(5.65.5) to the second-order linear difference equation

$$y_{n+2} - y_{n+1} + R y_n = 0, \ \ n = 0, 1, \ldots . \tag{5.65.9}$$

Eq.(5.65.9) can be solved explicitly, in terms of the characteristic roots

$$\lambda_1 = \frac{1 + \sqrt{1 - 4R}}{2} \quad \text{and} \quad \lambda_2 = \frac{1 - \sqrt{1 - 4R}}{2}. \tag{5.65.10}$$

Then by using (5.65.7) we obtain the solution of Eq.(5.65.5) provided that w_0 is chosen in such a way that

$$y_n \neq 0 \ \text{for all} \ n = 0, 1, \ldots .$$

The set of points w_0 for which $y_n = 0$ for some value of n is called the **forbidden set F** of Eq.(5.65.5).

There are very few rational equations for which we know something about the forbidden set of the equation. See, for example, [53] and [55].

In view of (5.65.10), the character of solutions of Eq.(5.65.5) and its forbidden set F depend on whether

$$R < \frac{1}{4}, \ \ R = \frac{1}{4}, \ \text{or} \ R > \frac{1}{4}.$$

The character of solutions of Eq.(5.65.1) and its forbidden set F are easily inferred by means of the change of variables (5.65.4).

Case 1: $R < \frac{1}{4}$. Then the general solution of Eq.(5.65.9) is

$$y_n = c_1 \lambda_1^n + c_2 \lambda_2^n, \ \ n = 0, 1, \ldots$$

and in view of the initial conditions (5.65.8)

$$c_1 = \frac{w_1 - \lambda_1}{\lambda_1 - \lambda_2} \quad \text{and} \quad c_2 = \frac{\lambda_1 - w_0}{\lambda_1 - \lambda_2}.$$

Thus, the forbidden set of Eq.(5.65.5) is the set of w_0 such that

$$y_n = \frac{w_1 - \lambda_1}{\lambda_1 - \lambda_2} \cdot \lambda_1^n + \frac{\lambda_1 - w_0}{\lambda_1 - \lambda_2} \cdot \lambda_2^n = 0, \ \ \text{for} \ n = 1, 2, \ldots$$

that is, the set

$$F = \{\frac{\lambda_1^n \lambda_2 - \lambda_1 \lambda_2^n}{\lambda_1 - \lambda_2} \ : \ n = 1, 2, \ldots\}.$$

When

$$w_0 \notin F,$$

$$w_n = \frac{(w_1 - \lambda_2)\lambda_1^{n+1} - (w_0 - \lambda_1)\lambda_2^{n+1}}{(w_1 - \lambda_2)\lambda_1^n - (w_0 - \lambda_1)\lambda_2^n}, \quad n = 0, 1, \dots .$$

Hence,

$$w_n = \lambda_2 \text{ if } w_0 = \lambda_2$$

and

$$\lim_{n\to\infty} w_n = \lambda_1 \text{ if } w_0 \neq \lambda_2.$$

Case 2: $R = \frac{1}{4}$. Then the general solution of Eq.(5.65.9) is

$$y_n = c_1(\frac{1}{2})^n + c_2 n(\frac{1}{2})^n, \quad n = 0, 1, \dots$$

and in view of (5.65.8)

$$c_1 = 1 \quad \text{and} \quad c_2 = 2w_0 - 1.$$

Thus, the forbidden set of Eq.(5.65.5) is the set of w_0 such that

$$y_n = (\frac{1}{2})^n + (2w_0 - 1)n(\frac{1}{2})^n = 0, \quad \text{for } n = 1, 2, \dots$$

that is

$$F = \{\frac{n-1}{2n} : n = 1, 2, \dots\}.$$

When

$$w_0 \notin F,$$

$$w_n = \frac{1 + (2w_0 - 1)(n+1)}{2 + 2(2w_0 - 1)n}, \quad n = 0, 1, \dots .$$

Hence, in this case

$$\lim_{n\to\infty} w_n = \frac{1}{2}.$$

Case 3: $R > \frac{1}{4}$. Here the characteristic roots of Eq.(5.65.10) are complex conjugate. Choose $\phi \in (0, \frac{\pi}{2})$ such that

$$\cos\phi = \frac{1}{2\sqrt{R}} \quad \text{and} \quad \sin\phi = \frac{\sqrt{4R-1}}{2\sqrt{R}}.$$

Then

$$y_n = R^{\frac{n}{2}}[c_1 \cos(n\phi) + c_2 \sin(n\phi)], \quad n = 0, 1, \dots$$

and by using (5.65.8) we find

$$c_1 = 1 \text{ and } c_2 = \frac{2w_0 - 1}{\sqrt{4R-1}}.$$

Note the forbidden set in this case is the set of all initial conditions w_0 such that

$$\cos(n\phi) + \frac{2w_0 - 1}{\sqrt{4R-1}} \cdot \sin(n\phi) = 0 \ \text{ for } \ n = 1, 2, \ldots$$

that is, the set of points

$$F = \{\frac{1 - \sqrt{4R-1}\cot(n\phi)}{2} \ : \ n = 1, 2, \ldots\}.$$

When

$$w_0 \notin F,$$

$$w_n = \sqrt{R} \cdot \frac{\sqrt{4R-1}\cos(n+1)\phi + (2w_0 - 1)\sin(n+1)\phi}{\sqrt{4R-1}\cos(n\phi) + (2w_0 - 1)\sin(n\phi)}, \quad n = 0, 1, \ldots .$$

We can rewrite this solution in a way that is easier to investigate the long-term behavior. Set

$$r = \sqrt{(4R-1) + (2w_0 - 1)^2}$$

and let $\theta \in (\frac{-\pi}{2}, \frac{\pi}{2})$ be such that

$$\cos\theta = \frac{\sqrt{4R-1}}{r} \quad \text{and} \quad \sin\theta = \frac{2w_0 - 1}{r}.$$

Then

$$w_n = \sqrt{R} \cdot \frac{\cos(n\phi + \phi - \theta)}{\sin(n\phi - \theta)} = \sqrt{R} \cdot \frac{\cos(n\phi - \theta)\cos\phi - \sin(n\phi - \theta)\sin\phi}{\cos(n\phi - \theta)}$$

$$= \sqrt{R} \cdot [\cos\phi - \sin\phi\tan(n\phi - \theta)]$$

$$= \frac{1}{2} - \frac{\sqrt{4R-1}}{2} \cdot \tan(n\phi - \theta).$$

From this it follows that if ϕ is a rational multiple of π, that is,

$$\phi = \frac{q}{p} \cdot \pi \in (0, \frac{\pi}{2})$$

where q and p are positive constants, then every solution of Eq.(5.65.5) with

$$w_0 \neq \frac{1 - \sqrt{4R-1}\cot(n\frac{q}{p}\pi)}{2} \quad \text{for } \ n = 1, 2, \ldots, p-1$$

is periodic with period p.

When ϕ is not a rational multiple of π, then no solution of Eq.(5.65.5) is periodic and for any $w_0 \notin F$, the set of limit points of the solution of Eq.(5.65.5) is dense in the real line. See [45].

Remark 5.65.1 *When the parameters α, β, A, and B of a Riccati equation are nonnegative, the Riccati number of the equation is less than or equal to $\frac{1}{4}$ and so every solution of a Riccati equation with nonnegative parameters converges to an equilibrium of the equation.*

Who introduced the name of this equation? What is the relationship of this equation to the Riccati differential equation?

Open Problem 5.65.1 *Assume that*

$$\{\alpha_n\},\ \{\beta_n\},\ \{A_n\},\ \{B_n\}$$

are periodic sequences of real numbers. Determine the ***forbidden set*** **F** *of the Riccati equation*

$$x_{n+1} = \frac{\alpha_n + \beta_n x_n}{A_n + B_n x_n}, \quad n = 0, 1, \ldots \tag{5.65.11}$$

and the character of solutions of Eq.(5.65.11) with $x_0 \notin F$.

The above problem with period-two sequences was investigated in [120].

5.66 Equation #66 : $x_{n+1} = \dfrac{\alpha + \beta x_n}{A + C x_{n-1}}$

The most substantial work on this equation was presented in [158]. See also [157] and [175]. Eq.(#66) can be written in the normalized form

$$x_{n+1} = \frac{\alpha + x_n}{A + x_{n-1}}, \quad n = 0, 1, \ldots \tag{5.66.1}$$

with positive parameters α, A and with arbitrary nonnegative initial conditions x_{-1}, x_0.

Eq.(5.66.1) has the unique equilibrium

$$\bar{x} = \frac{1 - A + \sqrt{(1-A)^2 + 4\alpha}}{2}.$$

The characteristic equation of the linearized equation of Eq.(5.66.1) about the equilibrium $\bar{x}$ is

$$\lambda^2 - \frac{1}{A + \bar{x}}\lambda + \frac{\bar{x}}{A + \bar{x}} = 0.$$

From this and Theorem 1.2.2 it follows that the equilibrium $\bar{x}$ of Eq.(5.66.1) is locally asymptotically stable for all positive values of the parameters.

The following conjecture, known for more than 15 years (see [157], [175] and [189]) has not been confirmed yet.

Conjecture 5.66.1 *Assume that*

$$\alpha, A \in (0, \infty).$$

Show that every positive solution of Eq.(5.66.1) has a finite limit.

In other words, the equilibrium $\bar{x}$ of Eq.(5.66.1) is globally asymptotically stable. To the best of our knowledge, any claims in the literature made prior to July 2007 that this conjecture has been confirmed are not correct.

For

$$\alpha < A,$$

the Conjecture is true and the proof is a straightforward consequence of Theorem 1.6.3.

For

$$A \geq 1,$$

the conjecture is true and the proof follows by Theorems 5.23.2 and 5.23.3.

Finally, for

$$\alpha = A$$

the conjecture is true and the proof follows from Theorem 5.26.1.

There are several publications in the literature where the Conjecture has also been confirmed is some subregions of parameters with

$$\alpha > A \text{ and } A < 1. \tag{5.66.2}$$

See [104], [157], [175], and the references cited therein. What is needed at this time is to confirm the conjecture in the "entire" region (5.66.2). This will confirm Conjecture 5.66.1. In this direction we offer the following conjecture which is more than what is needed to confirm Conjecture 5.66.1 in the entire region (5.66.2).

Conjecture 5.66.2 *Assume that the following conditions hold:*

(i) $f \in C[(0, \infty) \times (0, \infty), (0, \infty)]$.

(ii) $f(x, y)$ *is decreasing in* x *and strictly decreasing in* y.

(iii) $xf(x, x)$ *is strictly decreasing in* x.

(iv) The equation

$$x_{n+1} = x_n f(x_n, x_{n-1}), \quad n = 0, 1, \ldots \tag{5.66.3}$$

has a unique positive equilibrium $\bar{x}$*, which is locally asymptotically stable.*

Then $\bar{x}$ *is a global attractor of all positive solutions of Eq.(5.66.3).*

The following special case of Conjecture 5.66.2 would also confirm Conjecture 5.66.1.

Conjecture 5.66.3 *Confirm Conjecture 5.66.2 in the special case of the rational difference equation*

$$x_{n+1} = \frac{\alpha + \beta x_n + \gamma x_{n-1}}{A + Bx_n + Cx_{n-1}}, \quad n = 0, 1, \ldots$$

with nonnegative parameters and nonnegative initial conditions, that is, for the function

$$f(x, y) = \frac{1}{x} \cdot \frac{\alpha + \beta x + \gamma y}{A + Bx + Cy}$$

under the assumptions of Conjecture 5.66.2.

5.67 Equation #67 : $x_{n+1} = \dfrac{\alpha + \beta x_n}{A + Dx_{n-2}}$

Eq.(#67) can be written in the normalized form

$$x_{n+1} = \frac{\alpha + x_n}{A + x_{n-2}}, \quad n = 0, 1, \ldots \tag{5.67.1}$$

with positive parameters α, A and with arbitrary nonnegative initial conditions x_{-2}, x_{-1}, x_0.

Eq.(5.67.1) has the unique equilibrium

$$\bar{x} = \frac{1 - A + \sqrt{(1-A)^2 + 4\alpha}}{2}.$$

The characteristic equation of the linearized equation of Eq.(5.67.1) about the equilibrium $\bar{x}$ is

$$\lambda^3 - \frac{1}{A + \bar{x}}\lambda^2 + \frac{\bar{x}}{A + \bar{x}} = 0.$$

From this and Theorem 1.2.3 it follows that the equilibrium $\bar{x}$ of Eq.(5.67.1) is locally asymptotically stable when either

$$A \geq \frac{1}{2} \tag{5.67.2}$$

or

$$\frac{1}{3} < A < \frac{1}{2} \text{ and } \alpha < \frac{A^2(-A^2 + 3A - 1)}{(2A-1)^2} \tag{5.67.3}$$

and unstable when

$$\frac{1}{3} < A < \frac{1}{2} \text{ and } \alpha > \frac{A^2(-A^2 + 3A - 1)}{(2A-1)^2} \tag{5.67.4}$$

or

$$0 < A < \frac{1}{3}. \qquad (5.67.5)$$

By Theorems 5.23.2 and 5.23.3 it follows that the equilibrium $\bar{x}$ of Eq.(5.67.1) is globally asymptotically stable when

$$A \geq 1.$$

Conjecture 5.67.1 *Assume that either*

$$\frac{1}{2} \leq A < 1$$

or that (5.67.3) holds. Show that the equilibrium $\bar{x}$ of Eq.(5.67.1) is globally asymptotically stable.

Conjecture 5.67.2 *Show that Eq.(5.67.1) has solutions that do not converge to the equilibrium point $\bar{x}$ or to a periodic solution.*

5.68 Equation #68 : $x_{n+1} = \dfrac{\alpha + \beta x_n}{Bx_n + Cx_{n-1}}$

This equation was investigated in [175]. See also Theorem 2.1.1 where we established that every solution of the equation is bounded. Eq.(#68) can be written in the normalized form

$$x_{n+1} = \frac{\alpha + x_n}{Bx_n + x_{n-1}}, \quad n = 0, 1, \ldots \qquad (5.68.1)$$

with positive parameters α, B and with arbitrary positive initial conditions x_{-1}, x_0.

The only equilibrium of Eq.(5.68.1) is

$$\bar{x} = \frac{1 + \sqrt{1 + 4\alpha(B+1)}}{2(B+1)}.$$

The characteristic equation of the linearized equation of Eq.(5.68.1) about the equilibrium $\bar{x}$ is

$$\lambda^2 + \frac{B\bar{x} - 1}{\bar{x}(1+B)}\lambda + \frac{1}{1+B} = 0. \qquad (5.68.2)$$

From this and Theorem 1.2.2 it follows that the equilibrium $\bar{x}$ of Eq.(5.68.1) is locally asymptotically stable for all positive values of the parameters.

When

$$\alpha \geq \frac{1-B}{4B^2},$$

as we will see in Theorem 5.141.1, every solution of Eq.(5.68.1) converges to the equilibrium $\bar{x}$. For another proof of that result see [175].

Conjecture 5.68.1 *Show that for all positive values of the parameters the equilibrium $\bar{x}$ of Eq.(5.68.1) is globally asymptotically stable.*

5.69 Equation #69 : $x_{n+1} = \dfrac{\alpha + \beta x_n}{Bx_n + Dx_{n-2}}$

The boundedness character of solutions of this equation was investigated in [69]. See also Theorem 2.1.1 where we established that every solution of the equation is bounded. Eq.(#69) can be written in the normalized form

$$x_{n+1} = \frac{\alpha + x_n}{x_n + Dx_{n-2}}, \quad n = 0, 1, \ldots \tag{5.69.1}$$

with positive parameters α, D and with arbitrary positive initial conditions x_{-2}, x_{-1}, x_0.

The only equilibrium of Eq.(5.69.1) is

$$\bar{x} = \frac{1 + \sqrt{1 + 4\alpha(D+1)}}{2(D+1)}.$$

The characteristic equation of the linearized equation of Eq.(5.69.1) about the equilibrium $\bar{x}$ is

$$\lambda^3 + \frac{\bar{x} - 1}{\bar{x}(1+D)}\lambda^2 + \frac{D}{1+D} = 0. \tag{5.69.2}$$

From this and Theorem 1.2.3 it follows that the equilibrium $\bar{x}$ of Eq.(5.69.1) is locally asymptotically stable when

$$0 < D \leq 1 + \sqrt{2} \tag{5.69.3}$$

or

$$D > 1 + \sqrt{2} \text{ and } \alpha > \frac{D(D^2 - 2D - 1)}{(3D+1)^2} \tag{5.69.4}$$

and unstable when

$$D > 1 + \sqrt{2} \text{ and } \alpha < \frac{D(D^2 - 2D - 1)}{(3D+1)^2}.$$

When

$$\alpha \geq \frac{D-1}{4},$$

as we will see in Theorem 5.141.2, every solution of Eq.(5.69.1) converges to the equilibrium $\bar{x}$.

Conjecture 5.69.1 *Assume that*

$$D > 1 \text{ and } \frac{D(D^2-2D-1)}{(3D+1)^2} < \alpha < \frac{D-1}{4}.$$

Show that the equilibrium $\bar{x}$ of Eq.(5.69.1) is globally asymptotically stable.

Conjecture 5.69.2 *Assume that*

$$D > 1 \text{ and } \alpha < \frac{D(D^2-2D-1)}{(3D+1)^2}.$$

Show that Eq.(5.69.1) has solutions that do not converge to the equilibrium point $\bar{x}$ or to a periodic solution.

5.70 Equation #70 : $x_{n+1} = \dfrac{\alpha + \beta x_n}{Cx_{n-1} + Dx_{n-2}}$

This is an equation of paramount importance that has not been investigated yet. Eq.(#70) can be written in the normalized form

$$x_{n+1} = \frac{\alpha + x_n}{Cx_{n-1} + x_{n-2}}, \quad n = 0, 1, \ldots \tag{5.70.1}$$

with positive parameters α, C and with arbitrary positive initial conditions x_{-2}, x_{-1}, x_0.

In Section 4.7 we conjectured that Eq.(5.70.1) possesses a period-six trichotomy depending on whether

$$\alpha C^2 > 1, \quad \alpha C^2 = 1, \quad \text{or} \quad \alpha C^2 < 1\,.$$

The characteristic equation of the linearized equation of Eq.(5.70.1) about its unique equilibrium point $\bar{x}$,

$$\bar{x} = \frac{1 + \sqrt{1 + 4\alpha(C+1)}}{2(C+1)},$$

is

$$\lambda^2 - \frac{1}{(C+1)\bar{x}}\lambda^2 + \frac{C}{C+1}\lambda + \frac{1}{C+1} = 0. \tag{5.70.2}$$

From this and Theorem 1.2.3 it follows that the equilibrium $\bar{x}$ of Eq.(5.70.1) is locally asymptotically stable when

$$\alpha C^2 > 1$$

and unstable when

$$\alpha C^2 < 1.$$

When

$$\alpha C^2 = 1,$$

the three characteristic roots of Eq.(5.70.2) are

$$\lambda_1 = \frac{-1}{C+1}, \quad \lambda_2 = \frac{1 - i\sqrt{3}}{2}, \quad \text{and} \quad \lambda_3 = \frac{1 + i\sqrt{3}}{2}.$$

Please note that the dominant characteristic roots are sixth roots of unity. **Is this typical of periodic convergence?**

Conjecture 5.70.1 *Show that the solution of the equation*

$$x_{n+1} = \frac{1 + x_n}{x_{n-1} + x_{n-2}}, \quad n = 0, 1, \ldots \tag{5.70.3}$$

with initial conditions

$$x_{-2} = x_{-1} = x_0 = 2$$

converges to a prime period-six solution of Eq.(5.70.3).

5.71 Equation #71 : $x_{n+1} = \dfrac{\alpha + \gamma x_{n-1}}{A + Bx_n}$

This equation was investigated in [108]. Eq.(#71) possesses a period-two trichotomy depending on whether

$$\gamma < A, \quad \gamma = A, \quad \text{or} \quad \gamma > A.$$

This result is a special case of a more general period-two trichotomy result presented in Theorem 4.2.1.

When

$$\gamma > A,$$

it follows from Theorem 4.2.2 that every bounded solution of Eq.(#71) converges to the equilibrium.

Open Problem 5.71.1 *Can the character of solutions of the rational equation*

$$x_{n+1} = \frac{\alpha + \gamma x_{n-1}}{A + x_n}, \quad n = 0, 1, \ldots$$

be predicted from the characteristic roots of the linearized equation about the equilibrium point? Extend and generalize.

5.72 Equation #72 : $x_{n+1} = \dfrac{\alpha + \gamma x_{n-1}}{A + C x_{n-1}}$

This is a Riccati-type equation. For the character of solutions of Riccati equations see Section 5.65.

5.73 Equation #73 : $x_{n+1} = \dfrac{\alpha + \gamma x_{n-1}}{A + D x_{n-2}}$

This equation was investigated in [17] and [70]. Eq.(#73) possesses a period-two trichotomy depending on whether

$$\gamma < A, \quad \gamma = A, \quad \text{or} \quad \gamma > A.$$

This result is a special case of a more general period-two trichotomy result presented in Theorem 4.3.1.

Conjecture 5.73.1 *Assume that*

$$\gamma > A.$$

Show that every bounded solution of Eq.(#73) converges to the equilibrium.

Open Problem 5.73.1 *Can the character of solutions of the rational equation*

$$x_{n+1} = \frac{\alpha + \gamma x_{n-1}}{A + x_{n-2}}, \quad n = 0, 1, \ldots$$

be predicted from the characteristic roots of the linearized equation about the equilibrium point? Extend and generalize.

5.74 Equation #74 : $x_{n+1} = \dfrac{\alpha + \gamma x_{n-1}}{Bx_n + Cx_{n-1}}$

This equation was investigated in [162]. See also [175]. Eq.(#74) can be written in the normalized form

$$x_{n+1} = \frac{\alpha + x_{n-1}}{Bx_n + x_{n-1}}, \quad n = 0, 1, \ldots \tag{5.74.1}$$

with positive parameters α, B and with arbitrary positive initial conditions x_{-1}, x_0.

The only equilibrium of Eq.(5.74.1) is

$$\bar{x} = \frac{1 + \sqrt{1 + 4\alpha(B+1)}}{2(B+1)}.$$

The characteristic equation of the linearized equation of Eq.(5.74.1) about the equilibrium $\bar{x}$ is

$$\lambda^2 + \frac{B}{1+B}\lambda + \frac{\bar{x} - 1}{(B+1)\bar{x}} = 0. \tag{5.74.2}$$

From this and Theorem 1.2.2 it follows that the equilibrium $\bar{x}$ of Eq.(5.74.1) is locally asymptotically stable when

$$\alpha > \frac{B-1}{4} \tag{5.74.3}$$

and unstable when

$$\alpha < \frac{B-1}{4}. \tag{5.74.4}$$

When (5.74.4) holds, and only then, Eq.(5.74.1) has the unique period-two solution

$$\ldots, \frac{1 - \sqrt{1 - \frac{4\alpha}{B-1}}}{2}, \frac{1 + \sqrt{1 - \frac{4\alpha}{B-1}}}{2}, \ldots, \tag{5.74.5}$$

which is locally asymptotically stable. See [175].

When

$$\alpha \geq \frac{B-1}{4},$$

as we will see in Theorem 5.141.2, every solution of Eq.(5.74.1) converges to the equilibrium $\bar{x}$. For another proof of that result see [162] and [175].

When (5.74.4) holds as we will see in Theorem 5.145.1 every solution of Eq.(5.74.1) converges to a (not necessarily prime) period-two solution.

Open Problem 5.74.1 *Assume that (5.74.4) holds. Determine the set of all initial conditions* x_{-1}, x_0 *for which every solution of Eq.(5.74.1) converges to the prime period-two solution (5.74.5).*

Open Problem 5.74.2 *Assume that (5.74.4) holds. Determine the set of all initial conditions* x_{-1}, x_0 *for which every solution of Eq.(5.74.1) converges to the equilibrium* $\bar{x}$.

5.75 Equation #75 : $x_{n+1} = \dfrac{\alpha + \gamma x_{n-1}}{Bx_n + Dx_{n-2}}$

Eq.(#75) has unbounded solutions. This equation is part of a period-two trichotomy presented in Theorem 4.3.1.

Conjecture 5.75.1 *Show that every bounded solution of Eq.(#75) converges to the equilibrium of the equation.*

Open Problem 5.75.1 *Determine the set of all initial conditions of Eq.(#75) through which the solutions are unbounded.*

5.76 Equation #76 : $x_{n+1} = \dfrac{\alpha + \gamma x_{n-1}}{Cx_{n-1} + Dx_{n-2}}$

This equation was investigated in [102] where the special cases #74 and #76 were extended and unified. See also Theorem 2.1.1 where we established that every solution of this equation is bounded. Eq.(#76) can be written in the normalized form

$$x_{n+1} = \frac{\alpha + x_{n-1}}{x_{n-1} + Dx_{n-2}}, \quad n = 0, 1, ... \tag{5.76.1}$$

with positive parameters α, D and with arbitrary positive initial conditions x_{-2}, x_{-1}, x_0.

The only equilibrium of Eq.(5.76.1) is

$$\bar{x} = \frac{1 + \sqrt{1 + 4\alpha(D + 1)}}{2(D + 1)}.$$

The characteristic equation of the linearized equation of Eq.(5.76.1) about the equilibrium $\bar{x}$ is

$$\lambda^3 + \frac{\bar{x}-1}{\bar{x}(1+D)}\lambda + \frac{D}{1+D} = 0. \tag{5.76.2}$$

From this and Theorem 1.2.3 it follows that the equilibrium $\bar{x}$ of Eq.(5.76.1) is locally asymptotically stable when

$$D < 1 + 4\alpha \tag{5.76.3}$$

and unstable when

$$D > 1 + 4\alpha. \tag{5.76.4}$$

When (5.76.4) holds, Eq.(5.76.1) possesses the unique prime period-two solution

$$\ldots, \frac{1-\sqrt{1-\frac{4\alpha}{D-1}}}{2}, \frac{1+\sqrt{1-\frac{4\alpha}{D-1}}}{2}, \ldots,$$

which was shown in [102] to be locally asymptotically stable.

When

$$D \leq 1 + 4\alpha,$$

as we will see in Theorem 5.141.2, every solution of Eq.(5.76.1) converges to the equilibrium $\bar{x}$. For another proof of that result see [102].

Conjecture 5.76.1 *Assume that (5.76.4) holds. Show that every solution of Eq.(5.76.1) converges to a (not necessarily prime) period-two solution.*

5.77 Equation #77 : $x_{n+1} = \dfrac{\alpha + \delta x_{n-2}}{A + Bx_n}$

This equation was investigated in [49]. See also Theorem 2.5.1 where we established that every solution of the equation is bounded. Eq.(#77) can be written in the normalized form

$$x_{n+1} = \frac{\alpha + x_{n-2}}{A + x_n}, \quad n = 0, 1, \ldots \tag{5.77.1}$$

with positive parameters α, A and with arbitrary nonnegative initial conditions x_{-2}, x_{-1}, x_0.

The only equilibrium of Eq.(5.77.1) is

$$\bar{x} = \frac{1 - A + \sqrt{(1-A)^2 + 4\alpha}}{2}.$$

The characteristic equation of the linearized equation of Eq.(5.77.1) about the equilibrium $\bar{x}$ is

$$\lambda^3 + \frac{\bar{x}}{\bar{x}+A}\lambda^2 - \frac{1}{\bar{x}+A} = 0. \qquad (5.77.2)$$

From this and Theorem 1.2.3 it follows that the equilibrium $\bar{x}$ of Eq.(5.77.1) is locally asymptotically stable when

$$A \geq 1 \qquad (5.77.3)$$

or

$$A < 1 \text{ and } \alpha > \frac{2 - A + \sqrt{5A^2 - 4A^3}}{2} \qquad (5.77.4)$$

and unstable when

$$A < 1 \text{ and } \alpha < \frac{2 - A + \sqrt{5A^2 - 4A^3}}{2}.$$

By Theorems (5.23.2) and (5.23.3) it follows that when

$$A \geq 1,$$

the equilibrium $\bar{x}$ of Eq.(5.77.1) is globally asymptotically stable.

Conjecture 5.77.1 *Show that every solution of Eq.(5.77.1) converges to the positive equilibrium $\bar{x}$ when*

$$A < 1 \text{ and } \alpha > \frac{2 - A + \sqrt{5A^2 - 4A^3}}{2}.$$

Conjecture 5.77.2 *Assume that*

$$A < 1 \text{ and } \alpha < \frac{2 - A + \sqrt{5A^2 - 4A^3}}{2}.$$

Show that Eq.(5.77.1) has solutions that do not converge to the equilibrium point $\bar{x}$ or to a periodic solution.

5.78 Equation #78 : $x_{n+1} = \dfrac{\alpha + \delta x_{n-2}}{A + Cx_{n-1}}$

This equation was investigated in [49]. See also Theorem 2.7.1 where we established that every solution of the equation is bounded. This equation can be written in the normalized form

$$x_{n+1} = \frac{\alpha + x_{n-2}}{A + x_{n-1}}, \quad n = 0, 1, \ldots \qquad (5.78.1)$$

with positive parameters α, A and with arbitrary nonnegative initial conditions, x_{-2}, x_{-1}, x_0.

The only equilibrium of Eq.(5.78.1) is

$$\bar{x} = \frac{1 - A + \sqrt{(1-A)^2 + 4\alpha}}{2}.$$

The characteristic equation of the linearized equation of Eq.(5.78.1) about the equilibrium $\bar{x}$ is

$$\lambda^3 + \frac{\bar{x}}{\bar{x} + A}\lambda - \frac{1}{\bar{x} + A} = 0. \tag{5.78.2}$$

From this and Theorem 1.2.3 it follows that the equilibrium $\bar{x}$ is locally asymptotically stable when

$$A \geq 1 \tag{5.78.3}$$

or

$$A < 1 \text{ and } \alpha > \frac{(A-1)^2(A+1)}{A^2} \tag{5.78.4}$$

and unstable when

$$A < 1 \text{ and } \alpha < \frac{(A-1)^2(A+1)}{A^2}.$$

By Theorems (5.23.2) and (5.23.3) it follows that when

$$A \geq 1,$$

the equilibrium $\bar{x}$ of Eq.(5.78.1) is globally asymptotically stable.

Conjecture 5.78.1 *Show that every solution of Eq.(5.78.1) converges to the positive equilibrium $\bar{x}$ when*

$$A < 1 \textit{ and } \alpha > \frac{(A-1)^2(A+1)}{A^2}.$$

Conjecture 5.78.2 *Assume that*

$$A < 1 \textit{ and } \alpha < \frac{(A-1)^2(A+1)}{A^2}.$$

Show that Eq.(5.78.1) has solutions that do not converge to the equilibrium point $\bar{x}$ or to a periodic solution.

5.79 Equation #79 : $x_{n+1} = \dfrac{\alpha + \delta x_{n-2}}{A + D x_{n-2}}$

This is a Riccati-type equation. For the character of solutions of Riccati equations see Section 5.65.

5.80 Equation #80 : $x_{n+1} = \dfrac{\alpha + \delta x_{n-2}}{Bx_n + Cx_{n-1}}$

Eq.(#80) can be written in the normalized form

$$x_{n+1} = \frac{\alpha + x_{n-2}}{Bx_n + x_{n-1}}, \quad n = 0, 1, \ldots \tag{5.80.1}$$

with positive parameters α, B and with arbitrary positive initial conditions x_{-2}, x_{-1}, x_0.

The boundedness character of solutions of this equation was investigated in [69]. See also Theorem 3.4.1 where we established that the equation has unbounded solutions for all positive values of the parameters α and B.

Eq.(5.80.1) has the unique equilibrium

$$\bar{x} = \frac{1 + \sqrt{1 + 4\alpha(B+1)}}{2(B+1)}.$$

The characteristic equation of the linearized equation of Eq.(5.80.1) about the equilibrium $\bar{x}$ is

$$\lambda^3 + \frac{B}{B+1}\lambda^2 + \frac{1}{B+1}\lambda - \frac{1}{(B+1)\bar{x}} = 0\,. \tag{5.80.2}$$

From this and Theorem 1.2.3 it follows that the equilibrium $\bar{x}$ of Eq.(5.80.1) is locally asymptotically stable when

$$\alpha B > 1 \tag{5.80.3}$$

and unstable when

$$\alpha B < 1. \tag{5.80.4}$$

For the equilibrium $\bar{x}$ of Eq.(5.80.1),

Local Asymptotic Stabilty $\not\Rightarrow$ Global Asymptotic Stabilty.

More specifically, when the condition (5.80.3) is satisfied, the equilibrium $\bar{x}$ of Eq.(5.80.1) is locally asymptotically stable but not globally asymptotically stable. The reason is that for all positive values of the parameters α, B there exist initial conditions x_{-2}, x_{-1}, x_0 for which the solution of Eq.(5.80.1) is unbounded. See Theorem 3.4.1.

In addition to unbounded solutions, what other type's of solutions exist? Can there exist any periodic solutions? Can there exist any bounded solutions that are not periodic and do not converge to the equilibrium $\bar{x}$ or to a periodic solution?

It is interesting to note that when

$$B = 1,$$

unbounded solutions of Eq.(5.80.1) co-exist with periodic solutions. For example, when

$$\alpha \neq 1,$$

the sequence

$$\dots, 1, 1, \alpha, \dots$$

is a prime period-three solution of the equation

$$x_{n+1} = \frac{\alpha + x_{n-2}}{x_n + x_{n-1}}, \quad n = 0, 1, \dots .$$

Open Problem 5.80.1 *Determine all possible periodic solutions of Eq.(5.80.1).*

Conjecture 5.80.1 *Assume that (5.80.3) holds. Show that every bounded solution of Eq.(5.80.1) converges either to the equilibrium $\bar{x}$ or to a periodic solution.*

When (5.80.4) holds, numerical investigations indicate the existence of solutions of Eq.(5.80.1), which are bounded, not periodic, do not converge to the equilibrium $\bar{x}$, and do not converge to a periodic solution. For this type of solution we pose the following conjecture.

Conjecture 5.80.2 *Assume that (5.80.4) holds. Show that Eq.(5.80.1) has bounded solutions that do not converge to the equilibrium point $\bar{x}$ or to a periodic solution.*

5.81 Equation #81 : $x_{n+1} = \dfrac{\alpha + \delta x_{n-2}}{Bx_n + Dx_{n-2}}$

This equation was investigated in [90]. See also Section 2.1 where we established that every solution of the equation is bounded. This equation can be written in the normalized form

$$x_{n+1} = \frac{\alpha + x_{n-2}}{Bx_n + x_{n-2}}, \quad n = 0, 1, \dots \tag{5.81.1}$$

with positive parameters α, B and with arbitrary positive initial conditions, x_{-2}, x_{-1}, x_0.

The only equilibrium of Eq.(5.81.1) is

$$\bar{x} = \frac{1+\sqrt{1+4\alpha(B+1)}}{2(B+1)}.$$

The characteristic equation of the linearized equation of Eq.(5.81.1) about the equilibrium $\bar{x}$ is

$$\lambda^3 + \frac{B}{B+1}\lambda^2 + \frac{\bar{x}-1}{(B+1)\bar{x}} = 0. \tag{5.81.2}$$

From this and Theorem 1.2.3 it follows that the equilibrium $\bar{x}$ of Eq.(5.81.2) is locally asymptotically stable when

$$B \leq 1+\sqrt{2} \tag{5.81.3}$$

or

$$B > 1+\sqrt{2} \text{ and } \alpha > \frac{2+6B+2B^2+B^3-(1+2B)\sqrt{4+8B+5B^2}}{B^2(9+6B+B^2)} \tag{5.81.4}$$

and unstable when

$$B > 1+\sqrt{2} \text{ and } \alpha < \frac{2+6B+2B^2+B^3-(1+2B)\sqrt{4+8B+5B^2}}{B^2(9+6B+B^2)}. \tag{5.81.5}$$

In the next theorem we establish that when $\alpha > B$ (resp. $\alpha < B$), every solution of Eq.(5.81.1) eventually enters the invariant interval $[1, \frac{\alpha}{B}]$ (resp. $[\frac{\alpha}{B}, 1]$). See [90].

Theorem 5.81.1 *Assume that*

$$\alpha \neq B.$$

Then the following statements are true:

(a) The interval $[\min(\frac{\alpha}{B}, 1), \max(\frac{\alpha}{B}, 1)]$ *is invariant for every solution* $\{x_n\}$ *of Eq.(5.81.1).*

(b) Every solution $\{x_n\}$ *eventually enters the invariant interval* $[\min(\frac{\alpha}{B}, 1), \max(\frac{\alpha}{B}, 1)]$.

PROOF We will give the proof when $\alpha > B$. The proof when $\alpha < B$ is similar and will be omitted.

(*a*) Let $\{x_n\}$ be a solution of Eq.(5.81.1) with initial conditions x_{-2}, x_{-1}, x_0 such that

$$x_{-2,}, x_{-1}, x_0 \in [1, \frac{\alpha}{B}].$$

Then

$$1 = \frac{\alpha + \frac{\alpha}{B}}{B\frac{\alpha}{B} + \frac{\alpha}{B}} \leq x_1 = \frac{\alpha + x_{-2}}{Bx_0 + x_{-2}} \leq \frac{\alpha+1}{B\cdot 1+1} < \frac{\alpha}{B}$$

and by induction it follows that for all $n \geq 1$,

$$1 \leq x_n \leq \frac{\alpha}{B}.$$

(*b*) Let $\{x_n\}$ be a solution of Eq.(5.81.1). Observe that the following hold for $n \geq 1$ and $\alpha > B$:

$$x_{n+1} - 1 = \frac{\alpha - Bx_n}{Bx_n + x_{n-2}}, \tag{5.81.6}$$

and

$$x_{n+1} - 1 = B \cdot \frac{\alpha B(x_{n-1} - 1) + (\alpha - B)x_{n-3}}{(Bx_n + x_{n-2})(Bx_{n-1} + x_{n-3})}$$

$$> \frac{\alpha B^2}{(Bx_n + x_{n-2})(Bx_{n-1} + x_{n-3})} \cdot (x_{n-1} - 1), \tag{5.81.7}$$

and

$$x_{n+1} - \frac{\alpha}{B} = \frac{\alpha B \frac{Bx_{n-1} - \alpha}{Bx_{n-1} + x_{n-3}} + (B - \alpha)x_{n-2}}{Bx_n + x_{n-2}}$$

$$< \frac{\alpha B^2}{(Bx_n + x_{n-2})(Bx_{n-1} + x_{n-3})} \cdot (x_{n-1} - \frac{\alpha}{B}), \tag{5.81.8}$$

and

$$x_{n+1} - x_{n-1} = -\frac{Bx_{n-3}(x_{n-1} - \frac{\alpha}{B}) + (x_{n-1} - 1)(x_{n-3}x_{n-2} + Bx_{n-1}x_{n-2})}{B(\alpha + x_{n-3}) + x_{n-2}(Bx_{n-1} + x_{n-3})}. \tag{5.81.9}$$

Suppose that for some N sufficiently large

$$x_N > \frac{\alpha}{B} \quad \text{or} \quad x_N < 1. \tag{5.81.10}$$

We will prove that the solution $\{x_n\}$ eventually enters the invariant interval $[1, \frac{\alpha}{B}]$ when $x_N > \frac{\alpha}{B}$. The proof when $x_N < 1$ is similar and will be omitted. From (5.81.6), we see that

$$x_{N+1} < 1. \tag{5.81.11}$$

Also,

$$x_{N+2} = \frac{\alpha + x_{N-1}}{Bx_{N+1} + x_{N-1}} > \frac{\alpha + x_{N-1}}{B \cdot 1 + x_{N-1}} > 1$$

and so from (5.81.7), we obtain for all $k \geq 1$,

$$x_{N+2k} > 1. \tag{5.81.12}$$

Furthermore, we claim that for some $k \geq 1$,

$$x_{N+2k} < \frac{\alpha}{B}. \tag{5.81.13}$$

Suppose for the sake of contradiction that for all $k \geq 1$,

$$x_{N+2k} \geq \frac{\alpha}{B}.$$

From (5.81.6) we obtain that for $k \geq 1$,

$$x_{N+2k+1} \leq 1 < \frac{\alpha}{B} \leq x_{N+2k}.$$

From this and (5.81.9) it follows

$$x_{N+2k+3} \leq x_{N+2k+1} \leq 1 < \frac{\alpha}{B} \leq x_{N+2k} \leq x_{N+2k+2}, \quad k = 1, 2, \ldots$$

and so the two subsequences $\{x_{N+2k}\}_{k=1}^{\infty}$ and $\{x_{N+2k+1}\}_{k=1}^{\infty}$ both converge to finite limits. Let

$$\lim_{k \to \infty} x_{N+2k} = y, \quad \lim_{k \to \infty} x_{N+2k+1} = x.$$

Clearly,

$$0 \leq x \leq 1 < \frac{\alpha}{B} \leq y. \tag{5.81.14}$$

By taking limits on both sides of

$$x_{N+2k+3} = \frac{\alpha + x_{N+2k}}{Bx_{N+2k+2} + x_{N+2k}}$$

as $k \to \infty$ we find

$$x = \frac{\alpha + y}{(B+1)y} \quad \text{and} \quad y = \frac{\alpha + x}{(B+1)x},$$

from which it follows that $x = y$. This contradicts (5.81.14) and proves (5.81.13). Assume without loss of generality that (5.81.13) holds for $k = 1$. From this and (5.81.12) we have that

$$1 < x_{N+2} < \frac{\alpha}{B}.$$

From this, in view of (5.81.7) and (5.81.8), we obtain that for all $k \geq 1$,

$$1 < x_{N+2k} < \frac{\alpha}{B}. \tag{5.81.15}$$

From (5.81.6) and (5.81.15), we find that for all $k \geq 1$,

$$1 < x_{N+2k+1} < \frac{\alpha}{B}. \tag{5.81.16}$$

The proof is complete. ∎

In the next theorem we prove that the equilibrium $\bar{x}$ of Eq.(5.81.1) is globally asymptotically stable in some region of the parameters. For another proof see [90].

Theorem 5.81.2 *Assume that*

$$B \leq 4\alpha + 1. \tag{5.81.17}$$

Then the equilibrium $\bar{x}$ of Eq.(5.81.1) is globally asymptotically stable.

PROOF Note that

$$\frac{B-1}{4} > \frac{2+6B+2B^2+B^3-(1+2B)\sqrt{4+8B+5B^2}}{B^2(9+6B+B^2)}.$$

From this and in view of (5.81.17), (5.81.3), and (5.81.4), it follows that the equilibrium $\bar{x}$ of Eq.(5.81.1) is locally asymptotically stable. It suffices to show that the equilibrium is a global attractor of all solutions of Eq.(5.81.1). Let $\{x_n\}$ be a solution of Eq.(5.81.1). When

$$\alpha \neq B,$$

by employing, Theorem 1.6.5 and in view of Theorem 5.81.1, the result follows.

On the other hand, assume

$$\alpha = B.$$

From (5.81.9), we obtain that for all $n \geq 1$,

$$(x_{n+1} - x_{n-1})(x_{n-1} - 1) \leq 0.$$

From this it follows that for all $n \geq 0$,

$$1 \leq x_{2n+1} \leq x_{2n-1} \text{ or } x_{2n-1} \leq x_{2n+1} \leq 1$$

and

$$1 \leq x_{2n} \leq x_{2n-2} \text{ or } x_{2n-2} \leq x_{2n} \leq 1.$$

Hence, the subsequences of the even and odd terms both converge to finite limits. From this and the fact that Eq.(5.81.1) has no prime period-two solutions the result follows. The proof is complete. ∎

Conjecture 5.81.1 *Assume that (5.81.3) or (5.81.4) holds. Show that every solution of Eq.(5.81.1) converges to the equilibrium $\bar{x}$.*

Conjecture 5.81.2 *Show that Eq.(5.81.1) has solutions that do not converge to the equilibrium point $\bar{x}$ or to a periodic solution.*

5.82 Equation #82 : $x_{n+1} = \dfrac{\alpha + \delta x_{n-2}}{Cx_{n-1} + Dx_{n-2}}$

For some work on this equation see [125]. Eq.(#82) can be written in the normalized form

$$x_{n+1} = \frac{\alpha + x_{n-2}}{Cx_{n-1} + x_{n-2}}, \quad n = 0, 1, \ldots \tag{5.82.1}$$

with positive parameters α, C and with arbitrary positive initial conditions x_{-2}, x_{-1}, x_0.

Eq.(5.82.1) has the unique equilibrium

$$\bar{x} = \frac{1 + \sqrt{1 + 4\alpha(1+C)}}{2(1+C)}.$$

The characteristic equation of the linearized equation about the equilibrium $\bar{x}$ is

$$\lambda^3 + \frac{C}{C+1}\lambda + \frac{\bar{x} - 1}{(1+C)\bar{x}} = 0.$$

From this and Theorem 1.2.3 it follows that the equilibrium $\bar{x}$ of Eq.(5.82.1) is locally asymptotically stable when

$$C \leq \frac{1+\sqrt{5}}{2} \tag{5.82.2}$$

or

$$C > \frac{1+\sqrt{5}}{2} \quad \text{and} \quad \alpha > \frac{2 + 4C + C^2 - \sqrt{4 + 16C + 21C^2 + 9C^3}}{C^2} \tag{5.82.3}$$

and unstable when

$$C > \frac{1+\sqrt{5}}{2} \quad \text{and} \quad \alpha < \frac{2 + 4C + C^2 - \sqrt{4 + 16C + 21C^2 + 9C^3}}{C^2}.$$

The following theorem is a new result about the global attractivity of the equilibrium $\bar{x}$ of Eq.(5.82.1).

The following identities, which will be useful in the proof of the theorem that follows, hold for all $n \geq 0$:

$$x_{n+1} - 1 = \frac{\alpha - Cx_{n-2}}{Cx_{n-1} + x_{n-2}} \tag{5.82.4}$$

and

$$x_{n+1} - x_{n-3} = \frac{(\alpha - Cx_{n-3})x_{n-4} + x_{n-2}(x_{n-4} + Cx_{n-3})(1 - x_{n-3})}{C(\alpha + x_{n-4}) + x_{n-2}(Cx_{n-3} + x_{n-4})}. \tag{5.82.5}$$

Theorem 5.82.1 *Assume that*

$$C \leq 1 + 4\alpha. \tag{5.82.6}$$

Then the equilibrium $\bar{x}$ of Eq.(5.82.1) is globally asymptotically stable.

PROOF Note that

$$\frac{C-1}{4} > \frac{2 + 4C + C^2 - \sqrt{4 + 16C + 21C^2 + 9C^3}}{C^2}.$$

From this and in view of (5.82.2), (5.82.3), and (5.82.6), the equilibrium $\bar{x}$ of Eq.(5.82.1) is locally asymptotically stable. It suffices to show that the equilibrium is a global attractor of all solutions of Eq.(5.82.1). We consider the following two cases:

Case 1:

$$\alpha < C$$

and

Case 2:

$$\alpha \geq C.$$

We will give the proof in Case 1. The proof in Case 2 is similar and will be omitted. First we will establish that the interval $[\frac{\alpha}{C}, 1]$ is invariant. Let $\{x_n\}$ be a solution of Eq.(5.82.1) with initial conditions x_{-2}, x_{-1}, x_0 such that

$$x_{-2}, x_{-1}, x_0 \in [\frac{\alpha}{C}, 1] \,.$$

Then

$$\frac{\alpha}{C} < \frac{\alpha + \frac{\alpha}{C}}{C + \frac{\alpha}{C}} < x_1 = \frac{\alpha + x_{-2}}{Cx_{-1} + x_{-2}} < \frac{\alpha + 1}{C\frac{\alpha}{C} + 1} = 1$$

and by induction the result follows.

Next we will show that for $j \in \{0, 1, 2, 3\}$,

$$\{x_{4n+j}\} \text{ eventually enters the invariant interval } [\frac{\alpha}{C}, 1] \,. \tag{5.82.7}$$

We will show that the subsequence $\{x_{4n+1}\}$ eventually enters the invariant interval $[\frac{\alpha}{C}, 1]$. The proof that the other three subsequences eventually enter the invariant interval $[\frac{\alpha}{C}, 1]$ is similar and will be omitted. Suppose for the sake of contradiction that there exists N sufficiently large such that

$$x_{4N+1} < \frac{\alpha}{C} \text{ or } x_{4N+1} > 1.$$

We will give the proof in the case when $x_{4N+1} < \frac{\alpha}{C}$. The proof in the other case is similar and will be omitted. Then

$$x_{4N+3} = \frac{\alpha + x_{4N}}{Cx_{4N+1} + x_{4N}} > \frac{\alpha + x_{4N}}{C\frac{\alpha}{C} + x_{4N}} = 1$$

and from this it follows that

$$x_{4N+5} = \frac{\alpha + x_{4N+2}}{Cx_{4N+3} + x_{4N+2}} < \frac{\alpha + x_{4N+2}}{C \cdot 1 + x_{4N+2}} < 1. \tag{5.82.8}$$

To this end we claim that for some $k \geq 1$,

$$x_{4N+1+4k} > \frac{\alpha}{C}. \tag{5.82.9}$$

Otherwise, for all $k \geq 0$,

$$x_{4N+1+4k} \leq \frac{\alpha}{C}.$$

From (5.82.5) it follows that the subsequence $\{x_{4n+1}\}$ increases. By taking limits in (5.82.5) we get a contradiction.

Assume without loss of generality that (5.82.9) holds for $k = 1$. From this and from (5.82.8), we see that

$$\frac{\alpha}{C} < x_{4N+5} < 1.$$

From (5.82.4), we obtain that

$$x_{4N+7} < 1.$$

In addition,

$$x_{4N+7} > \frac{\alpha}{C}$$

because otherwise

$$x_{4N+7} = \frac{\alpha + x_{4N+4}}{Cx_{4N+5} + x_{4N+4}} \leq \frac{\alpha}{C},$$

implying that

$$x_{4N+5} > 1,$$

which is a contradiction. Similarly, we obtain

$$\frac{\alpha}{C} < x_{4N+9} < 1$$

and by induction, we find that for all $k \geq 1$,

$$\frac{\alpha}{C} < x_{4N+1+4k} < 1,$$

which proves (5.82.7).

The function

$$f(z_2, z_3) = \frac{\alpha + z_3}{Cz_2 + z_3}$$

clearly is strictly decreasing in z_2 and eventually strictly increasing in z_3. By employing Theorem 1.6.5 the result follows. ∎

Conjecture 5.82.1 *Assume that (5.82.2) or (5.82.3) holds. Show that every solution of Eq.(5.82.1) converges to the equilibrium $\bar{x}$.*

Conjecture 5.82.2 *Show that Eq.(5.82.1) has solutions that do not converge to the equilibrium point $\bar{x}$ or to a periodic solution.*

5.83 Equation #83 : $x_{n+1} = \dfrac{\beta x_n + \gamma x_{n-1}}{A + Bx_n}$

This equation was investigated in [179]. See also [175]. Eq.(#83) possesses a period-two trichotomy depending on whether

$$\gamma < \beta + A, \quad \gamma = \beta + A, \quad \text{or} \quad \gamma > \beta + A.$$

This result is a special case of a more general period-two trichotomy result presented in Theorem 4.2.1.

When

$$\gamma > \beta + A,$$

it follows from Theorem 4.2.2 that every positive and bounded solution of Eq.(#83) converges to the positive equilibrium, $\bar{x} = \frac{\beta+\gamma-A}{B}$.

Zero is always an equilibrium of the equation

$$x_{n+1} = \frac{\beta x_n + \gamma x_{n-1}}{A + x_n}, \quad n = 0, 1, \ldots \tag{5.83.1}$$

and when

$$\beta + \gamma > A,$$

Eq.(5.83.1) has, in addition to zero, the unique positive equilibrium

$$\bar{x} = \beta + \gamma - A.$$

By Theorems 5.23.2 and 5.23.4 it follows that the zero equilibrium of Eq.(5.84.1) is globally asymptotically stable when

$$\beta + \gamma \leq A.$$

When

$$A - \beta < \gamma < A + \beta,$$

the positive equilibrium $\bar{x}$ is locally asymptotically stable and by the period-two trichotomy Theorem 4.2.1 it is a global attractor of all positive solutions.

Open Problem 5.83.1 *Obtain "easily verifiable" conditions which determine the set of all positive initial conditions for which the solutions of Eq.(5.83.1) do exactly one of the following:*

(*i*) *converge to a prime period-two solution, when* $\gamma = \beta + A$

(*ii*) *converge to the positive equilibrium, when* $\gamma \geq \beta + A$

(*iii*) *are unbounded, when* $\gamma > \beta + A$.

5.84 Equation #84 : $x_{n+1} = \dfrac{\beta x_n + \gamma x_{n-1}}{A + C x_{n-1}}$

This equation was investigated in [180]. See also [175]. See also Theorem 2.1.1 where we established that every solution of the equation is bounded. Eq.(#84) can be written in the normalized form

$$x_{n+1} = \frac{\beta x_n + x_{n-1}}{A + x_{n-1}}, \quad n = 0, 1, \ldots \tag{5.84.1}$$

with positive parameters β, A and with arbitrary nonnegative initial conditions x_{-1}, x_0.

Zero is always an equilibrium point of Eq.(5.84.1). By Theorems 5.23.2 and 5.23.4 it follows that the zero equilibrium of Eq.(5.84.1) is globally asymptotically stable when

$$A \geq \beta + 1 \tag{5.84.2}$$

and unstable when

$$A < \beta + 1. \tag{5.84.3}$$

Furthermore, when Eq.(5.84.3) holds, Eq.(5.84.1) has also the unique positive equilibrium point

$$\bar{x} = \beta + 1 - A.$$

The characteristic equation of the linearized equation of Eq.(5.84.1) about the positive equilibrium $\bar{x}$ is

$$\lambda^2 - \frac{\beta}{\beta + 1}\lambda + \frac{\beta - A}{\beta + 1} = 0. \tag{5.84.4}$$

From this and Theorem 1.2.2 it follows that the positive equilibrium $\bar{x}$ of Eq.(5.84.1) is locally asymptotically stable for all positive values of the parameters, as long as $A < \beta + 1$.

In the next theorem we prove that the positive equilibrium $\bar{x}$ of Eq.(5.84.1) is a global attractor of all positive solutions provided that $\beta + 1 > A$. For another proof of this result see [175].

Theorem 5.84.1 *Assume that (5.84.3) holds. Then the positive equilibrium $\bar{x}$ of Eq.(5.84.1) is a global attractor of all positive solutions.*

PROOF Let $\{x_n\}$ be a positive solution of Eq.(5.84.1). From Theorem 2.1.1 we know that $\{x_n\}$ is bounded from above by a positive constant. We claim that $\{x_n\}$ is also bounded from below by a positive constant. Otherwise, there exists a sequence of indices $\{n_i\}$ such that

$$x_{n_i+1} \to 0, \quad \text{and} \quad x_{n_i+1} < x_j \quad \text{for all} \quad j < n_i + 1. \tag{5.84.5}$$

Then from (5.84.1), the subsequences $\{x_{n_i}\}$ and $\{x_{n_i-1}\}$ converge to zero. Hence, eventually,

$$x_{n_i}, x_{n_i-1} < \beta + 1 - A,$$

which implies that eventually,

$$x_{n_i+1} = \frac{\beta x_{n_i} + x_{n_i-1}}{A + x_{n_i-1}} > \frac{(\beta+1)\min(x_{n_i}, x_{n_i-1})}{A + (\beta + 1 - A)} = \min(x_{n_i}, x_{n_i-1}).$$

This contradicts (5.84.5) and establishes our claim that the solution is bounded from below by a positive constant. We divide the proof into the following three cases.

Case 1:

$$A - 1 < \beta < A.$$

We claim that the solution $\{x_n\}$ is eventually bounded from above by the positive constant $\frac{A}{\beta}$. Otherwise, for some N sufficiently large

$$x_{N+1} = \frac{\beta x_N + x_{N-1}}{A + x_{N-1}} \geq \frac{A}{\beta},$$

from which it follows that

$$x_N > (\frac{A}{\beta})^2.$$

Similarly, we find that

$$x_{N-1} > (\frac{A}{\beta})^3,$$

which eventually leads to a contradiction and proves our claim that the solution $\{x_n\}$ is eventually bounded from above by the positive constant $\frac{A}{\beta}$. The function

$$f(x_n, x_{n-1}) = \frac{\beta x_n + x_{n-1}}{A + x_{n-1}}$$

is eventually strictly increasing in x_n and x_{n-1}. The result follows by Theorem 1.6.7.

Case 2:

$$\beta > A.$$

We claim that the solution $\{x_n\}$ is eventually bounded from below by the positive constant $\frac{A}{\beta}$. Otherwise, for some N sufficiently large

$$x_{N+1} = \frac{\beta x_N + x_{N-1}}{A + x_{N-1}} \leq \frac{A}{\beta},$$

from which it follows that

$$x_N < (\frac{A}{\beta})^2.$$

Similarly, we find that

$$x_{N-1} < (\frac{A}{\beta})^3,$$

which eventually leads to a contradiction and proves our claim that the solution $\{x_n\}$ is eventually bounded from below by the positive constant $\frac{A}{\beta}$. Hence, for n sufficiently large,

$$x_{n+1} - 1 = \frac{\beta x_n - A}{A + x_{n-1}}$$

and so the solution $\{x_n\}$ is eventually bounded from below by 1. Using the change of variables

$$y_n = \frac{x_n - 1}{\beta},$$

Eq.(5.84.1) reduces to the equation

$$y_{n+1} = \frac{\frac{\beta - A}{\beta^2} + y_n}{\frac{A+1}{\beta} + y_{n-1}}, \quad n = 0, 1, \ldots .$$

The proof in this case is a straightforward consequence of Theorem 1.6.3.

Case 3:

$$\beta = A.$$

For all $n \geq 0$,

$$x_{n+1} - 1 = \beta \cdot \frac{x_n - 1}{\beta + x_{n-1}}$$

and

$$x_{n+1} - x_n = -x_{n-1} \cdot \frac{x_n - 1}{\beta + x_{n-1}}.$$

From this it follows that for $n \geq 0$,

$$x_n \leq x_{n+1} \leq 1 \text{ or } 1 \leq x_{n+1} \leq x_n,$$

from which the result follows. The proof is complete. ∎

Open Problem 5.84.1 *Let $\{\beta_n\}$ and $\{A_n\}$ be periodic sequences of nonnegative real numbers with prime period $k \geq 2$. Determine the global character of solutions of the difference equation*

$$x_{n+1} = \frac{\beta_n x_n + x_{n-1}}{A_n + x_{n-1}}, \quad n = 0, 1, \ldots .$$

Extend and generalize.

Open Problem 5.84.2 *Let $\{\beta_n\}$ and $\{A_n\}$ be convergent sequences of positive real numbers. Investigate the character of solutions of the difference equation*

$$x_{n+1} = \frac{\beta_n x_n + x_{n-1}}{A_n + x_{n-1}}, \quad n = 0, 1, \ldots .$$

Extend and generalize.

Open Problem 5.84.3 *Assume that β and A are given real numbers. Determine the "good" set G of the equation*

$$x_{n+1} = \frac{\beta x_n + x_{n-1}}{A + x_{n-1}}, \tag{5.84.6}$$

that is, the set of initial conditions

$$x_{-2}, x_{-1}, x_0 \in \Re$$

such that the equation (5.84.6) is well defined for all $n \geq 0$. Determine the character of solutions of Eq.(5.84.6) for all initial conditions in the "good" set G.

5.85 Equation #85 : $x_{n+1} = \dfrac{\beta x_n + \gamma x_{n-1}}{A + D x_{n-2}}$

Eq.(#85) can be written in the normalized form

$$x_{n+1} = \frac{x_n + \gamma x_{n-1}}{A + x_{n-2}}, \quad n = 0, 1, \ldots \tag{5.85.1}$$

with positive parameters γ, A and with arbitrary nonnegative initial conditions x_{-2}, x_{-1}, x_0.

The boundedness character of this equation was established in Theorem 3.2.1 where it was shown that when

$$\gamma > A + 1,$$

the equation has unbounded solutions. See also [49].

Zero is always an equilibrium point of Eq.(5.85.1). By Theorems 5.23.2 and 5.23.4 it follows that the zero equilibrium of Eq.(5.85.1) is globally asymptotically stable when

$$A \geq \gamma + 1 \tag{5.85.2}$$

and unstable when

$$A < \gamma + 1. \tag{5.85.3}$$

Furthermore, when Eq.(5.85.3) holds, Eq.(5.85.1) has also the unique positive equilibrium point

$$\bar{x} = \gamma + 1 - A.$$

The characteristic equation of the linearized equation of Eq.(5.85.1) about the positive equilibrium $\bar{x}$ is

$$\lambda^3 - \frac{1}{\gamma+1}\lambda^2 - \frac{\gamma}{\gamma+1}\lambda + \frac{\gamma+1-A}{\gamma+1} = 0. \tag{5.85.4}$$

From this and Theorem 1.2.3 it follows that the positive equilibrium $\bar{x}$ of Eq.(5.85.1) is locally asymptotically stable when

$$\sqrt{2A^2 - 3A + 1} - A < \gamma < A + 1 \tag{5.85.5}$$

and unstable when

$$\gamma < \sqrt{2A^2 - 3A + 1} - A.$$

Conjecture 5.85.1 *Show that every positive solution of Eq.(5.85.1) converges to the positive equilibrium $\bar{x}$ when*

$$\sqrt{2A^2 - 3A + 1} - A < \gamma < A + 1.$$

Conjecture 5.85.2 *Assume that*

$$\gamma > A + 1.$$

Show that every positive and bounded solution of Eq.(5.85.1) converges to the positive equilibrium $\bar{x}$.

Conjecture 5.85.3 *Assume that*

$$\gamma = A + 1.$$

Show that every solution of Eq.(5.85.1) converges to a (not necessarily prime) period-two solution.

Conjecture 5.85.4 *Assume that*

$$\gamma < \sqrt{2A^2 - 3A + 1} - A.$$

Show that Eq.(5.85.1) has bounded solutions that do not converge to an equilibrium point or to a periodic solution.

5.86 Equation #86 : $x_{n+1} = \dfrac{\beta x_n + \gamma x_{n-1}}{Bx_n + Cx_{n-1}}$

This equation was investigated in [175], [181], [184], and [205]. Eq.(#86) can be written in the normalized form

$$x_{n+1} = \frac{\beta x_n + x_{n-1}}{Bx_n + x_{n-1}}, \quad n = 0, 1, \ldots \tag{5.86.1}$$

with positive parameters β, B and with arbitrary positive initial conditions x_{-1}, x_0. We also assume that $\beta \neq B$ because otherwise the equation eventually becomes trivial.

The only equilibrium of Eq.(5.86.1) is

$$\bar{x} = \frac{\beta + 1}{B + 1}.$$

The characteristic equation of the linearized equation of Eq.(5.86.1) about the equilibrium $\bar{x}$ is

$$\lambda^2 - \frac{\beta - B}{(\beta + 1)(B + 1)}\lambda + \frac{\beta - B}{(\beta + 1)(B + 1)} = 0.$$

From this and Theorem 1.2.2 it follows that the equilibrium $\bar{x}$ of Eq.(5.86.1) is locally asymptotically stable when

$$\beta > B \tag{5.86.2}$$

or

$$\beta < B \text{ and } B < 3\beta + \beta B + 1 \tag{5.86.3}$$

and unstable when

$$\beta < B \text{ and } B > 3\beta + \beta B + 1. \tag{5.86.4}$$

When (5.86.4) holds, and only then, Eq.(5.86.1) possesses the unique prime period-two solution

$$\ldots, \frac{1-\beta-\sqrt{(1-\beta)^2-\frac{4\beta(1-\beta)}{B-1}}}{2}, \frac{1-\beta+\sqrt{(1-\beta)^2-\frac{4\beta(1-\beta)}{B-1}}}{2}, \ldots, \tag{5.86.5}$$

which is locally asymptotically stable. For the proof of this, see [175].

In the next theorem we present the global character of solutions of Eq.(5.86.1). See [181], [184], and [205].

Theorem 5.86.1 *The following statements are true:*
(a) The equilibrium $\bar{x}$ of Eq.(5.86.1) is globally asymptotically stable when (5.86.2) or (5.86.3) holds.

(b) Every solution of Eq.(5.86.1) converges to the equilibrium $\bar{x}$ of Eq.(5.86.1) when

$$\beta < B \text{ and } B = 3\beta + \beta B + 1. \tag{5.86.6}$$

(c) Every solution of Eq.(5.86.1) converges to a (not necessarily prime) period-two solution when (5.86.4) holds.

PROOF Let $\{x_n\}$ be a solution of Eq.(5.86.1) and assume that (5.86.2) holds. For all $n \geq 0$,

$$1 < x_{n+1} = \frac{\beta x_n + x_{n-1}}{Bx_n + x_{n-1}} = \frac{\beta}{B} \cdot \frac{\beta B x_n + B x_{n-1}}{\beta B x_n + \beta x_{n-1}} < \frac{\beta}{B},$$

which implies that the interval $[1, \frac{\beta}{B}]$ is invariant for the solution $\{x_n\}$. Furthermore, the solution $\{x_n\}$ satisfies the following equation:

$$x_{n+1} = \frac{\beta \cdot \frac{\beta x_{n-1}+x_{n-2}}{Bx_{n-1}+x_{n-2}} + x_{n-1}}{B \cdot \frac{\beta x_{n-1}+x_{n-2}}{Bx_{n-1}+x_{n-2}} + x_{n-1}}$$

$$= F(x_{n-1}, x_{n-2}) = \frac{\beta^2 x_{n-1} + Bx_{n-1}^2 + \beta x_{n-2} + x_{n-1}x_{n-2}}{\beta B x_{n-1} + Bx_{n-1}^2 + Bx_{n-2} + x_{n-1}x_{n-2}}. \tag{5.86.7}$$

Clearly,

$$F \in C([1, \frac{\beta}{B}]^2, [1, \frac{\beta}{B}]),$$

and

$$F_{x_{n-1}} = \frac{(B-\beta)\beta B x_{n-1}^2 + 2\beta B(B-\beta)x_{n-1}x_{n-2} + (B-\beta)x_{n-2}^2}{(\beta B x_{n-1} + Bx_{n-1}^2 + Bx_{n-2} + x_{n-1}x_{n-2})^2} < 0,$$

and

$$F_{x_{n-2}} = \frac{-x_{n-1}^2(B-\beta)^2}{(\beta B x_{n-1} + Bx_{n-1}^2 + Bx_{n-2} + x_{n-1}x_{n-2})^2} < 0,$$

and for each $m, M \in [1, \frac{\beta}{B}]$, the system

$$M = \frac{\beta^2 m + Bm^2 + \beta m + m^2}{\beta Bm + Bm^2 + Bm + m^2} \text{ and } m = \frac{\beta^2 M + BM^2 + \beta M + M^2}{\beta BM + BM^2 + BM + M^2}$$

has the unique solution $(m, M) = (\bar{x}, \bar{x})$. By employing Theorem 1.6.5 the result follows.

On the other hand, assume that

$$\beta < B.$$

Clearly, the function

$$\frac{\beta x_n + x_{n-1}}{Bx_n + x_{n-1}}$$

is strictly decreasing in x_n and strictly increasing in x_{n-1}. By employing Theorem 1.6.6, we find that the solution $\{x_n\}$ converges to a (not necessarily prime) period-two solution. Due to the fact that Eq.(5.86.1) possesses a prime period-two solution only when (5.86.4) holds, (a), (b), and (c) follow. The proof is complete. ▮

Open Problem 5.86.1 *Assume that β and B are given real numbers. Determine the "good" set of the equation*

$$x_{n+1} = \frac{\beta x_n + x_{n-1}}{Bx_n + x_{n-1}}, \tag{5.86.8}$$

that is, the set of initial conditions

$$x_{-1}, x_0 \in \Re$$

such that the equation (5.86.8) is well defined for all $n \geq 0$. Determine the character of solutions of Eq.(5.86.8) for all initial conditions in the "good" set G.

5.87 Equation #87 : $x_{n+1} = \dfrac{\beta x_n + \gamma x_{n-1}}{Bx_n + Dx_{n-2}}$

Eq.(#87) can be written in the normalized form

$$x_{n+1} = \frac{x_n + \gamma x_{n-1}}{x_n + Dx_{n-2}}, \quad n = 0, 1, \ldots \tag{5.87.1}$$

with positive parameters γ, D and with arbitrary positive initial conditions x_{-2}, x_{-1}, x_0.

For some work on this equation see [49]. See also Section 3.3 where we established that when

$$\gamma > 1,$$

the equation has unbounded solutions.

Eq.(5.87.1) has the unique equilibrium point

$$\bar{x} = \frac{\gamma + 1}{D + 1}.$$

The characteristic equation of the linearized equation of Eq.(5.87.1) about the equilibrium $\bar{x}$ is

$$\lambda^3 + \frac{\gamma - D}{(\gamma + 1)(D + 1)}\lambda^2 - \frac{\gamma}{\gamma + 1}\lambda + \frac{D}{D + 1} = 0.$$

From this and Theorem 1.2.3 it follows that the equilibrium $\bar{x}$ of Eq.(5.87.1) is locally asymptotically stable when

$$\frac{D^2 - 2D - 1}{D^2 + 5D + 2} < \gamma < 1 \tag{5.87.2}$$

and unstable when

$$\gamma < \frac{D^2 - 2D - 1}{D^2 + 5D + 2} \quad \text{or} \quad \gamma > 1.$$

When

$$\gamma = 1,$$

Eq.(5.87.1) has infinitely many prime period-two solutions of the form

$$\ldots, x, \frac{x}{(D + 1)x - 1}, x, \frac{x}{(D + 1)x - 1}, \ldots$$

with

$$x \in (\frac{1}{D + 1}, \infty) \text{ and } x \neq \frac{2}{D + 1}.$$

Conjecture 5.87.1 *Assume that (5.87.2) holds. Show that every solution of Eq.(5.87.1) converges to the equilibrium $\bar{x}$.*

Conjecture 5.87.2 *Assume that*

$$\gamma = 1.$$

Show that every solution of Eq.(5.87.1) converges to a (not necessarily prime) period-two solution.

Conjecture 5.87.3 *Assume that*

$$\gamma < \frac{D^2 - 2D - 1}{D^2 + 5D + 2}.$$

Show that Eq.(5.87.1) has bounded solutions that do not converge to the equilibrium point $\bar{x}$ or to a periodic solution.

Open Problem 5.87.1 *Assume that*

$$\gamma > 1.$$

(i) Determine the set of all initial conditions through which solutions of Eq.(5.87.1) are unbounded.

(ii) Determine the set of all initial conditions through which solutions of Eq.(5.87.1) converge to the equilibrium $\bar{x}$.

Conjecture 5.87.4 *Assume that*

$$\gamma > 1.$$

Show that every bounded solution of Eq.(5.87.1) converges to the equilibrium $\bar{x}$.

5.88 Equation #88 : $x_{n+1} = \dfrac{\beta x_n + \gamma x_{n-1}}{Cx_{n-1} + Dx_{n-2}}$

This equation was investigated in [14]. See also Theorem 2.8.1 where we have shown that every solution of Eq.(5.88.1) is bounded. Eq.(#88) can be written in the normalized form

$$x_{n+1} = \frac{\beta x_n + x_{n-1}}{x_{n-1} + Dx_{n-2}}, \quad n = 0, 1, \ldots \tag{5.88.1}$$

with positive parameters β, D and with arbitrary positive initial conditions x_{-2}, x_{-1}, x_0.

The only equilibrium of Eq.(5.88.1) is

$$\bar{x} = \frac{\beta + 1}{D + 1}.$$

The characteristic equation of the linearized equation of Eq.(5.88.1) about the equilibrium $\bar{x}$ is

$$\lambda^3 - \frac{\beta}{\beta+1}\lambda^2 + \frac{\beta - D}{(\beta+1)(D+1)}\lambda + \frac{D}{D+1} = 0.$$

From this and Theorem 1.2.3 it follows that the equilibrium $\bar{x}$ of Eq.(5.88.1) is locally asymptotically stable when

$$D \leq 1 \text{ or } D > 1 \text{ and } \frac{D-1}{D+3} < \beta < \frac{D^2+3D+1}{D^2} \tag{5.88.2}$$

and unstable when

$$D > 1 \text{ and } \beta < \frac{D-1}{D+3} \text{ or } \beta > \frac{D^2+3D+1}{D^2}.$$

Conjecture 5.88.1 *When (5.88.2) holds, every solution of Eq.(5.88.1) converges to the equilibrium $\bar{x}$.*

Conjecture 5.88.2 *Assume that*

$$D > 1 \text{ and } \beta > \frac{D^2+3D+1}{D^2}.$$

Show that Eq.(5.88.1) has solutions that do not converge to the equilibrium point $\bar{x}$ or to a periodic solution.

Open Problem 5.88.1 *Investigate the global character of solutions of Eq.(5.88.1) when*

$$\beta = \frac{D-1}{D+3} > 0.$$

When

$$D > 1 \text{ and } \beta < \frac{D-1}{D+3}, \tag{5.88.3}$$

Eq.(5.88.1) possesses a unique period-two solution of the form

$$\ldots, x, y, x, y, \ldots,$$

where x, y are the two positive solutions of the equation

$$(D-1)t^2 + (\beta - 1)(D-1)t + \beta(1-\beta) = 0.$$

Conjecture 5.88.3 *Show that when (5.88.3) holds the unique prime period-two solution of Eq.(5.88.1) is locally asymptotically stable.*

5.89 Equation #89 : $x_{n+1} = \dfrac{\beta x_n + \delta x_{n-2}}{A + Bx_n}$

This equation was investigated in [49]. See also Section 2.5 where we established that every solution of the equation is bounded. Eq.(#89) can be written in the normalized form

$$x_{n+1} = \frac{\beta x_n + \delta x_{n-2}}{1 + x_n}, \quad n = 0, 1, \ldots \tag{5.89.1}$$

with positive parameters β, δ and with arbitrary nonnegative initial conditions x_{-2}, x_{-1}, x_0.

Zero is always an equilibrium point of Eq.(5.89.1). By Theorems 5.23.2 and 5.23.4 it follows that the zero equilibrium of Eq.(5.89.1) is globally asymptotically stable when

$$\beta + \delta \leq 1 \tag{5.89.2}$$

and unstable when

$$\beta + \delta > 1. \tag{5.89.3}$$

Furthermore, when Eq.(5.89.3) holds, Eq.(5.89.1) has also the unique positive equilibrium point

$$\bar{x} = \beta + \delta - 1.$$

The characteristic equation of the linearized equation of Eq.(5.89.1) about the positive equilibrium, $\bar{x} = \beta + \delta - 1$, is

$$\lambda^3 + \frac{\delta - 1}{\beta + \delta}\lambda^2 - \frac{\delta}{\beta + \delta} = 0.$$

From this and Theorem 1.2.3 it follows that $\bar{x} = \beta + \delta - 1$ is locally asymptotically stable when

$$\beta > \sqrt{2\delta^2 - \delta} - \delta \tag{5.89.4}$$

and unstable when

$$\beta < \sqrt{2\delta^2 - \delta} - \delta.$$

When

$$\beta = \sqrt{2\delta^2 - \delta} - \delta,$$

one of the characteristic roots is real within the interval $(0, 1)$ and the other two characteristic roots are complex conjugates with magnitude equal to one.

In Section 2.5 we proved that every solution of Eq.(5.89.1) is bounded from above. Here we will also show that when (5.89.3) holds, every positive solution of Eq.(5.89.1) is also bounded from below by a positive constant. Assume for the sake of contradiction that there exists a positive solution $\{x_n\}$ of Eq.(5.89.1), which contains a subsequence $\{x_{n_i+1}\}$ such that

$$x_{n_i+1} \to 0 \text{ and } x_{n_i+1} < x_j, \;\; j < n_i + 1. \tag{5.89.5}$$

Clearly,

$$x_{n_i} \to 0 \text{ and } x_{n_i-2} \to 0.$$

Hence, there exists i_0 such that

$$x_{n_{i_0}} < \beta + \delta - 1.$$

Then

$$x_{n_{i_0}+1} = \frac{\beta x_{n_{i_0}} + \delta x_{n_{i_0}-2}}{1 + x_{n_{i_0}}} > (\beta+\delta)\frac{\min(x_{n_{i_0}}, x_{n_{i_0}-2})}{\beta+\delta} = \min(x_{n_{i_0}}, x_{n_{i_0}-2}).$$

This contradicts (5.89.5) and the proof is complete.

The next three theorems are new results about the global attractivity of the positive equilibrium $\bar{x}$ of Eq.(5.89.1).

Theorem 5.89.1 *Assume that*

$$\delta = 1.$$

Then every positive solution of Eq.(5.89.1) converges to β.

PROOF For all $n \geq 0$,

$$x_{n+1} - \beta = \frac{x_{n-2} - \beta}{1 + x_n}$$

and

$$x_{n+1} - x_{n-2} = x_n \cdot \frac{\beta - x_{n-2}}{1 + x_n}.$$

From this it follows that, for all $n \geq 0$,

$$\beta \leq x_{3n+1} \leq x_{3n-2}$$

or

$$x_{3n-2} \leq x_{3n+1} \leq \beta$$

and so the sequence $\{x_{3n+1}\}$ converges to a finite limit. Similarly, it follows that the sequences $\{x_{3n+2}\}$ and $\{x_{3n+3}\}$ converge to finite limits. Due to the fact that when $\delta = 1$, Eq.(5.89.1) has no prime period-three solutions, the result follows. The proof is complete. ∎

Theorem 5.89.2 *Assume that*

$$\beta > 1 - \delta > 0.$$

Then every positive solution of Eq.(5.89.1) converges to the positive equilibrium $\bar{x} = \beta + \delta - 1$.

PROOF We claim that eventually

$$x_n < \beta < \frac{1}{\delta} \cdot \beta. \tag{5.89.6}$$

Assume for the sake of contradiction that for some $N > 0$

$$x_{N+1} = \frac{\beta x_N + \delta x_{N-2}}{1 + x_N} \geq \beta.$$

From this it follows that

$$x_{N-2} \geq \frac{1}{\delta} \cdot \beta$$

and, similarly,

$$x_{N-5} \geq (\frac{1}{\delta})^2 \beta.$$

Inductively, we find

$$x_{N+1-3k} > (\frac{1}{\delta})^k \beta,$$

which is a contradiction and proves (5.89.6). Clearly, the function

$$f(x_n, x_{n-2}) = \frac{\beta x_n + \delta x_{n-2}}{1 + x_n}$$

is strictly increasing in x_n and x_{n-2}. By employing Theorem 1.6.7 the result follows. The proof is complete. ∎

Theorem 5.89.3 *Assume that*

$$\beta \geq \delta - 1 > 0.$$

Then every positive solution of Eq.(5.89.1) converges to the positive equilibrium $\bar{x} = \beta + \delta - 1$.

PROOF We claim that eventually

$$x_n > \beta > \frac{1}{\delta} \cdot \beta. \tag{5.89.7}$$

Assume for the sake of contradiction that for some $N > 0$

$$x_{N+1} = \frac{\beta x_N + \delta x_{N-2}}{1 + x_N} \leq \beta.$$

From this it follows that

$$x_{N-2} \leq \frac{1}{\delta} \cdot \beta$$

and, similarly,

$$x_{N-5} \leq (\frac{1}{\delta})^2 \cdot \beta.$$

Inductively, we find

$$x_{N+1-3k} \leq (\frac{1}{\delta})^k \cdot \beta,$$

which is a contradiction and proves (5.89.7). Clearly, the function

$$f(x_n, x_{n-2}) = \frac{\beta x_n + \delta x_{n-2}}{1 + x_n}$$

is strictly decreasing in x_n and strictly increasing in x_{n-2}. When

$$\beta > \delta - 1 > 0$$

by employing Theorem 1.6.7 the result follows. When

$$\beta = \delta - 1 > 0,$$

the Hypotheses of Theorem 1.6.8 are satisfied from which the result follows. The proof is complete. ∎

Conjecture 5.89.1 *Assume that*

$$\sqrt{2\delta^2 - \delta} - \delta < \beta < \delta - 1.$$

Show that every positive solution $\{x_n\}$ of Eq.(5.89.1) converges to the positive equilibrium $\bar{x} = \beta + \delta - 1$.

Conjecture 5.89.2 *Assume that*

$$\beta < \sqrt{2\delta^2 - \delta} - \delta.$$

Show that Eq.(5.89.1) has solutions that do not converge to an equilibrium point or to a periodic solution.

5.90 Equation #90 : $x_{n+1} = \dfrac{\beta x_n + \delta x_{n-2}}{A + C x_{n-1}}$

This equation was investigated in [49] . See also Section 2.7 where we established that every solution of the equation is bounded. Eq.(#90) can be written in the normalized form

$$x_{n+1} = \frac{\beta x_n + \delta x_{n-2}}{1 + x_{n-1}}, \quad n = 0, 1, \ldots \tag{5.90.1}$$

with positive parameters β, δ and with arbitrary nonnegative initial conditions x_{-2}, x_{-1}, x_0.

Zero is always an equilibrium point of Eq.(5.90.1). By Theorems 5.23.2 and 5.23.4, the zero equilibrium of Eq.(5.90.1) is globally asymptotically stable when

$$\beta + \delta \leq 1 \tag{5.90.2}$$

and unstable when

$$\beta + \delta > 1. \tag{5.90.3}$$

Furthermore, when Eq.(5.90.3) holds, Eq.(5.90.1) has also the unique positive equilibrium point

$$\bar{x} = \beta + \delta - 1.$$

The characteristic equation of the linearized equation of Eq.(5.90.1) about the positive equilibrium, $\bar{x} = \beta + \delta - 1$, is

$$\lambda^3 - \frac{\beta}{\beta + \delta}\lambda^2 + \frac{\beta + \delta - 1}{\beta + \delta}\lambda - \frac{\delta}{\beta + \delta} = 0.$$

From this and Theorem 1.2.3 it follows that $\bar{x} = \beta + \delta - 1$ is locally asymptotically stable when

$$\beta > \frac{\delta^2 - \delta}{\delta + 1} \tag{5.90.4}$$

and unstable when

$$\beta < \frac{\delta^2 - \delta}{\delta + 1}.$$

Conjecture 5.90.1 *Assume that*

$$\beta > \frac{\delta^2 - \delta}{\delta + 1}$$

Show that every positive solution $\{x_n\}$ of Eq.(5.90.1) converges to the positive equilibrium $\bar{x} = \beta + \delta - 1$.

Conjecture 5.90.2 *Assume that*

$$\beta < \frac{\delta^2 - \delta}{\delta + 1}.$$

Show that Eq.(5.90.1) has solutions that do not converge to an equilibrium point or to a periodic solution.

5.91 Equation #91 : $x_{n+1} = \dfrac{\beta x_n + \delta x_{n-2}}{A + D x_{n-2}}$

The boundedness character of this equation was investigated in [49]. See also Theorem 2.3.3 where we established that every solution of the equation is bounded. Eq.(#91) can be written in the normalized form

$$x_{n+1} = \frac{\beta x_n + x_{n-2}}{A + x_{n-2}}, \quad n = 0, 1, \ldots \tag{5.91.1}$$

with positive parameters β, A and with arbitrary nonnegative initial conditions x_{-2}, x_{-1}, x_0.

Zero is always an equilibrium of Eq.(5.91.1). By Theorems 5.23.2 and 5.23.4 it follows that when

$$A \geq \beta + 1$$

the zero equilibrium of Eq.(5.91.1) is globally asymptotically stable.

When

$$A < \beta + 1,$$

Eq.(5.91.1) has the positive equilibrium

$$\bar{x} = \beta + 1 - A.$$

The characteristic equation of the linearized equation of Eq.(5.91.1) about the positive equilibrium, $\bar{x} = \beta + 1 - A$, is

$$\lambda^3 - \frac{\beta}{\beta+1}\lambda^2 + \frac{\beta - A}{\beta + 1} = 0.$$

From this and Theorem 1.2.3 it follows that $\bar{x} = \beta + 1 - A$ is locally asymptotically stable when

$$\beta \leq 1 + \sqrt{2} \text{ and } A < \beta + 1 \tag{5.91.2}$$

or when

$$\beta > 1 + \sqrt{2} \text{ and } \frac{3\beta - \sqrt{5\beta^2 + 8\beta + 4}}{2} < A < \beta + 1 \tag{5.91.3}$$

and unstable when

$$\beta > 1+\sqrt{2} \text{ and } A < \frac{3\beta - \sqrt{5\beta^2+8\beta+4}}{2}.$$

The next theorem is a new result about the global attractivity of the positive equilibrium of Eq.(5.91.1).

Theorem 5.91.1 *Assume that*

$$\beta - 1 \le A < \beta + 1.$$

Then every positive solution of Eq.(5.91.1) converges to the positive equilibrium $\bar{x}$.

PROOF Let $\{x_n\}$ be a positive solution of Eq.(5.91.1). We consider the following three cases:

Case 1:

$$\beta - 1 \le A < \beta.$$

We claim that, eventually,

$$x_n > \frac{A}{\beta}.$$

Otherwise, there exists N sufficiently large such that

$$x_{N+1} = \frac{\beta x_N + x_{N-2}}{A + x_{N-2}} \le \frac{A}{\beta}.$$

From this it follows that

$$x_N < (\frac{A}{\beta})^2$$

and, similarly,

$$x_{N-1} < (\frac{A}{\beta})^3,$$

which eventually leads to a contradiction. Clearly, the function

$$f(x_n, x_{n-2}) = \frac{\beta x_n + x_{n-2}}{A + x_{n-2}}$$

is strictly increasing in x_n and eventually strictly decreasing in x_{n-2}. When

$$\beta - 1 < A < \beta,$$

by employing Theorem 1.6.7 the result follows. When

$$A = \beta - 1,$$

the Hypotheses of Theorem 1.6.8 are satisfied from which the result follows.

Case 2:

$$\beta < A < \beta + 1.$$

We claim that, eventually,

$$x_n < \frac{A}{\beta}.$$

Otherwise, there exists N sufficiently large such that

$$x_{N+1} = \frac{\beta x_N + x_{N-2}}{A + x_{N-2}} \geq \frac{A}{\beta}.$$

From this it follows that

$$x_N > (\frac{A}{\beta})^2$$

and, similarly,

$$x_{N-1} > (\frac{A}{\beta})^3,$$

which eventually leads to a contradiction. Clearly, the function

$$f(x_n, x_{n-2}) = \frac{\beta x_n + x_{n-2}}{A + x_{n-2}}$$

is strictly increasing in x_n and eventually strictly increasing in x_{n-2}. By employing Theorem 1.6.7 the result follows.

Case 3:

$$\beta = A.$$

For all $n \geq 0$,

$$x_{n+1} - 1 = \beta \cdot \frac{x_n - 1}{\beta + x_{n-2}} \text{ and } x_{n+1} - x_n = x_{n-2} \cdot \frac{1 - x_n}{\beta + x_{n-2}}.$$

Then, clearly, for all $n \geq 0$,

$$1 \leq x_{n+1} \leq x_n \text{ or } x_n \leq x_{n+1}, \leq 1$$

from which the result follows. The proof is complete. ∎

From Theorem 5.91.1 and in view of (5.91.2) and (5.91.3) it follows that when

$$A - 1 < \beta \leq 1,$$

the positive equilibrium $\bar{x}$ of Eq.(5.91.1) and with positive initial conditions is globally asymptotically stable.

Conjecture 5.91.1 *Assume that $\beta \in (1, \infty)$ and that*

$$\frac{3\beta - \sqrt{5\beta^2 + 8\beta + 4}}{2} < A < \beta - 1.$$

Show that every positive solution of Eq.(5.91.1) converges to the positive equilibrium $\bar{x}$.

Conjecture 5.91.2 *Assume that*

$$A < \frac{3\beta - \sqrt{5\beta^2 + 8\beta + 4}}{2}.$$

Show that Eq.(5.91.1) has solutions that do not converge to an equilibrium point or to a periodic solution.

5.92 Equation #92 : $x_{n+1} = \dfrac{\beta x_n + \delta x_{n-2}}{Bx_n + Cx_{n-1}}$

Eq.(#92) can be written in the normalized form

$$x_{n+1} = \frac{x_n + \delta x_{n-2}}{x_n + Cx_{n-1}}, \quad n = 0, 1, \ldots \tag{5.92.1}$$

with positive parameters δ, C and with arbitrary positive initial conditions x_{-2}, x_{-1}, x_0.

The boundedness character of solutions of this equation was investigated in [69]. See also Theorem 3.4.1 where we established that the Eq.(5.92.1) has unbounded solutions when

$$\delta > C.$$

From this and Theorem 5.221.1 it follows that every solution of Eq.(5.92.1) is bounded if and only if

$$\delta \leq C.$$

Eq.(5.92.1) has the unique equilibrium

$$\bar{x} = \frac{1+\delta}{1+C}.$$

The characteristic equation of the linearized equation of Eq.(5.92.1) about the equilibrium $\bar{x}$ is

$$\lambda^3 + \frac{\delta - C}{(1+\delta)(1+C)}\lambda^2 + \frac{C}{1+C}\lambda - \frac{\delta}{1+\delta} = 0. \tag{5.92.2}$$

From this and Theorem 1.2.3 it follows that the equilibrium $\bar{x}$ of Eq.(5.92.1) is locally asymptotically stable when

$$\delta < \frac{C+2+\sqrt{C^2+8C+8}}{2(1+C)} \tag{5.92.3}$$

and unstable when

$$\delta > \frac{C+2+\sqrt{C^2+8C+8}}{2(1+C)}. \tag{5.92.4}$$

It is interesting to note that for the equilibrium $\bar{x}$ of Eq.(5.92.1),

Local Asymptotic Stabilty $\not\Rightarrow$ Global Attractivity.

Indeed, for all positive values of C for which

$$C < \frac{C+2+\sqrt{C^2+8C+8}}{2(1+C)}$$

and for all values of δ such that

$$C < \delta < \frac{C+2+\sqrt{C^2+8C+8}}{2(1+C)},$$

the equilibrium $\bar{x}$ of Eq.(5.92.1) is locally asymptotically stable and at the same time the equation has unbounded solutions. In particular, for such initial conditions the equilibrium of the equation is not a global attractor.

Conjecture 5.92.1 *Assume that*

$$\delta \leq C.$$

Show that for the equilibrium $\bar{x}$ of Eq.(5.92.1),

Local Asymptotic Stabilty $\Longrightarrow$ Global Attractivity.

Open Problem 5.92.1 *Assume that*

$$\delta > C.$$

(*i*) *Determine the set of all initial conditions for which the solutions of Eq.(5.92.1) converge to the equilibrium $\bar{x}$.*

(*ii*) *Determine the set of all initial conditions for which the solutions of Eq.(5.92.1) are unbounded.*

(*iii*) *Determine all possible periodic solutions of Eq.(5.92.1).*

Conjecture 5.92.2 *Assume that (5.92.4) holds. Show that Eq.(5.92.1) has bounded solutions that do not converge to the equilibrium point $\bar{x}$ or to a periodic solution.*

5.93 Equation #93 : $x_{n+1} = \dfrac{\beta x_n + \delta x_{n-2}}{Bx_n + Dx_{n-2}}$

Eq.(#93) can be written in the normalized form

$$x_{n+1} = \frac{\beta x_n + x_{n-2}}{Bx_n + x_{n-2}}, \quad n = 0, 1, \ldots \tag{5.93.1}$$

with positive parameters β, B and with arbitrary positive initial conditions x_{-2}, x_{-1}, x_0. For some work on this equation see [141].

The only equilibrium of Eq.(5.93.1) is

$$\bar{x} = \frac{\beta + 1}{B + 1}.$$

The characteristic equation of the linearized equation of Eq.(5.93.1) about the equilibrium $\bar{x}$ is

$$\lambda^3 - \frac{\beta - B}{(\beta + 1)(B + 1)}\lambda^2 + \frac{\beta - B}{(\beta + 1)(B + 1)} = 0.$$

From this and Theorem 1.2.3 it follows that the equilibrium $\bar{x}$ of Eq.(5.93.1) is locally asymptotically stable when

$$-1 + \sqrt{2} \le B \le 1 + \sqrt{2}, \tag{5.93.2}$$

or

$$B < -1 + \sqrt{2} \text{ and } \beta < \frac{B\sqrt{2} + B + 1}{\sqrt{2} - B - 1}, \tag{5.93.3}$$

or

$$B > 1 + \sqrt{2} \text{ and } \beta > \frac{B\sqrt{2} - B - 1}{\sqrt{2} + B + 1} \tag{5.93.4}$$

and unstable when

$$B < -1 + \sqrt{2} \text{ and } \beta > \frac{B\sqrt{2} + B + 1}{\sqrt{2} - B - 1} \tag{5.93.5}$$

or

$$B > 1 + \sqrt{2} \text{ and } \beta < \frac{B\sqrt{2} - B - 1}{\sqrt{2} + B + 1}. \tag{5.93.6}$$

The next theorem is a new result about the global stability of the equilibrium $\bar{x}$ of Eq.(5.93.1).

Theorem 5.93.1 *Assume that*

$$1 \leq \beta < B, \quad or \ \beta < B \leq 1, \quad or \ \frac{B-1}{B+3} \leq \beta < 1 < B \tag{5.93.7}$$

or

$$B < \beta \leq 1, \quad or \ 1 \leq B < \beta, \quad or \ B < 1 < \frac{3B+1}{1-B} \leq \beta. \tag{5.93.8}$$

Then the equilibrium $\bar{x}$ of Eq.(5.93.1) is globally asymptotically stable.

PROOF When (5.93.7) or (5.93.8) holds, in view of (5.93.2), (5.93.3), and (5.93.4) the equilibrium $\bar{x}$ of Eq.(5.93.1) is locally asymptotically stable. It suffices to show that the equilibrium $\bar{x}$ is a global attractor of all solutions of Eq.(5.93.1).

Let $\{x_n\}$ be a solution of Eq.(5.93.1). For all $n \geq 0$,

$$\min(\frac{\beta}{B}, 1) < x_{n+1} = \frac{\beta x_n + x_{n-2}}{B x_n + x_{n-2}} < \max(\frac{\beta}{B}, 1)$$

and so the interval $[\min(\frac{\beta}{B}, 1), \max(\frac{\beta}{B}, 1)]$ is invariant for the solution $\{x_n\}$. When (5.93.7) holds, the function

$$f(x_n, x_{n-2}) = \frac{\beta x_n + x_{n-2}}{B x_n + x_{n-2}}$$

is strictly decreasing in x_n and strictly increasing in x_{n-2}, and for each $m, M \in [\frac{\beta}{B}, 1]$, the system

$$M = \frac{\beta m + M}{B m + M} \quad \text{and} \quad m = \frac{\beta M + m}{B M + m}$$

has the unique solution $(m, M) = (\bar{x}, \bar{x})$. By employing Theorem 1.6.4 the result follows.

When (5.93.8) holds, the function

$$f(x_n, x_{n-2}) = \frac{\beta x_n + x_{n-2}}{B x_n + x_{n-2}}$$

is strictly increasing in x_n and strictly decreasing in x_{n-2}, and for each $m, M \in [1, \frac{\beta}{B}]$, the system

$$M = \frac{\beta M + m}{B M + m} \quad \text{and} \quad m = \frac{\beta m + M}{B m + M}$$

has the unique solution $(m, M) = (\bar{x}, \bar{x})$. By employing Theorem 1.6.4 the result follows. The proof is complete. ∎

Conjecture 5.93.1 *Show that for the equilibrium $\bar{x}$ of Eq.(5.93.1),*

$$\textit{Local Asymptotic Stabilty} \Longrightarrow \textit{Global Attractivity.}$$

Conjecture 5.93.2 *Show that Eq.(5.93.1) has solutions that do not converge to the equilibrium point $\bar{x}$ or to a periodic solution.*

5.94 Equation #94 : $x_{n+1} = \dfrac{\beta x_n + \delta x_{n-2}}{C x_{n-1} + D x_{n-2}}$

Eq.(#94) can be written in the normalized form

$$x_{n+1} = \frac{\beta x_n + x_{n-2}}{C x_{n-1} + x_{n-2}}, \quad n = 0, 1, \ldots \tag{5.94.1}$$

with positive parameters β, C and with arbitrary positive initial conditions x_{-2}, x_{-1}, x_0.

The boundedness character of this equation was investigated in [152] where they established that every solution is bounded from above and from below by positive constants. See also Theorem 2.9.1 in Section 2.9.

Eq.(5.94.1) has the unique equilibrium point

$$\bar{x} = \frac{\beta + 1}{C + 1}.$$

The characteristic equation of the linearized equation of Eq.(5.94.1) about the equilibrium, $\bar{x} = \frac{\beta+1}{C+1}$, is

$$\lambda^3 - \frac{\beta}{\beta+1}\lambda^2 + \frac{C}{C+1}\lambda + \frac{\beta - C}{(\beta+1)(C+1)} = 0.$$

From this and Theorem 1.2.3 it follows that $\bar{x} = \frac{\beta+1}{C+1}$ is locally asymptotically stable when

$$\beta < 1 \text{ and } C < \frac{1 + 5\beta + (1+\beta)\sqrt{5 + 4\beta}}{2(1-\beta)}, \tag{5.94.2}$$

or

$$1 \leq \beta \leq 1 + \sqrt{2}, \tag{5.94.3}$$

or

$$\beta > 1 + \sqrt{2} \text{ and } C > \frac{1 + 5\beta - (1+\beta)\sqrt{5 + 4\beta}}{2(1-\beta)} \tag{5.94.4}$$

and unstable when

$$\beta < 1 \text{ and } C > \frac{1 + 5\beta + (1+\beta)\sqrt{5 + 4\beta}}{2(1-\beta)}$$

or

$$\beta > 1 + \sqrt{2} \text{ and } C < \frac{1 + 5\beta - (1 + \beta)\sqrt{5 + 4\beta}}{2(1 - \beta)}.$$

Conjecture 5.94.1 *Assume that (5.94.2), (5.94.3), or (5.94.4) holds. Show that every solution of Eq.(5.94.1) converges to the equilibrium $\bar{x}$.*

Conjecture 5.94.2 *Show that Eq.(5.94.1) has solutions that do not converge to the equilibrium point $\bar{x}$ or to a periodic solution.*

5.95 Equation #95 : $x_{n+1} = \dfrac{\gamma x_{n-1} + \delta x_{n-2}}{A + Bx_n}$

Eq.(#95) can be written in the normalized form

$$x_{n+1} = \frac{\gamma x_{n-1} + x_{n-2}}{A + x_n}, \quad n = 0, 1, \ldots \tag{5.95.1}$$

with positive parameters γ, A and with arbitrary nonnegative initial conditions x_{-2}, x_{-1}, x_0.

The boundedness character of this equation was established in Theorem 3.1.1 where it was shown that when

$$\gamma > A + 1,$$

the equation has unbounded solutions. The periodic character of this equation will be investigated in Theorem 5.195.2 where it will be shown that when

$$\gamma = A + 1,$$

every solution of the equation converges to a (not necessarily prime) period-two solution.

Zero is always an equilibrium point of Eq.(5.95.1). By Theorems 5.23.2 and 5.23.4 it follows that the zero equilibrium of Eq.(5.95.1) is globally asymptotically stable when

$$A \geq \gamma + 1 \tag{5.95.2}$$

and unstable when

$$A < \gamma + 1. \tag{5.95.3}$$

Furthermore, when Eq.(5.95.3) holds, Eq.(5.95.1) has also the unique positive equilibrium point

$$\bar{x} = \gamma + 1 - A.$$

The characteristic equation of the linearized equation of Eq.(5.95.1) about the positive equilibrium $\bar{x}$ is

$$\lambda^3 + \frac{\gamma + 1 - A}{\gamma + 1}\lambda^2 - \frac{\gamma}{\gamma + 1}\lambda - \frac{1}{\gamma + 1} = 0. \quad (5.95.4)$$

From this and Theorem 1.2.3 it follows that the positive equilibrium $\bar{x}$ of Eq.(5.95.1) is locally asymptotically stable when

$$\frac{\sqrt{3 - 2A} - 1}{2} < \gamma < A + 1 \quad (5.95.5)$$

and unstable when

$$\gamma < \frac{\sqrt{3 - 2A} - 1}{2}.$$

Open Problem 5.95.1 *Assume that*

$$\gamma > A + 1.$$

(a) Determine the set of all positive initial conditions through which the solutions of Eq.(5.95.1) converge to the positive equilibrium $\bar{x}$.

(b) Determine the set of all positive initial conditions through which the solutions of Eq.(5.95.1) are unbounded.

Open Problem 5.95.2 *Assume that*

$$\gamma = A + 1.$$

(a) Determine the set of all positive initial conditions through which the solutions of Eq.(5.95.1) converge to the positive equilibrium $\bar{x}$.

(b) Determine the set of all positive initial conditions through which the solutions of Eq.(5.95.1) converge to a prime period-two solution.

Conjecture 5.95.1 *Assume that (5.95.5) holds. Show that every positive solution of Eq.(5.95.1) converges to the positive equilibrium $\bar{x}$.*

Conjecture 5.95.2 *Assume that*

$$\gamma < \frac{\sqrt{3 - 2A} - 1}{2}.$$

Show that Eq.(5.95.1) has bounded solutions that do not converge to an equilibrium point $\bar{x}$ or to a periodic solution.

Conjecture 5.95.3 *Assume that*

$$\gamma > 1 + A.$$

Show that every positive and bounded solution of Eq.(5.95.1) converges to the positive equilibrium $\bar{x}$.

5.96 Equation #96 : $x_{n+1} = \dfrac{\gamma x_{n-1} + \delta x_{n-2}}{A + C x_{n-1}}$

This equation was investigated in [49]. See also Section 2.7 where we established that every solution of the equation is bounded. Eq.(#96) can be written in the normalized form

$$x_{n+1} = \frac{\gamma x_{n-1} + x_{n-2}}{A + x_{n-1}}, \quad n = 0, 1, \ldots \tag{5.96.1}$$

with positive parameters γ, A and with arbitrary nonnegative initial conditions x_{-2}, x_{-1}, x_0.

Zero is always an equilibrium point of Eq.(5.96.1). By Theorems 5.23.2 and 5.23.4 it follows that the zero equilibrium of Eq.(5.96.1) is globally asymptotically stable when

$$A \geq \gamma + 1 \tag{5.96.2}$$

and unstable when

$$A < \gamma + 1. \tag{5.96.3}$$

Furthermore, when Eq.(5.96.3) holds, Eq.(5.96.1) has also the unique positive equilibrium point

$$\bar{x} = \gamma + 1 - A.$$

The characteristic equation of the linearized equation of Eq.(5.96.1) about the positive equilibrium, $\bar{x} = \gamma + 1 - A$, is

$$\lambda^3 + \frac{1 - A}{\gamma + 1}\lambda - \frac{1}{\gamma + 1} = 0.$$

From this and Theorem 1.2.3 it follows that $\bar{x} = \gamma + 1 - A$ is locally asymptotically stable when

$$A > \frac{1 - \gamma - \gamma^2}{\gamma + 1} \tag{5.96.4}$$

and unstable when

$$A < \frac{1 - \gamma - \gamma^2}{\gamma + 1}.$$

In Section 2.7 we proved that every solution of Eq.(5.96.1) is bounded from above. Here we will also show that when (5.96.3) holds, every positive solution of Eq.(5.96.1) is also bounded from below by a positive constant. Assume for the sake of contradiction that there exists a positive solution $\{x_n\}$ of Eq.(5.96.1), which contains a subsequence $\{x_{n_i+1}\}$ such that

$$x_{n_i+1} \to 0 \text{ and } x_{n_i+1} < x_j, \ \ j < n_i + 1. \tag{5.96.5}$$

Clearly,

$$x_{n_i} \to 0 \text{ and } x_{n_i-2} \to 0.$$

Hence, there exists i_0 such that

$$x_{n_{i_0}} < \gamma + 1 - A.$$

Then

$$x_{n_{i_0}+1} = \frac{\gamma x_{n_{i_0}-1} + x_{n_{i_0}-2}}{A + x_{n_{i_0}-1}} > (\gamma+1)\frac{\min(x_{n_{i_0}-1}, x_{n_{i_0}-2})}{\gamma+1} = \min(x_{n_{i_0}-1}, x_{n_{i_0}-2}).$$

This contradicts (5.96.5) and the proof is complete.

The next three theorems are new results about the global attractivity of the positive equilibrium $\bar{x}$ of Eq.(5.96.1).

Theorem 5.96.1 *Assume that*

$$A = 1.$$

Then every positive solution of Eq.(5.96.1) converges to γ.

PROOF Let $\{x_n\}$ be a positive solution of Eq.(5.96.1). For all $n \geq 0$,

$$x_{n+1} - \gamma = \frac{x_{n-2} - \gamma}{1 + x_{n-1}}$$

and

$$x_{n+1} - x_{n-2} = x_{n-1} \cdot \frac{x_{n-2} - \gamma}{1 + x_{n-1}}.$$

From this it follows that, for all $n \geq 0$,

$$\gamma \leq x_{3n+1} \leq x_{3n-2} \text{ or } x_{3n-2} \leq x_{3n+1} \leq \gamma,$$

and so the sequence $\{x_{3n+1}\}$ converges to a finite limit. Similarly, it follows that the sequences $\{x_{3n+2}\}$ and $\{x_{3n+3}\}$ converge to finite limits. Due to the fact that Eq.(5.96.1) has no prime period-three solutions, the result follows. The proof is complete. ∎

Theorem 5.96.2 *Assume that*

$$\gamma > A - 1 > 0.$$

Then every positive solution of Eq.(5.96.1) converges to the positive equilibrium $\bar{x} = \gamma + 1 - A$.

PROOF We claim that, eventually,

$$x_n < \gamma. \tag{5.96.6}$$

Otherwise, there exists N sufficiently large such that

$$x_{N+1} = \frac{\gamma x_{N-1} + x_{N-2}}{A + x_{N-1}} \geq \gamma.$$

From this it follows that

$$x_{N-2} \geq A \cdot \gamma$$

and, similarly,

$$x_{N-5} \geq A^2\gamma.$$

Inductively, we find

$$x_{N+1-3k} \geq A^k\gamma,$$

which is a contradiction so our claim is established. Clearly, the function

$$f(x_{n-1}, x_{n-2}) = \frac{\gamma x_{n-1} + x_{n-2}}{A + x_{n-1}}$$

is strictly increasing in x_{n-1} and x_{n-2}. By employing Theorem 1.6.7 the result follows. The proof is complete. ∎

Theorem 5.96.3 *Assume that*

$$\gamma \geq 1 - A > 0.$$

Then every positive solution of Eq.(5.96.1) converges to the positive equilibrium $\bar{x} = \gamma + 1 - A$.

PROOF We claim that, eventually,

$$x_n > \gamma. \tag{5.96.7}$$

Otherwise, there exists N sufficiently large such that

$$x_{N+1} = \frac{\gamma x_{N-1} + x_{N-2}}{A + x_{N-1}} \leq \gamma.$$

From this it follows that

$$x_{N-2} \leq A \cdot \gamma$$

and, similarly,

$$x_{N-5} \leq A^2 \cdot \gamma.$$

Inductively, we find

$$x_{N+1-3k} \leq A^k \cdot \gamma,$$

which is a contradiction and proves (5.96.7). Clearly, the function

$$f(x_{n-1}, x_{n-2}) = \frac{\gamma x_{n-1} + x_{n-2}}{A + x_{n-1}}$$

is strictly decreasing in x_{n-1} and strictly increasing in x_{n-2}. When

$$\gamma > 1 - A > 0,$$

by employing Theorem 1.6.7, the result follows. When

$$\gamma = 1 - A > 0,$$

the Hypotheses of Theorem 1.6.8 are satisfied from which the result follows. The proof is complete. ∎

Conjecture 5.96.1 *Assume that*

$$\frac{1 - \gamma - \gamma^2}{\gamma + 1} < A < 1.$$

Show that every positive solution of Eq.(5.96.1) converges to the positive equilibrium $\bar{x} = \beta + \delta - 1$.

Conjecture 5.96.2 *Assume that*

$$A < \frac{1 - \gamma - \gamma^2}{\gamma + 1}.$$

Show that Eq.(5.96.1) has solutions that do not converge to an equilibrium point or to a periodic solution.

5.97 Equation #97 : $x_{n+1} = \dfrac{\gamma x_{n-1} + \delta x_{n-2}}{A + D x_{n-2}}$

This equation was investigated in [67] and [122]. Eq.(#97) possesses a period-two trichotomy depending on whether

$$\gamma < \delta + A, \quad \gamma = \delta + A, \quad \text{or} \quad \gamma > \delta + A.$$

This result is a special case of a more general period-two trichotomy result presented in Theorem 4.3.1.

Open Problem 5.97.1 *Assume*

$$\gamma = \delta + A.$$

Determine the set of all positive initial conditions for which every positive solution of the equation

$$x_{n+1} = \frac{\gamma x_{n-1} + \delta x_{n-2}}{A + x_{n-2}}, \quad n = 0, 1, \ldots \tag{5.97.1}$$

converges to the positive equilibrium 2δ.

Open Problem 5.97.2 *Assume*

$$\gamma > \delta + A.$$

Determine the set of all positive initial conditions for which every positive solution of Eq.(5.97.1) converges to the positive equilibrium.

Conjecture 5.97.1 *Assume*

$$\gamma > \delta + A.$$

Show that every positive and bounded solution of Eq.(5.97.1) converges to the positive equilibrium.

5.98 Equation #98 : $x_{n+1} = \frac{\gamma x_{n-1} + \delta x_{n-2}}{Bx_n + Cx_{n-1}}$

Eq.(#98) can be written in the normalized form

$$x_{n+1} = \frac{x_{n-1} + \delta x_{n-2}}{Bx_n + x_{n-1}}, \quad n = 0, 1, ... \tag{5.98.1}$$

with positive parameters δ, B and with arbitrary positive initial conditions x_{-2}, x_{-1}, x_0.

The boundedness character of Eq.(5.98.1) was investigated in [69]. See also Theorem 3.4.1 where we established the existence of unbounded solutions of Eq.(5.98.1) when

$$\delta > B.$$

From this and Theorem 5.221.1 it follows that every solution of Eq.(5.98.1) is bounded if and only if

$$\delta \leq B.$$

Eq.(5.98.1) has the unique equilibrium

$$\bar{x} = \frac{\delta + 1}{B + 1}.$$

The characteristic equation of the linearized equation of Eq.(5.98.1) about the equilibrium, $\bar{x} = \frac{\delta+1}{B+1}$, is

$$\lambda^3 + \frac{B}{B+1}\lambda^2 + \frac{\delta - B}{(\delta+1)(B+1)}\lambda - \frac{\delta}{\delta+1} = 0.$$

From this and Theorem 1.2.3 it follows that the equilibrium $\bar{x} = \frac{\delta+1}{B+1}$ is locally asymptotically stable when

$$\frac{B-1}{B+3} < \delta < \frac{1 + 2B + \sqrt{5 + 16B + 12B^2}}{2(B+1)} \tag{5.98.2}$$

and unstable when

$$\delta < \frac{B-1}{B+3} \quad \text{or} \quad \delta > \frac{1 + 2B + \sqrt{5 + 16B + 12B^2}}{2(B+1)}. \tag{5.98.3}$$

When

$$\delta = \frac{1 + 2B + \sqrt{5 + 16B + 12B^2}}{2(B+1)},$$

two characteristic roots are complex conjugate with magnitude equal to one and the third characteristic root lies in the interval $(0, 1)$. In particular, when

$$B = \delta = 1 + \sqrt{2},$$

two of the characteristic roots are eighth roots of unity and the third characteristic root lies in the interval $(0, 1)$.

It is interesting to note that for the equilibrium $\bar{x}$ of Eq.(5.98.1),

Local Asymptotic Stabilty $\nRightarrow$ Global Asymptotic Stabilty.

Indeed, for all positive values of B for which

$$B < \frac{1 + 2B + \sqrt{5 + 16B + 12B^2}}{2(B+1)},$$

and for all values of δ such that

$$B < \delta < \frac{1 + 2B + \sqrt{5 + 16B + 12B^2}}{2(B+1)},$$

the equilibrium $\bar{x}$ of Eq.(5.98.1) is locally asymptotically stable and at the same time the equation has unbounded solutions. In particular, for such initial conditions the equilibrium of the equation is not a global attractor.

When

$$\delta = \frac{B-1}{B+3},$$

two of the characteristic roots lie in the interval $(0,1)$ and the other root is equal to -1.

The sequence

$$\ldots, \phi, \psi, \phi, \psi, \ldots, \tag{5.98.4}$$

where ϕ and ψ are the two positive roots of the equation

$$(B-1)t^2 + (B-1)(\delta-1)t + \delta(1-\delta) = 0$$

is a unique prime period-two solution of Eq.(5.98.1) if and only if

$$\delta < \frac{B-1}{B+3}. \tag{5.98.5}$$

Conjecture 5.98.1 *Assume that*

$$0 < \frac{B-1}{B+3} < \delta \leq B$$

or

$$\delta \leq B \leq 1.$$

Show that for the equilibrium $\bar{x}$ of Eq.(5.98.1),

Local Asymptotic Stabilty $\Longrightarrow$ Global Attractivity.

Conjecture 5.98.2 *Assume that (5.98.5) is satisfied. Show that the unique prime period-two solution (5.98.4) of Eq.(5.98.1) is locally asymptotically stable.*

Conjecture 5.98.3 *Assume that (5.98.5) is satisfied. Show that every solution of Eq.(5.98.1) converges to a (not necessarily prime) period-two solution.*

Conjecture 5.98.4 *Assume that*

$$\delta = \frac{B-1}{B+3} > 0.$$

Show that every solution of Eq.(5.98.1) converges to the equilibrium $\bar{x}$.

Open Problem 5.98.1 *Assume that*

$$\delta > B.$$

(*i*) *Determine the set of all initial conditions for which the solutions of Eq.(5.98.1) converge to the equilibrium $\bar{x}$.*

(*ii*) *Determine the set of all initial conditions for which the solutions of Eq.(5.98.1) are unbounded.*

(*iii*) *Determine all possible periodic solutions of Eq.(5.98.1).*

Conjecture 5.98.5 *Assume that*

$$\delta > \frac{1+2B+\sqrt{5+16B+12B^2}}{2(B+1)}.$$

Show that Eq.(5.98.1) has bounded solutions that do not converge to the equilibrium point $\bar{x}$ or to a periodic solution.

5.99 Equation #99 : $x_{n+1} = \dfrac{\gamma x_{n-1} + \delta x_{n-2}}{Bx_n + Dx_{n-2}}$

For some work on this equation see [56]. Eq.(#99) can be written in the normalized form

$$x_{n+1} = \frac{\gamma x_{n-1} + x_{n-2}}{Bx_n + x_{n-2}}, \quad n = 0, 1, \ldots \tag{5.99.1}$$

with positive parameters γ, B and with arbitrary positive initial conditions x_{-2}, x_{-1}, x_0.

By Theorem 3.3.1 it follows that Eq.(5.99.1) has unbounded solutions when

$$\gamma > 1.$$

Eq.(5.99.1) has the unique equilibrium point

$$\bar{x} = \frac{\gamma+1}{B+1}.$$

The characteristic equation of the linearized equation of Eq.(5.99.1) about the equilibrium, $\bar{x} = \frac{\gamma+1}{B+1}$, is

$$\lambda^3 + \frac{B}{B+1}\lambda^2 - \frac{\gamma}{\gamma+1}\lambda + \frac{\gamma - B}{(\gamma+1)(B+1)} = 0.$$

From this and Theorem 1.2.3 it follows that $\bar{x} = \frac{\gamma+1}{B+1}$ is locally asymptotically stable when

$$0 < B \leq \sqrt{2}+1 \quad \text{and} \quad 0 < \gamma < 1 \tag{5.99.2}$$

or

$$B > \sqrt{2}+1 \quad \text{and} \quad \frac{(1+B)\sqrt{12B^2+16B+5} - 2B^2 - 9B - 3}{2(2B^2+5B+1)} < \gamma < 1. \tag{5.99.3}$$

When

$$B = 1 + \sqrt{2} \text{ and } \gamma = \frac{(1+B)\sqrt{12B^2+16B+5} - 2B^2 - 9B - 3}{2(2B^2+5B+1)},$$

one of the characteristic roots is real within the interval $(0, 1)$ and the other two characteristic roots are eighth roots of unity. When

$$B = 1 + \sqrt{2} \text{ and } \gamma = 1,$$

two of the characteristic roots are real within the interval $(-1, 1)$ and the other root is equal to -1.

Open Problem 5.99.1 *Determine whether the difference equation*

$$x_{n+1} = \frac{x_{n-1} + x_{n-2}}{x_n + x_{n-2}}, \quad n = 0, 1, \ldots$$

has any unbounded solutions.

Conjecture 5.99.1 *Assume that*

$$B \leq 1 + \sqrt{2}.$$

Then the following results hold:

(a) *When*

$$0 < \gamma < 1,$$

every solution of Eq.(5.99.1) converges to its equilibrium point $\bar{x}$.

(b) *When*

$$\gamma = 1,$$

every bounded solution of Eq.(5.99.1) converges to a (not necessarily prime) period-two solution.

Conjecture 5.99.2 *Assume that (5.99.2) or (5.99.3) holds. Show that every solution of Eq.(5.99.1) converges to its equilibrium point* $\bar{x}$.

Conjecture 5.99.3 *Assume that*

$$\gamma = 1.$$

Show that every bounded solution of Eq.(5.99.1) converges to a (not necessarily prime) period-two solution.

Conjecture 5.99.4 *Show that Eq.(5.99.1) has bounded solutions that do not converge to the equilibrium point* $\bar{x}$ *or to a periodic solution.*

Conjecture 5.99.5 *Assume that*

$$\gamma > 1.$$

Show that every bounded solution of Eq.(5.99.1) converges to the positive equilibrium.

5.100 Equation #100 : $x_{n+1} = \dfrac{\gamma x_{n-1} + \delta x_{n-2}}{Cx_{n-1} + Dx_{n-2}}$

This equation was investigated in [140]. See also [141]. Eq.(#100) can be written in the normalized form

$$x_{n+1} = \frac{\gamma x_{n-1} + x_{n-2}}{Cx_{n-1} + x_{n-2}}, \quad n = 0, 1, \ldots \tag{5.100.1}$$

with positive parameters γ and C and with arbitrary positive initial conditions x_{-2}, x_{-1}, x_0.

Eq.(5.100.1) has the unique equilibrium point

$$\bar{x} = \frac{\gamma + 1}{C + 1}.$$

The characteristic equation of the linearized equation about the equilibrium, $\bar{x} = \frac{\gamma+1}{C+1}$, is

$$\lambda^3 + \frac{C - \gamma}{(\gamma + 1)(C + 1)}\lambda + \frac{\gamma - C}{(\gamma + 1)(C + 1)} = 0.$$

From this and Theorem 1.2.3 it follows that $\bar{x} = \frac{\gamma+1}{C+1}$ is locally asymptotically stable when

$$C < 1 \text{ and } \gamma < \frac{3C + 1}{1 - C}, \tag{5.100.2}$$

or

$$1 \leq C \leq \frac{1 + \sqrt{5}}{2}, \tag{5.100.3}$$

or

$$C > \frac{1 + \sqrt{5}}{2} \text{ and } \gamma > \frac{\sqrt{5} - 1}{2} \cdot \frac{C - \frac{\sqrt{5}+1}{2}}{C + \frac{\sqrt{5}+3}{2}} \tag{5.100.4}$$

and unstable when

$$C < 1 \text{ and } \gamma > \frac{3C + 1}{1 - C} \tag{5.100.5}$$

or

$$C > \frac{1 + \sqrt{5}}{2} \text{ and } \gamma < \frac{\sqrt{5} - 1}{2} \cdot \frac{C - \frac{\sqrt{5}+1}{2}}{C + \frac{\sqrt{5}+3}{2}}. \tag{5.100.6}$$

When

$$\gamma = \frac{3C + 1}{1 - C},$$

two of the characteristic roots are complex conjugate with magnitude less than one and the other characteristic root is equal to -1.

When (5.100.5) holds, and only then, the sequence

$$\ldots, \phi, \psi, \phi, \psi, \ldots$$

is a unique prime period-two solution of Eq.(5.100.1), where ϕ and ψ are the two positive roots of the quadratic equation

$$C(C-1)w^2 + (\gamma - 1)(1-C)w + 1 - \gamma = 0.$$

In this case none of the characteristic roots is equal to -1, while in the case where

$$\gamma = \frac{3C+1}{1-C}$$

one of the characteristic roots is equal to -1 but the equation does not have prime period-two solutions. For some stability results and the local stability of the unique prime period-two solution of Eq.(5.100.1), see [140] and [141].

Conjecture 5.100.1 *Assume that*

$$C < 1 \text{ and } \gamma \leq \frac{3C+1}{1-C},$$

or

$$1 \leq C \leq \frac{1+\sqrt{5}}{2},$$

or

$$C > \frac{1+\sqrt{5}}{2} \text{ and } \gamma \geq \frac{\sqrt{5}-1}{2} \cdot \frac{C - \frac{\sqrt{5}+1}{2}}{C + \frac{\sqrt{5}+3}{2}}.$$

Show that every solution of Eq.(5.100.1) converges to its equilibrium point $\bar{x}$.

Conjecture 5.100.2 *Assume that*

$$C < 1 \text{ and } \gamma > \frac{3C+1}{1-C}.$$

Show that every solution of Eq.(5.100.1) converges to a (not necessarily prime) period-two solution.

Open Problem 5.100.1 *Show that Eq.(5.100.1) has solutions that do not converge to the equilibrium point $\bar{x}$ or to a periodic solution.*

5.101 Equation #101 : $x_{n+1} = \dfrac{\alpha}{A + Bx_n + Cx_{n-1}}$

This equation is a special case of a more general equation investigated in Section 5.17. It follows from Theorem 5.17.1 that the equilibrium of this equation is globally asymptotically stable.

Open Problem 5.101.1 *Investigate the global character of solutions of the equation*

$$x_{n+1} = \frac{1}{A_n + B_n x_n + x_{n-1}}, \quad n = 0, 1, \ldots \tag{5.101.1}$$

with periodic coefficients $\{A_n\}$ *and* $\{B_n\}$.

Open Problem 5.101.2 *Investigate the global character of solutions of Eq.(5.101.1) with convergent coefficients* $\{A_n\}$ *and* $\{B_n\}$.

5.102 Equation #102 : $x_{n+1} = \dfrac{\alpha}{A + Bx_n + Dx_{n-2}}$

This equation is a special case of a more general equation investigated in Section 5.17. It follows from Theorem 5.17.1 that the equilibrium of this equation is globally asymptotically stable.

Open Problem 5.102.1 *Investigate the global character of solutions of the equation*

$$x_{n+1} = \frac{1}{A_n + B_n x_n + x_{n-2}}, \quad n = 0, 1, \ldots \tag{5.102.1}$$

with periodic coefficients $\{A_n\}$ *and* $\{B_n\}$.

Open Problem 5.102.2 *Investigate the global character of solutions of Eq.(5.102.1) with convergent coefficients* $\{A_n\}$ *and* $\{B_n\}$.

5.103 Equation #103 : $x_{n+1} = \dfrac{\alpha}{A + Cx_{n-1} + Dx_{n-2}}$

This equation is a special case of a more general equation investigated in Section 5.17. It follows from Theorem 5.17.1 that the equilibrium of this equation is globally asymptotically stable.

Open Problem 5.103.1 *Investigate the global character of solutions of the equation*

$$x_{n+1} = \frac{1}{A_n + C_n x_{n-1} + x_{n-2}}, \quad n = 0, 1, \ldots \tag{5.103.1}$$

with periodic coefficients $\{A_n\}$ *and* $\{C_n\}$.

Open Problem 5.103.2 *Investigate the global character of solutions of Eq.(5.103.1) with convergent coefficients $\{A_n\}$ and $\{C_n\}$.*

5.104 Equation #104 : $x_{n+1} = \dfrac{\alpha}{Bx_n + Cx_{n-1} + Dx_{n-2}}$

This equation is a special case of a more general equation investigated in Section 5.17. It follows from Theorem 5.17.2 that the equilibrium of this equation is globally asymptotically stable.

Open Problem 5.104.1 *Investigate the global character of solutions of the equation*

$$x_{n+1} = \frac{1}{B_n x_n + C_n x_{n-1} + x_{n-2}}, \quad n = 0, 1, \ldots \tag{5.104.1}$$

with periodic coefficients $\{B_n\}$ and $\{C_n\}$.

Open Problem 5.104.2 *Investigate the global character of solutions of Eq.(5.104.1) with convergent coefficients $\{B_n\}$ and $\{C_n\}$.*

5.105 Equation #105 : $x_{n+1} = \dfrac{\beta x_n}{A + Bx_n + Cx_{n-1}}$

Eq.(#105) can be written in the normalized form

$$x_{n+1} = \frac{\beta x_n}{1 + Bx_n + x_{n-1}}, \quad n = 0, 1, \ldots \tag{5.105.1}$$

with positive parameters β, B and with arbitrary nonnegative initial conditions x_{-1}, x_0. Zero is always an equilibrium point of Eq.(5.105.1). By Theorems 5.23.2 and 5.23.4 it follows that the zero equilibrium of Eq.(5.105.1) is globally asymptotically stable when

$$\beta \leq 1 \tag{5.105.2}$$

and unstable when

$$\beta > 1. \tag{5.105.3}$$

Furthermore when (5.105.3) holds, Eq.(5.105.1) has also the unique positive equilibrium point

$$\bar{x} = \frac{\beta - 1}{B + 1}.$$

The characteristic equation of the linearized equation of Eq.(5.105.1) about the positive equilibrium, $\bar{x} = \frac{\beta-1}{B+1}$, is

$$\lambda^2 - \frac{\beta + B}{\beta(B+1)}\lambda + \frac{\beta - 1}{\beta(B+1)} = 0.$$

From this and Theorem 1.2.2 it follows that, $\bar{x} = \frac{\beta-1}{B+1}$, is locally asymptotically stable when

$$\beta > 1$$

and unstable when

$$\beta < 1.$$

It is a straightforward consequence of Theorem 1.6.3 that when

$$\beta > 1,$$

every positive solution of Eq.(5.105.1) converges to the positive equilibrium $\bar{x}$.

Open Problem 5.105.1 *Investigate the global character of solutions of the equation*

$$x_{n+1} = \frac{\beta_n x_n}{1 + B_n x_n + x_{n-1}}, \quad n = 0, 1, \ldots \tag{5.105.4}$$

with periodic coefficients $\{\beta_n\}$ *and* $\{B_n\}$.

Open Problem 5.105.2 *Investigate the global character of solutions of Eq.(5.105.4) with convergent coefficients* $\{\beta_n\}$ *and* $\{B_n\}$.

5.106 Equation #106 : $x_{n+1} = \dfrac{\beta x_n}{A + Bx_n + Dx_{n-2}}$

Eq.(#106) can be written in the normalized form

$$x_{n+1} = \frac{\beta x_n}{1 + Bx_n + x_{n-2}}, \quad n = 0, 1, \ldots \tag{5.106.1}$$

with positive parameters β, B and with arbitrary nonnegative initial conditions x_{-2}, x_{-1}, x_0.

Zero is always an equilibrium point of Eq.(5.106.1). By Theorems 5.23.2 and 5.23.4 it follows that the zero equilibrium of Eq.(5.106.1) is globally asymptotically stable when

$$\beta \leq 1 \tag{5.106.2}$$

and unstable when

$$\beta > 1. \tag{5.106.3}$$

When (5.106.3) holds, Eq.(5.106.1) has also the unique positive equilibrium point

$$\bar{x} = \frac{\beta - 1}{B + 1}.$$

The characteristic equation of the linearized equation of Eq.(5.106.1) about the positive equilibrium $\bar{x}$ is

$$\lambda^3 - \frac{\beta + B}{\beta(B+1)}\lambda^2 + \frac{\beta - 1}{\beta(B+1)} = 0.$$

From this and Theorem 1.2.3 it follows that $\bar{x}$ is locally asymptotically stable when

$$B \geq \sqrt{2} - 1 \text{ and } \beta > 1 \tag{5.106.4}$$

or

$$B < \sqrt{2} - 1 \text{ and } 1 < \beta < \frac{3 - B + \sqrt{5 + 6B - 3B^2 - 4B^3}}{2(1 - 2B - B^2)} \tag{5.106.5}$$

and unstable when

$$B < \sqrt{2} - 1 \text{ and } \beta > \frac{3 - B + \sqrt{5 + 6B - 3B^2 - 4B^3}}{2(1 - 2B - B^2)}. \tag{5.106.6}$$

The following theorem about the global behavior of solutions of Eq.(5.106.1) was established in [157].

Theorem 5.106.1 *[157] Assume that*

$$0 < B < 1 \text{ and } 1 < \beta \leq 2.$$

Then every positive solution of Eq.(5.106.1) converges to its positive equilibrium point $\bar{x}$.

The following theorem is a new result about the global character of solutions of Eq.(5.106.1).

Theorem 5.106.2 *Assume that*

$$\beta > 1 \quad \text{and} \quad B \geq 1.$$

Then every positive solution of Eq.(5.106.1) converges to its positive equilibrium point $\bar{x}$.

PROOF Let $\{x_n\}$ be a positive solution of Eq.(5.106.1). Then, clearly,

$$x_{n+1} = \frac{\beta x_n}{1 + Bx_n + x_{n-2}} < \frac{\beta}{B}, \quad \text{for } n \geq 1. \tag{5.106.7}$$

Next, we claim that the solution $\{x_n\}$ is also bounded from below by a positive constant. Otherwise, there exists a sequence of indices $\{n_i\}$ such that

$$x_{n_i+1} \to 0$$

and

$$x_{n_i+1} < x_j, \quad \text{for all } j < n_i + 1. \tag{5.106.8}$$

Then, clearly, from (5.106.1),

$$x_{n_i} \to 0$$

and also

$$x_{n_i-2} \to 0.$$

Hence, eventually,

$$x_{n_i}, x_{n_i-2} < \frac{\beta - 1}{B + 1}$$

which implies that, eventually,

$$x_{n_i+1} = \frac{\beta x_{n_i}}{1 + Bx_{n_i} + x_{n_i-2}} > \frac{\beta x_{n_i}}{1 + (\beta - 1)} = x_{n_i}.$$

This contradicts (5.106.8) and establishes our claim that the solution $\{x_n\}$ is also bounded from below. Clearly, the function

$$f(x_n, x_{n-2}) = \frac{\beta x_n}{1 + Bx_n + x_{n-2}}$$

is strictly increasing in x_n and strictly decreasing in x_{n-2}. When

$$\beta > 1 \text{ and } B > 1,$$

the result follows by employing Theorem 1.6.7. When

$$\beta > 1 \text{ and } B = 1,$$

the Hypotheses of Theorem 1.6.8 are satisfied from which the result follows. The proof is complete. ∎

Conjecture 5.106.1 *Assume that (5.106.4) or (5.106.5) holds. Show that every positive solution of Eq.(5.106.1) converges to the positive equilibrium* $\bar{x}$.

Conjecture 5.106.2 *Show that Eq.(5.106.1) has solutions which do not converge to an equilibrium point or to a periodic solution.*

Open Problem 5.106.1 *Investigate the global character of the equation*

$$x_{n+1} = \frac{\beta_n x_n}{1 + B_n x_n + x_{n-2}}, \quad n = 0, 1, \ldots \tag{5.106.9}$$

with periodic coefficients $\{\beta_n\}$ *and* $\{B_n\}$.

Open Problem 5.106.2 *Investigate the global character of solutions of Eq.(5.106.9) with convergent coefficients* $\{\beta_n\}$ *and* $\{B_n\}$.

Eq.(#106) is a special case of the more general $(k+1)^{st}$-order rational difference equation

$$x_{n+1} = \frac{x_n}{A + \sum_{i=0}^{k} B_i x_{n-i}}, \quad n = 0, 1, \ldots, \tag{5.106.10}$$

with nonnegative parameters and with arbitrary nonnegative initial conditions $x_{-k}, \ldots, x_0$ such that the denominator is always positive. For some work on this equation see [157]. For $k = 0$ this is a special case of the **Riccati difference** equation, which in mathematical biology is also known as the **Holt-Beverton model**. For $k = 1$ and $B_0 = 0$, this is **Pielou's equation**, which is the discrete analog of the delay **logistic equation**

$$N'(t) = rN(t)[1 - \frac{N(t-\tau)}{P}], \ t \geq 0. \tag{5.106.11}$$

Actually in her books [196,197], Pielou proposed the equation

$$x_{n+1} = \frac{\alpha x_n}{1 + x_{n-k}}, \quad n = 0, 1, \ldots \tag{5.106.12}$$

as the discrete analogue of the delay logistic equation (5.106.11). One arrives at Eq.(5.106.11) from the logistic differential equation

$$N'(t) = rN(t)[1 - \frac{N(t)}{P}], \ t \geq 0 \tag{5.106.13}$$

by assuming that there is a delay τ in the response of the growth rate per individual to density changes. Pielou arrived at her model (5.106.12) as follows: The solution of Eq.(5.106.13) is

$$N(t) = \frac{P}{1 + (\frac{P}{N_0 - 1})e^{-rt}}$$

and Pielou observed that $N(t)$ satisfies the first-order difference equation

$$N(t+1) = \frac{\alpha N(t)}{1+\beta N(t)} \tag{5.106.14}$$

with

$$\alpha = e^r > 1 \text{ and } \beta = \frac{e^r - 1}{P} > 0.$$

Now from (5.106.14), Pielou arrived at her model (5.106.12) by assuming, as in the continuous case, that there should be a delay k in the response of the growth rate per individual to density changes.

When

$$A = 0 \text{ or } A \geq 1,$$

Eq.(5.106.10) has a unique equilibrium point. When

$$A = 0,$$

the unique equilibrium is

$$\bar{x} = \frac{1}{\sum_{i=0}^{k} B_i}$$

and when

$$A \geq 1,$$

zero is the only equilibrium. When

$$0 < A < 1,$$

Eq.(5.106.10) has two equilibrium points, namely, the zero equilibrium and the positive equilibrium

$$\bar{x} = \frac{1-A}{\sum_{i=0}^{k} B_i}.$$

The characteristic equation of the linearized equation about the zero equilibrium point, that exists, provided that $A > 0$, is

$$\lambda^{k+1} - \frac{1}{A}\lambda^k = 0. \tag{5.106.15}$$

The characteristic equation of the linearized equation about the positive equilibrium point which exists provided that $0 \leq A < 1$, is

$$\lambda^{k+1} - \frac{AB_0 + \sum_{i=1}^{k} B_i}{\sum_{i=0}^{k} B_i}\lambda^k + \frac{1-A}{\sum_{i=0}^{k} B_i}\sum_{i=1}^{k} B_i \lambda^{k-i} = 0. \tag{5.106.16}$$

From (5.106.15) it follows that the zero equilibrium point is locally asymptotically stable when

$$A > 1$$

and unstable when

$$0 < A < 1.$$

Furthermore, it follows from Eq.(5.106.10) that

$$x_{n+1} \leq \frac{1}{A} x_n, \quad \text{for } n \geq 0$$

and so when

$$A \geq 1,$$

the zero equilibrium point of Eq.(5.106.10) is globally asymptotically stable.

By using Theorems 1.2.2, 1.2.3, and 1.2.4 we can determine the region of local asymptotic stability of the positive equilibrium of Eq.(5.106.10) for the values of

$$k \in \{1, 2, 3\}.$$

Unfortunately, we do not know the local asymptotic stability of Eq.(5.106.10) for

$$k \geq 4.$$

Open Problem 5.106.3 *Determine the region of parameters of Eq.(5.106.16) where all roots of the equation lie inside the unit disk.*

Theorem 5.106.3 *Assume that*

$$0 \leq A < 1 \quad \text{and} \quad \sum_{i=1}^{k} B_i \leq B_0.$$

Then every positive solution of Eq.(5.106.10) converges to its positive equilibrium point.

PROOF Let $\{x_n\}_{n=-k}^{\infty}$ be any positive solution of Eq.(5.106.10). Then

$$x_{n+1} \leq \frac{x_n}{A + B_0 x_n}, \quad \text{for } n \geq 1$$

and by using (the comparison result) Theorem 1.4.1, we find

$$\limsup_{n \to \infty} x_{n+1} \leq \frac{1 - A}{B_0}.$$

Let $\epsilon > 0$ and assume without loss of generality that, for $n \geq 0$,

$$x_n < \frac{1 - A + \epsilon}{B_0}.$$

We claim that

$$\liminf_{n \to \infty} x_n > 0. \tag{5.106.17}$$

Assume for the sake of contradiction that (5.106.17) is not true. Then there exists a sequence of indices $\{n_j\}$ such that

$$\lim_{j\to\infty} x_{n_j+1} = 0 \quad \text{and} \quad x_{n_j+1} < x_t, \quad \text{for} \quad t < n_j + 1. \tag{5.106.18}$$

At this point we will give two different proofs of (5.106.17). The reason we present the two different proofs is that the first proof is valid only when

$$0 \le A < 1 \quad \text{and} \quad \sum_{i=1}^{k} B_i < B_0$$

but it extends our theorem to a more general equation. (See Theorem 5.106.4). The second proof is valid when

$$0 \le A < 1 \quad \text{and} \quad \sum_{i=1}^{k} B_i \le B_0$$

as needed, but we cannot use it to prove Theorem 5.106.4.

First Proof: Clearly,

$$\lim_{j\to\infty} x_{n_j} = 0$$

and

$$\sum_{i=1}^{k} B_i \frac{1-A+\epsilon}{B_0} + B_0 x_{n_j} > \sum_{i=1}^{k} B_i x_{n_j-i} + B_0 x_{n_j} > 1 - A,$$

from which it follows that

$$B_0 x_{n_j} > (1-A)\frac{B_0 - \sum_{i=1}^{k} B_i}{B_1} - \epsilon \frac{\sum_{i=1}^{k} B_i}{B_0}$$

and

$$B_0 \liminf_{j\to\infty} x_{n_j} \ge (1-A)\frac{B_0 - \sum_{i=1}^{k} B_i}{B_0} > 0$$

and this contradiction establishes (5.106.17).

Second Proof: Clearly,

$$x_{n_j}, \ldots, x_{n_j-k} \to 0.$$

Hence, eventually,

$$x_{n_j}, \ldots, x_{n_j-k} < \frac{1-A}{\sum_{i=0}^{k} B_i},$$

which implies that, eventually,

$$x_{n_j+1} = \frac{x_{n_j}}{A + \sum_{i=0}^{k} B_i x_{n_j-i}} > \frac{x_{n_j}}{A + \sum_{i=0}^{k} B_i \frac{1-A}{\sum_{i=0}^{k} B_i}} = x_{n_i}.$$

This contradicts (5.106.18) and establishes (5.106.17).

Clearly, the function

$$f(x_n, \ldots, x_{n-k}) = \frac{x_n}{A + \sum_{i=0}^{k} B_i x_{n-i}}$$

is strictly increasing in x_n and strictly decreasing in all other arguments. When

$$0 \leq A < 1 \text{ and } \sum_{i=1}^{k} B_i < B_0,$$

the result follows by employing Theorem 1.6.7. When

$$0 \leq A < 1 \text{ and } \sum_{i=1}^{k} B_i = B_0,$$

the Hypotheses of Theorem 1.6.8 are satisfied from which the result follows. The proof is complete. ∎

One can easily see that Theorem 5.106.3 has the following straightforward generalization.

Theorem 5.106.4 *Assume that for some* $l \in \{1, 2, \ldots, k\}$

$$0 \leq A < 1 \quad \textit{and} \quad \sum_{i=0, i \neq l}^{k} B_i < B_l.$$

Then every positive solution of the equation

$$x_{n+1} = \frac{x_{n-l}}{A + B_l x_{n-l} + \sum_{i=0, i \neq l}^{k} B_i x_{n-i}}, \quad n = 0, 1, \ldots$$

converges to its positive equilibrium point

$$\bar{x} = \frac{1 - A}{\sum_{i=0}^{k} B_i}.$$

5.107 Equation #107 : $x_{n+1} = \dfrac{\beta x_n}{A + C x_{n-1} + D x_{n-2}}$

The boundedness character of solutions of this equation was investigated in [69]. See also Theorem 2.3.1 where we established that every solution of the equation is bounded. Eq.(#107) can be written in the normalized form

$$x_{n+1} = \frac{\beta x_n}{1 + C x_{n-1} + x_{n-2}}, \quad n = 0, 1, \ldots \tag{5.107.1}$$

with positive parameters β, C and with arbitrary nonnegative initial conditions x_{-2}, x_{-1}, x_0. Zero is always an equilibrium point of Eq.(5.107.1). By Theorems 5.23.2 and 5.23.4 it follows that the zero equilibrium of Eq.(5.107.1) is globally asymptotically stable when

$$\beta \leq 1 \tag{5.107.2}$$

and unstable when

$$\beta > 1. \tag{5.107.3}$$

When (5.107.3) holds, Eq.(5.107.1) has also the unique positive equilibrium point

$$\bar{x} = \frac{\beta - 1}{C + 1}.$$

The characteristic equation of the linearized equation of Eq.(5.107.1) about the positive equilibrium, $\bar{x} = \frac{\beta-1}{C+1}$, is

$$\lambda^3 - \lambda^2 + \frac{C(\beta - 1)}{\beta(C + 1)}\lambda + \frac{\beta - 1}{\beta(C + 1)} = 0.$$

From this and Theorem 1.2.3 it follows that $\bar{x} = \frac{\beta-1}{C+1}$ is locally asymptotically stable when

$$1 < \beta \leq \frac{3 + \sqrt{5}}{2} \tag{5.107.4}$$

or

$$\beta > \frac{3 + \sqrt{5}}{2} \quad \text{and} \quad C > \frac{-\beta + (\beta - 1)\sqrt{\beta}}{\beta} \tag{5.107.5}$$

and unstable when

$$\beta > \frac{3 + \sqrt{5}}{2} \quad \text{and} \quad C < \frac{-\beta + (\beta - 1)\sqrt{\beta}}{\beta}. \tag{5.107.6}$$

When

$$\beta = \frac{3 + \sqrt{5}}{2} \quad \text{and} \quad C = \frac{-\beta + (\beta - 1)\sqrt{\beta}}{\beta},$$

two of the characteristic roots are 10th roots of unity and the third one is inside the interval $(-1, 0)$.

The following theorem about the global stability of solutions of Eq.(5.107.1) was established in [157].

Theorem 5.107.1 *[157] Assume that*

$$1 < \beta \leq 2.$$

Then every positive solution of Eq.(5.107.1) converges to its positive equilibrium point $\bar{x}$.

Conjecture 5.107.1 *Assume that (5.107.4) or (5.107.5) holds. Show that every positive solution of Eq.(5.107.1) converges to its positive equilibrium point* $\bar{x}$.

Conjecture 5.107.2 *Assume that*

$$\beta > \frac{3+\sqrt{5}}{2} \quad \text{and } C < \frac{-\beta + (\beta-1)\sqrt{\beta}}{\beta}.$$

Show that Eq.(5.107.1) has solutions that do not converge to an equilibrium or to a periodic solution.

Open Problem 5.107.1 *Investigate the global character of solutions of the equation*

$$x_{n+1} = \frac{\beta_n x_n}{1 + C_n x_{n-1} + x_{n-2}}, \quad n = 0, 1, \ldots \tag{5.107.7}$$

with periodic coefficients $\{\beta_n\}$ *and* $\{C_n\}$.

Open Problem 5.107.2 *Investigate the global character of solutions of Eq.(5.107.7) with convergent coefficients* $\{\beta_n\}$ *and* $\{C_n\}$.

5.108 Equation #108 : $x_{n+1} = \dfrac{\beta x_n}{Bx_n + Cx_{n-1} + Dx_{n-2}}$

Eq.(#108) can be written in the normalized form

$$x_{n+1} = \frac{x_n}{x_n + Cx_{n-1} + Dx_{n-2}}, \quad n = 0, 1, \ldots \tag{5.108.1}$$

with positive parameters C, D and with arbitrary positive initial conditions x_{-2}, x_{-1}, x_0.

Eq.(5.108.1) has the unique equilibrium point

$$\bar{x} = \frac{1}{C + D + 1}.$$

The characteristic equation of the linearized equation of Eq.(5.108.1) about the equilibrium, $\bar{x} = \frac{1}{C+D+1}$, is

$$\lambda^3 - \frac{C+D}{1+C+D}\lambda^2 + \frac{C}{1+C+D}\lambda + \frac{D}{1+C+D} = 0.$$

From this and Theorem 1.2.3 it follows that $\bar{x} = \frac{1}{1+C+D}$ is locally asymptotically stable when

$$C > D^2 - 2D - 1 \tag{5.108.2}$$

and unstable when

$$C < D^2 - 2D - 1. \quad (5.108.3)$$

It is noteworthy that when

$$D = 1 + \sqrt{2} \text{ and } C = 0,$$

two of the characteristic roots are eighth roots of unity and the other root is inside the interval $(-1, 0)$.

The following theorem, which is a new result about the global character of solutions of Eq.(5.108.1), is an immediate application of Theorem 5.106.3.

Theorem 5.108.1 *Assume that*

$$C + D \leq 1.$$

Then every solution of Eq.(5.108.1) converges to its equilibrium point $\bar{x}$.

Conjecture 5.108.1 *Assume that (5.108.2) holds. Show that every solution of Eq.(5.108.1) converges to its equilibrium point.*

Conjecture 5.108.2 *Assume that*

$$C < D^2 - 2D - 1.$$

Show that Eq.(5.108.1) has solutions that do not converge to the equilibrium point $\bar{x}$ *or to a periodic solution.*

Open Problem 5.108.1 *Investigate the global character of solutions of Eq.(5.108.1) with periodic coefficients.*

5.109 Equation #109 : $x_{n+1} = \dfrac{\gamma x_{n-1}}{A + Bx_n + Cx_{n-1}}$

For some work on this equation see [175]. Here we present a detailed account on the character of its solutions and confirm Conjectures 7.5.1 and 7.5.2 in [175]. Actually Conjecture 7.5.1 was confirmed by Hristo Voulov in his talk at the annual AMS meeting in Santiago in 2002 by using Theorem 1.6.4. Here we present a very simple and direct proof based on Theorem 1.6.6.

Eq.(#109) can be written in the normalized form

$$x_{n+1} = \frac{x_{n-1}}{A + Bx_n + x_{n-1}}, \quad n = 0, 1, \ldots . \quad (5.109.1)$$

We will allow the parameter A to be nonnegative so that Eq.(5.109.1) also includes the special case #32. The parameter B is assumed positive; otherwise, this is a Riccati-type difference equation. The initial conditions x_{-1}, x_0 of Eq.(5.109.1) are arbitrary nonnegative real numbers such that the denominator is always positive.

When

$$A = 0 \quad \text{or} \quad A \geq 1,$$

Eq.(5.109.1) has a unique equilibrium point.

When

$$A = 0,$$

the unique equilibrium of Eq.(5.109.1) is

$$\bar{x} = \frac{1}{1+B}$$

and when

$$A \geq 1,$$

zero is the only equilibrium.

When

$$0 < A < 1,$$

Eq.(5.109.1) has two equilibrium points, namely, the zero equilibrium and the unique positive equilibrium point

$$\bar{x} = \frac{1-A}{1+B}.$$

The characteristic equation of the linearized equation of Eq.(5.109.1) about the zero equilibrium point, which exists as long as $A > 0$, is

$$\lambda^2 - \frac{1}{A} = 0. \tag{5.109.2}$$

The characteristic equation of the linearized equation of Eq.(5.109.1) about the positive equilibrium $\bar{x} = \frac{1-A}{1+B}$, which exists as long as $0 \leq A < 1$, is

$$\lambda^2 + \frac{B(1-A)}{1+B}\lambda - \frac{A+B}{1+B} = 0. \tag{5.109.3}$$

From (5.109.2) and Theorem 1.2.2 it follows that the zero equilibrium is locally asymptotically stable when

$$A > 1$$

and unstable when

$$0 < A < 1.$$

Also, it follows from Theorem 1.2.2 that the positive equilibrium $\bar{x} = \frac{1-A}{1+B}$ is locally asymptotically stable when

$$0 \leq A < 1 \quad \text{and} \quad B < 1$$

and unstable when

$$0 \leq A < 1 \quad \text{and} \quad B > 1.$$

In addition to its equilibrium points, Eq.(5.109.1) has period-two solutions. If

$$\ldots, \phi, \psi, \phi, \psi, \ldots$$

is a prime period-two solution of Eq.(5.109.1) then, clearly,

$$\phi = \frac{\phi}{A + B\psi + \phi} \quad \text{and} \quad \psi = \frac{\psi}{A + B\phi + \psi}.$$

It follows that prime period-two solutions exist if and only if

$$0 \leq A < 1.$$

Furthermore, when

$$B \neq 1,$$

the only prime period-two solution of Eq.(5.109.1) is

$$\ldots, 0, 1 - A, 0, 1 - A, \ldots \tag{5.109.4}$$

and when

$$B = 1,$$

all prime period-two solutions are

$$\ldots, \phi, \psi, \phi, \psi, \ldots$$

with

$$\phi + \psi = 1 - A$$
$$\phi \neq \psi \quad \text{and} \quad \phi, \psi \in [0, 1 - A].$$

To investigate the local asymptotic stability of the unique period-two solution (5.109.4) when $B \neq 1$, we set

$$u_n = x_{n-1} \quad \text{and} \quad v_n = x_n.$$

Then

$$u_{n+1} = v_n \quad \text{and} \quad v_{n+1} = \frac{u_n}{A + Bv_n + u_n}.$$

That is, Eq.(5.109.1) is equivalent to the map

$$T(u, v) = \left(v, \frac{u}{A + Bv + u} \right).$$

Then

$$T^2(u,v) = T\left(v, \frac{u}{A+Bv+u}\right) = (f(u,v), g(u,v))$$

with

$$f(u,v) = \frac{u}{A+Bv+u}$$

and

$$g(u,v) = \frac{v}{A+Bf(u,v)v+v}.$$

Observe that the period-two solution (5.109.4) is a fixed point of the second iterate T^2 of the map T. The Jacobian determinant of T^2 at the period-two solution of (5.109.4) is

$$J_{T^2}\binom{0}{1-A} = \begin{vmatrix} \frac{1}{A+B(1-A)} & 0 \\ \\ \frac{-B(1-A)}{A+B(1-A)} & A \end{vmatrix}.$$

It follows that both eigenvalues of T^2 are inside the unit disk, and so the period-two solution (5.109.4) is locally asymptotically stable when

$$0 \le A < 1 \quad \text{and} \quad B > 1$$

and unstable when

$$0 \le A < 1 \quad \text{and} \quad B < 1.$$

Next we establish the following result, which confirms Conjectures 7.5.1 and 7.5.2 in [175].

Theorem 5.109.1 *(a) Assume that*

$$A \ge 1.$$

Then the zero equilibrium of Eq.(5.109.1) is globally asymptotically stable.

(b) Assume

$$0 \le A < 1.$$

Then every solution of Eq.(5.109.1) converges to a (not necessarily prime) period-two solution.

(c) Assume

$$0 \le A < 1 \quad \text{and} \quad B < 1.$$

Then every positive solution of Eq.(5.109.1) converges to its positive equilibrium point $\bar{x}$. Actually, if we only allow positive initial conditions, the positive equilibrium is globally asymptotically stable.

PROOF (*a*) This is a consequence of the inequality

$$x_{n+1} \leq \frac{1}{A} x_{n-1}$$

and the fact that there are no period-two solutions when $A \geq 1$.

(*b*) This is a consequence of Theorem 1.6.6 and the fact that every solution of Eq.(5.109.1) is bounded.

(*c*) Let $\{x_n\}_{n=-1}^{\infty}$ be a positive solution of Eq.(5.109.1). Then in this case (as Voulov pointed out) Theorem 1.6.4 applies, from which the result follows. The result also follows from Theorem 5.106.4. We will also present a third proof based on the "simple" Theorem 1.6.6. By this result and the fact that every solution of Eq.(5.109.1) is bounded from above, it follows that the subsequences $\{x_{2n}\}$ and $\{x_{2n+1}\}$ of the solution converge monotonically to finite limits L_E and L_O. From Eq.(5.109.1)

$$x_{2n+2} = \frac{x_{2n}}{A + Bx_{2n+1} + x_{2n}}, \quad n = 0, 1, \ldots \tag{5.109.5}$$

and

$$x_{2n+1} = \frac{x_{2n-1}}{A + Bx_{2n} + x_{2n-1}}, \quad n = 0, 1, \ldots . \tag{5.109.6}$$

Note that if both limits L_E and L_O are positive, then by taking limits in (5.109.5) and (5.109.6), as $n \to \infty$, and by simplifying we find

$$\begin{aligned} BL_O + L_E &= 1 - A \\ L_O + BL_E &= 1 - A. \end{aligned}$$

Hence,

$$L_O = L_E = \frac{1 - A}{1 + B},$$

which is exactly what we want to establish. To complete the proof it suffices to show that none of the limits L_E or L_O could be zero. Assume for the sake of contradiction that

$$L_O = 0.$$

Then either $L_E = 0$ or $L_E > 0$. In particular, when

$$L_E > 0$$

from (5.109.5) we see that

$$L_E = 1 - A.$$

Let $\epsilon > 0$ be such that

$$\epsilon B < (1 - A)(1 - B).$$

Then for n sufficiently large

$$x_{2n} < L_E + \epsilon \leq 1 - A + \epsilon$$

and

$$x_{2n+1} = \frac{x_{2n-1}}{A + Bx_{2n} + x_{2n-1}} \geq \frac{x_{2n-1}}{A + B(1 - A + \epsilon) + x_{2n-1}}.$$

Therefore, by (the comparison result) Theorem 1.4.1,

$$x_{2n+1} \geq y_{2n+1},$$

where

$$y_{2n+1} = \frac{y_{2n-1}}{A + B(1 - A + \epsilon) + y_{2n-1}}, \quad n = 1, 2, \ldots \tag{5.109.7}$$

with

$$y_1 > 0.$$

But (5.109.7) is a Riccati equation with positive initial condition and so the $\lim_{n\to\infty} y_{2n+1}$ is the positive equilibrium of Eq.(5.109.7), which is

$$\bar{y} = (1 - A)(1 - B) - B\epsilon.$$

Then

$$0 = \lim_{n\to\infty} x_{2n+1} \geq \bar{y} > 0$$

and this contradiction completes the proof of the theorem. ∎

Open Problem 5.109.1 *Investigate the global character of solutions of Eq.(5.109.1) with periodic coefficients.*

Eq.(#109) is a special case of the more general $(k+1)^{st}$-order rational difference equation

$$x_{n+1} = \frac{x_{n-1}}{A + \sum_{i=0}^{k} B_i x_{n-i}}, \quad n = 0, 1, \ldots \tag{5.109.8}$$

with nonnegative parameters and with arbitrary nonnegative initial conditions $x_{-k}, \ldots, x_0$ such that the denominator is always positive.

When

$$A = 0 \quad \text{or} \quad A \geq 1,$$

Eq.(5.109.8) has a unique equilibrium point.

When

$$A = 0,$$

the unique equilibrium is

$$\bar{x} = \frac{1}{\sum_{i=0}^{k} B_i}$$

and when

$$A \geq 1,$$

zero is the only equilibrium. When

$$0 < A < 1,$$

Eq.(5.109.8) has two equilibrium points, namely, the zero equilibrium and the unique positive equilibrium point

$$\bar{x} = \frac{1-A}{\sum_{i=0}^{k} B_i}.$$

The characteristic equation of the linearized equation of Eq.(5.109.8) about the zero equilibrium point, which exists provided that $A > 0$, is

$$\lambda^{k+1} - \frac{1}{A}\lambda^{k-1} = 0. \tag{5.109.9}$$

The characteristic equation of the linearized equation of Eq.(5.109.8) about the positive equilibrium $\bar{x} = \frac{1-A}{\sum_{i=0}^{k} B_i}$, which exists provided that $0 \leq A < 1$, is

$$\lambda^{k+1} + \frac{B_0(1-A)}{\sum_{i=0}^{k} B_i}\lambda^k - \frac{AB_1 + \sum_{i=0, i\neq 1}^{k} B_i}{\sum_{i=0}^{k} B_i}\lambda^{k-1} + \frac{(1-A)}{\sum_{i=0}^{k} B_i}\sum_{i=2}^{k} B_i\lambda^{k-i} = 0. \tag{5.109.10}$$

From (5.109.9) it follows that the zero equilibrium is locally asymptotically stable when

$$A > 1$$

and unstable when

$$0 < A < 1.$$

In addition to its equilibrium points, Eq.(5.109.8) has period-two solutions. If

$$\ldots, \phi, \psi, \phi, \psi, \ldots$$

is a period-two solution of Eq.(5.109.8) then, clearly,

$$\phi = \frac{\phi}{A + \sum_{i=0}^{S} B_{2i+1}\phi + \sum_{i=0}^{T} B_{2i}\psi} \quad \text{and} \quad \psi = \frac{\psi}{A + \sum_{i=0}^{S} B_{2i+1}\psi + \sum_{i=0}^{T} B_{2i}\phi},$$

where

$$T = \frac{k}{2} \quad \text{and} \quad S = \frac{k-2}{2}, \quad \text{if } k \text{ is even}$$

and

$$S = T = \frac{k-1}{2}, \quad \text{if } k \text{ is odd.}$$

It follows that prime period-two solutions exist if and only if

$$0 \leq A < 1.$$

Furthermore, when

$$\sum_{i=0}^{T} B_{2i} \neq \sum_{i=0}^{S} B_{2i+1},$$

$$\ldots, 0, \frac{1-A}{\sum_{i=0}^{k} B_{2i}}, 0, \frac{1-A}{\sum_{i=0}^{k} B_{2i}}, \ldots \tag{5.109.11}$$

is the only period-two solution of Eq.(5.109.8) and when

$$\sum_{i=0}^{T} B_{2i} = \sum_{i=0}^{S} B_{2i+1},$$

all prime period-two solutions are

$$\ldots, \phi, \psi, \phi, \psi, \ldots$$

with

$$\phi + \psi = \frac{1-A}{\sum_{i=0}^{k} B_{2i}}$$

$$\phi \neq \psi \quad \text{and} \quad \phi, \psi \in \left[0, \frac{1-A}{\sum_{i=0}^{k} B_{2i}}\right].$$

We are now ready to establish the following global asymptotic stability result.

Theorem 5.109.2 (*a*) *Assume that*

$$A \geq 1.$$

Then the zero equilibrium of Eq.(5.109.8) is globally asymptotically stable.

(*b*) *Assume*

$$0 \leq A < 1 \text{ and } \sum_{i=0, i\neq 1}^{k} B_i < B_1.$$

Then every positive solution of Eq.(5.109.8) converges to its positive equilibrium point.

PROOF (*a*) This is a consequence of the inequality

$$x_{n+1} \leq \frac{1}{A} x_{n-1}$$

and the fact that there are no period-two solutions when $A \geq 1$.

(*b*) The result follows from Theorem 5.106.4. ∎

5.110 Equation #110 : $x_{n+1} = \dfrac{\gamma x_{n-1}}{A + Bx_n + Dx_{n-2}}$

Eq.(#110) possesses a period-two trichotomy depending on whether

$$\gamma < A, \quad \gamma = A, \quad \text{or} \quad \gamma > A.$$

This result is a special case of a more general period-two trichotomy result presented in Theorem 4.3.1.

Open Problem 5.110.1 *Investigate the global character of solutions of Eq.(#110) with periodic coefficients.*

Open Problem 5.110.2 *Investigate the global character of solutions of the equation*

$$x_{n+1} = \frac{\gamma_n x_{n-1}}{A_n + B_n x_n + x_{n-2}}, \quad n = 0, 1, \ldots$$

with convergent coefficients $\{\gamma_n\}$, $\{A_n\}$, *and* $\{B_n\}$. *Extend and generalize.*

Conjecture 5.110.1 *Assume that*

$$\gamma > A.$$

Show that every positive and bounded solution of Eq.(#110) converges to the positive equilibrium.

5.111 Equation #111 : $x_{n+1} = \dfrac{\gamma x_{n-1}}{A + Cx_{n-1} + Dx_{n-2}}$

Eq.(#111) can be written in the normalized form

$$x_{n+1} = \frac{x_{n-1}}{A + x_{n-1} + Dx_{n-2}}, \quad n = 0, 1, \ldots . \tag{5.111.1}$$

We will allow the parameter A to be nonnegative so that Eq.(5.111.1) will also include the special case #34. The parameter D is positive and the initial conditions x_{-2}, x_{-1}, x_0 are arbitrary nonnegative real numbers such that the denominator is always positive. When

$$0 \le A < 1 \text{ and } D < 1,$$

this equation is also a special case of a more general equation whose positive equilibrium is a global attractor of all positive solutions of the equation. See Theorem 5.106.4.

As in the special case of Eq.(#109), when

$$A = 0 \quad \text{or} \quad A \geq 1,$$

Eq.(5.111.1) has a unique equilibrium point.

When

$$A = 0,$$

the unique equilibrium is

$$\bar{x} = \frac{1}{1+D}$$

and when

$$A \geq 1,$$

zero is the only equilibrium. When

$$0 < A < 1,$$

Eq.(5.111.1) has two equilibrium points, namely, the zero equilibrium and the unique positive equilibrium point

$$\bar{x} = \frac{1-A}{1+D}.$$

The characteristic equation of the linearized equation of Eq.(5.111.1) about the zero equilibrium point, which exists as long as $A > 0$, is

$$\lambda^3 - \frac{1}{A}\lambda = 0. \tag{5.111.2}$$

The characteristic equation of the linearized equation of Eq.(5.111.1) about the positive equilibrium $\bar{x} = \frac{1-A}{D+1}$, which exists as long as $0 \leq A < 1$, is

$$\lambda^3 - \frac{D+A}{1+D}\lambda + \frac{D(1-A)}{1+D} = 0. \tag{5.111.3}$$

From (5.111.2) and Theorem 1.2.3 it follows that the zero equilibrium is locally asymptotically stable when

$$A > 1$$

and unstable when

$$0 < A < 1.$$

Also, it follows from Theorem 1.2.3 that the positive equilibrium $\bar{x} = \frac{1-A}{1+D}$ is locally asymptotically stable when

$$0 \leq A < 1 \quad \text{and} \quad D < 1$$

and unstable when

$$0 \le A < 1 \quad \text{and} \quad D > 1.$$

In addition to its equilibrium points, Eq.(5.111.1) has period-two solutions. If

$$\ldots, \phi, \psi, \phi, \psi, \ldots$$

is a period-two solution of Eq.(5.111.1) then, clearly,

$$\phi = \frac{\phi}{A + \phi + D\psi} \quad \text{and} \quad \psi = \frac{\psi}{A + \psi + D\phi}.$$

It follows that prime period-two solutions exist if and only if

$$0 \le A < 1.$$

Furthermore, when

$$D \ne 1$$

the only prime period-two solution of Eq.(5.111.1) is

$$\ldots, 0, 1 - A, 0, 1 - A, \ldots \tag{5.111.4}$$

and one can see that it is LAS when $D > 1$.

On the other hand, when

$$D = 1,$$

all prime period-two solutions are

$$\ldots, \phi, \psi, \phi, \psi, \ldots$$

with

$$\phi + \psi = 1 - A$$
$$\phi \ne \psi \quad \text{and} \quad \phi, \psi \in [0, 1 - A].$$

We are now ready to present the following global asymptotic stability result.

Theorem 5.111.1 (a) *Assume that*

$$A \ge 1.$$

Then the zero equilibrium of Eq.(5.111.1) is globally asymptotically stable.

(b) *Assume*

$$0 \le A < 1 \text{ and } D < 1.$$

Then every positive solution of Eq.(5.111.1) converges to its positive equilibrium point. Actually, if we only allow positive initial conditions, the positive equilibrium is globally asymptotically stable.

PROOF (*a*) This is a consequence of the inequality

$$x_{n+1} \leq \frac{1}{A} x_{n-1}$$

and the fact that there are no period-two solutions when $A \geq 1$.

(*b*) The proof follows from Theorem 5.106.4. ∎

Conjecture 5.111.1 *Assume that*

$$A < 1 \quad \text{and} \quad D \geq 1.$$

Show that every solution $\{x_n\}$ of Eq.(5.111.1) converges to a (not necessarily prime) period-two solution.

Open Problem 5.111.1 *Investigate the global character of solutions of Eq.(5.111.1) with periodic coefficients.*

Open Problem 5.111.2 *Investigate the global character of solutions of the equation*

$$x_{n+1} = \frac{\gamma_n x_{n-1}}{A_n + C_n x_{n-1} + x_{n-2}}, \quad n = 0, 1, \ldots$$

with convergent coefficients $\{\gamma_n\}$, $\{A_n\}$, and $\{C_n\}$. Extend and generalize.

5.112 Equation #112 : $x_{n+1} = \dfrac{\gamma x_{n-1}}{Bx_n + Cx_{n-1} + Dx_{n-2}}$

Eq.(#112) can be written in the normalized form

$$x_{n+1} = \frac{x_{n-1}}{Bx_n + x_{n-1} + Dx_{n-2}}, \quad n = 0, 1, \ldots \tag{5.112.1}$$

with positive parameters B, D and with arbitrary nonnegative initial conditions x_{-2}, x_{-1}, x_0 such that the denominator is always positive.

Eq.(5.112.1) has the unique equilibrium point

$$\bar{x} = \frac{1}{B + D + 1}.$$

The characteristic equation of the linearized equation of Eq.(5.112.1) about the equilibrium, $\bar{x} = \frac{1}{B+D+1}$, is

$$\lambda^3 + \frac{B}{B+D+1}\lambda^2 - \frac{B+D}{B+D+1}\lambda + \frac{D}{B+D+1} = 0. \tag{5.112.2}$$

From this and Theorem 1.2.3 it follows that the equilibrium $\bar{x} = \frac{1}{B+D+1}$ is locally asymptotically stable when

$$B + D < 1$$

and unstable when

$$B + D > 1.$$

In addition to its equilibrium point, the sequence

$$\ldots, \phi, 1 - \phi, \phi, 1 - \phi, \ldots$$

is a period-two solution of Eq.(5.112.1) for all positive values of the parameters B and D with

$$\phi \in [0, 1].$$

Furthermore, when

$$B + D \neq 1,$$

the only prime period-two solution of Eq.(5.112.1) is

$$\ldots, 0, 1, 0, 1, \ldots \tag{5.112.3}$$

and one can see that it is LAS when $B + D > 1$.

We are now ready to establish the following result.

Theorem 5.112.1 *Assume that*

$$B + D < 1.$$

Then every solution of Eq.(5.112.1) converges to its equilibrium point $\bar{x}$.

PROOF The proof follows from Theorem 5.106.4. ∎

Conjecture 5.112.1 *Assume that*

$$B + D \geq 1.$$

Show that every solution of Eq.(5.112.1) converges to a (not necessarily prime) period-two solution.

Open Problem 5.112.1 *Investigate the global character of solutions of Eq.(5.112.1) with periodic coefficients.*

Open Problem 5.112.2 *Investigate the global character of solutions of the equation*

$$x_{n+1} = \frac{\gamma_n x_{n-1}}{B_n x_n + C_n x_{n-1} + x_{n-2}}, \quad n = 0, 1, \ldots$$

with convergent coefficients $\{\gamma_n\}$, $\{B_n\}$, and $\{C_n\}$. Extend and generalize.

5.113 Equation #113 : $x_{n+1} = \dfrac{\delta x_{n-2}}{A + Bx_n + Cx_{n-1}}$

This equation was investigated in [60] and [146]. See also Section 4.4 where we established that this equation possesses a period-three trichotomy depending on whether

$$\delta < A, \quad \delta = A, \quad \text{or} \quad \delta > A.$$

Open Problem 5.113.1 *Determine the set of all initial conditions*

$$x_{-2}, x_{-1}, x_0 \in (0, \infty)$$

for which the solutions of the equation

$$x_{n+1} = \frac{2x_{n-2}}{1 + x_n + x_{n-1}}, \quad n = 0, 1, \ldots \tag{5.113.1}$$

are bounded.

5.114 Equation #114 : $x_{n+1} = \dfrac{\delta x_{n-2}}{A + Bx_n + Dx_{n-2}}$

For the global character of solutions of Eq.(#114) see Section 5.136.

Open Problem 5.114.1 *Determine the set of all initial conditions*

$$x_{-2}, x_{-1}, x_0 \in (0, \infty)$$

such that the solutions of the equation

$$x_{n+1} = \frac{x_{n-2}}{1 + x_n + x_{n-2}}, \quad n = 0, 1, \ldots \tag{5.114.1}$$

converge to zero.

Conjecture 5.114.1 *Show that for the positive equilibrium $\bar{x}$ of Eq.(#114) and with positive initial conditions,*

Local Asymptotic Stability $\Rightarrow$ Global Attractivity.

Conjecture 5.114.2 *Show that Eq.(5.114.1) has solutions that do not converge to an equilibrium point or to a periodic solution.*

5.115 Equation #115 : $x_{n+1} = \dfrac{\delta x_{n-2}}{A + Cx_{n-1} + Dx_{n-2}}$

For the global character of solutions of Eq.(#115) see Section 5.136.

Open Problem 5.115.1 *Determine the set of all initial conditions*

$$x_{-2}, x_{-1}, x_0 \in (0, \infty)$$

for which the solutions of the equation

$$x_{n+1} = \frac{x_{n-2}}{1 + x_{n-1} + x_{n-2}}, \quad n = 0, 1, \ldots \tag{5.115.1}$$

converge to zero.

Conjecture 5.115.1 *Show that for the positive equilibrium $\bar{x}$ of Eq.(#115) and with positive initial conditions,*

Local Asymptotic Stability $\Rightarrow$ ***Global Attractivity.***

Conjecture 5.115.2 *Show that Eq.(5.115.1) has solutions that do not converge to an equilibrium point or to a periodic solution.*

5.116 Equation #116 : $x_{n+1} = \dfrac{\delta x_{n-2}}{Bx_n + Cx_{n-1} + Dx_{n-2}}$

For the global character of solutions of Eq.(#116) see Section 5.136.

Open Problem 5.116.1 *Determine the set of all initial conditions*

$$x_{-2}, x_{-1}, x_0 \in (0, \infty)$$

for which the solutions of the equation

$$x_{n+1} = \frac{x_{n-2}}{x_n + x_{n-1} + x_{n-2}}, \quad n = 0, 1, \ldots \tag{5.116.1}$$

converge to $\frac{1}{3}$.

Conjecture 5.116.1 *Show that for the equilibrium $\bar{x}$ of Eq.(#114),*

Local Asymptotic Stability $\Rightarrow$ ***Global Attractivity.***

Conjecture 5.116.2 *Show that Eq.(5.116.1) has solutions that do not converge to the equilibrium point or to a periodic solution.*

5.117 Equation #117 : $x_{n+1} = \dfrac{\alpha + \beta x_n + \gamma x_{n-1}}{A}$

The equation in this special case is linear.

5.118 Equation #118 : $x_{n+1} = \dfrac{\alpha + \beta x_n + \gamma x_{n-1}}{B x_n}$

Eq.(#118) possesses a period-two trichotomy depending on whether

$$\gamma < \beta, \quad \gamma = \beta, \quad \text{or} \quad \gamma > \beta.$$

This result is a special case of a more general period-two trichotomy result presented in Theorem 4.2.1.

When

$$\gamma > \beta,$$

it follows from Theorem 4.2.2 that every bounded solution of Eq.(#118) converges to the equilibrium.

Open Problem 5.118.1 *Investigate the global character of solutions of Eq.(#118) with periodic coefficients.*

Open Problem 5.118.2 *Investigate the global character of solutions of the equation*

$$x_{n+1} = \frac{\alpha_n + \beta_n x_n + \gamma_n x_{n-1}}{x_n}, \quad n = 0, 1, \ldots$$

with convergent coefficients $\{\alpha_n\}$, $\{\beta_n\}$, *and* $\{\gamma_n\}$. *Extend and generalize.*

5.119 Equation #119 : $x_{n+1} = \dfrac{\alpha + \beta x_n + \gamma x_{n-1}}{C x_{n-1}}$

The change of variables, $x_n = y_n + \frac{\gamma}{C}$, transforms Eq.(#119) to Eq.(#66).

Conjecture 5.119.1 *Show that every solution of the equation*

$$x_{n+1} = \frac{\alpha + \beta x_n + x_{n-1}}{x_{n-1}}, \quad n = 0, 1, \ldots \tag{5.119.1}$$

has a finite limit.

5.120 Equation #120 : $x_{n+1} = \dfrac{\alpha + \beta x_n + \gamma x_{n-1}}{D x_{n-2}}$

For some work on this equation see [157]. In this section we allow the parameter α to be nonnegative and so the results presented here are also true for the special case #56. Eq.(#120) can be written in the normalized form,

$$x_{n+1} = \frac{\alpha + \beta x_n + x_{n-1}}{x_{n-2}}, \quad n = 0, 1, \ldots \tag{5.120.1}$$

with positive parameter β and with arbitrary positive initial conditions x_{-2}, x_{-1}, x_0.

Eq.(5.120.1) has the unique equilibrium

$$\bar{x} = \frac{\beta + 1 + \sqrt{(\beta+1)^2 + 4\alpha}}{2}.$$

The characteristic equation of the linearized equation of Eq.(5.120.1) about the equilibrium $\bar{x}$ is

$$\lambda^3 - \frac{\beta}{\bar{x}}\lambda^2 - \frac{1}{\bar{x}}\lambda + 1 = 0. \tag{5.120.2}$$

From this and Theorem 1.2.3 it follows that the equilibrium $\bar{x}$ of Eq.(5.120.1) is unstable when

$$\beta \neq 1.$$

When

$$\beta = 1,$$

that is, for the equation

$$x_{n+1} = \frac{\alpha + x_n + x_{n-1}}{x_{n-2}}, \quad n = 0, 1, \ldots, \tag{5.120.3}$$

the equilibrium $\bar{x}$ is nonhyperbolic, with one characteristic root equal to -1 and the other two complex conjugates given by:

$$\frac{2 + \sqrt{1+\alpha} \pm i\sqrt{4\sqrt{1+\alpha} + 3(1+\alpha)}}{2(1 + \sqrt{1+\alpha})}$$

both with magnitude equal to one. Eq.(5.120.3), which is called **Todd's equation**, possesses the following invariant (see [122], [157], or [175]):

$$(\alpha + x_{n-2} + x_{n-1} + x_n)(1 + \frac{1}{x_{n-2}})(1 + \frac{1}{x_{n-1}})(1 + \frac{1}{x_n}) = \text{constant}. \tag{5.120.4}$$

Todd's equation can be extended to the more general $(k+1)^{\text{st}}$-order rational equation

$$x_{n+1} = \frac{\alpha + x_n + \cdots + x_{n-(k-1)}}{x_{n-k}}, \quad n = 0, 1, \ldots, \tag{5.120.5}$$

which possesses the invariant

$$(\alpha + x_{n-k} + \cdots + x_n)(1 + \frac{1}{x_{n-k}}) \cdots (1 + \frac{1}{x_n}) = \text{constant}. \tag{5.120.6}$$

From this it follows that every solution of Eq.(5.120.5) is bounded from above and below by positive constants when

$$\alpha > 0.$$

When $\alpha = 0$ it also follows from (5.120.6) that the solution is bounded from above. We now claim that the solution is also bounded from below. Otherwise, there exists a solution $\{x_n\}$ of Eq.(5.120.5) such that

$$x_{n_i+1} \to 0 \quad \text{and} \quad x_{n_i+1} < x_j \ \text{ for } \ j < n_i + 1.$$

Then, clearly,

$$x_{n_i} \to 0$$

$$\ldots$$

and

$$x_{n_i-k} \to 0.$$

Also, from

$$x_{n_i+1} = \frac{x_{n_i} + \cdots + x_{n_i-(k-1)}}{x_{n_i-k}} < x_{n_i}$$

it follows that

$$x_{n_i}(1 - x_{n_i-k}) + \cdots + x_{n_i-(k-1)} < 0,$$

which is a contradiction.

In the special case of Eq.(5.120.3) where

$$\alpha = \beta = 1,$$

that is, for the equation

$$x_{n+1} = \frac{1 + x_n + x_{n-1}}{x_{n-2}}, \quad n = 0, 1, \ldots, \tag{5.120.7}$$

the three characteristic roots are eighth roots of unity. In this case one can also see that every solution of Eq.(5.120.7) is periodic with period eight. See [122], [157], or [175].

Conjecture 5.120.1 *Assume that*

$$\alpha \neq 1 \quad or \quad \beta \neq 1.$$

Show that Eq.(5.120.1) has bounded solutions that do not converge to the equilibrium point $\bar{x}$ or to a periodic solution.

5.121 Equation #121 : $x_{n+1} = \dfrac{\alpha + \beta x_n + \delta x_{n-2}}{A}$

The equation in this special case is linear.

5.122 Equation #122 : $x_{n+1} = \dfrac{\alpha + \beta x_n + \delta x_{n-2}}{B x_n}$

The change of variables, $x_n = y_n + \frac{\beta}{B}$, transforms Eq.(#122) to Eq.(#77). See Section 5.77.

Conjecture 5.122.1 *Show that for the equilibrium $\bar{x}$ of Eq.(#122),*

Local Asymptotic Stability ⇒ Global Asymptotic Stability.

Conjecture 5.122.2 *Show that Eq.(#122) has solutions that do not converge to the equilibrium point or to a periodic solution.*

5.123 Equation #123 : $x_{n+1} = \dfrac{\alpha + \beta x_n + \delta x_{n-2}}{C x_{n-1}}$

In this section we allow the parameter α to be nonnegative and so this equation also includes the special case #59.

We have conjectured in Section 4.5 that Eq.(#123) possesses a period-four trichotomy depending on whether

$$\delta < \beta, \quad \delta = \beta, \quad \text{or} \quad \delta > \beta.$$

See [59] and [150]. The only part of this conjecture that has not been established yet is when $\delta > \beta$.

Conjecture 5.123.1 *Assume that*

$$\alpha \geq 0 \text{ and } \beta > 1.$$

Show that the equilibrium of the equation

$$x_{n+1} = \frac{\alpha + \beta x_n + x_{n-2}}{x_{n-1}}, \quad n = 0, 1, \ldots$$

is globally asymptotically stable.

5.124 Equation #124 : $x_{n+1} = \dfrac{\alpha + \beta x_n + \delta x_{n-2}}{D x_{n-2}}$

The change of variables, $x_n = y_n + \frac{\delta}{D}$, transforms Eq.(#124) to Eq.(#67). See Section 5.67.

Conjecture 5.124.1 *Show that for the equilibrium $\bar{x}$ of Eq.(#124),*

Local Asymptotic Stability $\Rightarrow$ ***Global Asymptotic Stability.***

Conjecture 5.124.2 *Show that Eq.(#124) has solutions which do not converge to the equilibrium point or to a periodic solution.*

5.125 Equation #125 : $x_{n+1} = \dfrac{\alpha + \gamma x_{n-1} + \delta x_{n-2}}{A}$

The equation in this special case is linear.

5.126 Equation #126 : $x_{n+1} = \dfrac{\alpha + \gamma x_{n-1} + \delta x_{n-2}}{B x_n}$

This equation was investigated in [46]. Eq.(#126) can be written in the normalized form

$$x_{n+1} = \frac{\alpha + \gamma x_{n-1} + x_{n-2}}{x_n}, \quad n = 0, 1, \ldots \tag{5.126.1}$$

with positive parameters α, γ and with arbitrary positive initial conditions x_{-2}, x_{-1}, x_0.

It follows from Theorem 3.1.1 that when

$$\gamma > 1,$$

Eq.(5.126.1) possesses unbounded solutions. It was also shown in [46], that when

$$\alpha = \gamma = 1 \tag{5.126.2}$$

and by choosing initial conditions x_{-2}, x_{-1}, and x_0 such that

$$x_0 \leq x_{-2} \leq 1 \text{ and } x_{-1} \in (0, \infty),$$

then

$$\lim_{n\to\infty} x_{2n} \in [0, 1] \text{ and } \lim_{n\to\infty} x_{2n+1} = \infty.$$

Furthermore, it follows from the work in [46] that when (5.126.2) holds, every bounded solution of Eq.(5.126.1) converges to a (not necessarily prime) period-two solution. This result is also true for

$$\gamma = 1 \text{ and } \alpha > 0.$$

The proof is an immediate consequence of the following identity:

$$x_{n+2} - x_n = \frac{1}{x_{n+1}} (x_n - x_{n-2}), \quad n = 0, 1, \ldots .$$

It was shown by R. Nigmatulin (personal communication with G. Ladas) that when

$$\gamma > 1,$$

every unbounded solution $\{x_n\}_{n=-2}^{\infty}$ of Eq.(5.126.1) is such that the subsequences of the even and odd terms converge one of them to zero and the other to ∞.

It was also shown (by R. Nigmatulin) that when

$$\alpha = \gamma = 1,$$

and by choosing initial conditions x_{-2}, x_{-1}, and x_0 such that

$$x_{-2} = x_{-1} = x_0 = 1,$$

then

$$x_{2n} = 1 \text{ and } x_{2n+1} = 2n + 1 \to \infty.$$

The only equilibrium of Eq.(5.126.1) is

$$\bar{x} = \frac{\gamma + 1 + \sqrt{(\gamma + 1)^2 + 4\alpha}}{2}.$$

The characteristic equation of the linearized equation of Eq.(5.126.1) about the equilibrium $\bar{x}$ is

$$\lambda^3 + \lambda^2 - \frac{\gamma}{\bar{x}}\lambda - \frac{1}{\bar{x}} = 0. \quad (5.126.3)$$

From this and Theorem 1.2.3 it follows that the equilibrium $\bar{x}$ of Eq.(5.126.1) is locally asymptotically stable when

$$\alpha \geq 1 \text{ and } \gamma < 1 \quad (5.126.4)$$

or

$$\alpha < 1 \text{ and } (1-\alpha)\cdot\frac{-1+\sqrt{3+2\alpha}}{2(\alpha+1)} < \gamma < 1 \quad (5.126.5)$$

and unstable when

$$\alpha < 1 \text{ and } \gamma < (1-\alpha)\cdot\frac{-1+\sqrt{3+2\alpha}}{2(\alpha+1)}.$$

Conjecture 5.126.1 *Assume that (5.126.4) or (5.126.5) holds. Show that the equilibrium of (5.126.1) is globally asymptotically stable.*

Open Problem 5.126.1 *Show that Eq.(5.126.1) has bounded solutions that do not converge to the equilibrium point $\bar{x}$ or to a periodic solution.*

Conjecture 5.126.2 *Assume that*

$$\gamma > 1.$$

Show that every bounded solution of Eq.(5.126.1) converges to the equilibrium $\bar{x}$.

Open Problem 5.126.2 *Assume that*

$$\gamma > 1.$$

(a) Determine the set of all initial conditions through which the solutions of Eq.(5.126.1) converge to the equilibrium $\bar{x}$.

(b) Determine the set of all initial conditions through which the solutions of Eq.(5.126.1) are unbounded.

Open Problem 5.126.3 *Assume that*

$$\gamma = 1.$$

(a) Determine the set of all initial conditions through which the solutions of Eq.(5.126.1) converge to the equilibrium $\bar{x}$.

(b) Determine the set of all initial conditions through which the solutions of Eq.(5.126.1) converge to a prime period-two solution.

(c) Determine the set of all initial conditions through which the solutions of Eq.(5.126.1) are unbounded.

5.127 Equation #127 : $x_{n+1} = \dfrac{\alpha + \gamma x_{n-1} + \delta x_{n-2}}{C x_{n-1}}$

The change of variables, $x_n = y_n + \frac{\gamma}{C}$, transforms Eq.(#127) to Eq.(#78). See Section 5.78.

Conjecture 5.127.1 *Show that for the equilibrium $\bar{x}$ of Eq.(#127),*

Local Asymptotic Stability $\Rightarrow$ ***Global Asymptotic Stability.***

Conjecture 5.127.2 *Show that Eq.(#127) has solutions that do not converge to the equilibrium point or to a periodic solution.*

5.128 Equation #128 : $x_{n+1} = \dfrac{\alpha + \gamma x_{n-1} + \delta x_{n-2}}{D x_{n-2}}$

Eq.(#128) possesses a period-two trichotomy depending on whether

$$\gamma < \delta, \quad \gamma = \delta, \quad \text{or} \quad \gamma > \delta.$$

This result is a special case of a more general period-two trichotomy result presented in Theorem 4.3.1.

Conjecture 5.128.1 *Assume that*

$$\gamma > \delta.$$

Show that every bounded solution of Eq.(#128) converges to the equilibrium.

Open Problem 5.128.1 *Investigate the global character of solutions of Eq.(#128) with periodic coefficients.*

5.129 Equation #129 : $x_{n+1} = \dfrac{\beta x_n + \gamma x_{n-1} + \delta x_{n-2}}{A}$

The equation in this special case is linear.

5.130 Equation #130 : $x_{n+1} = \dfrac{\beta x_n + \gamma x_{n-1} + \delta x_{n-2}}{Bx_n}$

The change of variables, $x_n = y_n + \frac{\beta}{B}$, transforms Eq.(#130) to Eq.(#95). See Section 5.95.

Conjecture 5.130.1 *Assume that*

$$\gamma > \beta + \delta.$$

Show that every bounded solution of Eq.(#130) converges to the equilibrium.

Conjecture 5.130.2 *Assume that*

$$\gamma < \beta + \delta.$$

Show that for the equilibrium $\bar{x}$ of Eq.(#130),

Local Asymptotic Stability ⇒ Global Asymptotic Stability.

Conjecture 5.130.3 *Assume that*

$$\gamma = \beta + \delta.$$

Show that every solution of Eq.(#130) converges to a (not necessarily prime) period-two solution.

Conjecture 5.130.4 *Assume that*

$$\gamma < \beta + \delta.$$

Show that Eq.(#130) has bounded solutions that do not converge to the equilibrium point or to a periodic solution.

5.131 Equation #131 : $x_{n+1} = \dfrac{\beta x_n + \gamma x_{n-1} + \delta x_{n-2}}{Cx_{n-1}}$

The change of variables, $x_n = y_n + \frac{\gamma}{C}$, transforms Eq.(#131) to Eq.(#172). See Section 5.172.

Conjecture 5.131.1 *Show that for the equilibrium $\bar{x}$ of Eq.(#131),*

Local Asymptotic Stability ⇒ Global Asymptotic Stability.

Conjecture 5.131.2 *Show that Eq.(#131) has solutions that do not converge to the equilibrium point or to a periodic solution.*

5.132 Equation #132 : $x_{n+1} = \dfrac{\beta x_n + \gamma x_{n-1} + \delta x_{n-2}}{D x_{n-2}}$

Eq.(#132) can be written in the normalized form

$$x_{n+1} = \frac{\beta x_n + \gamma x_{n-1} + x_{n-2}}{x_{n-2}}, \quad n = 0, 1, \ldots \tag{5.132.1}$$

with positive parameters β, γ and with arbitrary positive initial conditions x_{-2}, x_{-1}, x_0.

The boundedness character of this equation was investigated in [49]. See also Theorem 3.2.1 where we established that the equation has unbounded solutions when

$$\gamma > \beta + 1.$$

Conjecture 5.132.1 *Assume that*

$$\gamma > \beta + 1.$$

Show that every bounded solution of Eq.(5.132.1) converges to the equilibrium.

Conjecture 5.132.2 *Assume that*

$$\gamma < \beta + 1.$$

Show that for the equilibrium $\bar{x}$ of Eq.(5.132.1),

Local Asymptotic Stability $\Rightarrow$ ***Global Asymptotic Stability.***

Conjecture 5.132.3 *Assume that*

$$\gamma = \beta + 1.$$

Show that every solution of Eq.(5.132.1) converges to a (not necessarily prime) period-two solution.

Conjecture 5.132.4 *Assume that*

$$\gamma < \beta + 1.$$

Show that Eq.(5.132.1) has bounded solutions that do not converge to the equilibrium point or to a periodic solution.

5.133 Equation #133 : $x_{n+1} = \dfrac{\alpha}{A + Bx_n + Cx_{n-1} + Dx_{n-2}}$

This equation is a special case of a more general equation investigated in Section 5.17. It follows from Theorem 5.17.1 that the equilibrium of this equation is globally asymptotically stable.

5.134 Equation #134 : $x_{n+1} = \dfrac{\beta x_n}{A + Bx_n + Cx_{n-1} + Dx_{n-2}}$

This equation is a special case of a more general equation that we investigated in Section 5.106.

Conjecture 5.134.1 *Show that for the positive equilibrium $\bar{x}$ of Eq.(#134) and with positive initial conditions,*

Local Asymptotic Stability $\Rightarrow$ ***Global Asymptotic Stability.***

Conjecture 5.134.2 *Show that Eq.(#134) has solutions that do not converge to an equilibrium point or to a periodic solution.*

5.135 Equation #135 : $x_{n+1} = \dfrac{\gamma x_{n-1}}{A + Bx_n + Cx_{n-1} + Dx_{n-2}}$

This equation is a special case of a more general equation that we investigated in Section 5.109.

Conjecture 5.135.1 *Show that for the positive equilibrium $\bar{x}$ of Eq.(#135) and with positive initial conditions,*

Local Asymptotic Stability $\Rightarrow$ ***Global Asymptotic Stability.***

Conjecture 5.135.2 *Show that Eq.(#135) has solutions that do not converge to an equilibrium point or to a periodic solution.*

5.136 Equation #136 : $x_{n+1} = \dfrac{\delta x_{n-2}}{A + Bx_n + Cx_{n-1} + Dx_{n-2}}$

This equation was investigated in [60]. The equation

$$x_{n+1} = \frac{x_{n-2}}{A + Bx_n + Cx_{n-1} + x_{n-2}}, \quad n = 0, 1, \ldots \tag{5.136.1}$$

with nonnegative parameters A, B, C such that

$$A + B + C > 0$$

and with arbitrary nonnegative initial conditions x_{-2}, x_{-1}, x_0 such that the denominator is always positive contains the special case #37, which is a Riccati-type equation and the following six cases:

#39, #40, #114, #115, #116, #136.

When $A \geq 1$, zero is the only equilibrium point of Eq.(5.136.1) and, clearly, zero in this case is globally asymptotically stable.

When $A \in (0, 1)$, Eq.(5.136.1) has two equilibrium points, namely, the zero equilibrium point, which is unstable, and the positive equilibrium point

$$\bar{x} = \frac{1 - A}{B + C + 1}.$$

When $A = 0$, $\bar{x} = \frac{1}{B+C+1}$ is the only equilibrium of Eq.(5.136.1). The positive equilibrium point $\bar{x}$ of Eq.(5.136.1) is locally asymptotically stable when

$$A \in [0, 1), \quad B < \frac{2 - A + \sqrt{A^2 + 8}}{2} \quad \text{and} \quad C < \frac{1 - 2B + \sqrt{5 + 4A + 4B(1 - A)}}{2}. \tag{5.136.2}$$

Conjecture 5.136.1 *Assume that (5.136.2) holds. Show that the positive equilibrium of Eq.(5.136.1) is a global attractor of all positive solutions.*

Conjecture 5.136.2 *Show that Eq.(5.136.1) has solutions that do not converge to an equilibrium point or to a periodic solution.*

The following global result was established in [60].

Theorem 5.136.1 *Assume that*

$$A, B + C \in [0, 1).$$

Then every positive solution of Eq.(5.136.1) converges to the positive equilibrium $\bar{x}$.

PROOF The proof follows from Theorem 5.106.4. ∎

By using the identity

$$a^3 + b^3 + c^3 - 3abc = \frac{1}{2}(a+b+c)\left[(a-b)^2 + (b-c)^2 + (c-a)^2\right]$$

one can show that Eq.(5.136.1) has positive prime period-three solutions if and only if

$$0 \leq A < 1 \text{ and } B = C = 1. \tag{5.136.3}$$

All other possible prime period-three solutions of Eq.(5.136.1) are of the form

$$\ldots, 0, 0, \phi, 0, 0, \phi, \ldots \tag{5.136.4}$$

with $\phi \in (0, \infty)$ or of the form

$$\ldots, 0, \phi, \psi, 0, \phi, \psi, \ldots \tag{5.136.5}$$

with $\phi, \psi \in (0, \infty)$.

One can see that Eq.(5.136.1) has prime period-three solutions of the form (5.136.4) if and only if

$$A \in (0, 1) \text{ or } A = 0 \text{ and } B, C \in (0, \infty). \tag{5.136.6}$$

Furthermore, when (5.136.6) holds, Eq.(5.136.1) has a unique prime period-three solution of the form (5.136.4) with $\phi = 1 - A$.

Also, Eq.(5.136.1) has prime period-three solutions of the form (5.136.5) if and only if

$$0 \leq A < 1 \text{ and } B = C = 1, \tag{5.136.7}$$

or

$$0 \leq A < 1 \text{ and } B, C \in (1, \infty), \tag{5.136.8}$$

or

$$0 \leq A < 1 \text{ and } B, C \in [0, 1). \tag{5.136.9}$$

Furthermore, when (5.136.7) holds, the values of ϕ, ψ in (5.136.5) are all positive numbers ϕ and ψ such that

$$\phi + \psi = 1 - A$$

and Eq.(5.136.1) has infinitely many period-three solutions.

When (5.136.8) or (5.136.9) holds, the values of ϕ, ψ in (5.136.5) are

$$\phi = \frac{(1-A)(1-C)}{1-BC} \text{ and } \psi = \frac{(1-A)(1-B)}{1-BC}. \tag{5.136.10}$$

and Eq.(5.136.1) has a unique prime period-three solution in this case.

When (5.136.3) holds, the positive prime period-three solutions of Eq.(5.136.1)

$$\ldots, \phi, \psi, \omega, \ldots \tag{5.136.11}$$

are given by

$$\phi + \psi + \omega = 1 - A$$

with

$$\phi, \psi, \omega \in (0, 1) \text{ and } (\phi, \psi, \omega) \neq \left(\frac{1-A}{3}, \frac{1-A}{3}, \frac{1-A}{3} \right).$$

In view of the above we see that when $A \in [0, 1)$, all prime period-three solutions of the equation

$$x_{n+1} = \frac{x_{n-2}}{A + x_n + x_{n-1} + x_{n-2}}, \quad n = 0, 1, \ldots \tag{5.136.12}$$

are of the form (5.136.11), where ϕ, ψ, ω are all solutions of the equation

$$\phi + \psi + \omega = 1 - A$$

with

$$\phi, \psi, \omega \in [0, 1] \text{ and } (\phi, \psi, \omega) \neq \left(\frac{1-A}{3}, \frac{1-A}{3}, \frac{1-A}{3} \right).$$

The following result shows that when

$$0 \le A < 1,$$

every solution of Eq.(5.136.12) converges to a (not necessarily prime) period-three solution.

Theorem 5.136.2 *Assume that $A \in [0, 1)$. Then every solution of Eq.(5.136.12) converges to a (not necessarily prime) period-three solution.*

PROOF Note that

$$x_{n+1} - x_{n-2} = \frac{x_{n-2}}{A + x_n + x_{n-1} + x_{n-2}} \left(1 - A - x_n - x_{n-1} - x_{n-2}\right), \quad n = 0, 1, \ldots .$$

Set

$$J_n = 1 - A - x_n - x_{n-1} - x_{n-2}.$$

Then for $n \geq 0$,

$$J_{n+1} = 1 - A - x_n - x_{n-1} - \frac{x_{n-2}}{A + x_n + x_{n-1} + x_{n-2}} = \frac{A + x_n + x_{n-1}}{A + x_n + x_{n-1} + x_{n-2}} J_n.$$

Hence, the signum of J_n is constant, from which the result follows because every solution of Eq.(5.136.12) is bounded. ∎

Open Problem 5.136.1 *Determine the set of all positive initial conditions through which the solutions of Eq.(5.136.12) converge to a prime period-three solution.*

Open Problem 5.136.2 *Assume that (5.136.6) is satisfied. Determine the global character of Eq.(5.136.1), and in particular, determine the basin of attraction of the period-three solution*

$$\ldots, 0, 0, 1-A, \ldots .$$

Open Problem 5.136.3 *Assume that (5.136.8) or (5.136.9) is satisfied. Determine, in each case, the global character of solutions of Eq.(5.136.1), and in particular, determine the basin of attraction of each of the period-three solutions*

$$\ldots, 0, 0, 1-A, \ldots$$

$$\ldots, \frac{(1-A)(1-C)}{1-BC}, \frac{(1-A)(1-B)}{1-BC}, 0, \ldots .$$

It is not difficult to see that when (5.136.6) holds, the basin of attraction of the period-three solution

$$\ldots, 0, 0, 1-A, \ldots$$

includes all solutions of Eq.(5.136.1) with two of the three initial conditions x_{-2}, x_{-1}, x_0 equal to zero and the third positive.

When (5.136.8) or (5.136.9) holds, every solution of Eq.(5.136.1) with one of the three initial conditions x_{-2}, x_{-1}, x_0 equal to zero and the other two positive converges to a period-three solution.

Conjecture 5.136.3 *Assume that*

$$0 \leq A < 1 \text{ and } B, C \in (1, \infty).$$

Then every positive solution of Eq.(5.136.1) converges to a period-three solution of Eq.(5.136.1).

Open Problem 5.136.4 *Investigate the behavior of solutions of Eq.(#136) when*

$$\text{either } B \neq 1 \text{ or } C \neq 1.$$

Open Problem 5.136.5 *Assume that k is a positive integer and $A, B_0, \ldots, B_k \in [0, \infty)$. Determine the global stability of the periodic solutions of the difference equation*

$$x_{n+1} = \frac{x_{n-k}}{A + B_0 x_n + \ldots + B_k x_{n-k}}, \quad n = 0, 1, \ldots .$$

Conjecture 5.136.4 *Show that Eq.(5.136.1) has solutions that do not converge to an equilibrium point or to a periodic solution.*

5.137 Equation #137 : $x_{n+1} = \dfrac{\alpha + \beta x_n + \gamma x_{n-1} + \delta x_{n-2}}{A}$

The equation in this special case is linear.

5.138 Equation #138 : $x_{n+1} = \dfrac{\alpha + \beta x_n + \gamma x_{n-1} + \delta x_{n-2}}{B x_n}$

This equation is a special case of a more general equation that we investigate in Section 5.195.

Conjecture 5.138.1 *Assume that*

$$\gamma > \beta + \delta.$$

Show that every bounded solution of Eq.(#138) converges to the equilibrium.

Conjecture 5.138.2 *Assume that*

$$\gamma < \beta + \delta.$$

Show that for the equilibrium $\bar{x}$ of Eq.(#138),

Local Asymptotic Stability $\Rightarrow$ ***Global Asymptotic Stability.***

Conjecture 5.138.3 *Assume that*

$$\gamma < \beta + \delta.$$

Show that Eq.(#138) has bounded solutions that do not converge to the equilibrium point or to a periodic solution.

5.139 Equation #139 : $x_{n+1} = \dfrac{\alpha + \beta x_n + \gamma x_{n-1} + \delta x_{n-2}}{C x_{n-1}}$

The change of variables, $x_n = y_n + \frac{\gamma}{C}$, transforms Eq.(#139) to Eq.(#172). See Section 5.172.

Conjecture 5.139.1 *Show that for the equilibrium $\bar{x}$ of Eq.(#139),*

$$\textit{Local Asymptotic Stability} \Rightarrow \textit{Global Asymptotic Stability.}$$

Conjecture 5.139.2 *Show that Eq.(#139) has solutions that do not converge to the equilibrium point or to a periodic solution.*

5.140 Equation #140 : $x_{n+1} = \dfrac{\alpha + \beta x_n + \gamma x_{n-1} + \delta x_{n-2}}{D x_{n-2}}$

The change of variables, $x_n = y_n + \frac{\delta}{D}$, transforms Eq.(#140) to Eq.(#167). See Section 5.167.

Conjecture 5.140.1 *Assume that*

$$\gamma > \beta + \delta.$$

Show that every bounded solution of Eq.(#140) converges to the equilibrium.

Conjecture 5.140.2 *Assume that*

$$\gamma < \beta + \delta.$$

Show that for the equilibrium $\bar{x}$ of Eq.(#140),

$$\textit{Local Asymptotic Stability} \Rightarrow \textit{Global Asymptotic Stability.}$$

Conjecture 5.140.3 *Assume that*

$$\gamma = \beta + \delta.$$

Show that every solution of Eq.(#140) converges to a (not necessarily prime) period-two solution.

Conjecture 5.140.4 *Assume that*

$$\gamma < \beta + \delta.$$

Show that Eq.(#140) has bounded solutions that do not converge to the equilibrium point or to a periodic solution.

5.141 Equation #141 : $x_{n+1} = \dfrac{\alpha + \beta x_n}{A + Bx_n + Cx_{n-1}}$

This equation was investigated in [176]. See also [175]. Eq.(#141) can be written in the normalized form

$$x_{n+1} = \frac{\alpha + x_n}{A + Bx_n + x_{n-1}}, \quad n = 0, 1, \ldots \tag{5.141.1}$$

with positive parameters α and B and with arbitrary nonnegative initial conditions x_{-1}, x_0. Throughout this section we allow the parameter A to be nonnegative so that we also include the special case #68.

Eq.(5.141.1) has the unique equilibrium

$$\bar{x} = \frac{1 - A + \sqrt{(1-A)^2 + 4\alpha(1+B)}}{2(B+1)}.$$

The characteristic equation of the linearized equation of Eq.(5.141.1) about the equilibrium $\bar{x}$ is

$$\lambda^2 + \frac{B\bar{x} - 1}{A + (1+B)\bar{x}}\lambda + \frac{\bar{x}}{A + (1+B)\bar{x}} = 0.$$

From this and Theorem 1.2.2 it follows that the equilibrium $\bar{x}$ of Eq.(5.141.1) is locally asymptotically stable for all the values of the parameters α, A, B.

From Theorems 5.23.2 and 5.23.3 it follows that when

$$A \geq 1,$$

the equilibrium $\bar{x}$ of Eq.(5.141.1) is globally asymptotically stable. In this section we will show that in a subregion of

$$0 \leq A < 1$$

every solution of Eq.(5.141.1) converges to the equilibrium $\bar{x}$.

The following identity will be useful in the sequel.

$$x_{n+1} - x_{n-3} = \frac{(\alpha A - A^2 x_{n-3} - A x_{n-3}^2) + x_n(A + x_{n-3})(1 - Bx_{n-3})}{(A + Bx_n)(A + Bx_{n-2} + x_{n-3}) + \alpha + x_{n-2}}$$

$$+\frac{x_{n-2}(\alpha B - (AB+1)x_{n-3}) + x_n x_{n-2} B(1 - Bx_{n-3})}{(A + Bx_n)(A + Bx_{n-2} + x_{n-3}) + \alpha + x_{n-2}}. \tag{5.141.2}$$

Clearly,

$$x_{n-3} \geq \frac{1}{B} \text{ and } \alpha < \frac{A}{B} + \frac{1}{B^2}$$

implies that

$$\alpha A - A^2 x_{n-3} - A x_{n-3}^2 \leq 0,$$

and

$$(A + x_{n-3})(1 - B x_{n-3}) \leq 0,$$

and

$$\alpha B - (AB + 1) x_{n-3} < 0.$$

Also,

$$x_{n-3} \leq \frac{1}{B} \text{ and } \alpha > \frac{A}{B} + \frac{1}{B^2}$$

implies that

$$\alpha A - A^2 x_{n-3} - A x_{n-3}^2 \geq 0,$$

and

$$(A + x_{n-3})(1 - B x_{n-3}) \geq 0,$$

and

$$\alpha B - (AB + 1) x_{n-3} > 0.$$

Set

$$f(x_n, x_{n-1}) = \frac{\alpha + x_n}{A + B x_n + x_{n-1}}.$$

Theorem 5.141.1 *Let $\{x_n\}$ be any solution of Eq.(5.141.1). Then the following statements are true:*
(i) When

$$0 \leq A < 1 \text{ and } \frac{(1-B)(1-A)^2}{4B^2} \leq \alpha < \frac{A}{B} + \frac{1}{B^2} \tag{5.141.3}$$

then the solution $\{x_n\}$ eventually enters the interval $\left[\alpha B - A, \frac{1}{B}\right]$ and the function $f(x_n, x_{n-1})$ is eventually strictly increasing in x_n and strictly decreasing in x_{n-1}. Furthermore, the solution $\{x_n\}$ converges to the equilibrium $\bar{x}$.
(ii) When

$$0 \leq A < 1 \text{ and } \alpha > \frac{A}{B} + \frac{1}{B^2}, \tag{5.141.4}$$

the solution $\{x_n\}$ eventually enters the interval $[\frac{1}{B}, \alpha B - A]$ and the function $f(x_n, x_{n-1})$ is eventually strictly decreasing in x_n and x_{n-1}. Furthermore, the solution $\{x_n\}$ converges to the equilibrium $\bar{x}$.
(iii) When

$$0 \leq A < 1 \text{ and } \alpha = \frac{A}{B} + \frac{1}{B^2}, \tag{5.141.5}$$

then the solution $\{x_n\}$ converges to the equilibrium $\bar{x}$.

PROOF Let $\{x_n\}$ be a solution of Eq.(5.141.1) with nonnegative initial conditions. We claim that

$$\left[\min(\alpha B - A, \frac{1}{B}), \max(\alpha B - A, \frac{1}{B})\right]$$

is an attracting interval for the solution $\{x_n\}$ of Eq.(5.141.1).

We will prove that when (5.141.3) or (5.141.4) holds, all four subsequences of the solution $\{x_n\}$, of the form $\{x_{4n+j}\}_{j=0}^{3}$, lie eventually within the interval

$$\left[\min(\alpha B - A, \frac{1}{B}), \max(\alpha B - A, \frac{1}{B})\right].$$

We will give the proof when (5.141.3) holds. The proof when (5.141.4) holds is similar and will be omitted. Furthermore, we will give the proof for the subsequence $\{x_{4n+1}\}$. The proof for all the other subsequences is similar and will be omitted.

Suppose for the sake of contradiction that there exists N sufficiently large such that

$$x_{4N+1} < \alpha B - A \quad \text{or} \quad x_{4N+1} > \frac{1}{B}.$$

We will give the proof in the case where $x_{4N+1} < \alpha B - A$. The proof in the other case is similar and will be omitted. Then from

$$x_{4N+1} < \alpha B - A$$

it follows that

$$x_{4N+3} = \frac{\alpha + x_{4N+2}}{A + Bx_{4N+2} + x_{4N+1}} > \frac{\alpha + x_{4N+2}}{A + Bx_{4N+2} + \alpha B - A} = \frac{1}{B} > \alpha B - A.$$

From this it follows that

$$x_{4N+5} = \frac{\alpha + x_{4N+4}}{A + Bx_{4N+4} + x_{4N+3}} < \frac{\alpha + x_{4N+4}}{A + Bx_{4N+4} + \alpha B - A} = \frac{1}{B}. \tag{5.141.6}$$

We claim that for some $k \geq 1$,

$$x_{4N+4k+1} \geq \alpha B - A. \tag{5.141.7}$$

Otherwise, for all $k \geq 1$,

$$x_{4N+4k+1} < \alpha B - A.$$

Then, clearly, for all $k \geq 1$,

$$x_{4N+4k+3} = \frac{\alpha + x_{4N+4k+2}}{A + Bx_{4N+4k+2} + x_{4N+4k+1}} > \frac{\alpha + x_{4N+4k+2}}{A + Bx_{4N+4k+2} + \alpha B - A} = \frac{1}{B}.$$

From (5.141.2) it follows that the subsequence $\{x_{4N+4k+1}\}$ decreases. By taking limits in (5.141.2) we get a contradiction that proves (5.141.7). Assume without loss of generality that (5.141.7) holds for $k = 1$. From this and (5.141.6) we see that

$$\alpha B - A < x_{4N+5} < \frac{1}{B}.$$

Then

$$x_{4N+7} = \frac{\alpha + x_{4N+6}}{A + Bx_{4N+6} + x_{4N+6}} < \frac{\alpha + x_{4N+6}}{A + Bx_{4N+6} + \alpha B - A} < \frac{1}{B}$$

and

$$x_{4N+7} = \frac{\alpha + x_{4N+6}}{A + Bx_{4N+6} + x_{4N+5}} > \frac{\alpha + x_{4N+6}}{A + Bx_{4N+6} + \frac{1}{B}} > \frac{\alpha}{A + \frac{1}{B}} > \alpha B - A$$

and the result follows by induction.

When (5.141.3) holds, and due to the fact that the solution $\{x_n\}$ eventually enters the interval $[\alpha B - A, \frac{1}{B}]$, we see that the function

$$f(x_n, x_{n-1}) = \frac{\alpha + x_n}{A + Bx_n + x_{n-1}}$$

is eventually strictly increasing in x_n and strictly decreasing in x_{n-1}. Furthermore, for each $m, M \in [\alpha B - A, \frac{1}{B}]$, in view of (5.141.3), the system

$$M = \frac{\alpha + M}{A + BM + m} \quad \text{and} \quad m = \frac{\alpha + m}{A + Bm + M}$$

has the unique solution $(m, M) = (\bar{x}, \bar{x})$. Hence, the result follows by Theorem 1.6.5.

When (5.141.4) holds, and due to the fact that the solution $\{x_n\}$ eventually enters the interval $[\frac{1}{B}, \alpha B - A]$, we see that the function

$$f(x_n, x_{n-1}) = \frac{\alpha + x_n}{A + Bx_n + x_{n-1}}$$

is strictly decreasing in x_n and eventually strictly decreasing in x_{n-1}. Furthermore, for each $m, M \in [\frac{1}{B}, \alpha B - A]$, the system

$$M = \frac{\alpha + m}{A + (B+1)m} \quad \text{and} \quad m = \frac{\alpha + M}{A + (B+1)M}$$

has the unique solution $(m, M) = (\bar{x}, \bar{x})$. Hence, the result follows by Theorem 1.6.5.

Finally, assume that (5.141.5) holds. Then, clearly, for all $n \geq 0$,

$$x_{n+1} - \frac{1}{B} = \frac{1}{B} \cdot \frac{\frac{1}{B} - x_{n-1}}{A + Bx_n + x_{n-1}}$$

from which it follows that each one of the four subsequences $\{x_{4n+j}\}$, $j \in \{0,1,2,3\}$ is either above $\frac{1}{B}$, or below $\frac{1}{B}$, or identically equal to $\frac{1}{B}$. In view of (5.141.2) all four subsequences converge monotonically to finite limits. In addition, from (5.141.2) we see that for all $n \geq 3$,

$$x_{n+1} = x_{n-3} \text{ if and only if } x_{n-3} = \frac{1}{B}.$$

Hence, all four subsequences converge to $\frac{1}{B}$. The proof is complete. ∎

The following theorem extends the result of Theorem 5.141.1 to the more general rational equation

$$x_{n+1} = \frac{\alpha + x_{n-m}}{A + Mx_{n-m} + Lx_{n-l}}, \quad n = 0, 1, \ldots \tag{5.141.8}$$

with $l, m \in \{0, 1, \ldots\}$, with positive parameters α, M, L, and with arbitrary nonnegative initial conditions.

The proof, as in the case of Theorem 5.141.1, is based on the identity:

$$x_{n+1} - x_{n-2l-1}$$
$$= \frac{(\alpha A - A^2 x_{n-2l-1} - ALx_{n-2l-1}^2) + x_{n-m}(A + Lx_{n-2l-1})(1 - Mx_{n-2l-1})}{(A + Mx_{n-m})(A + Mx_{n-l-m-1} + Lx_{n-2l-1}) + L\alpha + Lx_{n-l-m-1}}$$
$$+ \frac{x_{n-l-m-1}(\alpha M - (AM + L)x_{n-2l-1}) + x_{n-m}x_{n-l-m-1}M(1 - Mx_{n-2l-1})}{(A + Mx_{n-m})(A + Mx_{n-l-m-1} + Lx_{n-2l-1}) + L\alpha + Lx_{n-l-m-1}}. \tag{5.141.9}$$

Theorem 5.141.2 *Let $\{x_n\}$ be any solution of Eq.(5.141.8). Then the following statements are true:*
(i) When

$$0 \leq A < 1 \text{ and } \frac{(L - M)(1 - A)^2}{4M^2} \leq \alpha < \frac{A}{M} + \frac{L}{M^2}, \tag{5.141.10}$$

the solution $\{x_n\}$ eventually enters the interval $\left[\frac{\alpha M - A}{L}, \frac{1}{M}\right]$ and the function $f(x_{n-m}, x_{n-l})$ is eventually strictly increasing in x_{n-m} and strictly decreasing in x_{n-l}. Furthermore, the solution $\{x_n\}$ converges to the equilibrium.
(ii) When

$$0 \leq A < 1 \text{ and } \alpha > \frac{A}{M} + \frac{L}{M^2}, \tag{5.141.11}$$

the solution $\{x_n\}$ eventually enters the interval $\left[\frac{1}{M}, \frac{\alpha M - A}{L}\right]$ and the function $f(x_{n-m}, x_{n-l})$ is eventually strictly decreasing in x_{n-m} and x_{n-l}. Furthermore, the solution $\{x_n\}$ converges to the equilibrium.
(iii) When

$$0 \leq A < 1 \text{ and } \alpha = \frac{A}{M} + \frac{L}{M^2}, \tag{5.141.12}$$

the solution $\{x_n\}$ converges to the equilibrium.

PROOF The proof is similar to the proof of Theorem 5.141.1 and will be omitted. ∎

Conjecture 5.141.1 *Assume that*

$$\alpha, A, B \in (0, \infty).$$

Show that every solution of Eq.(5.141.1) has a finite limit.

In other words, the equilibrium $\bar{x}$ of Eq.(5.141.1) is globally asymptotically stable. To the best of our knowledge any claims in the literature, made prior to July 2007, that the conjecture has been confirmed are not correct. For some partial results see [175] and [176].

5.142 Equation #142 : $x_{n+1} = \dfrac{\alpha + \beta x_n}{A + Bx_n + Dx_{n-2}}$

Eq.(#142) can be written in the normalized form

$$x_{n+1} = \frac{\alpha + x_n}{A + x_n + Dx_{n-2}}, \quad n = 0, 1, \ldots \tag{5.142.1}$$

with positive parameters α and D and with arbitrary nonnegative initial conditions x_{-2}, x_{-1}, x_0. Throughout this section we will allow the parameter A to be nonnegative so that we also include the special case #69.

Eq.(5.142.1) has the unique equilibrium

$$\bar{x} = \frac{1 - A + \sqrt{(1 - A)^2 + 4\alpha(1 + D)}}{2(1 + D)}.$$

The characteristic equation of the linearized equation of Eq.(5.142.1) about the equilibrium $\bar{x}$ is

$$\lambda^3 + \frac{\bar{x} - 1}{A + (1 + D)\bar{x}}\lambda^2 + \frac{D\bar{x}}{A + (1 + D)\bar{x}} = 0.$$

From this and Theorem 1.2.3 it follows that the equilibrium $\bar{x}$ of Eq.(5.142.1) is locally asymptotically stable when

$$A \geq 1, \tag{5.142.2}$$

or

$$0 \leq A < 1 \text{ and } D < \frac{4A(A + 1)}{(2A - 1)^2}, \tag{5.142.3}$$

or

$$0 \leq A < 1, \ \ D \geq \frac{4A(A+1)}{(2A-1)^2}, \ \ \text{and} \ \ \alpha > \alpha^* \tag{5.142.4}$$

and unstable when

$$0 \leq A < 1, \ \ D \geq \frac{4A(A+1)}{(2A-1)^2}, \ \ \text{and} \ \ \alpha < \alpha^*,$$

where

$$\alpha^* = \frac{2A - D + 5AD - 4A^2D - 2D^2 + D^3 - 4AD^3 + 4A^2D^3}{2(9D^2 + 6D + 1)}$$

$$\frac{-(1 + A + 2D + AD - D^2 + 2AD^2)\sqrt{D}\sqrt{4A^2D - 4AD + D - 4A^2 - 4A}}{2(9D^2 + 6D + 1)}.$$

By Theorems 5.23.2 and 5.23.3 it follows that when

$$A \geq 1,$$

the equilibrium of Eq.(5.142.1) is globally asymptotically stable.

From Theorem 5.141.2 it follows that when

$$0 \leq A < 1 \ \text{and} \ \frac{(D-1)(1-A)^2}{4} \leq \alpha, \tag{5.142.5}$$

every solution of Eq.(5.142.1) converges to the equilibrium $\bar{x}$.

Conjecture 5.142.1 *Assume that*

$$\alpha^* < \alpha < \frac{(D-1)(1-A)^2}{4}.$$

Show that every solution of Eq.(5.142.1) converges to the equilibrium $\bar{x}$.

Conjecture 5.142.2 *Show that Eq.(5.142.1) has solutions that do not converge to the equilibrium point $\bar{x}$ or to a periodic solution.*

5.143 Equation #143 : $x_{n+1} = \dfrac{\alpha + \beta x_n}{A + Cx_{n-1} + Dx_{n-2}}$

The boundedness character of solutions of this equation was investigated in [69]. See also Theorem 2.3.1 where we established that every solution of the equation is bounded. Eq.(#143) can be written in the normalized form

$$x_{n+1} = \frac{\alpha + x_n}{A + Cx_{n-1} + x_{n-2}}, \quad n = 0, 1, \ldots \tag{5.143.1}$$

with positive parameters α, A, C and with arbitrary nonnegative initial conditions x_{-2}, x_{-1}, x_0.

Eq.(5.143.1) has the unique equilibrium

$$\bar{x} = \frac{1 - A + \sqrt{(1-A)^2 + 4\alpha(1+C)}}{2(1+C)}.$$

The characteristic equation of the linearized equation of Eq.(5.143.1) about the equilibrium $\bar{x}$ is

$$\lambda^3 + \frac{-1}{A + (1+C)\bar{x}}\lambda^2 + \frac{C\bar{x}}{A + (C+1)\bar{x}}\lambda + \frac{\bar{x}}{A + (1+C)\bar{x}} = 0.$$

From this and Theorem 1.2.3 it follows that the equilibrium $\bar{x}$ of Eq.(5.143.1) is locally asymptotically stable when

$$(A^2 + A + 2\alpha)C^2 + 2(A^2 + \alpha)C + 3A - 1 + 2AC$$
$$+[(A(C+1)^2 + A - 1]\sqrt{(1-A)^2 + 4\alpha(C+1)} > 0 \qquad (5.143.2)$$

and unstable when

$$(A^2 + A + 2\alpha)C^2 + 2(A^2 + \alpha)C + 3A - 1 + 2AC$$
$$+[(A(C+1)^2 + A - 1]\sqrt{(1-A)^2 + 4\alpha(C+1)} < 0$$

By Theorems 5.23.2 and 5.23.3 it follows that when

$$A \geq 1,$$

the equilibrium $\bar{x}$ of Eq.(5.143.1) is globally asymptotically stable.

Conjecture 5.143.1 *Assume that (5.143.2) holds. Show that the equilibrium $\bar{x}$ of Eq.(5.143.1) is globally asymptotically stable.*

Conjecture 5.143.2 *Show that Eq.(5.143.1) has solutions that do not converge to the equilibrium point $\bar{x}$ or to a periodic solution.*

5.144 Equation #144 : $x_{n+1} = \dfrac{\alpha + \beta x_n}{Bx_n + Cx_{n-1} + Dx_{n-2}}$

The boundedness character of solutions of this equation was investigated in [69]. See also Theorem 2.1.1 where we established that every solution of the equation is bounded. Eq.(#144) can be written in the normalized form

$$x_{n+1} = \frac{\alpha + x_n}{x_n + Cx_{n-1} + Dx_{n-2}}, \quad n = 0, 1, \ldots \qquad (5.144.1)$$

with positive parameters α, C, D and with arbitrary positive initial conditions x_{-2}, x_{-1}, x_0.

Eq.(5.144.1) has the unique equilibrium

$$\bar{x} = \frac{1 + \sqrt{1 + 4\alpha(1 + C + D)}}{2(1 + C + D)}.$$

The characteristic equation of the linearized equation of Eq.(5.144.1) about the equilibrium $\bar{x}$ is

$$\lambda^3 + \frac{\bar{x} - 1}{(1 + C + D)\bar{x}}\lambda^2 + \frac{C}{1 + C + D}\lambda + \frac{D}{1 + C + D} = 0.$$

From this and Theorem 1.2.3 it follows that the equilibrium $\bar{x}$ of Eq.(5.144.1) is locally asymptotically stable when

$$D < 1, \ \ C < \frac{-1 - D + \sqrt{-3D^2 + 6D + 1}}{2}, \tag{5.144.2}$$

and

$$\alpha > \frac{D(C + C^2 + 2CD + 2D^2)}{(C + C^2 - D + CD + D^2)^2}. \tag{5.144.3}$$

Conjecture 5.144.1 *Assume that (5.144.2) and (5.144.3) hold. Show that every solution of Eq.(5.144.1) converges to the equilibrium $\bar{x}$.*

Conjecture 5.144.2 *Show that Eq.(5.144.1) has solutions that do not converge to the equilibrium point $\bar{x}$ or to a periodic solution.*

5.145 Equation #145 : $x_{n+1} = \dfrac{\alpha + \gamma x_{n-1}}{A + Bx_n + Cx_{n-1}}$

This equation was investigated in [175]. Eq.(#145) can be written in the normalized form

$$x_{n+1} = \frac{\alpha + x_{n-1}}{A + Bx_n + x_{n-1}}, \quad n = 0, 1, \ldots \tag{5.145.1}$$

with positive parameters α and B and with arbitrary nonnegative initial conditions x_{-1}, x_0. Throughout this section we allow the parameter A to be nonnegative so that we also include the special case #74.

Eq.(5.145.1) has the unique equilibrium

$$\bar{x} = \frac{1 - A + \sqrt{(1 - A)^2 + 4\alpha(1 + B)}}{2(1 + B)}.$$

The characteristic equation of the linearized equation of Eq.(5.145.1) about the equilibrium $\bar{x}$ is

$$\lambda^2 + \frac{B\bar{x}}{A+(1+B)\bar{x}}\lambda + \frac{\bar{x}-1}{A+(1+B)\bar{x}} = 0.$$

From this and Theorem 1.2.2 it follows that the equilibrium $\bar{x}$ of Eq.(5.145.1) is locally asymptotically stable when

$$\bar{x} > \frac{1-A}{2},$$

which is equivalent to

$$A \geq 1, \tag{5.145.2}$$

or

$$0 \leq A < 1 \text{ and } B \leq 1, \tag{5.145.3}$$

or

$$0 \leq A < 1, \;\; B > 1, \;\; \text{and } \alpha > \frac{(B-1)(1-A)^2}{4} \tag{5.145.4}$$

and unstable when

$$0 \leq A < 1, \;\; B > 1, \;\; \text{and } \alpha < \frac{(B-1)(1-A)^2}{4}. \tag{5.145.5}$$

By Theorems 5.23.2 and 5.23.3 it follows that when

$$A \geq 1,$$

the equilibrium $\bar{x}$ of Eq.(5.145.1) is globally asymptotically stable.

By Theorem 5.141.2 it follows that when

$$0 \leq A < 1 \text{ and } \alpha \geq \frac{(B-1)(1-A)^2}{4}, \tag{5.145.6}$$

every solution of Eq.(5.145.1) converges to the equilibrium $\bar{x}$.

When (5.145.5) holds, Eq.(5.145.1) has the unique prime period-two solution

$$\ldots, \frac{1-A-\sqrt{(1-A)^2-\frac{4\alpha}{B-1}}}{2}, \frac{1-A+\sqrt{(1-A)^2-\frac{4\alpha}{B-1}}}{2}, \ldots, \tag{5.145.7}$$

which is locally asymptotically stable. See [175].

The following theorem is a new result about the global behavior of solutions of Eq.(5.145.1) when (5.145.5) holds.

Theorem 5.145.1 *Assume that (5.145.5) holds. Then every solution of Eq.(5.145.1) converges to a (not necessarily prime) period-two solution.*

PROOF Let $\{x_n\}$ be a solution of Eq.(5.145.1). Due to the fact that

$$\frac{(B-1)(1-A)^2}{4} < B + A,$$

it follows from (5.145.5) that

$$\alpha < B + A.$$

From this and Theorem 5.141.2 (*i*) it follows that the function

$$f(x_n, x_{n-1}) = \frac{\alpha + x_{n-1}}{A + Bx_n + x_{n-1}}$$

increases in x_{n-1} and decreases in x_n. By Theorem 1.6.6 it follows that the subsequences of the even and odd terms are eventually monotonic and because the solution is bounded these subsequences converge to finite limits. The proof is complete. ∎

Open Problem 5.145.1 *Assume that (5.145.5) holds.*
(i) Determine the set of all initial conditions x_{-1}, x_0 *for which every solution of Eq.(5.145.1) converges to the equilibrium* $\bar{x}$.

(ii) Determine the set of all initial conditions x_{-1}, x_0 *for which every solution of Eq.(5.145.1) converges to (5.145.7).*

5.146 Equation #146 : $x_{n+1} = \dfrac{\alpha + \gamma x_{n-1}}{A + Bx_n + Dx_{n-2}}$

This equation was investigated in [72]. When

$$\gamma + A + B > 0,$$

Eq.(#146) possesses a period-two trichotomy depending on whether

$$\gamma < A, \quad \gamma = A, \quad \text{or} \quad \gamma > A.$$

The precise result that allows for the parameters α, A, and B to be nonnegative was presented in Theorem 4.3.1.

What is it that makes Eq.(#146) possess a period-two trichotomy?

Could the period-two trichotomy of Eq.(#146) be predicted from the linearized equation of Eq.(#146) and its dominant characteristic root?

Open Problem 5.146.1 *Investigate the global character of solutions of the equation*

$$x_{n+1} = \frac{\alpha_n + \gamma_n x_{n-1}}{1 + B_n x_n + x_{n-2}}, \quad n = 0, 1, \ldots \tag{5.146.1}$$

with periodic coefficients $\{\alpha_n\}$, $\{\gamma_n\}$, *and* $\{B_n\}$.

Open Problem 5.146.2 *Investigate the global character of solutions of Eq.(5.146.1) with convergent coefficients* $\{\alpha_n\}$, $\{\gamma_n\}$, *and* $\{B_n\}$.

Conjecture 5.146.1 *Assume that*

$$\gamma > A.$$

Show that every bounded solution of Eq.(#146) converges to the equilibrium.

5.147 Equation #147 : $x_{n+1} = \dfrac{\alpha + \gamma x_{n-1}}{A + C x_{n-1} + D x_{n-2}}$

Eq.(#147) can be written in the normalized form

$$x_{n+1} = \frac{\alpha + x_{n-1}}{A + x_{n-1} + D x_{n-2}}, \quad n = 0, 1, \ldots \tag{5.147.1}$$

with positive parameters α and D and with arbitrary nonnegative initial conditions x_{-2}, x_{-1}, x_0. Throughout this section we allow the parameter A to be nonnegative so that we also include the special case #76.

Eq.(5.147.1) has the unique equilibrium

$$\bar{x} = \frac{1 - A + \sqrt{(1 - A)^2 + 4\alpha(D + 1)}}{2(D + 1)}.$$

The characteristic equation of the linearized equation of Eq.(5.147.1) about the equilibrium $\bar{x}$ is

$$\lambda^3 + \frac{\bar{x} - 1}{A + (D + 1)\bar{x}}\lambda + \frac{D\bar{x}}{A + (D + 1)\bar{x}} = 0. \tag{5.147.2}$$

From this and Theorem 1.2.3 it follows that the equilibrium $\bar{x}$ of Eq.(5.147.1) is locally asymptotically stable when

$$A \geq 1, \text{ or } 0 \leq A < 1 \text{ and } D < 1 + \frac{4\alpha}{(1 - A)^2} \tag{5.147.3}$$

and unstable when

$$0 \leq A < 1 \text{ and } D > 1 + \frac{4\alpha}{(1-A)^2}. \tag{5.147.4}$$

By Theorems 5.23.2 and 5.23.3 it follows that when

$$A \geq 1,$$

the equilibrium $\bar{x}$ of Eq.(5.147.1) is globally asymptotically stable.

From Theorem 5.141.2 it follows that when

$$0 \leq A < 1 \text{ and } D \leq 1 + \frac{4\alpha}{(1-A)^2}, \tag{5.147.5}$$

every solution of Eq.(5.147.1) converges to the equilibrium $\bar{x}$.

When (5.147.4) holds, Eq.(5.147.1) has the unique period-two solution

$$\ldots, \frac{1 - A - \sqrt{(1-A)^2 - \frac{4\alpha}{D-1}}}{2}, \frac{1 - A + \sqrt{(1-A)^2 - \frac{4\alpha}{D-1}}}{2}, \ldots. \tag{5.147.6}$$

Conjecture 5.147.1 *Show that the period-two cycle (5.147.6) is locally asymptotically stable.*

Conjecture 5.147.2 *Assume that (5.147.4) holds. Show that every solution of Eq.(5.147.1) converges to a (not necessarily prime) period-two solution.*

Open Problem 5.147.1 *Assume that (5.147.4) holds.*
(i) Determine the set of all initial conditions x_{-2}, x_{-1}, x_0 for which every solution of Eq.(5.147.1) converges to the equilibrium $\bar{x}$.

(ii) Determine the set of all initial conditions x_{-2}, x_{-1}, x_0 for which every solution of Eq.(5.147.1) converges to (5.147.6).

5.148 Equation #148 : $x_{n+1} = \dfrac{\alpha + \gamma x_{n-1}}{Bx_n + Cx_{n-1} + Dx_{n-2}}$

The boundedness character of solutions of this equation was investigated in [69]. See also Theorem 2.1.1 where we established that every solution of the equation is bounded. Eq.(#148) can be written in the normalized form

$$x_{n+1} = \frac{\alpha + x_{n-1}}{Bx_n + x_{n-1} + Dx_{n-2}}, \quad n = 0, 1, \ldots \tag{5.148.1}$$

with positive parameters α, B, D and with arbitrary positive initial conditions x_{-2}, x_{-1}, x_0.

Eq.(5.148.1) has the unique equilibrium

$$\bar{x} = \frac{1 + \sqrt{1 + 4\alpha(1 + B + D)}}{2(1 + B + D)}.$$

The characteristic equation of the linearized equation of Eq.(5.148.1) about the equilibrium $\bar{x}$ is

$$\lambda^3 + \frac{B}{1 + B + D}\lambda^2 + \frac{\bar{x} - 1}{(1 + B + D)\bar{x}}\lambda + \frac{D}{1 + B + D} = 0.$$

From this and Theorem 1.2.3 it follows that the equilibrium $\bar{x}$ is locally asymptotically stable when

$$\alpha > \frac{B + D - 1}{4}, \quad D > 1, \quad \text{and} \quad B \geq \frac{D^2 + D + 1}{D - 1},$$

or when

$$D \leq 1$$

and

$$\frac{B + D - 1}{4} < \alpha < \frac{(3B + 1)D^2 + (4B^2 + 5B + 1)D + B^3 + 2B^2 + 6}{(D^2 + D(1 - B) + B + 1)^2}.$$

When

$$\alpha < \frac{B + D - 1}{4}, \tag{5.148.2}$$

and only then, Eq.(5.148.1) has a unique prime period-two solution of the form

$$\ldots, x, y, x, y, \ldots,$$

where x, y are the positive solutions of the quadratic equation

$$t^2 - t + \frac{\alpha}{B + D - 1} = 0.$$

Conjecture 5.148.1 *Assume that (5.148.2) holds. Show that the unique prime period-two solution of Eq.(5.148.1) is locally asymptotically stable.*

Conjecture 5.148.2 *Show that every solution of Eq.(5.148.1) converges to a (not necessarily prime) period-two solution.*

5.149 Equation #149 : $x_{n+1} = \dfrac{\alpha + \delta x_{n-2}}{A + Bx_n + Cx_{n-1}}$

Eq.(#149) can be written in the normalized form

$$x_{n+1} = \frac{\alpha + x_{n-2}}{A + Bx_n + x_{n-1}}, \quad n = 0, 1, \ldots \tag{5.149.1}$$

with positive parameters α, A, B and with arbitrary nonnegative initial conditions x_{-2}, x_{-1}, x_0.

The boundedness character of solutions of this equation was investigated in [69]. See also Theorem 3.4.1 where we established that the equation has unbounded solutions when

$$A < 1. \tag{5.149.2}$$

Actually, as we will see later, (5.149.2) is a necessary and sufficient condition for Eq.(5.149.1) to have unbounded solutions.

Eq.(5.149.1) has the unique equilibrium

$$\bar{x} = \frac{1 - A + \sqrt{(1-A)^2 + 4\alpha(B+1)}}{2(B+1)}.$$

The characteristic equation of the linearized equation of Eq.(5.149.1) about the equilibrium $\bar{x}$ is

$$\lambda^3 + \frac{B\bar{x}}{A + (B+1)\bar{x}}\lambda^2 + \frac{\bar{x}}{A + (B+1)\bar{x}}\lambda - \frac{1}{A + (B+1)\bar{x}} = 0. \tag{5.149.3}$$

From this and Theorem 1.2.3 it follows that the equilibrium $\bar{x}$ of Eq.(5.149.1) is locally asymptotically stable when

$$A \geq 1,$$

or

$$0 < A < 1 \text{ and } \alpha \geq \frac{(1-A)^2(1+A)}{A^2}, \tag{5.149.4}$$

or

$$0 < A < 1, \text{ and } 0 < \alpha < \frac{(1-A)^2(1+A)}{A^2}, \text{ and } B > \frac{2 - A - A\sqrt{5 + 4\alpha - 4A}}{2\alpha} \tag{5.149.5}$$

and unstable when

$$0 < A < 1, \; 0 < \alpha < \frac{(1-A)^2(1+A)}{A^2}, \text{ and } B < \frac{2 - A - A\sqrt{5 + 4\alpha - 4A}}{2\alpha}. \tag{5.149.6}$$

By Theorems 5.23.2 and 5.23.3 it follows that when

$$A \geq 1,$$

the equilibrium $\bar{x}$ of Eq.(5.149.1) is globally asymptotically stable.

For the equilibrium $\bar{x}$ of Eq.(5.149.1),

Local Asymptotic Stabilty $\not\Rightarrow$ Global Asymptotic Stabilty.

More specifically, when the condition (5.149.5) is satisfied, the equilibrium of Eq.(5.149.1) is locally asymptotically stable but not globally asymptotically stable. The reason is that when the condition (5.149.5) is satisfied, there exist initial conditions x_{-2}, x_{-1}, x_0 for which the solution of Eq.(5.149.1) is unbounded. See Theorem 3.4.1.

In addition to unbounded solutions, what other type's of solutions exist? Can there exist any periodic solutions? Can there exist any bounded solutions that are not periodic and do not converge to the equilibrium point $\bar{x}$ or to a periodic solution?

Note that when

$$0 \leq A < 1 \text{ and } B = 1,$$

unbounded solutions of Eq.(5.149.1) coexist with periodic solutions. For example, when

$$A \neq 1 - \sqrt{\alpha},$$

the sequence

$$\ldots, 1-A, 1-A, \frac{\alpha}{1-A}, \ldots$$

is a prime period-three solution of the equation

$$x_{n+1} = \frac{\alpha + x_{n-2}}{A + x_n + x_{n-1}}, \quad n = 0, 1, \ldots .$$

Conjecture 5.149.1 *Show that Eq.(5.149.1) has bounded solutions that do not converge to the equilibrium point $\bar{x}$ or to a periodic solution.*

Open Problem 5.149.1 *Investigate the global character of solutions of the equation*

$$x_{n+1} = \frac{\alpha_n + x_{n-2}}{A_n + B_n x_n + x_{n-1}}, \quad n = 0, 1, \ldots \tag{5.149.7}$$

with periodic coefficients $\{\alpha_n\}$, $\{A_n\}$, and $\{B_n\}$.

Open Problem 5.149.2 *Investigate the global character of solutions of Eq.(5.149.7) with convergent coefficients $\{\alpha_n\}$, $\{A_n\}$, and $\{B_n\}$.*

5.150 Equation #150 : $x_{n+1} = \dfrac{\alpha + \delta x_{n-2}}{A + Bx_n + Dx_{n-2}}$

Eq.(#150) can be written in the normalized form

$$x_{n+1} = \frac{\alpha + x_{n-2}}{A + Bx_n + x_{n-2}}, \quad n = 0, 1, \ldots \tag{5.150.1}$$

with positive parameters α, A, B and with arbitrary nonnegative initial conditions x_{-2}, x_{-1}, x_0.

Eq.(5.150.1) has the unique equilibrium

$$\bar{x} = \frac{1 - A + \sqrt{(1-A)^2 + 4\alpha(1+B)}}{2(1+B)}.$$

The characteristic equation of the linearized equation of Eq.(5.150.1) about the equilibrium $\bar{x}$ is

$$\lambda^3 + \frac{B\bar{x}}{A + (1+B)\bar{x}}\lambda^2 + \frac{\bar{x} - 1}{A + (1+B)\bar{x}} = 0.$$

From this and Theorem 1.2.3 it follows that the equilibrium $\bar{x}$ of Eq.(5.150.1) is locally asymptotically stable when

$$A \geq 1, \tag{5.150.2}$$

or

$$0 < A < 1 \text{ and } 0 < B \leq 1, \tag{5.150.3}$$

or

$$0 < A < 1, \ \ B > 1, \ \text{ and } \alpha > \frac{s - \sqrt{s^2 - 4t}}{2B^2(B+3)^2}, \tag{5.150.4}$$

where

$$s = 4(1+A)(A + 3B + 1) - B^2(AB - 2B + 4A^2 - 9A - 4)$$

and

$$t = B^2(B+3)^2(A-1)^2(A+1)(B^2 + AB - 2B - A - 1)$$

and unstable when

$$0 < A < 1, \ \ B > 1 \text{ and } \alpha < \frac{s - \sqrt{s^2 - 4t}}{2B^2(B+3)^2}. \tag{5.150.5}$$

By Theorems 5.23.2 and 5.23.3 it follows that when

$$A \geq 1,$$

the equilibrium $\bar{x}$ of Eq.(5.150.1) is globally asymptotically stable.

From Theorem 5.141.2 it follows that when

$$0 < A < 1 \text{ and } \alpha \geq \frac{(B-1)(1-A)^2}{4}, \tag{5.150.6}$$

every solution of Eq.(5.150.1) converges to the equilibrium $\bar{x}$.

Open Problem 5.150.1 *Assume that*

$$0 < A < 1, \quad B > 1, \quad \textit{and} \quad \frac{s - \sqrt{s^2 - 4t}}{2B^2(B+3)^2} < \alpha < \frac{(B-1)(1-A)^2}{4}.$$

Show that every solution of Eq.(5.150.1) converges to the equilibrium $\bar{x}$.

Conjecture 5.150.1 *Show that Eq.(5.150.1) has solutions that do not converge to the equilibrium point $\bar{x}$ or to a periodic solution.*

5.151 Equation #151 : $x_{n+1} = \dfrac{\alpha + \delta x_{n-2}}{A + Cx_{n-1} + Dx_{n-2}}$

Eq.(#151) can be written in the normalized form

$$x_{n+1} = \frac{\alpha + x_{n-2}}{A + Cx_{n-1} + x_{n-2}}, \quad n = 0, 1, \ldots \tag{5.151.1}$$

with positive parameters α, A, C and with arbitrary nonnegative initial conditions x_{-2}, x_{-1}, x_0.

By Theorems 5.23.2 and 5.23.3 it follows that when

$$A \geq 1,$$

the equilibrium of Eq.(5.151.1) is globally asymptotically stable.

Also, when

$$0 < A < 1 \text{ and } \alpha \geq \frac{(C-1)(1-A)^2}{4}, \tag{5.151.2}$$

by Theorem 5.141.2, every solution of Eq.(5.151.1) converges to the equilibrium $\bar{x}$.

Conjecture 5.151.1 *Show that for the equilibrium $\bar{x}$ of Eq.(5.151.1),*

Local Asymptotic Stability* $\Rightarrow$ *Global Asymptotic Stability.

Conjecture 5.151.2 *Assume that*

$$0 < A < 1 \text{ and } \alpha < \frac{(C-1)(1-A)^2}{4}.$$

Show that Eq.(5.151.1) has solutions that do not converge to the equilibrium point or to a periodic solution.

5.152 Equation #152 : $x_{n+1} = \dfrac{\alpha + \delta x_{n-2}}{Bx_n + Cx_{n-1} + Dx_{n-2}}$

The boundedness character of solutions of this equation was investigated in [69]. See also Theorem 2.1.1 where we established that every solution of the equation is bounded. Eq.(#152) can be written in the normalized form

$$x_{n+1} = \frac{\alpha + x_{n-2}}{Bx_n + Cx_{n-1} + x_{n-2}}, \quad n = 0, 1, \ldots \tag{5.152.1}$$

with positive parameters α, B, C and with arbitrary positive initial conditions x_{-2}, x_{-1}, x_0.

Conjecture 5.152.1 *Show that for the equilibrium $\bar{x}$ of Eq.(5.152.1),*

Local Asymptotic Stability $\Rightarrow$ ***Global Asymptotic Stability.***

Conjecture 5.152.2 *Show that Eq.(5.152.1) has solutions that do not converge to the equilibrium point or to a periodic solution.*

5.153 Equation #153 : $x_{n+1} = \dfrac{\beta x_n + \gamma x_{n-1}}{A + Bx_n + Cx_{n-1}}$

Eq.(#153) can be written in the normalized form

$$x_{n+1} = \frac{\beta x_n + x_{n-1}}{A + Bx_n + x_{n-1}}, \quad n = 0, 1, \ldots \tag{5.153.1}$$

with positive parameters β, A, B and with arbitrary nonnegative initial conditions x_{-1}, x_0.

Zero is always an equilibrium of Eq.(5.153.1). By Theorems 5.23.2 and 5.23.4 it follows that when

$$A \geq \beta + 1,$$

the zero equilibrium of Eq.(5.153.1) is globally asymptotically stable. When

$$A < \beta + 1, \tag{5.153.2}$$

the zero equilibrium is unstable.

When (5.153.2) holds, Eq.(5.153.1) has the unique positive equilibrium point

$$\bar{x} = \frac{1+\beta-A}{1+B}.$$

The characteristic equation of the linearized equation of Eq.(5.153.1) about the positive equilibrium $\bar{x}$ is

$$\lambda^2 + \frac{B-AB-\beta}{(\beta+1)(B+1)}\lambda + \frac{\beta-A-B}{(1+B)(\beta+1)} = 0.$$

From this and Theorem 1.2.2 it follows that the positive equilibrium $\bar{x}$ of Eq.(5.153.1), which exists provided that (5.153.2) is satisfied, is locally asymptotically stable when

$$\beta > \frac{(B-1)(1-A)}{B+3} \tag{5.153.3}$$

and unstable when

$$\beta < \frac{(B-1)(1-A)}{B+3}. \tag{5.153.4}$$

When (5.153.2) and (5.153.3) both hold, it has been recently established (see [135]) that every positive solution of Eq.(5.153.1) converges to the positive equilibrium.

When (5.153.4) holds, and only then, Eq.(5.153.1) has the unique prime period-two solution

$$\ldots, \frac{1-\beta-A-\sqrt{(1-\beta-A)^2-\frac{4\beta(1-\beta-A)}{B-1}}}{2},$$

$$\frac{1-\beta-A+\sqrt{(1-\beta-A)^2-\frac{4\beta(1-\beta-A)}{B-1}}}{2}, \ldots. \tag{5.153.5}$$

Conjecture 5.153.1 *Show that the period-two cycle (5.153.5) is locally asymptotically stable.*

Conjecture 5.153.2 *Assume that (5.153.2) and (5.153.4) holds. Show that every solution of Eq.(5.153.1) converges to a (not necessarily prime) period-two solution.*

5.154 Equation #154 : $x_{n+1} = \dfrac{\beta x_n + \gamma x_{n-1}}{A + Bx_n + Dx_{n-2}}$

The boundedness character of this equation was investigated in [49]. Eq.(#154) can be written in the normalized form

$$x_{n+1} = \frac{x_n + \gamma x_{n-1}}{A + x_n + Dx_{n-2}}, \quad n = 0, 1, \ldots \tag{5.154.1}$$

with positive parameters γ, A, D and with arbitrary nonnegative initial conditions x_{-2}, x_{-1}, x_0.

By Theorem 3.3.1 it follows that Eq.(5.154.1) has unbounded solutions when

$$\gamma > 1 + A.$$

By Theorems 5.23.2 and 5.23.4 it follows that when

$$A \geq 1 + \gamma,$$

the zero equilibrium of Eq.(5.154.1) is globally asymptotically stable.

When

$$A < 1 + \gamma,$$

Eq.(5.154.1) has the unique positive equilibrium point

$$\bar{x} = \frac{1 + \gamma - A}{1 + D}.$$

Conjecture 5.154.1 *Assume that*

$$\gamma > 1 + A.$$

Show that every positive and bounded solution of Eq.(5.154.1) converges to the positive equilibrium $\bar{x}$.

Conjecture 5.154.2 *Assume that*

$$A - 1 < \gamma < 1 + A.$$

Show that for the positive equilibrium $\bar{x}$ of Eq.(5.154.1) and with positive initial conditions,

Local Asymptotic Stability $\Rightarrow$ Global Attractivity.

Conjecture 5.154.3 *Assume that*

$$\gamma = 1 + A.$$

Show that every solution of Eq.(5.154.1) converges to a (not necessarily prime) period-two solution.

Conjecture 5.154.4 *Assume that*

$$A - 1 < \gamma < 1 + A.$$

Show that Eq.(5.154.1) has bounded solutions that do not converge to an equilibrium point or to a periodic solution.

5.155 Equation #155 : $x_{n+1} = \dfrac{\beta x_n + \gamma x_{n-1}}{A + Cx_{n-1} + Dx_{n-2}}$

The boundedness character of solutions of this equation was investigated in [69]. See also Theorem 2.3.1 where we established that every solution of the equation is bounded. Eq.(#155) can be written in the normalized form

$$x_{n+1} = \frac{\beta x_n + x_{n-1}}{A + x_{n-1} + Dx_{n-2}}, \quad n = 0, 1, \ldots \tag{5.155.1}$$

with positive parameters β, A, D and with arbitrary nonnegative initial conditions x_{-2}, x_{-1}, x_0.

By Theorems 5.23.2 and 5.23.4 it follows that when

$$A \geq 1 + \beta,$$

the zero equilibrium of Eq.(5.155.1) is globally asymptotically stable.

When

$$A < 1 + \beta,$$

Eq.(5.155.1) has the unique positive equilibrium point

$$\bar{x} = \frac{1 + \beta - A}{1 + D}.$$

Conjecture 5.155.1 *Show that for the positive equilibrium $\bar{x}$ of Eq.(#155) and with positive initial conditions,*

Local Asymptotic Stability $\Rightarrow$ Global Attractivity.

Conjecture 5.155.2 *Assume that*

$$A < 1 + \beta.$$

Show that Eq.(5.155.1) has solutions that do not converge to an equilibrium point or to a periodic solution.

Conjecture 5.155.3 *It follows from the work in Section 4.2 that Eq.(5.155.1) has a unique prime period-two solution. Show that this period-two solution of Eq.(5.155.1) is locally asymptotically stable.*

5.156 Equation #156 : $x_{n+1} = \dfrac{\beta x_n + \gamma x_{n-1}}{Bx_n + Cx_{n-1} + Dx_{n-2}}$

Eq.(#156) can be written in the normalized form

$$x_{n+1} = \frac{x_n + \gamma x_{n-1}}{x_n + Cx_{n-1} + Dx_{n-2}}, \quad n = 0, 1, \ldots \tag{5.156.1}$$

with positive parameters γ, C, D and with arbitrary positive initial conditions x_{-2}, x_{-1}, x_0.

Open Problem 5.156.1 *Investigate the periodic nature of solutions and the global stability of the equilibrium of Eq.(5.156.1).*

Conjecture 5.156.1 *Show that for the equilibrium $\bar{x}$ of Eq.(5.156.1),*

Local Asymptotic Stability $\Rightarrow$ ***Global Attractivity.***

Conjecture 5.156.2 *It follows from the work in Section 4.2 that Eq.(5.156.1) has a unique prime period-two solution. Show that this period-two solution of Eq.(5.156.1) is locally asymptotically stable.*

Conjecture 5.156.3 *Show that Eq.(5.156.1) has solutions that do not converge to the equilibrium point or to a periodic solution.*

5.157 Equation #157 : $x_{n+1} = \dfrac{\beta x_n + \delta x_{n-2}}{A + Bx_n + Cx_{n-1}}$

Eq.(#157) can be written in the normalized form

$$x_{n+1} = \frac{\beta x_n + x_{n-2}}{A + Bx_n + x_{n-1}} \tag{5.157.1}$$

with positive parameters β, B and with arbitrary nonnegative initial conditions x_{-2}, x_{-1}, x_0. We will assume that the parameter A is nonnegative so that this section also includes the study of the special case #92.

The boundedness character of this equation was investigated in [69]. See also Theorem 3.4.1 where we established that the equation has unbounded solutions when

$$\beta < B(1 - A). \tag{5.157.2}$$

By Theorem 5.221.1 it follows that every solution of Eq.(5.157.1) is bounded if and only if

$$\beta \geq B(1-A).$$

Eq.(5.157.1) has one or two equilibrium points. When

$$\beta + 1 \leq A,$$

zero is the unique equilibrium of Eq.(5.157.1).

When

$$A = 0,$$

Eq.(5.157.1) has the unique positive equilibrium

$$\bar{x} = \frac{\beta+1}{B+1}.$$

When

$$\beta + 1 > A \text{ and } A > 0,$$

Eq.(5.157.1) has two equilibrium points, namely, the zero equilibrium and the positive equilibrium $\bar{x}$.

It follows from Theorems 5.23.2 and 5.23.4 that the zero equilibrium is globally asymptotically stable when

$$A \geq \beta + 1.$$

The characteristic equation of the linearized equation of Eq.(5.157.1) about the positive equilibrium $\bar{x}$ is

$$\lambda^3 + \frac{B\bar{x}-\beta}{A+(B+1)\bar{x}}\lambda^2 + \frac{\bar{x}}{A+(B+1)\bar{x}}\lambda - \frac{1}{A+(B+1)\bar{x}} = 0. \tag{5.157.3}$$

From this and Theorem 1.2.3 it follows that the positive equilibrium $\bar{x}$ of Eq.(5.157.1), which exists as long as $A < \beta + 1$, is locally asymptotically stable when

$$1 \leq A < \beta + 1 \tag{5.157.4}$$

or

$$0 < A < 1 \text{ and } \beta > \frac{-1-A-2B+\sqrt{(1+A)^2+4B(2+(2-A)B)}}{2B} = \beta^*. \tag{5.157.5}$$

For the positive equilibrium $\bar{x}$ of Eq.(5.157.1),

Local Asymptotic Stability $\not\Rightarrow$ Global Attractivity.

Indeed, for all values of β, A, and B for which

$$1 > A \geq 0,\ 1 > B \geq \frac{\sqrt{5}-1}{2} \text{ or } \frac{1}{2} \leq B \leq \frac{\sqrt{5}-1}{2} \text{ and } 1 > A > \frac{B^2+B-1}{B^2-B} \tag{5.157.6}$$

we have

$$\beta^* < \beta < B(1-A)$$

and so the positive equilibrium of Eq.(5.157.1) is locally asymptotically stable and, in the same region, by Theorem 3.4.1, the equation has unbounded solutions. Hence, the positive equilibrium of the equation cannot be a global attractor in the entire region of its local asymptotic stability.

Open Problem 5.157.1 *Assume that (5.157.6) holds.*
(a) Determine the set of all positive initial conditions x_{-2}, x_{-1}, x_0 through which the solutions of Eq.(5.157.1) converge to the positive equilibrium point $\bar{x}$.

(b) Determine the set of all positive initial conditions x_{-2}, x_{-1}, x_0 through which the solutions of Eq.(5.157.1) are unbounded.

When

$$\frac{1}{2} < B < \frac{\sqrt{5}-1}{2} \quad \text{and} \quad A < \frac{B^2+B-1}{B^2-B} \quad \text{or} \quad 0 < B < \frac{1}{2} \quad \text{and} \quad 0 \le A < 1, \tag{5.157.7}$$

it follows that

$$\beta^* > \beta > B(1-A).$$

In this case numerical investigations indicate chaotic behavior of solutions of Eq.(5.157.1).

Conjecture 5.157.1 *Show that Eq.(5.157.1) has bounded solutions that do not converge to an equilibrium point or to a periodic solution.*

Lemma 5.157.1 *Assume that $A = 1$. Let $\{x_n\}$ be a solution of Eq.(5.157.1) for which there exists $N > 0$ such that*

$$0 < x_{N-1}, x_{N-2}, x_N < \frac{\beta}{B}. \tag{5.157.8}$$

Then

$$x_{N+1} < \frac{\beta}{B}.$$

PROOF In view of Eq.(5.157.1), we get

$$x_{N+1} = \frac{\beta x_N + x_{N-2}}{1 + Bx_N + x_{N-1}} < \frac{\frac{\beta}{B}(\beta+1)}{1+\beta} = \frac{\beta}{B}.$$

The proof is complete. ∎

Lemma 5.157.2 *Assume that $A = 1$. Let $\{x_n\}$ be a solution of Eq.(5.157.1) such that*

$$x_{N+1} \geq \frac{\beta}{B}. \tag{5.157.9}$$

Then

$$x_{N-1} < \frac{\beta}{B}. \tag{5.157.10}$$

PROOF Suppose for the sake of contradiction that

$$x_{N-1} \geq \frac{\beta}{B}.$$

Then in view of Eq.(5.157.1), we get

$$x_{N-2} \geq \frac{\beta}{B}(1 + \frac{\beta}{B}) \text{ and } x_{N-4} \geq \frac{\beta}{B},$$

which implies that

$$x_{N-5} \geq \frac{\beta}{B}(1 + \frac{\beta}{B})^2 \text{ and } x_{N-7} \geq \frac{\beta}{B}.$$

Inductively, we get

$$x_{N-3k-2} > \frac{\beta}{B}(1 + \frac{\beta}{B})^{k+1}, \quad k = 0, 1, \ldots,$$

which is a contradiction and the proof is complete. ▮

Theorem 5.157.1 *Assume that $A \geq 1$. Let $\{x_n\}$ be a solution of Eq.(5.157.1). Then there exists $N > 0$ such that for all $n \geq N$,*

$$x_n < \frac{\beta}{B}. \tag{5.157.11}$$

PROOF We will consider two cases. First assume that

$$A = 1.$$

In view of Lemma 5.157.1 it suffices to show that there exists $N > 0$ such that (5.157.8) holds. For the sake of contradiction and in view of (5.157.9) and (5.157.10) assume that there exists $N > 0$ such that for all $n \geq 0$

$$0 < x_{3n+N}, x_{3n+N+1} < \frac{\beta}{B} < x_{3n+N-1}.$$

From Eq.(5.157.1) we get

$$x_{3n+N+2} = \frac{\beta x_{3n+N+1} + x_{3n+N-1}}{1 + Bx_{3n+N+1} + x_{3n+N}} < x_{3(n-1)+N+2}$$

and so the subsequence $\{x_{3n+N+2}\}$ converges. Set

$$S = \limsup_{n\to\infty} x_n.$$

Then, clearly,

$$S = \lim_{n\to\infty} x_{3n+N+2} \geq \frac{\beta}{B}.$$

In addition, there exists a sequence of indices $\{n_i\}$ and positive numbers $\{l_t\}_{t=0,t\neq 2}^{4}$ such that

$$l_t = \lim_{i\to\infty} x_{3n_i+N+t}.$$

Then, clearly,

$$l_3 = \frac{\beta S + l_0}{1 + BS + l_1} > 0 \text{ and } l_4 = \frac{\beta l_3 + l_1}{1 + Bl_3 + l_2} > 0$$

and

$$S = l_5 = \frac{\beta l_4 + S}{1 + Bl_4 + l_3} = \frac{\beta l_4}{Bl_4 + l_3} < \frac{\beta}{B},$$

which is a contradiction and the proof of (5.157.11) is complete when $A = 1$.

When

$$A > 1,$$

assume for the sake of contradiction that there exists N sufficiently large such that

$$x_{N+1} = \frac{\beta x_N + x_{N-2}}{A + Bx_N + x_{N-1}} > \frac{\beta}{B}, \tag{5.157.12}$$

from which it follows that

$$x_{N-2} = \frac{\beta x_{N-3} + x_{N-5}}{A + Bx_{N-3} + x_{N-4}} > \frac{\beta}{B}A$$

and

$$x_{N-5} > A^2 \frac{\beta}{B}.$$

Inductively, we have that

$$x_{N-3k-2} > A^{k+1}\frac{\beta}{B}, \quad k = 0, 1, \ldots,$$

which is a contradiction and so the proof of (5.157.11) is complete. ∎

Lemma 5.157.3 *Assume that* $1 \leq A < \beta + 1$. *Let* $\{x_n\}$ *be a positive solution of Eq.(5.157.1). Then*

$$S = \limsup_{n\to\infty} x_n > 0. \tag{5.157.13}$$

PROOF Assume for the sake of contradiction that $S = 0$. Choose two positive numbers ϵ and m such that

$$0 < m = \frac{A + (B+1)\epsilon}{\beta + 1} < 1. \tag{5.157.14}$$

There exists N sufficiently large such that

$$x_{N+1} = \frac{\beta x_N + x_{N-2}}{A + Bx_N + x_{N-1}} < \epsilon,$$

which implies

$$\min\{x_{N-2}, x_N\} < \epsilon m$$

and

$$\min\{x_{N-5}, x_{N-3}, x_{N-1}\} < \epsilon m^2$$

and eventually leads to a contradiction. The proof is complete. ∎

Theorem 5.157.2 *Assume that*

$$1 \leq B < \infty \text{ and } 1 \leq A < \beta + 1. \tag{5.157.15}$$

Then every positive solution of Eq.(5.157.1) converges to the positive equilibrium $\bar{x}$ of Eq.(5.157.1).

PROOF From Theorem 5.157.1 we know that the solution $\{x_n\}$ of Eq.(5.157.1) is bounded from above by the positive constant $\frac{\beta}{B}$. Let

$$S = \limsup_{n\to\infty} x_n < \infty$$

and

$$I = \liminf_{n\to\infty} x_n \geq 0.$$

Then, clearly,

$$S \leq \frac{(\beta+1)S}{A + BS}$$

and due to the fact that $S > 0$,

$$S \leq \frac{\beta + 1 - A}{B}.$$

Assume for the sake of contradiction that $S = \frac{\beta+1-A}{B}$. There exists a sequence of indices $\{n_i\}$ and positive numbers $\{l_t\}_{t=0}^{3}$ such that

$$S = \lim_{i\to\infty} x_{n_i+1} \text{ and } l_i = \lim_{i\to\infty} x_{n_i-t}.$$

Then

$$S = \frac{\beta l_0 + l_{-2}}{A + Bl_0 + l_{-1}},$$

which implies $l_0 = l_{-2} = S$ and $l_{-1} = 0$. From

$$l_0 = S = \frac{\beta l_{-1} + l_{-3}}{A + Bl_{-1} + l_{-2}} = \frac{l_{-3}}{A + S} \leq \frac{S}{A + S}$$

it follows that

$$S \leq 1 - A \leq 0,$$

which is a contradiction and so

$$S < \frac{\beta + 1 - A}{B}. \tag{5.157.16}$$

There exist $\epsilon > 0$, $m > 0$, and $N > 0$ with

$$0 < m < \min\{x_{N-2}, x_{N-1}, x_N\},$$

$$S + \epsilon < \min\{\frac{\beta + 1 - A}{B}, \beta + 1 - A - Bm\}, \text{ for } B \geq 1,$$

and

$$x_n < S + \epsilon < \beta + 1 - A - Bm, \text{ for } n \geq N - 2.$$

Then, clearly,

$$x_{N+1} = \frac{\beta x_N + x_{N-2}}{A + Bx_N + x_{N-1}} > \frac{(\beta + 1)m}{A + Bm + \beta + 1 - A - Bm} = m$$

and, inductively, we obtain

$$x_n > m, \text{ for } n \geq N - 2$$

from which it follows that $I > 0$. Then, clearly,

$$S \leq \frac{(\beta + 1)S}{A + BS + I} \text{ and } I \geq \frac{(\beta + 1)I}{A + BI + S}$$

from which it follows

$$BS + I \leq \beta + 1 - A \leq BI + S$$

and

$$(B - 1)(S - I) \leq 0.$$

We divide the proof into the following two cases:

Case 1:

$$B > 1.$$

In this case $I = S$.

Case 2:

$$B = 1.$$

Clearly,

$$S = \frac{\beta l_0 + l_{-2}}{A + l_0 + l_{-1}} \leq \frac{(\beta + 1)S}{A + S + I}$$

and so

$$S \leq \beta + 1 - A - I.$$

Assume for the sake of contradiction that

$$S = \beta + 1 - A - I \text{ and } S > I.$$

Then, clearly,

$$l_0 = l_{-2} = S \text{ and } l_{-1} = I.$$

In addition,

$$l_0 = S = \frac{\beta l_{-1} + l_{-3}}{A + l_{-1} + l_{-2}} < \frac{(\beta + 1)S}{A + S + I}$$

from which it follows that

$$S < \beta + 1 - A - I,$$

a contradiction, and so

$$S < \beta + 1 - A - I \text{ or } S = I.$$

Assume that $S < \beta + 1 - A - I$. There exists a sequence of indices $\{n_i\}$ and positive numbers $\{m_t\}_{t=0}^{3}$ such that

$$I = \lim_{j \to \infty} x_{n_j+1} \text{ and } m_t = \lim_{j \to \infty} x_{n_j-t}.$$

Then, clearly,

$$I = \frac{\beta m_0 + m_{-2}}{A + m_0 + m_{-1}} \geq \frac{(\beta + 1)I}{A + I + S}$$

and so $S \geq \beta + 1 - A - I$ which is a contradiction. Hence, $S = I$. The proof is complete. ∎

Conjecture 5.157.2 *Assume that*

$$1 \leq A < \beta + 1 \text{ and } 0 < B < 1.$$

Show that every positive solution of Eq.(5.157.1) converges to the positive equilibrium of Eq.(5.157.1).

Theorem 5.157.3 *Assume that*

$$\beta > B(1 - A), \ 1 \geq A \geq 0, \text{ and } B \geq 1. \tag{5.157.17}$$

Then $(1-A, \frac{\beta}{B})$ *is an invariant interval for all positive solutions of Eq.(5.157.1).*

PROOF Assume that $\{x_n\}$ is a solution of Eq.(5.157.1) with initial conditions x_{-2}, x_{-1}, x_0 such that

$$1 - A < x_{-2}, x_{-1}, x_0 < \frac{\beta}{B}.$$

Then

$$x_1 - \frac{\beta}{B} = \frac{Bx_{-2} - A\beta - \beta x_{-1}}{B(A + Bx_0 + x_{-1})} < \frac{\beta(1 - A - x_{-1})}{B(A + Bx_0 + x_{-1})} < 0.$$

In addition,

$$x_1 - (1-A) = \frac{\beta x_0 + x_{-2}}{A + Bx_0 + x_{-1}} - (1-A) = \frac{[\beta - (1-A)B]x_0 + x_{-2} - (1-A)(A + x_{-1})}{A + Bx_0 + x_{-1}}$$

$$> \frac{(1-A)[\beta - (1-A)B + 1 - A - \frac{\beta}{B}]}{A + Bx_0 + x_{-1}} = \frac{(1-A)(B-1)[\beta - (1-A)B]}{A + Bx_0 + x_{-1}} > 0.$$

Inductively, the result follows. The proof is complete. ∎

Theorem 5.157.4 *Assume that*

$$\beta > B(1 - A), \quad 1 \geq A \geq 0, \quad \text{and} \quad B \geq 1. \tag{5.157.18}$$

Then every solution of Eq.(5.157.1) with initial conditions in the invariant interval $(1 - A, \frac{\beta}{B})$ *converges to the positive equilibrium* $\bar{x}$ *of Eq.(5.157.1).*

PROOF We will divide the proof into the following two cases:
Case 1:

$$B = 1.$$

Let

$$S = \limsup_{n\to\infty} x_n \quad \text{and} \quad I = \liminf_{n\to\infty} x_n.$$

Then in view of Eq.(5.157.1) we get

$$S \leq \frac{(\beta + 1)S}{A + S + I},$$

which implies that

$$S \leq \beta + 1 - A - I.$$

Assume for the sake of contradiction that

$$S = \beta + 1 - A - I \quad \text{and} \quad S > I.$$

There exists a sequence of indices $\{n_i\}$ and positive numbers $\{l_t\}_{t=0}^{3}$ such that

$$S = \lim_{i\to\infty} x_{n_i+1} \quad \text{and} \quad l_t = \lim_{i\to\infty} x_{n_i-t}.$$

Then

$$S = \frac{\beta l_0 + l_{-2}}{A + l_0 + l_{-1}}$$

from which it follows that

$$l_0 = l_{-2} = S \text{ and } l_{-1} = I.$$

From

$$l_0 = S = \frac{\beta l_{-1} + l_{-3}}{A + l_{-1} + l_{-2}} \frac{\beta I + l_{-3}}{A + I + S} < \frac{(\beta + 1)S}{1 + S + I}$$

we obtain

$$S < \beta + 1 - A - I,$$

a contradiction, and so

$$S < \beta + 1 - A \text{ or } S = I.$$

Assume for the sake of contradiction that

$$S < \beta + 1 - A.$$

There exists a sequence of indices $\{n_i\}$ and positive numbers $\{m_t\}_{t=0}^{3}$ such that

$$I = \lim_{j\to\infty} x_{n_j+1} \text{ and } m_t = \lim_{j\to\infty} x_{n_j-t}.$$

Then, clearly,

$$I = \frac{\beta m_0 + m_{-2}}{A + m_0 + m_{-1}} \geq \frac{(\beta + 1)I}{A + I + S}$$

and so

$$S \geq \beta + 1 - A - I,$$

which is a contradiction. Hence, $S = I$.
Case 2:

$$B > 1.$$

In this case the result follows from Theorem 1.6.5 applied in the invariant interval $(1 - A, \frac{\beta}{B})$. The only Hypothesis of Theorem 1.6.5 that needs to be verified is whether the system

$$\begin{cases} M = \frac{(\beta+1)M}{A+BM+m} \\ m = \frac{(\beta+1)m}{A+Bm+M} \end{cases}$$

has a unique solution. This is clear because $B > 1$. The proof is complete. ∎

Open Problem 5.157.2 *Assume that (5.157.18) holds. Show that every positive solution of Eq.(5.157.1) converges to the positive equilibrium point* $\bar{x}$.

When $B = 1$, using an appropriate change of variables Eq.(#157) becomes

$$x_{n+1} = \frac{x_n + \delta x_{n-2}}{A + x_n + x_{n-1}}, \quad n = 0, 1, \ldots \tag{5.157.19}$$

with positive parameters $\delta > 0$, $A \geq 0$, and with arbitrary nonnegative initial conditions x_{-2} ,x_{-1}, x_0 such that the denominator is always positive.

Theorem 5.157.5 *Eq.(5.157.19) possesses a unique prime period-three solution of the form*

$$\ldots, p, q, r, p, q, r, \ldots$$

if and only if

$$\delta > A + 1.$$

Furthermore, p, q, r are the three positive solutions of the cubic equation

$$-Lx^3 + 2(L^2+L+1)x^2 - (L^3+3L^2+3L+2)x + L(L^2+L+1) = 0, \tag{5.157.20}$$

where $L = \delta - A - 1$. In fact, if p is one of the solutions of (5.157.20), the other two solutions are

$$q = \delta - A - \frac{1}{1 + p + A - \delta}, \quad r = \delta - A - \frac{p + A - \delta + 1}{p + A - \delta}. \tag{5.157.21}$$

PROOF Let

$$x_{-2} = p, \quad x_{-1} = q, \quad x_0 = r$$

where p, q, r are not all equal. Then the triple p, q, r is a prime period-three solution of Eq.(5.157.19) if and only if they satisfy the system of equations

$$\begin{aligned} r + \delta p &= Ap + rp + qp \\ p + \delta q &= Aq + qp + qr \\ q + \delta r &= Ar + qr + rp \end{aligned} \tag{5.157.22}$$

and $A + p + q, A + p + r > 0, A + q + r > 0$. Using the change of variables

$$P = p - \delta + A, \quad Q = q - \delta + A, \quad R = r - \delta + A$$

we have

$$\begin{aligned} R - P &= R(P - Q) \\ R - Q &= Q(P - R) \\ P - Q &= P(Q - R). \end{aligned} \tag{5.157.23}$$

In view of (5.157.23), we find

$$Q = -\frac{1}{1 + P}, \quad R = -\frac{P + 1}{P}$$

Substituting $P = p - \delta + A$, $Q = q - \delta + A$ and $R = r - \delta + A$, we see that

$$q = \delta - A - \frac{1}{1 + p + A - \delta}, \quad r = \delta - A - \frac{p + A - \delta + 1}{p + A - \delta}.$$

Finally, in view of (5.157.22), we get

$$\frac{f(p)}{(p + A - \delta)(p + A - \delta + 1)}$$

$$= \frac{-Lp^3 + 2(L^2 + L + 1)p^2 - (L^3 + 3L^2 + 3L + 2)p + L(L^2 + L + 1)}{(p + A - \delta)(p + A - \delta + 1)} = 0.$$

It holds that $(p + A - \delta)(p + A - \delta + 1) = 0$ if and only if $p = q = r$. Therefore, $f(p) = 0$. Similarly, $f(q) = f(r) = 0$. Also, Eq.(5.157.20) has three distinct positive solutions if and only if

$$\delta > A + 1.$$

The proof is complete. ∎

5.158 Equation #158 : $x_{n+1} = \dfrac{\beta x_n + \delta x_{n-2}}{A + Bx_n + Dx_{n-2}}$

Eq.(#158) can be written in the normalized form

$$x_{n+1} = \frac{x_n + \delta x_{n-2}}{A + x_n + Dx_{n-2}}, \quad n = 0, 1, \ldots \tag{5.158.1}$$

with positive parameters δ, A, D and with arbitrary nonnegative initial conditions x_{-2}, x_{-1}, x_0.

By Theorems 5.23.2 and 5.23.4 it follows that when

$$A \geq 1 + \delta,$$

the zero equilibrium of Eq.(5.158.1) is globally asymptotically stable.

When

$$A < 1 + \delta,$$

Eq.(5.158.1) has the unique positive equilibrium point

$$\bar{x} = \frac{1 + \delta - A}{1 + D}.$$

Conjecture 5.158.1 *Show that for the positive equilibrium $\bar{x}$ of Eq.(5.158.1) and with positive initial conditions,*

$$\textbf{\textit{Local Asymptotic Stability}} \Rightarrow \textbf{\textit{Global Attractivity.}}$$

Conjecture 5.158.2 *Assume that*

$$A < 1 + \delta.$$

Show that Eq.(5.158.1) has solutions that do not converge to an equilibrium point or to a periodic solution.

5.159 Equation #159 : $x_{n+1} = \dfrac{\beta x_n + \delta x_{n-2}}{A + Cx_{n-1} + Dx_{n-2}}$

The boundedness character of solutions of this equation was investigated in [69]. See also Theorem 2.1.1 where we established that every solution of the equation is bounded. Eq.(#159) can be written in the normalized form

$$x_{n+1} = \frac{\beta x_n + x_{n-2}}{A + Cx_{n-1} + x_{n-2}}, \quad n = 0, 1, \ldots \tag{5.159.1}$$

with positive parameters β, A, C and with arbitrary nonnegative initial conditions x_{-2}, x_{-1}, x_0.

By Theorems 5.23.2 and 5.23.4 it follows that when

$$A \geq 1 + \beta,$$

the zero equilibrium of Eq.(5.159.1) is globally asymptotically stable.

When

$$A < 1 + \beta,$$

Eq.(5.159.1) has the unique positive equilibrium point

$$\bar{x} = \frac{1 + \beta - A}{1 + C}.$$

Conjecture 5.159.1 *Show that for the positive equilibrium $\bar{x}$ of Eq.(5.159.1) and with positive initial conditions,*

$$\textbf{\textit{Local Asymptotic Stability}} \Rightarrow \textbf{\textit{Global Attractivity.}}$$

Conjecture 5.159.2 *Assume that*

$$A < 1 + \beta.$$

Show that Eq.(5.159.1) has solutions that do not converge to an equilibrium point or to a periodic solution.

5.160 Equation #160 : $x_{n+1} = \dfrac{\beta x_n + \delta x_{n-2}}{Bx_n + Cx_{n-1} + Dx_{n-2}}$

Eq.(#160) can be written in the normalized form

$$x_{n+1} = \frac{\beta x_n + x_{n-2}}{Bx_n + Cx_{n-1} + x_{n-2}}, \quad n = 0, 1, \ldots \tag{5.160.1}$$

with positive parameters δ, C, D and with arbitrary positive initial conditions x_{-2}, x_{-1}, x_0.

Open Problem 5.160.1 *Investigate the periodic nature of solutions and the global stability of the equilibrium of Eq.(5.160.1).*

Conjecture 5.160.1 *Show that for the equilibrium $\bar{x}$ of Eq.(5.160.1),*

Local Asymptotic Stability ⇒ Global Attractivity.

Conjecture 5.160.2 *Show that Eq.(5.160.1) has solutions that do not converge to the equilibrium point or to a periodic solution.*

5.161 Equation #161 : $x_{n+1} = \dfrac{\gamma x_{n-1} + \delta x_{n-2}}{A + Bx_n + Cx_{n-1}}$

Eq.(#161) can be written in the normalized form

$$x_{n+1} = \frac{x_{n-1} + \delta x_{n-2}}{A + Bx_n + x_{n-1}}, \quad n = 0, 1, \ldots \tag{5.161.1}$$

with positive parameters δ, A, B and with arbitrary nonnegative initial conditions x_{-2}, x_{-1}, x_0.

The boundedness character of solutions of this equation was investigated in [69]. See also Theorem 3.4.1 where we established that the equation has unbounded solutions when

$$\delta > A + B.$$

By Theorem 5.221.1 it follows that every solution of Eq.(5.161.1) is bounded if

$$\delta \leq A + B.$$

By Theorems 5.23.2 and 5.23.4 it follows that when

$$A \geq \delta + 1,$$

the zero equilibrium of Eq.(5.161.1) is globally asymptotically stable.

When

$$A < \delta + 1,$$

Eq.(5.161.1) has the unique positive equilibrium point

$$\bar{x} = \frac{1 + \delta - A}{1 + B}.$$

The characteristic equation of the linearized equation of Eq.(5.161.1) about the positive equilibrium $\bar{x}$ is

$$\lambda^3 + \frac{B(\delta + 1 - A)}{(B + 1)(\delta + 1)}\lambda^2 + \frac{\delta - A - B}{(\delta + 1)(B + 1)}\lambda - \frac{\delta}{\delta + 1} = 0.$$

From this and Theorem 1.2.3 it follows that the positive equilibrium $\bar{x}$ of Eq.(5.161.1) is locally asymptotically stable when

$$\frac{(B - 1)(1 - A)}{B + 3} < \delta$$

$$< \frac{1 + A + 2B + AB + \sqrt{(2B + AB + A + 1)^2 + 4(B + 1)(1 + A + 2B)}}{2(B + 1)}.$$

It is now easy to see that for the positive equilibrium $\bar{x}$ of Eq.(5.161.1),

Local Asymptotic Stabilty $\not\Rightarrow$ Global Attractivity.

Indeed, for all positive values of A, B for which

$$A + B < \frac{1 + A + 2B + AB + \sqrt{(2B + AB + A + 1)^2 + 4(B + 1)(1 + A + 2B)}}{2(B + 1)}$$

and for all values of δ such that

$$A + B < \delta$$

$$< \frac{1 + A + 2B + AB + \sqrt{(2B + AB + A + 1)^2 + 4(B + 1)(1 + A + 2B)}}{2(B + 1)},$$

the positive equilibrium $\bar{x}$ of Eq.(5.161.1) is locally asymptotically stable and, in the same region, by Theorem 3.4.1, the equation has unbounded solutions. Hence, the positive equilibrium of the equation cannot be a global attractor in the entire region of its local asymptotic stability.

Conjecture 5.161.1 *Assume that*

$$A - 1 < \delta \leq A + B.$$

Show that for the positive equilibrium $\bar{x}$ of Eq.(5.161.1) and with positive initial conditions,

Local Asymptotic Stabilty $\Longrightarrow$ Global Attractivity.

Conjecture 5.161.2 *It follows from the work in Section 4.2 that Eq.(5.161.1) has a unique prime period-two solution. Show that this period-two solution of Eq.(5.161.1) is locally asymptotically stable.*

Conjecture 5.161.3 *Show that Eq.(5.161.1) has bounded solutions that do not converge to an equilibrium point or to a periodic solution.*

5.162 Equation #162 : $x_{n+1} = \dfrac{\gamma x_{n-1} + \delta x_{n-2}}{A + Bx_n + Dx_{n-2}}$

Eq.(#162) can be written in the normalized form

$$x_{n+1} = \frac{\gamma x_{n-1} + x_{n-2}}{A + Bx_n + x_{n-2}}, \quad n = 0, 1, \ldots \tag{5.162.1}$$

with positive parameters γ, A, B and with arbitrary nonnegative initial conditions x_{-2}, x_{-1}, x_0.

By Theorems 5.23.2 and 5.23.4 it follows that when

$$A \geq 1 + \gamma,$$

the zero equilibrium of Eq.(5.162.1) is globally asymptotically stable.

When

$$A < 1 + \gamma,$$

Eq.(5.162.1) has the unique positive equilibrium point

$$\bar{x} = \frac{1 + \gamma - A}{1 + B}.$$

Conjecture 5.162.1 *Assume that*

$$\gamma > 1 + A.$$

Show that every positive and bounded solution of Eq.(5.162.1) converges to the positive equilibrium.

Conjecture 5.162.2 *Assume that*

$$\gamma < 1 + A.$$

Show that for the positive equilibrium $\bar{x}$ of Eq.(5.162.1) and with positive initial conditions,

Local Asymptotic Stability $\Rightarrow$ Global Attractivity.

Conjecture 5.162.3 *Assume that*

$$\gamma = 1 + A.$$

Show that every bounded solution of Eq.(5.162.1) converges to a (not necessarily prime) period-two solution.

Conjecture 5.162.4 *Assume that*

$$A - 1 < \gamma < 1 + A.$$

Show that Eq.(5.162.1) has bounded solutions that do not converge to an equilibrium point or to a periodic solution.

5.163 Equation #163 : $x_{n+1} = \dfrac{\gamma x_{n-1} + \delta x_{n-2}}{A + Cx_{n-1} + Dx_{n-2}}$

Eq.(#163) can be written in the normalized form

$$x_{n+1} = \frac{x_{n-1} + \delta x_{n-2}}{A + x_{n-1} + Dx_{n-2}}, \quad n = 0, 1, \ldots \tag{5.163.1}$$

with positive parameters δ, A, D and with arbitrary nonnegative initial conditions x_{-2}, x_{-1}, x_0.

By Theorems 5.23.2 and 5.23.4 it follows that when

$$A \geq 1 + \delta,$$

the zero equilibrium of Eq.(5.163.1) is globally asymptotically stable.

When

$$A < 1 + \delta,$$

Eq.(5.163.1) has the unique positive equilibrium point

$$\bar{x} = \frac{1 + \delta - A}{1 + D}.$$

Conjecture 5.163.1 *Show that for the positive equilibrium $\bar{x}$ of Eq.(5.163.1) and with positive initial conditions,*

Local Asymptotic Stability* $\Rightarrow$ *Global Attractivity.

Conjecture 5.163.2 *Assume that*

$$A < 1 + \delta.$$

Show that Eq.(5.163.1) has solutions that do not converge to an equilibrium point or to a periodic solution.

Conjecture 5.163.3 *It follows from the work in Section 4.2 that Eq.(5.163.1) has a unique prime period-two solution. Show that this period-two solution of Eq.(5.163.1) is locally asymptotically stable.*

5.164 Equation #164 : $x_{n+1} = \dfrac{\gamma x_{n-1} + \delta x_{n-2}}{Bx_n + Cx_{n-1} + Dx_{n-2}}$

Eq.(#164) can be written in the normalized form

$$x_{n+1} = \frac{x_{n-1} + \delta x_{n-2}}{Bx_n + x_{n-1} + Dx_{n-2}}, \quad n = 0, 1, \ldots \tag{5.164.1}$$

with positive parameters δ, B, D and with arbitrary positive initial conditions x_{-2}, x_{-1}, x_0.

Open Problem 5.164.1 *Investigate the periodic nature of solutions and the global stability of the equilibrium of Eq.(5.164.1).*

Conjecture 5.164.1 *Show that for the equilibrium $\bar{x}$ of Eq.(5.164.1),*

Local Asymptotic Stability $\Rightarrow$ Global Attractivity.

Conjecture 5.164.2 *Show that Eq.(5.164.1) has solutions that do not converge to the equilibrium point or to a periodic solution.*

Conjecture 5.164.3 *It follows from the work in Section 4.2 that Eq.(5.164.1) has a unique prime period-two solution. Show that this period-two solution of Eq.(5.164.1) is locally asymptotically stable.*

5.165 Equation #165 : $x_{n+1} = \dfrac{\alpha + \beta x_n + \gamma x_{n-1}}{A + Bx_n}$

This equation was investigated in [16], [108], [112], and [179]. Eq.(#165) possesses a period-two trichotomy depending on whether

$$\gamma < \beta + A, \quad \gamma = \beta + A, \quad \text{or} \quad \gamma > \beta + A.$$

The precise result that allows the parameters α, β, and A to be nonnegative was presented in Theorem 4.2.1.

When

$$\gamma > \beta + A,$$

it follows from Theorem 4.2.2 that every positive and bounded solution of Eq.(#165) converges to the positive equilibrium.

Open Problem 5.165.1 *Investigate the global character of solutions of Eq.(#165) with periodic coefficients.*

5.166 Equation #166 : $x_{n+1} = \dfrac{\alpha + \beta x_n + \gamma x_{n-1}}{A + Cx_{n-1}}$

This equation was investigated in [175]. Eq.(#166) can be written in the normalized form

$$x_{n+1} = \frac{\alpha + \beta x_n + x_{n-1}}{A + x_{n-1}}, \quad n = 0, 1, \ldots \tag{5.166.1}$$

with positive parameters α, β, A and with arbitrary nonnegative initial conditions x_{-1}, x_0.

Eq.(5.166.1) has the unique equilibrium

$$\bar{x} = \frac{\beta + 1 - A + \sqrt{(\beta + 1 - A)^2 + 4\alpha}}{2}.$$

The characteristic equation of the linearized equation of Eq.(5.166.1) about the equilibrium $\bar{x}$ is

$$\lambda^2 - \frac{\beta}{A + \bar{x}}\lambda + \frac{\bar{x} - 1}{\bar{x}} = 0.$$

From this and Theorem 1.2.2 it follows that the equilibrium $\bar{x}$ of Eq.(5.166.1) is locally asymptotically stable for all values of the parameters.

By Theorems 5.23.2 and 5.23.4 it follows that when

$$A \geq \beta + 1,$$

the equilibrium $\bar{x}$ of Eq.(5.166.1) is globally asymptotically stable.

The following theorem about the global attractivity of the equilibrium $\bar{x}$ Eq.(5.166.1) is a new result. See also [175].

Theorem 5.166.1 *Assume that*

$$\alpha \leq A + \beta(2A + 1) + 2(A + 1)^2. \tag{5.166.2}$$

Then the equilibrium of Eq.(5.166.1) is globally asymptotically stable.

PROOF It suffices to show that the equilibrium $\bar{x}$ of Eq.(5.166.1) is a global attractor of all solutions. Let $\{x_n\}$ be a solution of Eq.(5.166.1). We divide the proof into the following four cases:

Case 1:

$$A \leq \alpha \leq A + \beta(2A+1) + 2(A+1)^2.$$

By using the change of variables $x_n = \beta y_n + 1$, Eq.(5.166.1) takes the form

$$y_{n+1} = \frac{\frac{\alpha+\beta-A}{\beta^2} + y_n}{\frac{A+1}{\beta} + y_{n-1}}, \quad n = 0, 1, \ldots . \tag{5.166.3}$$

The equation in this case is a special case of Eq.(#66). For the proof in this case see [157], Theorem 3.4.1(f), p. 73.

The following identity will be useful in the sequel:

$$x_{n+1} - 1 = \beta \cdot \frac{x_n - \frac{A-\alpha}{\beta}}{A + x_{n-1}}, \quad n = 0, 1, \ldots . \tag{5.166.4}$$

Case 2:

$$\alpha < A < \beta + \alpha.$$

Clearly, from (5.166.4), we obtain that either

$$x_n \leq 1, \quad \text{for all } n \geq 0 \tag{5.166.5}$$

or, eventually,

$$x_n > 1 > \frac{A-\alpha}{\beta}. \tag{5.166.6}$$

When (5.166.5) holds, from (5.166.4), we see that, for all $n \geq 1$,

$$x_n < \frac{A-\alpha}{\beta}.$$

Clearly, the function

$$f(x_n, x_{n-1}) = \frac{\alpha + \beta x_n + x_{n-1}}{A + x_{n-1}}$$

is strictly increasing in x_n and x_{n-1}. By employing Theorem 1.6.7 the result follows.

When (5.166.6) holds, by using the change of variables, $x_n = \beta y_n + 1$, Eq.(5.166.1) reduces to Eq.(5.166.3), which is included in Eq.(#66). For the proof in this case see [157], Corollary 3.4.1(e), p. 73.

Case 3:

$$A > \beta + \alpha.$$

Clearly, from (5.166.4), we obtain that either

$$x_n > 1, \text{ for all } n \geq 0 \tag{5.166.7}$$

or, eventually,

$$x_n \leq 1 < \frac{A - \alpha}{\beta}. \tag{5.166.8}$$

When (5.166.7) holds, from (5.166.4), we see that, for all $n \geq 1$,

$$x_n > \frac{A - \alpha}{\beta}.$$

Clearly, the function

$$f(x_n, x_{n-1}) = \frac{\alpha + \beta x_n + x_{n-1}}{A + x_{n-1}}$$

is strictly increasing in x_n and strictly decreasing in x_{n-1}. By employing Theorem 1.6.7 the result follows.

When (5.166.8) holds, clearly the function

$$f(x_n, x_{n-1}) = \frac{\alpha + \beta x_n + x_{n-1}}{A + x_{n-1}}$$

is strictly increasing in x_n and x_{n-1}. By employing Theorem 1.6.7 the result follows.

Case 4:

$$\alpha < A = \beta + \alpha.$$

Observe that, for all $n \geq 0$,

$$x_{n+1} - x_n = \frac{(\alpha + x_{n-1})(1 - x_n)}{\alpha + \beta + x_{n-1}}.$$

From this and from (5.166.4), we obtain that, for all $n \geq 0$,

$$x_n \leq x_{n+1} \leq 1 \text{ or } 1 \leq x_{n+1} \leq x_n,$$

from which the result follows. The proof is complete. ∎

Open Problem 5.166.1 *Assume that*

$$\alpha > A + \beta(2A + 1) + 2(A + 1)^2.$$

Show that the equilibrium $\bar{x}$ of Eq.(5.166.1) is globally asymptotically stable.

5.167 Equation #167 : $x_{n+1} = \dfrac{\alpha + \beta x_n + \gamma x_{n-1}}{A + D x_{n-2}}$

Eq.(#167) can be written in the normalized form

$$x_{n+1} = \frac{\alpha + x_n + \gamma x_{n-1}}{A + x_{n-2}}, \quad n = 0, 1, \ldots \tag{5.167.1}$$

with positive parameters α, γ, A and with arbitrary nonnegative initial conditions x_{-2}, x_{-1}, x_0.

By Theorems 5.23.2 and 5.23.3 it follows that when

$$A \geq \gamma + 1,$$

the equilibrium of Eq.(5.167.1) is globally asymptotically stable.

Conjecture 5.167.1 *Assume that*

$$\gamma > 1 + A.$$

Show that every bounded solution of Eq.(5.167.1) converges to the equilibrium.

Conjecture 5.167.2 *Assume that*

$$\gamma = 1 + A.$$

Show that every solution of Eq.(5.167.1) converges to a (not necessarily prime) period-two solution.

Conjecture 5.167.3 *Assume that*

$$\gamma < 1 + A.$$

Show that for the equilibrium $\bar{x}$ of Eq.(5.167.1),

Local Asymptotic Stability* $\Rightarrow$ *Global Attractivity.

Conjecture 5.167.4 *Assume that*

$$A - 1 < \gamma < 1 + A.$$

Show that Eq.(5.167.1) has bounded solutions that do not converge to the equilibrium point or to a periodic solution.

5.168 Equation #168 : $x_{n+1} = \dfrac{\alpha + \beta x_n + \gamma x_{n-1}}{Bx_n + Cx_{n-1}}$

Eq.(#168) can be written in the normalized form

$$x_{n+1} = \frac{\alpha + x_n + \gamma x_{n-1}}{x_n + Cx_{n-1}}, \quad n = 0, 1, \ldots \tag{5.168.1}$$

with positive parameters α, γ, C and with arbitrary positive initial conditions x_{-1}, x_0.

Open Problem 5.168.1 *Investigate the periodic nature of solutions and the global stability of the equilibrium of Eq.(5.168.1).*

Conjecture 5.168.1 *Show that for the equilibrium $\bar{x}$ of Eq.(5.168.1),*

Local Asymptotic Stability ⇒ Global Attractivity.

Conjecture 5.168.2 *It follows from the work in Section 4.2 that Eq.(5.168.1) has a unique prime period-two solution. Show that this period-two solution of Eq.(5.168.1) is locally asymptotically stable.*

Conjecture 5.168.3 *Show that every solution of Eq.(5.168.1) converges to a (not necessarily prime) period-two solution.*

5.169 Equation #169 : $x_{n+1} = \dfrac{\alpha + \beta x_n + \gamma x_{n-1}}{Bx_n + Dx_{n-2}}$

Eq.(#169) can be written in the normalized form

$$x_{n+1} = \frac{\alpha + x_n + \gamma x_{n-1}}{x_n + Dx_{n-2}}, \quad n = 0, 1, \ldots \tag{5.169.1}$$

with positive parameters α, γ, D and with arbitrary positive initial conditions x_{-2}, x_{-1}, x_0.

The boundedness character of solutions of this equation was investigated in [49]. See also Theorem 3.3.1 where we established that the equation has unbounded solutions when

$$\gamma > 1.$$

Conjecture 5.169.1 *Assume that*

$$\gamma > 1.$$

Show that every bounded solution of Eq.(5.169.1) converges to the equilibrium.

Conjecture 5.169.2 *Assume that*

$$\gamma < 1.$$

Show that for the equilibrium $\bar{x}$ of Eq.(5.169.1),

Local Asymptotic Stability* $\Rightarrow$ *Global Attractivity.

Conjecture 5.169.3 *Assume that*

$$\gamma = 1.$$

Show that every solution of Eq.(5.169.1) converges to a (not necessarily prime) period-two solution.

Conjecture 5.169.4 *Assume that*

$$\gamma < 1.$$

Show that Eq.(5.169.1) has bounded solutions that do not converge to the equilibrium point or to a periodic solution.

5.170 Equation #170 : $x_{n+1} = \dfrac{\alpha + \beta x_n + \gamma x_{n-1}}{C x_{n-1} + D x_{n-2}}$

The boundedness character of solutions of this equation was investigated in [49]. See also Theorem 2.8.1 where we established that every solution of the equation is bounded. Eq.(#170) can be written in the normalized form

$$x_{n+1} = \frac{\alpha + \beta x_n + x_{n-1}}{x_{n-1} + D x_{n-2}}, \quad n = 0, 1, \ldots \tag{5.170.1}$$

with positive parameters α, β, D and with arbitrary positive initial conditions x_{-2}, x_{-1}, x_0.

Open Problem 5.170.1 *Investigate the periodic nature of solutions and the global stability of the equilibrium of Eq.(5.170.1).*

Conjecture 5.170.1 *Show that for the equilibrium $\bar{x}$ of Eq.(5.170.1),*

$$\textbf{\textit{Local Asymptotic Stability}} \Rightarrow \textbf{\textit{Global Attractivity.}}$$

Conjecture 5.170.2 *Show that Eq.(5.170.1) has solutions that do not converge to the equilibrium point or to a periodic solution.*

Conjecture 5.170.3 *It follows from the work in Section 4.2 that Eq.(5.170.1) has a unique prime period-two solution. Show that this period-two solution of Eq.(5.170.1) is locally asymptotically stable.*

5.171 Equation #171 : $x_{n+1} = \dfrac{\alpha + \beta x_n + \delta x_{n-2}}{A + Bx_n}$

The boundedness character of this equation was investigated in [49]. See also Theorem 2.5.1 where we established that every solution of the equation is bounded. Eq.(#171) can be written in the normalized form

$$x_{n+1} = \frac{\alpha + \beta x_n + x_{n-2}}{A + x_n}, \quad n = 0, 1, \ldots \tag{5.171.1}$$

with positive parameters α, β, A and with arbitrary nonnegative initial conditions x_{-2}, x_{-1}, x_0.

The only equilibrium of Eq.(5.171.1) is

$$\bar{x} = \frac{\beta + 1 - A + \sqrt{(\beta + 1 - A)^2 + 4\alpha}}{2}.$$

The characteristic equation of the linearized equation of Eq.(5.171.1) about the equilibrium $\bar{x}$ is

$$\lambda^3 + \frac{\bar{x} - \beta}{A + \bar{x}}\lambda^2 - \frac{1}{A + \bar{x}} = 0.$$

From this and Theorem 1.2.3 it follows that the equilibrium $\bar{x}$ of Eq.(5.171.1) is locally asymptotically stable when

$$\beta \geq 1 - A, \tag{5.171.2}$$

or

$$\sqrt{2 - A} - 1 \leq \beta < 1 - A, \tag{5.171.3}$$

or

$$\beta < \sqrt{2 - A} - 1 \text{ and } \alpha > \frac{2 + 2\beta A - 3\beta - A - (A + 1)\sqrt{5 - 4(A + \beta)}}{2} \tag{5.171.4}$$

and unstable when

$$\beta < \sqrt{2-A} - 1 \text{ and } \alpha \le \frac{2 + 2\beta A - 3\beta - A - (A+1)\sqrt{5 - 4(A+\beta)}}{2}.$$

Theorem 5.171.1 *Assume that*

$$\beta + 1 \le A.$$

Then every solution of Eq.(5.171.1) converges to the equilibrium $\bar{x}$.

PROOF Let

$$S = \limsup_{n\to\infty} x_n \text{ and } I = \liminf_{n\to\infty} x_n.$$

We divide the proof into the following two cases:

Case 1:

$$\beta + 1 < A.$$

Then, clearly,

$$S \le \frac{\alpha + (\beta+1)S}{A+I}$$

and

$$I \ge \frac{\alpha + (\beta+1)I}{A+S}.$$

From these two inequalities it follows that

$$\alpha + (\beta + 1 - A)I \le SI \le \alpha + (\beta + 1 - A)S.$$

From this it follows that $S = I$.

Case 2:

$$\beta + 1 = A.$$

Then

$$S \le \frac{\alpha + (\beta+1)S}{\beta+1+I}$$

and

$$I \ge \frac{\alpha + (\beta+1)I}{\beta+1+S}.$$

From these two inequalities it follows that

$$\alpha \le SI \le \alpha$$

and so

$$SI = \alpha.$$

Let $\{x_{n_i+1}\}$, $\{x_{n_i}\}$, $\{x_{n_i-2}\}$, be three subsequences of the solution $\{x_n\}$ such that

$$\lim_{i\to\infty} x_{n_i+1} = S$$

and

$$\lim_{i\to\infty} x_{n_i} = l_0, \quad \lim_{i\to\infty} x_{n_i-2} = l_{-2}.$$

Then

$$S = \frac{\alpha + \beta l_0 + l_{-2}}{\beta + 1 + l_0}.$$

We will show that

$$l_0 = S = I.$$

Suppose for the sake of contradiction

$$l_0 \neq S \text{ or } l_0 \neq I.$$

Then, clearly,

$$S < \frac{\alpha + (\beta+1)S}{\beta + 1 + I}.$$

From this it follows that

$$SI < \alpha,$$

which is a contradiction. The proof is complete. ∎

Theorem 5.171.2 *Assume that*

$$\beta \geq 1 - A > 0.$$

Then every solution of Eq.(5.171.1) converges to the equilibrium $\bar{x}$.

PROOF We will make use of the fact we proved in Lemma 2.5.1 of Section 2.5 that every solution of Eq.(5.171.1) is bounded from below by β.

Let

$$S = \limsup_{n\to\infty} x_n \text{ and } I = \liminf_{n\to\infty} x_n.$$

We divide the proof into the following two cases:

Case 1:

$$\beta > 1 - A.$$

Then, clearly,

$$S \leq \frac{\alpha + \beta I + S}{A + I}$$

and

$$I \geq \frac{\alpha + \beta S + I}{A + S}.$$

From these two inequalities it follows that

$$\alpha + \beta S + (1 - A)I \leq SI \leq \alpha + \beta I + (1 - A)S.$$

From this it follows that

$$(\beta + 1 - A)(S - I) \leq 0$$

and so $S = I$.

Case 2:

$$\beta = 1 - A > 0.$$

Then

$$S \leq \frac{\alpha + \beta I + S}{1 - \beta + I}$$

and

$$I \geq \frac{\alpha + \beta S + I}{1 - \beta + S}.$$

From these two inequalities it follows that

$$\alpha + \beta I + \beta S \leq SI \leq \alpha + \beta S + \beta I.$$

It also holds that

$$I > \beta,$$

otherwise,

$$\alpha + \beta^2 + \beta S = \beta S,$$

which is not true. Hence,

$$S = \frac{\alpha + \beta I}{I - \beta} \text{ and } I = \frac{\alpha + \beta S}{S - \beta}.$$

There exists a sequence of indices $\{n_i\}$ and positive numbers $\{l_{t+1}\}_{t=-4}^{1}$ such that

$$\lim_{i \to \infty} x_{n_i + 1 - t} = l_{t+1}$$

with

$$l_1 = S.$$

Then

$$l_1 = \frac{\alpha + \beta l_0 + l_{-2}}{1 - \beta + l_0}.$$

From this it follows that

$$l_0 = I \text{ and } l_{-2} = S$$

because otherwise

$$S < \frac{\alpha + \beta I}{I - \beta},$$

which is a contradiction. Similarly,

$$l_{-1} = S \text{ and } l_{-3} = I$$

and

$$l_{-2} = I \text{ and } l_{-4} = S.$$

Hence,

$$I = l_{-2} = S.$$

The proof is complete. ▮

Conjecture 5.171.1 *Assume that (5.171.2) or (5.171.3) or (5.171.4) holds. Show that every solution of Eq.(5.171.1) converges to the equilibrium $\bar{x}$.*

Conjecture 5.171.2 *Show that Eq.(5.171.1) has solutions that do not converge to the equilibrium point $\bar{x}$ or to a periodic solution.*

5.172 Equation #172 : $x_{n+1} = \dfrac{\alpha + \beta x_n + \delta x_{n-2}}{A + C x_{n-1}}$

The boundedness character of this equation was investigated in [49]. See also Theorem 2.7.1 where we established that every solution of the equation is bounded. Eq.(#172) can be written in the normalized form

$$x_{n+1} = \frac{\alpha + \beta x_n + \delta x_{n-2}}{1 + x_{n-1}}, \quad n = 0, 1, \ldots \tag{5.172.1}$$

with positive parameters α, β, δ and with arbitrary nonnegative initial conditions x_{-2}, x_{-1}, x_0.

Eq.(5.172.1) has the unique equilibrium

$$\bar{x} = \frac{\beta + \delta - 1 + \sqrt{(\beta + \delta - 1)^2 + 4\alpha}}{2}.$$

The characteristic equation of the linearized equation of Eq.(5.172.1) about the equilibrium $\bar{x}$ is

$$\lambda^3 - \frac{\beta}{1 + \bar{x}}\lambda^2 + \frac{\bar{x}}{1 + \bar{x}}\lambda - \frac{\delta}{1 + \bar{x}} = 0.$$

From this and Theorem 1.2.3 it follows that the equilibrium $\bar{x}$ of Eq.(5.172.1) is locally asymptotically stable when

$$\beta \geq \delta - 1 \tag{5.172.2}$$

or

$$\beta < \delta - 1 \text{ and } \alpha > (1+\beta\delta-\delta^2)(\beta+\delta+\beta\delta-\delta^2) \tag{5.172.3}$$

and unstable when

$$\beta < \delta - 1 \text{ and } \alpha < (1+\beta\delta-\delta^2)(\beta+\delta+\beta\delta-\delta^2).$$

Theorem 5.172.1 *Assume that*

$$0 < \beta \leq 1 - \delta.$$

Then every solution of Eq.(5.172.1) converges to the equilibrium $\bar{x}$.

PROOF Let

$$S = \limsup_{n\to\infty} x_n \text{ and } I = \limsup_{n\to\infty} x_n.$$

Clearly,

$$S \leq \frac{\alpha+(\beta+\delta)S}{1+I}$$

and

$$I \leq \frac{\alpha+(\beta+\delta)I}{1+S}.$$

Combining the two inequalities, we find that

$$\alpha + (\beta+\delta-1)I \leq SI \leq \alpha + (\beta+\delta-1)S. \tag{5.172.4}$$

We divide the proof into the following two cases:

Case 1:

$$0 < \beta < 1 - \delta.$$

From (5.172.4) it follows that

$$S = I.$$

Case 2:

$$0 < \beta = 1 - \delta.$$

From (5.172.4) it follows that

$$S = \frac{\alpha}{I}.$$

There exists a sequence of indices $\{n_i\}$ and positive numbers $\{l_{-t}\}_{t=0}^{2}$ such that

$$\lim_{i\to\infty} x_{n_i+1} = S \text{ and } l_{-t} = \lim_{i\to\infty} x_{n_i-t}.$$

Then

$$S = \frac{\alpha + \beta l_0 + (1-\beta)l_{-2}}{1 + l_{-1}}.$$

From this it follows that

$$l_0 = l_{-2} = S$$

and

$$l_{-1} = I$$

otherwise,

$$S \neq \frac{\alpha}{I},$$

which is a contradiction. Similarly,

$$l_{-1} = l_{-3} = S$$

and

$$l_{-2} = I.$$

Hence, $S = I$. The proof is complete. ▮

Conjecture 5.172.1 *Assume that (5.172.2) or (5.172.3) holds. Show that every solution of Eq.(5.172.1) converges to the equilibrium $\bar{x}$.*

Conjecture 5.172.2 *Show that Eq.(5.172.1) has solutions that do not converge to the equilibrium point $\bar{x}$ or to a periodic solution.*

5.173 Equation #173 : $x_{n+1} = \dfrac{\alpha + \beta x_n + \delta x_{n-2}}{A + D x_{n-2}}$

The boundedness character of solutions of this equation was investigated in [49]. See also Theorem 2.3.3 where we established that every solution of the equation is bounded. Eq.(#173) can be written in the normalized form

$$x_{n+1} = \frac{\alpha + \beta x_n + x_{n-2}}{A + x_{n-2}}, \quad n = 0, 1, \ldots \tag{5.173.1}$$

with positive parameters α, β, A and with arbitrary nonnegative initial conditions x_{-2}, x_{-1}, x_0.

By Theorems 5.23.2 and 5.23.4 it follows that when

$$\beta + 1 \leq A,$$

the equilibrium of Eq.(5.173.1) is globally asymptotically stable.

Conjecture 5.173.1 *Show that for the equilibrium $\bar{x}$ of Eq.(5.173.1),*

$$\textbf{\textit{Local Asymptotic Stability}} \Rightarrow \textbf{\textit{Global Attractivity.}}$$

Conjecture 5.173.2 *Assume that*

$$\beta + 1 > A.$$

Show that Eq.(5.173.1) has solutions that do not converge to the equilibrium point or to a periodic solution.

5.174 Equation #174 : $x_{n+1} = \dfrac{\alpha + \beta x_n + \delta x_{n-2}}{Bx_n + Cx_{n-1}}$

Eq.(#174) can be written in the normalized form

$$x_{n+1} = \frac{\alpha + x_n + \delta x_{n-2}}{x_n + Cx_{n-1}}, \quad n = 0, 1, \ldots \tag{5.174.1}$$

with positive parameters α, δ, C and with arbitrary positive initial conditions x_{-2}, x_{-1}, x_0.

The boundedness character of solutions of this equation was investigated in [69]. See also Theorems 3.4.1 where we established that Eq.(5.174.1) has unbounded solutions when

$$\delta > C.$$

From Theorem 5.221.1 it follows that every solution of Eq.(5.174.1) is bounded when

$$\delta < C.$$

Eq.(5.174.1) has the unique equilibrium

$$\bar{x} = \frac{1 + \delta + \sqrt{(\delta + 1)^2 + 4\alpha(1 + C)}}{2(1 + C)}.$$

The characteristic equation of the linearized equation of Eq.(5.174.1) about the equilibrium $\bar{x}$ is

$$\lambda^3 + \frac{\bar{x} - 1}{(1 + C)\bar{x}}\lambda^2 + \frac{C}{1 + C}\lambda - \frac{\delta}{(1 + C)\bar{x}} = 0. \tag{5.174.2}$$

From this and Theorem 1.2.3 it follows that the equilibrium $\bar{x}$ of Eq.(5.174.1) is locally asymptotically stable when

$$0 < \delta \leq \frac{C + 2 + \sqrt{C^2 + 8C + 8}}{2(1 + C)} \tag{5.174.3}$$

or

$$\delta > \frac{C + 2 + \sqrt{C^2 + 8C + 8}}{2(1 + C)} \tag{5.174.4}$$

and

$$\alpha > \frac{-3\delta - 2C\delta + 2\delta^2 + 2C\delta^2 - \sqrt{\delta}\sqrt{-4 - 4C + 5\delta + 4C\delta}}{2(C + 1)} \tag{5.174.5}$$

and unstable when

$$\delta > \frac{C + 2 + \sqrt{C^2 + 8C + 8}}{2(1 + C)} \tag{5.174.6}$$

and

$$\alpha < \frac{-3\delta - 2C\delta + 2\delta^2 + 2C\delta^2 - \sqrt{\delta}\sqrt{-4 - 4C + 5\delta + 4C\delta}}{2(C + 1)}. \tag{5.174.7}$$

It is interesting to note that for the equilibrium $\bar{x}$ of Eq.(5.174.1),

Local Asymptotic Stabilty $\not\Rightarrow$ Global Asymptotic Stabilty.

Indeed, for all positive values of C for which

$$C < \frac{C + 2 + \sqrt{C^2 + 8C + 8}}{2(1 + C)}$$

and for all values of δ such that

$$C < \delta < \frac{C + 2 + \sqrt{C^2 + 8C + 8}}{2(1 + C)},$$

the equilibrium $\bar{x}$ of Eq.(5.174.1) is locally asymptotically stable and at the same time the equation has unbounded solutions. In particular, for such initial conditions the equilibrium $\bar{x}$ of the equation is not a global attractor.

Conjecture 5.174.1 *Assume that*

$$\delta \leq C.$$

Show that for the equilibrium $\bar{x}$ of Eq.(5.174.1),

Local Asymptotic Stabilty $\Longrightarrow$ Global Attractivity.

Open Problem 5.174.1 *Assume that*

$$\delta > C.$$

(*i*) *Determine the set of all initial conditions for which the solutions of Eq.(5.174.1) converge to the equilibrium.*

(*ii*) *Determine the set of all initial conditions for which the solutions of Eq.(5.174.1) are unbounded.*

(*iii*) *Determine all possible periodic solutions of Eq.(5.174.1).*

Conjecture 5.174.2 *Show that Eq.(5.174.1) has bounded solutions that do not converge to the equilibrium point* $\bar{x}$ *or to a periodic solution.*

5.175 Equation #175 : $x_{n+1} = \dfrac{\alpha + \beta x_n + \delta x_{n-2}}{B x_n + D x_{n-2}}$

The boundedness character of solutions of this equation was investigated in [49]. See also Theorem 2.1.1 where we established that every solution of the equation is bounded. Eq.(#175) can be written in the normalized form

$$x_{n+1} = \frac{\alpha + \beta x_n + x_{n-2}}{B x_n + x_{n-2}}, \quad n = 0, 1, \ldots \tag{5.175.1}$$

with positive parameters α, β, B and with arbitrary positive initial conditions x_{-2}, x_{-1}, x_0.

Open Problem 5.175.1 *Investigate the periodic nature of solutions and the global stability of the equilibrium of Eq.(5.175.1).*

Conjecture 5.175.1 *Show that for the equilibrium* $\bar{x}$ *of Eq.(5.175.1),*

Local Asymptotic Stability $\Rightarrow$ ***Global Attractivity.***

Conjecture 5.175.2 *Show that Eq.(5.175.1) has solutions that do not converge to the equilibrium point or to a periodic solution.*

5.176 Equation #176 : $x_{n+1} = \dfrac{\alpha + \beta x_n + \delta x_{n-2}}{C x_{n-1} + D x_{n-2}}$

Eq.(#176) can be written in the normalized form

$$x_{n+1} = \frac{\alpha + \beta x_n + x_{n-2}}{C x_{n-1} + x_{n-2}}, \quad n = 0, 1, \ldots \qquad (5.176.1)$$

with positive parameters α, β, C and with arbitrary positive initial conditions x_{-2}, x_{-1}, x_0.

The boundedness character of this equation was investigated in [152] where it was established that every solution is bounded from above and from below by positive constants. See also Theorem 2.9.1 in Section 2.9.

Open Problem 5.176.1 *Investigate the periodic nature of solutions and the global stability of the equilibrium of Eq.(5.176.1).*

Conjecture 5.176.1 *Show that for the equilibrium $\bar{x}$ of Eq.(5.176.1),*

Local Asymptotic Stability* $\Rightarrow$ *Global Attractivity.

Conjecture 5.176.2 *Show that Eq.(5.176.1) has solutions that do not converge to the equilibrium point or to a periodic solution.*

5.177 Equation #177 : $x_{n+1} = \dfrac{\alpha + \gamma x_{n-1} + \delta x_{n-2}}{A + B x_n}$

This equation is a special case of a more general equation that is investigated in Section 5.195.

Conjecture 5.177.1 *Assume that*

$$\gamma > \delta + A.$$

Show that every bounded solution of Eq.(#177) converges to the equilibrium.

Conjecture 5.177.2 *Assume that*

$$\gamma = \delta + A.$$

Show that every solution of Eq.(#177) converges to a (not necessarily prime) period-two solution.

Conjecture 5.177.3 *Assume that*

$$\gamma < \delta + A.$$

Show that for the equilibrium $\bar{x}$ of Eq.(#177),

Local Asymptotic Stability ⇒ Global Attractivity.

Conjecture 5.177.4 *Assume that*

$$\gamma < \delta + A.$$

Show that Eq.(#177) has bounded solutions that do not converge to the equilibrium point or to a periodic solution.

5.178 Equation #178 : $x_{n+1} = \dfrac{\alpha + \gamma x_{n-1} + \delta x_{n-2}}{A + C x_{n-1}}$

The boundedness character of solutions of this equation was investigated in [49]. See also Theorem 2.7.1 where we established that every solution of the equation is bounded. Eq.(#178) can be written in the normalized form

$$x_{n+1} = \frac{\alpha + \gamma x_{n-1} + \delta x_{n-2}}{1 + x_{n-1}}, \quad n = 0, 1, \ldots \tag{5.178.1}$$

with positive parameters α, γ, δ and with arbitrary nonnegative initial conditions x_{-2}, x_{-1}, x_0.

Eq.(5.178.1) has the unique equilibrium point

$$\bar{x} = \frac{\gamma + \delta - 1 + \sqrt{(\gamma + \delta - 1)^2 + 4\alpha}}{2}.$$

The characteristic equation of the linearized equation of Eq.(5.178.1) about the equilibrium $\bar{x}$ is

$$\lambda^3 + \frac{\bar{x} - \gamma}{1 + \bar{x}}\lambda - \frac{\delta}{1 + \bar{x}} = 0.$$

From this and Theorem 1.2.3 it follows that the equilibrium $\bar{x}$ of Eq.(5.178.1) is locally asymptotically stable when

$$\gamma \geq \delta - 1 \tag{5.178.2}$$

or

$$\gamma < \delta - 1 \text{ and } \alpha > \frac{(1 + \gamma - \delta^2)(\gamma + \gamma^2 + \delta + \gamma\delta - \delta^2)}{(1 + \gamma)^2} \tag{5.178.3}$$

and unstable when

$$\gamma < \delta - 1 \text{ and } \alpha < \frac{(1+\gamma-\delta^2)(\gamma+\gamma^2+\delta+\gamma\delta-\delta^2)}{(1+\gamma)^2}.$$

By Theorems 5.23.2 and 5.23.4 it follows that when

$$\gamma + \delta \leq 1,$$

the equilibrium of Eq.(5.178.1) is globally asymptotically stable. This global stability condition is improved by the next theorem. Also, the next two theorems, which we present here for the first time, establish the global stability of the equilibrium $\bar{x}$ of Eq.(5.178.1) when (5.178.2) holds.

Theorem 5.178.1 *Assume that*

$$\delta \leq 1.$$

Then every solution of Eq.(5.178.1) converges to the equilibrium $\bar{x}$.

PROOF Let

$$S = \limsup_{n\to\infty} x_n \text{ and } I = \liminf_{n\to\infty} x_n.$$

We divide the proof into the following two cases:

Case 1:

$$\delta < 1.$$

We will show that, eventually,

$$x_n < \gamma < \frac{1}{\delta} \cdot \gamma. \tag{5.178.4}$$

Assume for the sake of contradiction that there exists N such that

$$x_{N+1} = \frac{\alpha + \gamma x_{N-1} + \delta x_{N-2}}{1 + x_{N-1}} > \gamma.$$

From this it follows that

$$x_{N-2} > \frac{1}{\delta} \cdot \gamma > \gamma$$

and, similarly,

$$x_{N-5} > (\frac{1}{\delta})^2 \cdot \gamma > \gamma.$$

Inductively, we find that

$$x_{N+1-3k} > (\frac{1}{\delta})^k,$$

which is a contradiction and so (5.178.4) holds. With the use of (5.178.4), we find that the function

$$f(x_{n-1}, x_{n-2}) = \frac{\alpha + \gamma x_{n-1} + \delta x_{n-2}}{1 + x_{n-1}}$$

is strictly increasing in x_{n-1} and x_{n-2}. Then, clearly,

$$S \leq \frac{\alpha + (\gamma + \delta)S}{1 + S} \text{ and } I \geq \frac{\alpha + (\gamma + \delta)I}{1 + I},$$

from which it follows that

$$S \leq \bar{x} \leq I.$$

Case 2:

$$\delta = 1.$$

Observe that

$$x_{n+1} - x_{n-1} = \frac{\alpha + (\gamma - x_{n-2})x_{n-1}}{1 + x_{n-1}}$$

and

$$x_{n+1} - \gamma = \frac{\alpha + x_{n-2} - \gamma}{1 + x_{n-1}}.$$

Therefore, either

$$x_n > \gamma,$$

eventually, in which case from

$$S \leq \frac{\alpha + \gamma I + S}{1 + I} \quad \text{and} \quad I \geq \frac{\alpha + \gamma S + I}{1 + S}$$

it follows that

$$S = I,$$

or there exists a subsequence, of the form $\{x_{3n}\}$ or $\{x_{3n+1}\}$ or $\{x_{3n+2}\}$, which is less that γ and increasing. Assume without loss of generality that

$$x_{3n} < x_{3n+3} < \gamma.$$

From this it follows that $\{x_{3n}\}$ converges to a positive finite limit l_0:

$$x_{3n} \to l_0 \in (0, \infty).$$

From

$$x_{3n+3} = \frac{\alpha + \gamma x_{3n+1} + x_{3n}}{1 + x_{3n+1}}$$

it follows that

$$x_{3n+1} \to l_1 \in (0, \infty).$$

Similarly,

$$x_{3n+2} \to l_2 \in (0, \infty).$$

When $\delta = 1$, Eq.(5.178.1) does not have prime period-three solutions and so

$$l_0 = l_1 = l_2 = \bar{x}.$$

The proof is complete. ∎

Theorem 5.178.2 *Assume that*

$$\gamma \geq \delta - 1 > 0.$$

Then every solution of Eq.(5.178.1) converges to the equilibrium $\bar{x}$.

PROOF We will make use of the fact we proved, in Lemma 2.7.1 of Section 2.7, that every solution of Eq.(5.178.1) is bounded from below by γ.

Let

$$S = \limsup_{n\to\infty} x_n, \quad I = \liminf_{n\to\infty} x_n.$$

Then, clearly,

$$S \leq \frac{\alpha + \gamma I + \delta S}{1 + I}$$

and

$$I \geq \frac{\alpha + \gamma S + \delta I}{1 + S}.$$

From these two inequalities it follows that

$$\alpha + \gamma S + (\delta - 1)I \leq SI \leq \alpha + \gamma I + (\delta - 1)S.$$

We divide the proof in the following two cases:

Case 1:

$$\gamma > \delta - 1.$$

In this case, clearly,

$$(\gamma + 1 - \delta)(S - I) \leq 0$$

and so $S = I$.

Case 2:

$$\gamma = \delta - 1 > 0.$$

In this case

$$SI = \alpha + \gamma S + \gamma I$$

There exists a sequence of indices $\{n_i\}$ and positive numbers $\{l_{-t}\}_{t=0}^{5}$ such that

$$S = \lim_{i\to\infty} x_{n_i+1} \text{ and } l_{-t} = \lim_{i\to\infty} x_{n_i-t}$$

From

$$l_1 = \frac{\alpha + \gamma l_{-1} + (\gamma+1)l_{-2}}{1 + l_{-1}}$$

we have

$$l_{-1} = I \text{ and } l_{-2} = S.$$

Similarly,

$$l_{-3} = S \text{ and } l_{-4} = I$$

and

$$l_{-4} = S \text{ and } l_{-5} = I.$$

Therefore,

$$S = I = l_{-4}.$$

The proof is complete. ∎

Conjecture 5.178.1 *Assume that (5.178.3) holds. Show that every solution of Eq.(5.178.1) converges to the equilibrium $\bar{x}$.*

Conjecture 5.178.2 *Show that Eq.(5.178.1) has solutions that do not converge to the equilibrium point $\bar{x}$ or to a periodic solution.*

5.179 Equation #179 : $x_{n+1} = \dfrac{\alpha + \gamma x_{n-1} + \delta x_{n-2}}{A + Dx_{n-2}}$

This equation was investigated in [17], [70], [128], and [206]. Eq.(#179) possesses a period-two trichotomy depending on whether

$$\gamma < \delta + A, \quad \gamma = \delta + A, \quad \text{or} \quad \gamma > \delta + A.$$

The precise result that allows the parameters α, δ, and A to be nonnegative was presented in Theorem 4.3.1.

Conjecture 5.179.1 *Assume that*

$$\gamma > \delta + A.$$

Show that every bounded solution of Eq.(#179) converges to the equilibrium.

Open Problem 5.179.1 *Investigate the global character of solutions of Eq.(#179) with periodic coefficients.*

5.180 Equation #180 : $x_{n+1} = \dfrac{\alpha + \gamma x_{n-1} + \delta x_{n-2}}{Bx_n + Cx_{n-1}}$

Eq.(#180) can be written in the normalized form

$$x_{n+1} = \frac{\alpha + x_{n-1} + \delta x_{n-2}}{Bx_n + x_{n-1}}, \quad n = 0, 1, \ldots \tag{5.180.1}$$

with positive parameters α, δ, B and with arbitrary positive initial conditions x_{-2}, x_{-1}, x_0.

The boundedness character of solutions of this equation was investigated in [69]. See also Theorem 3.4.1 where we established that Eq.(5.180.1) has unbounded solutions when

$$\delta > B.$$

From Theorem 5.221.1 it follows that every solution of Eq.(5.180.1) is bounded if

$$\delta < B.$$

Eq.(5.180.1) has the unique equilibrium

$$\bar{x} = \frac{1 + \delta + \sqrt{(\delta + 1)^2 + 4\alpha(1 + B)}}{2(1 + B)}.$$

The characteristic equation of the linearized equation of Eq.(5.180.1) about the equilibrium $\bar{x}$ is

$$\lambda^3 + \frac{B}{1+B}\lambda^2 + \frac{\bar{x} - 1}{(1+B)\bar{x}}\lambda - \frac{\delta}{(1+B)\bar{x}} = 0\,. \tag{5.180.2}$$

From this and Theorem 1.2.3 it follows that the equilibrium $\bar{x}$ of Eq.(5.180.1) is locally asymptotically stable when

$$0 < \delta \leq \frac{1 + 2B + \sqrt{5 + 16B + 12B^2}}{2(1+B)} \tag{5.180.3}$$

or

$$\delta > \frac{1 + 2B + \sqrt{5 + 16B + 12B^2}}{2(1+B)} \tag{5.180.4}$$

and

$$\alpha > \frac{1 + 3B + 2B^2 - B\delta - 2B^2\delta + 2B\delta^2 + 2B^2\delta^2}{2(B^2 + B^3)}$$

$$\frac{-(1+2B)\sqrt{1 + 2B + B^2 - 2B\delta - 2B^2\delta + 4B\delta^2 + 5B^2\delta^2}}{2(B^2 + B^3)} \tag{5.180.5}$$

and unstable when

$$\delta > \frac{1+2B+\sqrt{5+16B+12B^2}}{2(1+B)} \tag{5.180.6}$$

and

$$\alpha < \frac{1+3B+2B^2-B\delta-2B^2\delta+2B\delta^2+2B^2\delta^2}{2(B^2+B^3)}$$

$$\frac{-(1+2B)\sqrt{1+2B+B^2-2B\delta-2B^2\delta+4B\delta^2+5B^2\delta^2}}{2(B^2+B^3)}. \tag{5.180.7}$$

It is interesting to note that for the equilibrium $\bar{x}$ of Eq.(5.180.1),

Local Asymptotic Stabilty $\not\Rightarrow$ Global Asymptotic Stabilty.

Indeed, for all positive values of B for which

$$B < \frac{1+2B+\sqrt{5+16B+12B^2}}{2(1+B)}$$

and for all values of δ such that

$$B < \delta < \frac{1+2B+\sqrt{5+16B+12B^2}}{2(1+B)},$$

the equilibrium $\bar{x}$ of Eq.(5.180.1) is locally asymptotically stable and at the same time the equation has unbounded solutions. In particular, for such initial conditions the equilibrium $\bar{x}$ of the equation is not a global attractor.

Conjecture 5.180.1 *Assume that*

$$\delta \leq B.$$

Show that for the equilibrium $\bar{x}$ of Eq.(5.180.1),

Local Asymptotic Stabilty $\Longrightarrow$ Global Attractivity.

Open Problem 5.180.1 *Assume that*

$$\delta > B.$$

(*i*) *Determine the set of all initial conditions for which the solutions of Eq.(5.180.1) converge to the equilibrium $\bar{x}$.*

(*ii*) *Determine the set of all initial conditions for which the solutions of Eq.(5.180.1) are unbounded.*

(*iii*) *Determine all possible periodic solutions of Eq.(5.180.1).*

Conjecture 5.180.2 *Show that Eq.(5.180.1) has bounded solutions that do not converge to the equilibrium point $\bar{x}$ or to a periodic solution.*

Conjecture 5.180.3 *It follows from the work in Section 4.2 that Eq.(5.180.1) has a unique prime period-two solution. Show that this period-two solution of Eq.(5.180.1) is locally asymptotically stable.*

5.181 Equation #181 : $x_{n+1} = \dfrac{\alpha + \gamma x_{n-1} + \delta x_{n-2}}{Bx_n + Dx_{n-2}}$

This equation can be written in the normalized form

$$x_{n+1} = \frac{\alpha + \gamma x_{n-1} + x_{n-2}}{Bx_n + x_{n-2}}, \quad n = 0, 1, \ldots \tag{5.181.1}$$

with positive parameters α, γ, B and with arbitrary positive initial conditions x_{-2}, x_{-1}, x_0.

The boundedness character of solutions of this equation was investigated in [49]. See also Theorem 3.3.1 where we established that the equation has unbounded solutions when

$$\gamma > 1.$$

Conjecture 5.181.1 *Assume that*

$$\gamma > 1.$$

Show that every bounded solution of Eq.(5.181.1) converges to the equilibrium.

Conjecture 5.181.2 *Assume that*

$$\gamma = 1.$$

Show that every bounded solution of Eq.(5.181.1) converges to a (not necessarily prime) period-two solution.

Conjecture 5.181.3 *Assume that*

$$\gamma < 1.$$

Show that for the equilibrium $\bar{x}$ of Eq.(5.181.1),

$$\textbf{\textit{Local Asymptotic Stability}} \Rightarrow \textbf{\textit{Global Attractivity.}}$$

Conjecture 5.181.4 *Assume that*

$$\gamma < 1.$$

Show that Eq.(5.181.1) has bounded solutions that do not converge to the equilibrium point or to a periodic solution.

5.182 Equation #182 : $x_{n+1} = \dfrac{\alpha + \gamma x_{n-1} + \delta x_{n-2}}{Cx_{n-1} + Dx_{n-2}}$

The boundedness character of solutions of this equation was investigated in [69]. See also Theorem 2.1.1 where we established that every solution of the equation is bounded. Eq.(#182) can be written in the normalized form

$$x_{n+1} = \frac{\alpha + \gamma x_{n-1} + x_{n-2}}{Cx_{n-1} + x_{n-2}}, \quad n = 0, 1, \ldots \tag{5.182.1}$$

with positive parameters α, γ, C and with arbitrary positive initial conditions x_{-2}, x_{-1}, x_0.

Open Problem 5.182.1 *Investigate the periodic nature of solutions and the global stability of the equilibrium of Eq.(5.182.1).*

Conjecture 5.182.1 *Show that for the equilibrium $\bar{x}$ of Eq.(5.182.1),*

Local Asymptotic Stability* $\Rightarrow$ *Global Attractivity.

Conjecture 5.182.2 *Show that Eq.(5.182.1) has solutions that do not converge to the equilibrium point or to a periodic solution.*

Conjecture 5.182.3 *It follows from the work in Section 4.2 that Eq.(5.182.1) has a unique prime period-two solution. Show that this period-two solution of Eq.(5.182.1) is locally asymptotically stable.*

5.183 Equation #183 : $x_{n+1} = \dfrac{\beta x_n + \gamma x_{n-1} + \delta x_{n-2}}{A + Bx_n}$

This equation is a special case of a more general equation that is investigated in Section 5.195.

Conjecture 5.183.1 *Assume that*

$$\gamma > \beta + \delta + A.$$

Show that every positive and bounded solution of Eq.(#183) converges to the positive equilibrium.

Conjecture 5.183.2 *Assume that*

$$\gamma < \beta + \delta + A.$$

Show that for the positive equilibrium $\bar{x}$ of Eq.(#183) and with positive initial conditions,

Local Asymptotic Stability $\Rightarrow$ ***Global Attractivity.***

Conjecture 5.183.3 *Assume that*

$$\gamma < \beta + \delta + A.$$

Show that Eq.(#183) has bounded solutions that do not converge to an equilibrium point or to a periodic solution.

5.184 Equation #184 : $x_{n+1} = \dfrac{\beta x_n + \gamma x_{n-1} + \delta x_{n-2}}{A + C x_{n-1}}$

The boundedness character of solutions of this equation was investigated in [49]. See also Theorem 2.7.1 where we established that every solution of the equation is bounded. Eq.(#184) can be written in the normalized form

$$x_{n+1} = \frac{\beta x_n + x_{n-1} + \delta x_{n-2}}{A + x_{n-1}}, \quad n = 0, 1, \ldots \tag{5.184.1}$$

with positive parameters β, δ, A and with arbitrary nonnegative initial conditions x_{-2}, x_{-1}, x_0.

By Theorems 5.23.2 and 5.23.4 it follows that when

$$A \geq \beta + \delta + 1,$$

the zero equilibrium of Eq.(5.184.1) is globally asymptotically stable.

When

$$A < \beta + \delta + 1,$$

Eq.(5.184.1) has the unique positive equilibrium point

$$\bar{x} = \beta + \delta + 1 - A.$$

Conjecture 5.184.1 *Assume that*

$$A < \beta + \delta + 1.$$

Show that Eq.(5.184.1) has solutions that do not converge to an equilibrium point or to a periodic solution.

Conjecture 5.184.2 *Show that for the positive equilibrium $\bar{x}$ of Eq.(5.184.1) and with positive initial conditions,*

Local Asymptotic Stability ⇒ Global Attractivity.

5.185 Equation #185 : $x_{n+1} = \dfrac{\beta x_n + \gamma x_{n-1} + \delta x_{n-2}}{A + Dx_{n-2}}$

Eq.(#185) can be written in the normalized form

$$x_{n+1} = \frac{\beta x_n + \gamma x_{n-1} + x_{n-2}}{A + x_{n-2}}, \quad n = 0, 1, \ldots \tag{5.185.1}$$

with positive parameters β, γ, A and with arbitrary nonnegative initial conditions x_{-2}, x_{-1}, x_0.

The boundedness character of solutions of this equation was investigated in [49]. See also Theorem 3.2.1 where we established that the equation has unbounded solutions when

$$\gamma > \beta + 1 + A.$$

By Theorems 5.23.2 and 5.23.4 it follows that when

$$A \geq \beta + \gamma + 1,$$

the zero equilibrium of Eq.(5.185.1) is globally asymptotically stable.

When

$$A < \beta + \gamma + 1,$$

Eq.(5.185.1) has the unique positive equilibrium point

$$\bar{x} = \beta + \gamma + 1 - A.$$

Conjecture 5.185.1 *Assume that*

$$\gamma > \beta + 1 + A.$$

Show that every bounded and positive solution of Eq.(5.185.1) converges to the positive equilibrium $\bar{x}$.

Conjecture 5.185.2 *Show that for the positive equilibrium $\bar{x}$ of Eq.(5.185.1) and with positive initial conditions,*

$$\textit{Local Asymptotic Stability} \Rightarrow \textit{Global Attractivity.}$$

Conjecture 5.185.3 *Assume that*

$$\gamma = \beta + 1 + A.$$

Show that every solution of Eq.(5.185.1) converges to a (not necessarily prime) period-two solution.

Conjecture 5.185.4 *Assume that*

$$A - \beta - 1 < \gamma < \beta + 1 + A.$$

Show that Eq.(5.185.1) has bounded solutions that do not converge to an equilibrium point or to a periodic solution.

5.186 Equation #186 : $x_{n+1} = \dfrac{\beta x_n + \gamma x_{n-1} + \delta x_{n-2}}{Bx_n + Cx_{n-1}}$

Eq.(#186) can be written in the normalized form

$$x_{n+1} = \frac{x_n + \gamma x_{n-1} + \delta x_{n-2}}{x_n + Cx_{n-1}}, \quad n = 0, 1, \ldots \tag{5.186.1}$$

with positive parameters γ, δ, C and with arbitrary positive initial conditions x_{-2}, x_{-1}, x_0.

The boundedness character of solutions of this equation was investigated in [69]. See also Theorem 3.4.1 where we established that Eq.(5.186.1) has unbounded solutions when

$$\delta > \frac{\gamma}{C} + C.$$

By Theorem 5.221.1 it follows that every solution of Eq.(5.186.1) is bounded when

$$\delta < \frac{\gamma}{C} + C.$$

Eq.(5.186.1) has the unique equilibrium

$$\bar{x} = \frac{1 + \gamma + \delta}{1 + C}.$$

The characteristic equation of the linearized equation of Eq.(5.186.1) about the equilibrium $\bar{x}$ is

$$\lambda^3+\frac{\gamma+\delta-C}{(1+\gamma+\delta)(1+C)}\lambda^2+\frac{C+C\delta-\gamma}{(1+C)(1+\gamma+\delta)}\lambda-\frac{\delta}{1+\gamma+\delta}=0. \quad (5.186.2)$$

From this and Theorem 1.2.3 it follows that the equilibrium $\bar{x}$ of Eq.(5.186.1) is locally asymptotically stable when

$$\delta<\frac{2+2\gamma+C+\gamma C}{2(1+C)}$$

$$\frac{+\sqrt{8+20\gamma+12\gamma^2+C^2+8C+24\gamma C+16\gamma^2C+6\gamma C^2+5\gamma^2C^2}}{2(1+C)} \quad (5.186.3)$$

and unstable when

$$\delta>\frac{2+2\gamma+C+\gamma C}{2(1+C)}$$

$$\frac{+\sqrt{8+20\gamma+12\gamma^2+C^2+8C+24\gamma C+16\gamma^2C+6\gamma C^2+5\gamma^2C^2}}{2(1+C)}. \quad (5.186.4)$$

It is interesting to note that for the equilibrium $\bar{x}$ of Eq.(5.186.1),

Local Asymptotic Stabilty $\not\Rightarrow$ Global Asymptotic Stabilty.

Indeed, for all positive values of γ, C for which

$$\gamma+C<\frac{2+2\gamma+C+\gamma C}{2(1+C)}$$

$$\frac{+\sqrt{8+20\gamma+12\gamma^2+C^2+8C+24\gamma C+16\gamma^2C+6\gamma C^2+5\gamma^2C^2}}{2(1+C)}$$

and for all values of δ such that

$$\gamma+C<\delta<\frac{2+2\gamma+C+\gamma C}{2(1+C)}$$

$$\frac{+\sqrt{8+20\gamma+12\gamma^2+C^2+8C+24\gamma C+16\gamma^2C+6\gamma C^2+5\gamma^2C^2}}{2(1+C)},$$

the equilibrium $\bar{x}$ of Eq.(5.186.1) is locally asymptotically stable and at the same time the equation has unbounded solutions. In particular, for such initial conditions the equilibrium of the equation is not a global attractor.

Conjecture 5.186.1 *Assume that*

$$\delta \leq \frac{\gamma}{C} + C.$$

Show that for the equilibrium $\bar{x}$ *of Eq.(5.186.1),*

$$\textbf{Local Asymptotic Stabilty} \Longrightarrow \textbf{Global Attractivity}.$$

Open Problem 5.186.1 *Assume that*

$$\delta > \frac{\gamma}{C} + C.$$

(i) *Determine the set of all initial conditions for which the solutions of Eq.(5.186.1) converge to the equilibrium* $\bar{x}$.

(ii) *Determine the set of all initial conditions for which the solutions of Eq.(5.186.1) are unbounded.*

(iii) *Determine all possible periodic solutions of Eq.(5.186.1).*

Conjecture 5.186.2 *Show that Eq.(5.186.1) has bounded solutions that do not converge to the equilibrium point* $\bar{x}$ *or to a periodic solution.*

Conjecture 5.186.3 *It follows from the work in Section 4.2 that Eq.(5.186.1) has a unique prime period-two solution. Show that this period-two solution of Eq.(5.186.1) is locally asymptotically stable.*

5.187 Equation #187 : $x_{n+1} = \dfrac{\beta x_n + \gamma x_{n-1} + \delta x_{n-2}}{Bx_n + Dx_{n-2}}$

Eq.(#187) can be written in the normalized form

$$x_{n+1} = \frac{x_n + \gamma x_{n-1} + \delta x_{n-2}}{x_n + Dx_{n-2}}, \quad n = 0, 1, \ldots \tag{5.187.1}$$

with positive parameters γ, δ, D and with arbitrary positive initial conditions x_{-2}, x_{-1}, x_0.

The boundedness character of solutions of this equation was investigated in [49]. See also Theorem 3.3.1 where we established that the equation has unbounded solutions when

$$\gamma > 1 + \delta.$$

Open Problem 5.187.1 *Investigate the periodic nature of solutions and the global stability of the equilibrium of Eq.(5.187.1).*

Conjecture 5.187.1 *Assume that*

$$\gamma > 1 + \delta.$$

Show that every bounded solution of Eq.(5.187.1) converges to the equilibrium.

Conjecture 5.187.2 *Assume that*

$$\gamma < 1 + \delta.$$

Show that for the equilibrium $\bar{x}$ of Eq.(5.187.1),

Local Asymptotic Stability* $\Rightarrow$ *Global Attractivity.

Conjecture 5.187.3 *Assume that*

$$\gamma = 1 + \delta.$$

Show that every solution of Eq.(5.187.1) converges to a (not necessarily prime) period-two solution.

Conjecture 5.187.4 *Assume that*

$$\gamma < 1 + \delta.$$

Show that Eq.(5.187.1) has bounded solutions that do not converge to the equilibrium point or to a periodic solution.

5.188 Equation #188 : $x_{n+1} = \dfrac{\beta x_n + \gamma x_{n-1} + \delta x_{n-2}}{C x_{n-1} + D x_{n-2}}$

The boundedness character of solutions of this equation was investigated in [69]. See also Theorem 2.3.2 where we established that every solution of this equation is bounded. Eq.(#188) can be written in the normalized form

$$x_{n+1} = \frac{\beta x_n + x_{n-1} + \delta x_{n-2}}{x_{n-1} + D x_{n-2}}, \quad n = 0, 1, \ldots \tag{5.188.1}$$

with positive parameters β, δ, D and with arbitrary positive initial conditions x_{-2}, x_{-1}, x_0.

Open Problem 5.188.1 *Investigate the periodic nature of solutions and the global stability of the equilibrium of Eq.(5.188.1).*

Conjecture 5.188.1 *Show that for the equilibrium $\bar{x}$ of Eq.(5.188.1),*

$$\textbf{\textit{Local Asymptotic Stability}} \Rightarrow \textbf{\textit{Global Attractivity.}}$$

Conjecture 5.188.2 *Show that Eq.(5.188.1) has solutions that do not converge to the equilibrium point or to a periodic solution.*

Conjecture 5.188.3 *It follows from the work in Section 4.2 that Eq.(5.188.1) has a unique prime period-two solution. Show that this period-two solution of Eq.(5.188.1) is locally asymptotically stable.*

5.189 Equation #189 : $x_{n+1} = \dfrac{\alpha + \beta x_n}{A + Bx_n + Cx_{n-1} + Dx_{n-2}}$

Eq.(#189) can be written in the normalized form

$$x_{n+1} = \frac{\alpha + x_n}{A + x_n + Cx_{n-1} + Dx_{n-2}}, \quad n = 0, 1, \ldots \tag{5.189.1}$$

with positive parameters α, A, C, D and with arbitrary nonnegative initial conditions x_{-2}, x_{-1}, x_0.

By Theorems 5.23.2 and 5.23.3 it follows that when

$$A \geq 1,$$

the equilibrium of Eq.(5.189.1) is globally asymptotically stable.

Conjecture 5.189.1 *Show that for the equilibrium $\bar{x}$ of Eq.(5.189.1),*

$$\textbf{\textit{Local Asymptotic Stabilty}} \Longrightarrow \textbf{\textit{Global Attractivity.}}$$

Conjecture 5.189.2 *Assume that*

$$A < 1.$$

Show that Eq.(5.189.1) has solutions that do not converge to the equilibrium point or to a periodic solution.

5.190 Equation #190 : $x_{n+1} = \dfrac{\alpha + \gamma x_{n-1}}{A + Bx_n + Cx_{n-1} + Dx_{n-2}}$

Eq.(#190) can be written in the normalized form

$$x_{n+1} = \frac{\alpha + x_{n-1}}{A + Bx_n + x_{n-1} + Dx_{n-2}}, \quad n = 0, 1, \ldots \tag{5.190.1}$$

with positive parameters α, A, B, D and with arbitrary nonnegative initial conditions x_{-2}, x_{-1}, x_0.

By Theorems 5.23.2 and 5.23.3 it follows that when

$$A \geq 1,$$

the equilibrium of Eq.(5.190.1) is globally asymptotically stable.

Conjecture 5.190.1 *Show that for the equilibrium $\bar{x}$ of Eq.(5.190.1),*

Local Asymptotic Stability* $\Rightarrow$ *Global Attractivity.

Conjecture 5.190.2 *It follows from the work in Section 4.2 that Eq.(5.190.1) has a unique prime period-two solution. Show that this period-two solution of Eq.(5.190.1) is locally asymptotically stable.*

Conjecture 5.190.3 *Assume that*

$$A < 1.$$

Show that Eq.(5.190.1) has solutions that do not converge to the equilibrium point or to a periodic solution.

5.191 Equation #191 : $x_{n+1} = \dfrac{\alpha + \delta x_{n-2}}{A + Bx_n + Cx_{n-1} + Dx_{n-2}}$

Eq.(#191) can be written in the normalized form

$$x_{n+1} = \frac{\alpha + x_{n-2}}{A + Bx_n + Cx_{n-1} + x_{n-2}}, \quad n = 0, 1, \ldots \tag{5.191.1}$$

with positive parameters α, A, B, C and with arbitrary nonnegative initial conditions x_{-2}, x_{-1}, x_0.

By Theorems 5.23.2 and 5.23.3 it follows that when

$$A \geq 1,$$

the equilibrium of Eq.(5.191.1) is globally asymptotically stable.

Conjecture 5.191.1 *Show that for the equilibrium $\bar{x}$ of Eq.(5.191.1),*

$$\textbf{\textit{Local Asymptotic Stability}} \Rightarrow \textbf{\textit{Global Attractivity.}}$$

Conjecture 5.191.2 *Assume that*

$$A < 1.$$

Show that Eq.(5.191.1) has solutions that do not converge to the equilibrium point or to a periodic solution.

5.192 Equation #192 : $x_{n+1} = \dfrac{\beta x_n + \gamma x_{n-1}}{A + Bx_n + Cx_{n-1} + Dx_{n-2}}$

Eq.(#192) can be written in the normalized form

$$x_{n+1} = \frac{x_n + \gamma x_{n-1}}{A + x_n + Cx_{n-1} + Dx_{n-2}}, \quad n = 0, 1, \ldots \tag{5.192.1}$$

with positive parameters γ, A, C, D and with arbitrary nonnegative initial conditions x_{-2}, x_{-1}, x_0.

By Theorems 5.23.2 and 5.23.4 it follows that when

$$A \geq \gamma + 1,$$

the zero equilibrium of Eq.(5.192.1) is globally asymptotically stable.

When

$$A < \gamma + 1,$$

Eq.(5.192.1) has the unique positive equilibrium point

$$\bar{x} = \frac{1 + \gamma - A}{1 + C + D}.$$

Conjecture 5.192.1 *Show that for the positive equilibrium $\bar{x}$ of Eq.(5.192.1) and with positive initial conditions,*

$$\textbf{\textit{Local Asymptotic Stability}} \Rightarrow \textbf{\textit{Global Attractivity.}}$$

Conjecture 5.192.2 *Assume that*

$$A < \gamma + 1.$$

Show that Eq.(5.192.1) has solutions that do not converge to an equilibrium point or to a periodic solution.

Conjecture 5.192.3 *It follows from the work in Section 4.2 that Eq.(5.192.1) has a unique prime period-two solution. Show that this period-two solution of Eq.(5.192.1) is locally asymptotically stable.*

5.193 Equation #193 : $x_{n+1} = \dfrac{\beta x_n + \delta x_{n-2}}{A + Bx_n + Cx_{n-1} + Dx_{n-2}}$

Eq.(#193) can be written in the normalized form

$$x_{n+1} = \frac{x_n + \delta x_{n-2}}{A + x_n + Cx_{n-1} + Dx_{n-2}}, \quad n = 0, 1, \ldots \tag{5.193.1}$$

with positive parameters δ, A, C, D and with arbitrary nonnegative initial conditions x_{-2}, x_{-1}, x_0.

By Theorems 5.23.2 and 5.23.4 it follows that when

$$A \geq \delta + 1,$$

the zero equilibrium of Eq.(5.193.1) is globally asymptotically stable.

When

$$A < \delta + 1,$$

Eq.(5.193.1) has the unique positive equilibrium point

$$\bar{x} = \frac{1 + \delta - A}{1 + C + D}.$$

Conjecture 5.193.1 *Show that for the positive equilibrium $\bar{x}$ of Eq.(5.193.1) and with positive initial conditions,*

$$\textit{Local Asymptotic Stability} \Rightarrow \textit{Global Attractivity.}$$

Conjecture 5.193.2 *Assume that*

$$A < \delta + 1.$$

Show that Eq.(5.193.1) has solutions that do not converge to an equilibrium point or to a periodic solution.

5.194 Equation #194 : $x_{n+1} = \frac{\gamma x_{n-1} + \delta x_{n-2}}{A + Bx_n + Cx_{n-1} + Dx_{n-2}}$

Eq.(#194) can be written in the normalized form

$$x_{n+1} = \frac{x_{n-1} + \delta x_{n-2}}{A + Bx_n + x_{n-1} + Dx_{n-2}}, \quad n = 0, 1, \ldots \tag{5.194.1}$$

with positive parameters δ, A, B, D and with arbitrary nonnegative initial conditions x_{-2}, x_{-1}, x_0.

By Theorems 5.23.2 and 5.23.4 it follows that when

$$A \geq \delta + 1,$$

the zero equilibrium of Eq.(5.194.1) is globally asymptotically stable.

When

$$A < \delta + 1,$$

Eq.(5.194.1) has the unique positive equilibrium point

$$\bar{x} = \frac{1 + \delta - A}{1 + B + D}.$$

Conjecture 5.194.1 *Show that for the positive equilibrium $\bar{x}$ of Eq.(5.194.1) and with positive initial conditions,*

Local Asymptotic Stability ⇒ Global Attractivity.

Conjecture 5.194.2 *Assume that*

$$A < \delta + 1.$$

Show that Eq.(5.194.1) has solutions that do not converge to an equilibrium point or to a periodic solution.

Conjecture 5.194.3 *It follows from the work in Section 4.2 that Eq.(5.194.1) has a unique prime period-two solution. Show that this period-two solution of Eq.(5.194.1) is locally asymptotically stable.*

5.195 Equation #195 : $x_{n+1} = \frac{\alpha + \beta x_n + \gamma x_{n-1} + \delta x_{n-2}}{A + Bx_n}$

In this section we allow the parameters α, β, γ, δ, A to be nonnegative and so the results we present here are also true in several special cases of Eq.(5.0.1).

Eq.(#195) can be written in the normalized form

$$x_{n+1} = \frac{\alpha + \beta x_n + \gamma x_{n-1} + \delta x_{n-2}}{A + x_n}, \quad n = 0, 1, \ldots \tag{5.195.1}$$

with nonnegative parameters $\alpha, \beta, \gamma, \delta, A$ and with arbitrary nonnegative initial conditions x_{-2}, x_{-1}, x_0.

The boundedness character of this equation was investigated in [67]. See also Theorem 3.1.1 where we established that the equation has unbounded solutions when

$$\gamma > \beta + \delta + A.$$

Eq.(5.195.1) has one or two equilibrium points, namely, the zero equilibrium and/or the positive equilibrium

$$\bar{x} = \frac{\beta + \gamma + \delta - A + \sqrt{(\beta + \gamma + \delta - A)^2 + 4\alpha}}{2}.$$

The characteristic equation of the linearized equation of Eq.(5.195.1) about the zero equilibrium, as long as it exists, is

$$\lambda^3 - \frac{\beta}{A}\lambda^2 - \frac{\gamma}{A}\lambda - \frac{\delta}{A} = 0. \tag{5.195.2}$$

The characteristic equation of the linearized equation of Eq.(5.195.1) about the positive equilibrium $\bar{x}$, as long as it exists, is

$$\lambda^3 + \frac{\bar{x} - \beta}{A + \bar{x}}\lambda^2 - \frac{\gamma}{A + \bar{x}}\lambda - \frac{\delta}{A + \bar{x}} = 0. \tag{5.195.3}$$

By Theorems 5.23.2 and 5.23.4 it follows that the zero equilibrium of Eq.(5.195.1) is globally asymptotically stable when

$$\alpha = 0 \text{ and } \beta + \gamma + \delta \leq A.$$

When

$$\alpha = 0 \text{ and } \beta + \gamma + \delta > A,$$

Eq.(5.195.1) has the unique positive equilibrium

$$\bar{x} = \beta + \gamma + \delta - A.$$

The following theorem about the global stability of the positive equilibrium of Eq.(5.195.1) was established in [68].

Theorem 5.195.1 *Assume that*

$$\delta = A \text{ and } 0 \leq \gamma \leq \beta.$$

Then the positive equilibrium

$$\bar{x} = \beta + \gamma$$

of the equation

$$x_{n+1} = \frac{\beta x_n + \gamma x_{n-1} + A x_{n-2}}{A + x_n}, \quad n = 0, 1, \ldots \tag{5.195.4}$$

is globally asymptotically stable.

PROOF From Eq.(5.195.3) and Theorem 1.2.3 the local stability of the positive equilibrium of Eq.(5.195.4) follows. It remains to show that $\bar{x}$ is a global attractor of all positive solutions of Eq.(5.195.4).

When

$$\gamma = 0,$$

the proof follows from Theorem 5.89.1.

Assume that

$$0 < \gamma \leq \beta$$

and let $\{x_n\}$ be a solution of Eq.(5.195.4). We claim that, eventually,

$$x_n > \beta.$$

Suppose for the sake of contradiction that there exists N, sufficiently large, such that

$$x_{N+1} = \frac{\beta x_N + \gamma x_{N-1} + A x_{N-2}}{A + x_N} \leq \beta.$$

Then, clearly,

$$\min\{x_{N-1}, x_{N-2}\} \leq \frac{\beta A}{\gamma + A}.$$

Similarly,

$$\min\{x_{N-3}, x_{N-4}, x_{N-5}\} \leq \frac{\beta A^2}{(\gamma + A)^2}.$$

Sufficient repetition of this argument leads to a contradiction and proves our claim that, eventually,

$$x_n > \beta.$$

From this it follows that for some $N > 0$, sufficiently large, there exists a positive number

$$L \in (\beta, \beta + \gamma)$$

such that

$$L < x_{N-2}, x_{N-1}, x_N < U = \frac{\gamma L}{L - \beta}.$$

We claim that

$$x_n \in [L, U], \text{ for } n \geq N - 2.$$

Indeed,

$$L = \frac{\beta U + (\gamma + A)L}{A + U} < x_{N+1} = \frac{\beta x_N + \gamma x_{N-1} + x_{N-2}}{1 + x_N} < \frac{\beta L + (\gamma + A)U}{A + L} = U$$

and the proof follows by induction. Set

$$S = \limsup_{n\to\infty} x_{n+1} \text{ and } I = \liminf_{n\to\infty} x_{n+1}.$$

Then, clearly,

$$S \leq \frac{\beta I + (\gamma + A)S}{A + I} \text{ and } I \leq \frac{\beta S + (\gamma + A)I}{A + S}.$$

From this it follows that

$$\beta S + \gamma I \leq SI \leq \beta I + \gamma S. \tag{5.195.5}$$

We divide the proof into the following two cases:

Case 1:

$$0 < \gamma < \beta.$$

By Eq.(5.195.5) we find that

$$(\gamma - \beta)(S - I) \leq 0.$$

Hence, $S = I$.

Case 2:

$$0 < \gamma = \beta.$$

By (5.195.5) it follows that

$$S = \frac{\beta I}{I - \beta}.$$

There exists a sequence of indices $\{n_i\}$ and positive numbers $\{L_{-t}\}_{t=0}^{2}$ such that

$$S = \lim_{i\to\infty} x_{n_i+1} \text{ and } L_{-t} = \lim_{i\to\infty} x_{n_i-t}.$$

By taking limits in Eq.(5.195.4) we find that

$$S = \frac{\beta L_0 + \gamma L_{-1} + AL_{-2}}{A + L_0}.$$

From this it follows that

$$L_0 = I \text{ and } L_{-1} = L_{-2} = S,$$

otherwise,

$$S < \frac{\beta I}{I - \beta},$$

which is a contradiction. Similarly,

$$L_{-1} = S \text{ and } L_{-2} = L_{-3} = I.$$

Hence,

$$I = L_{-1} = S.$$

The proof is complete. ∎

Conjecture 5.195.1 *Show that the special case of Eq.(5.195.1) given by*

$$x_{n+1} = \frac{\beta x_n + \gamma x_{n-1} + x_{n-2}}{1 + x_n}, \quad n = 0, 1, \ldots \tag{5.195.6}$$

with $\beta, \gamma \in (0, \infty)$ has a period-two trichotomy character. More precisely, show that the following three statements are true:

(*a*) *Every solution of Eq.(5.195.6) has a finite limit if and only if*

$$\gamma < \beta + 2.$$

(*b*) *Every solution of Eq.(5.195.6) converges to a (not necessarily prime) period-two solution if and only if*

$$\gamma = \beta + 2.$$

(*c*) *Eq.(5.195.6) has unbounded solutions if and only if*

$$\gamma > \beta + 2.$$

By Theorems 5.23.2 and 5.23.3 it follows that when

$$\alpha > 0 \text{ and } \beta + \gamma + \delta \leq A,$$

the positive equilibrium of Eq.(5.195.1) is globally asymptotically stable.

Conjecture 5.195.2 *Show that for the positive equilibrium $\bar{x}$ of Eq.(5.195.1) and with positive initial conditions,*

Local Asymptotic Stability ⇒ Global Attractivity.

Conjecture 5.195.3 *Assume that*

$$\alpha > 0 \text{ and } A - \beta - \delta < \gamma < \beta + \delta + A.$$

Show that Eq.(5.195.1) has bounded solutions that do not converge to the equilibrium point or to a periodic solution.

For the remainder of this section we assume that

$$\gamma = \beta + \delta + A$$

and

$$\beta + A > 0. \tag{5.195.7}$$

In [67] it was established that every solution of Eq.(5.195.1) converges to a (not necessarily prime) period-two solution. The restriction (5.195.7) cannot be relaxed in this period-two convergence result. In fact, when $\beta = A = 0$ the resulting equation

$$x_{n+1} = \frac{\alpha + \gamma x_{n-1} + \delta x_{n-2}}{x_n}, \quad n = 0, 1, \ldots \tag{5.195.8}$$

with

$$\alpha \geq 0 \text{ and } \delta > 0$$

has unbounded solutions. See [46] and [76]. In particular, every solution of Eq.(5.195.8) with

$$x_{-2} = x_0 \leq \delta$$

is such that

$$x_{2n} = x_0, \text{ for } n \geq 0$$

and

$$x_{2n+1} = \frac{\delta}{x_0} x_{2n-1} + \left(\delta + \frac{\alpha}{x_0}\right) \to \infty, \text{ as } n \to \infty.$$

Theorem 5.195.2 *Assume that*

$$\gamma = \beta + \delta + A \text{ and } \beta + A > 0.$$

Then every solution of Eq.(5.195.1) converges to a (not necessarily prime) period-two solution.

The proof of Theorem 5.195.2 is very long and in order to simplify it we first establish several lemmas describing the character of solutions of Eq.(5.195.1). We begin by stating several identities that follow from Eq.(5.195.1) and will be used throughout this section. They are all valid for $n \geq 0$.

$$x_{2n+1} = \frac{\alpha + \beta x_{2n} + (\beta + \delta + A)\, x_{2n-1} + \delta x_{2n-2}}{A + x_{2n}} \tag{5.195.9}$$

$$= \frac{\alpha}{A + x_{2n}} + \beta \frac{x_{2n}}{A + x_{2n}} + (\beta + \delta + A) \frac{x_{2n-1}}{A + x_{2n}} + \delta \frac{x_{2n-2}}{A + x_{2n}}.$$

$$x_{2n+2} = \frac{\alpha + \beta x_{2n+1} + (\beta + \delta + A)\, x_{2n} + \delta x_{2n-1}}{A + x_{2n+1}}$$

$$= \frac{\alpha}{A + x_{2n+1}} + \beta \frac{x_{2n+1}}{A + x_{2n+1}} + (\beta + \delta + A) \frac{x_{2n}}{A + x_{2n+1}} + \delta \frac{x_{2n-1}}{A + x_{2n+1}}.$$

$$x_{n+2} - x_n = \frac{\beta + A}{A + x_{n+1}} (x_{n+1} - x_{n-1}) + \frac{\delta}{A + x_{n+1}} (x_n - x_{n-2}). \quad (5.195.10)$$

$$x_{2n+2} - x_{2n} = \frac{\beta + A}{A + x_{2n+1}} (x_{2n+1} - x_{2n-1}) + \frac{\delta}{A + x_{2n+1}} (x_{2n} - x_{2n-2}). \quad (5.195.11)$$

$$x_{2n+3} - x_{2n+1} = \frac{\beta + A}{A + x_{2n+2}} (x_{2n+2} - x_{2n}) + \frac{\delta}{A + x_{2n+2}} (x_{2n+1} - x_{2n-1}). \quad (5.195.12)$$

$$x_{2n+1} - x_{2n-1} = \frac{\alpha + \beta x_{2n} + (\beta + \delta - x_{2n}) x_{2n-1} + \delta x_{2n-2}}{A + x_{2n}}. \quad (5.195.13)$$

$$x_{2n+2} - x_{2n} = \frac{\alpha + \beta x_{2n+1} + (\beta + \delta - x_{2n+1}) x_{2n} + \delta x_{2n-1}}{A + x_{2n+1}}. \quad (5.195.14)$$

$$\frac{x_{2n+1}}{x_{2n-1}} = \frac{\alpha}{A + x_{2n}} \cdot \frac{1}{x_{2n-1}} + \beta \frac{x_{2n}}{A + x_{2n}} \cdot \frac{1}{x_{2n-1}} + \frac{\beta + \delta + A}{A + x_{2n}} + \delta \frac{x_{2n-2}}{A + x_{2n}} \cdot \frac{1}{x_{2n-1}}. \quad (5.195.15)$$

$$\frac{x_{2n+2}}{x_{2n}} = \frac{\alpha}{A + x_{2n+1}} \cdot \frac{1}{x_{2n}} + \beta \frac{x_{2n+1}}{A + x_{2n+1}} \cdot \frac{1}{x_{2n}} + \frac{\beta + \delta + A}{A + x_{2n+1}} + \delta \frac{x_{2n-1}}{A + x_{2n+1}} \cdot \frac{1}{x_{2n}}. \quad (5.195.16)$$

Among the above equations, the identity described by (5.195.10) is at the heart of the period-two convergence of solutions of Eq.(5.195.1). Its proof is a consequence of Eq.(5.195.1) as follows. Note that

$$x_{n+1} x_n = \alpha + \beta x_n + (\beta + \delta + A) x_{n-1} + \delta x_{n-2} - A x_{n+1}$$

and so,

$$x_{n+2} - x_n = \frac{\alpha + \beta x_{n+1} + (\beta + \delta + A) x_n + \delta x_{n-1}}{A + x_{n+1}} - x_n$$

$$= \frac{\alpha + \beta x_{n+1} + (\beta + \delta) x_n + \delta x_{n-1} - \alpha - \beta x_n - (\beta + \delta + A) x_{n-1} - \delta x_{n-2} + A x_{n+1}}{A + x_{n+1}}$$

$$= \frac{(\beta + A) x_{n+1} - (\beta + A) x_{n-1} + \delta (x_n - x_{n-2})}{A + x_{n+1}}$$

$$= \frac{\beta + A}{A + x_{n+1}} (x_{n+1} - x_{n-1}) + \frac{\delta}{A + x_{n+1}} (x_n - x_{n-2}).$$

From (5.195.10), and more precisely from its equivalent versions (5.195.11) and (5.195.12), it is now clear that the following result is true about the subsequences of the even terms $\{x_{2n}\}_{n=-1}^{\infty}$ and the odd terms $\{x_{2n+1}\}_{n=-1}^{\infty}$ of every solution $\{x_n\}_{n=-2}^{\infty}$ of Eq.(5.195.1).

Lemma 5.195.1 *The two subsequences $\{x_{2n}\}$ and $\{x_{2n+1}\}$ of every solution of Eq.(5.195.1) are either both eventually monotonically increasing, or they are both eventually monotonically decreasing, or one of them is monotonically increasing and the other is monotonically decreasing.*

In the sequel we will denote the limits of the subsequences of the even and odd terms of a solution of Eq.(5.195.1) by L_0 and L_1, respectively. That is,

$$L_0 = \lim_{n\to\infty} x_{2n} \quad \text{and} \quad L_1 = \lim_{n\to\infty} x_{2n+1}.$$

Each of these limits may only have one of the following three values:

$$0, \ \infty, \ \text{or a positive real number.}$$

Now let us look for all period-two solutions

$$\ldots, \phi, \ \psi, \ \ldots$$

of Eq.(5.195.1). From Eq.(5.195.1) we have

$$\phi = \frac{\alpha + \beta\psi + (\beta + \delta + A)\,\phi + \delta\psi}{A + \psi}$$

and so

$$\phi\psi = \alpha + (\beta + \delta)\,(\phi + \psi)\,,$$

which implies that

$$\phi\,[\psi - (\beta + \delta)] = \alpha + (\beta + \delta)\,\psi$$

and

$$\psi\,[\phi - (\beta + \delta)] = \alpha + (\beta + \delta)\,\phi.$$

When $\phi \neq \psi$ we have a prime period-two solution, while when $\phi = \psi$ we see that ϕ is the equilibrium $\overline{x}$ of Eq.(5.195.1). Note also that in all cases

$$\overline{x}, \ \phi, \ \psi, \ \in (\beta + \delta, \infty)\,,$$

provided that $\alpha+\beta+\delta > 0$. For the sake of completeness and unification, our proof here of the period-two convergence of Eq.(5.195.1) includes all previous special cases of Eq.(5.195.1). When

$$\alpha = \beta = \delta = 0, \tag{5.195.17}$$

Eq.(5.195.1) reduces to

$$x_{n+1} = \frac{Ax_{n-1}}{A + x_n}, \quad \text{for } n = 0, 1, \ldots \tag{5.195.18}$$

with $A > 0$, from which it is clear that

$$x_{n+1} \leq x_{n-1}.$$

Therefore, in this case, every solution of Eq.(5.195.18) converges to a (not necessarily prime) period-two solution of the form

$$\ldots, \phi, 0, \ldots \tag{5.195.19}$$

with $\phi \geq 0$. This completes the proof of the Theorem when (5.195.17) holds.

When $\delta = 0$, that is, for the equation

$$x_{n+1} = \frac{\alpha + \beta x_n + (\beta + A)\, x_{n-1}}{A + x_n}, \quad n = 0, 1, \ldots \tag{5.195.20}$$

it follows from (5.195.10) that

$$x_{n+2} - x_n = \frac{\beta + A}{A + x_{n+1}} \left(x_{n+1} - x_{n-1}\right), \quad \text{for } n \geq 0. \tag{5.195.21}$$

From (5.195.21) we see that for every solution of Eq.(5.195.20) exactly one of the following statements is true for all $n \geq 0$:

$$x_{n+1} < x_{n-1}$$

$$x_{n+1} = x_{n-1}$$

$$x_{n+1} > x_{n-1}.$$

Clearly, all bounded solutions of Eq.(5.195.20) converge to a period-two solution. As in [175, p. 40], assume for the sake of contradiction that there is an unbounded solution, that is, a solution such that:

$$\lim_{n \to \infty} x_{2n} = \infty$$

while $\{x_{2n+1}\}$ is increasing. The case where

$$\lim_{n \to \infty} x_{2n+1} = \infty$$

and $\{x_{2n}\}$ is increasing is similar and will be omitted. Choose $N \geq 0$ such that

$$\frac{\beta + A}{A + x_{2N+1}} \cdot \frac{\beta + A}{A + x_{2N}} < 1.$$

Define

$$\rho = \frac{\beta + A}{A + x_{2N+1}} \cdot \frac{\beta + A}{A + x_{2N}} \quad \text{and} \quad \sigma = \left(x_{2N} - x_{2N-2}\right).$$

Then

$$\frac{\beta + A}{A + x_{2n+1}} \cdot \frac{\beta + A}{A + x_{2n}} < \rho, \quad \text{for } n > N.$$

Hence, from (5.195.21) we find that

$$x_{2N+2} - x_{2N} = \frac{\beta + A}{A + x_{2N+1}} (x_{2N+1} - x_{2N-1})$$

$$= \frac{\beta + A}{A + x_{2N+1}} \cdot \frac{\beta + A}{A + x_{2N}} (x_{2N} - x_{2N-2}) = \sigma \rho$$

and

$$x_{2N+4} - x_{2N+2} = \frac{\beta + A}{A + x_{2N+3}} (x_{2N+3} - x_{2N+1})$$

$$= \frac{\beta + A}{A + x_{2N+3}} \cdot \frac{\beta + A}{A + x_{2N+2}} (x_{2N+2} - x_{2N}) < \sigma \rho^2.$$

It follows by induction that for $\mu = 1, 2, \ldots$

$$x_{2N+2(\mu+1)} - x_{2N+2\mu} < \sigma \rho^{\mu+1}$$

and so by summing up

$$x_{2N+2(\mu+1)} < x_{2N+2} + \frac{\sigma \rho^2}{1 - \rho}, \quad \text{for } \mu = 1, 2, \ldots.$$

This contradicts the assumption that the subsequence of the even terms converges to ∞, and completes the proof of the Theorem when $\delta = 0$. Therefore, in the sequel we will assume that

$$\delta > 0.$$

Returning to the period-two solutions of Eq.(5.195.1) the following result is now clear.

Lemma 5.195.2 *All prime period-two solutions*

$$\ldots, \phi, \psi, \ldots$$

of Eq.(5.195.1) are given by

$$\phi = \frac{(\beta + \delta)\,\psi + \alpha}{\psi - (\beta + \delta)}$$

with

$$\phi \neq \psi \quad \textit{and} \quad \phi, \psi \in (\beta + \delta, \infty)\,.$$

Clearly, when both L_0 and L_1 are positive numbers, the sequence

$$\ldots, L_0, L_1, \ldots$$

is a period-two solution of Eq.(5.195.1) and as we showed before

$$L_0 L_1 = \alpha + (\beta + \delta)\,(L_1 + L_2)\,.$$

In particular,

$$L_0, L_1 \in (\beta + \delta, \infty).$$

When $L_0 = L_1$, then the solution converges to the equilibrium $\overline{x}$ of Eq.(5.195.1) and when $L_0 \neq L_1$ the solution converges to a prime period-two solution of Eq.(5.195.1). Note that Eq.(5.195.1) has a huge set of prime period-two solutions, as described by Lemma 5.195.2.

The following Lemma shows that neither L_0 nor L_1 can be zero.

Lemma 5.195.3 *Neither of the subsequences* $\{x_{2n}\}$ *and* $\{x_{2n+1}\}$ *may converge to zero.*

PROOF Assume for the sake of contradiction that

$$L_0 = 0.$$

The case where $L_1 = 0$ is similar and will be omitted. Now note that when $L_0 = 0$, A must be positive. Otherwise, $A = 0, \beta > 0$, and so clearly $x_{n+2} \geq \beta > 0$.

There are three possibilities for L_1: L_1 may be zero, ∞, or a positive number. We will show that each of them leads to a contradiction.

If $L_1 = 0$, then both subsequences are eventually decreasing to zero and (5.195.13), with n sufficiently large, implies that

$$0 \geq \frac{\alpha + \beta x_{2n} + (\beta + \delta - x_{2n})\, x_{2n-1} + \delta x_{2n-2}}{A + x_{2n}} > 0,$$

which is impossible.

If $L_1 \in (0, \infty)$, by taking limits in (5.195.9) we see that

$$L_1 = \frac{\alpha + (\beta + \delta + A)\, L_1}{A} > L_1,$$

which is also impossible.

Finally, if $L_1 = \infty$, by taking limits in (5.195.15) we find

$$\lim_{n \to \infty} \left(\frac{x_{2n+1}}{x_{2n-1}} \right) = \frac{\beta + \delta + A}{A}$$

and from (5.195.16) we obtain

$$1 \geq \left(\frac{x_{2n+2}}{x_{2n}} \right) \geq \delta \frac{1}{\frac{A}{x_{2n-1}} + \left(\frac{x_{2n+1}}{x_{2n-1}} \right)} \cdot \frac{1}{x_{2n}}$$

which leads to a contradiction, as $n \to \infty$. The proof is complete. ∎

It follows from Lemma 5.195.3 that L_0 and L_1 are positive numbers or ∞. The next results establish that neither L_0 nor L_1 may be below

$$\beta + \delta.$$

Lemma 5.195.4

$$L_0,\ L_1 \in [\beta+\delta, \infty].$$

PROOF Assume for the sake of contradiction that

$$L_0 < \beta + \delta.$$

The proof when $L_1 \leq \beta + \delta$ is similar and will be omitted. There are two possibilities for L_1: L_1 may be positive or $L_1 = \infty$. We will show that each of them leads to a contradiction.

If $L_1 \in (0, \infty)$ then

$$\ldots, L_0, L_1, \ldots$$

is a period-two solution of Eq.(5.195.1) and so, by Lemma 5.195.2, $L_0 > \beta+\delta$, which is a contradiction. On the other hand, if $L_1 = \infty$, then from (5.195.15)

$$\lim_{n\to\infty}\left(\frac{x_{2n+1}}{x_{2n-1}}\right) = \frac{\beta+\delta+A}{A+L_0}$$

and so from (5.195.16),

$$1 = \lim_{n\to\infty}\left(\frac{x_{2n+2}}{x_{2n}}\right) = \frac{\beta}{L_0} + \delta\frac{1}{0+\frac{\beta+\delta+A}{A+L_0}}\cdot\frac{1}{L_0},$$

that is,

$$L_0 = \beta + \delta,$$

which is also a contradiction. ∎

Lemma 5.195.5 *(i) If $L_0 \in (0,\infty)$ and $L_1 = \infty$, then $L_0 = \beta + \delta$.*

(ii) If $L_0 = \infty$ and $L_1 \in (0,\infty)$, then $L_1 = \beta + \delta$.

PROOF We will prove (i). The proof of (ii) is similar and will be omitted. From (5.195.15) we obtain

$$\lim_{n\to\infty}\left(\frac{x_{2n+1}}{x_{2n-1}}\right) = \frac{\beta+\delta+A}{A+L_0}$$

and so from (5.195.16),

$$1 = \frac{\beta}{L_0} + \frac{\delta}{L_0}\cdot\frac{A+L_0}{\beta+\delta+A}.$$

Hence,

$$L_0 = \beta + \delta$$

and the proof is complete. ∎

Lemma 5.195.6 *It is not possible that both L_0 and L_1 are equal to ∞.*

PROOF Otherwise, from (5.195.15)

$$1 \leq \lim_{n\to\infty}\left(\frac{x_{2n+1}}{x_{2n-1}}\right) = 0,$$

which is impossible. ∎

Lemma 5.195.7 *Every solution of Eq.(5.195.1) is eventually bounded from below by $(\beta+\delta)$.*

PROOF Assume for the sake of contradiction that there exists a solution of Eq.(5.195.1) not bounded from below by $(\beta+\delta)$. Then in view of the previous lemmas the only thing that the solution can do is that one of the subsequences $\{x_{2n}\}$ and $\{x_{2n+1}\}$ eventually increases to $(\beta+\delta)$ while the other eventually increases to ∞. We will assume that

$$\lim_{n\to\infty} x_{2n} = \beta+\delta \text{ and } \lim_{n\to\infty} x_{2n+1} = \infty$$

with both subsequences being eventually increasing. The case where the even and odd subsequences are interchanged is similar and will be omitted.

Let $\epsilon \in (0, \beta+\delta)$ be given and sufficiently small, and let $N \geq 0$ be such that

$$\beta+\delta-\epsilon < x_{2n} < \beta+\delta, \text{ for } n \geq N.$$

Then for any $N_0 \geq N$ and sufficiently large we have

$$x_{2N_0+2} < \beta+\delta,$$

which implies that

$$x_{2N_0+2} = \frac{\alpha+\beta x_{2N_0+1}+(\beta+\delta+A)\,x_{2N_0}+\delta x_{2N_0-1}}{A+x_{2N_0+1}} < \beta+\delta. \quad (5.195.22)$$

Define

$$R_0 = \frac{(\beta+\delta+A)\,(\beta+\delta-\epsilon)+\alpha-(\beta+\delta)\,A}{\delta}.$$

Then (5.195.22) implies that

$$\alpha+\beta x_{2N_0+1}+(\beta+\delta+A)\,x_{2N_0}+\delta x_{2N_0-1} < (\beta+\delta)\,A+(\beta+\delta)\,x_{2N_0+1}$$

and so

$$x_{2N_0+1} > x_{2N_0-1}+\frac{\beta+\delta+A}{\delta}x_{2N_0}+\frac{\alpha-(\beta+\delta)\,A}{\delta}$$

$$> x_{2N_0-1}+\frac{(\beta+\delta+A)}{\delta}\,(\beta+\delta-\epsilon)+\frac{\alpha-(\beta+\delta)\,A}{\delta}.$$

Hence,

$$x_{2N_0+1} > x_{2N_0-1} + R_0$$

and by using Eq.(5.195.1),

$$\frac{\alpha + \beta x_{2N_0} + (\beta + \delta + A)\, x_{2N_0-1} + \delta x_{2N_0-2}}{A + x_{2N_0}} > x_{2N_0-1} + R_0.$$

Therefore,

$$\alpha + \beta x_{2N_0} + (\beta + \delta + A)\, x_{2N_0-1} + \delta x_{2N_0-2} > A x_{2N_0-1} + x_{2N_0} x_{2N_0-1} + R_0 \left(A + x_{2N_0}\right)$$

and so

$$\left(\beta + \delta - x_{2N_0}\right) x_{2N_0-1} > R_0 \left(A + x_{2N_0}\right) - \alpha - \beta x_{2N_0} - \delta x_{2N_0-2},$$

that is,

$$x_{2N_0-1} > \frac{R_0 \left(A + x_{2N_0}\right) - \alpha - \beta x_{2N_0} - \delta x_{2N_0-2}}{\beta + \delta - x_{2N_0}}.$$

Hence,

$$(\beta + \delta)\, x_{2N_0-1} - x_{2N_0} x_{2N_0-1} > R_0 \left(A + x_{2N_0}\right) - \alpha - \beta x_{2N_0} - \delta x_{2N_0-2}$$

or, equivalently,

$$\begin{aligned}
(\beta + \delta)\, x_{2N_0-1} - \alpha - \beta x_{2N_0-1} - (\beta + \delta + A)\, x_{2N_0-2} - \delta x_{2N_0-3} + A x_{2N_0} \\
> \quad R_0 \left(A + x_{2N_0}\right) - \alpha - \beta x_{2N_0} - \delta x_{2N_0-2}.
\end{aligned}$$

Thus,

$$\begin{aligned}
x_{2N_0-1} &> x_{2N_0-3} + \frac{R_0}{\delta} \left(A + x_{2N_0}\right) + \frac{1}{\delta} (\beta + A) \left(x_{2N_0-2} - x_{2N_0}\right) \\
&> \quad x_{2N_0-3} + \frac{R_0}{\delta} (A + \beta + \delta - \epsilon) - \frac{1}{\delta} (\beta + A) (\beta + \delta)
\end{aligned}$$

and so from Eq.(5.195.1) we see that

$$\begin{aligned}
&\frac{\alpha + \beta x_{2N_0-2} + (\beta + \delta + A)\, x_{2N_0-3} + \delta x_{2N_0-4}}{A + x_{2N_0-2}} \\
&> \quad x_{2N_0-3} + \frac{R_0}{\delta} (A + \beta + \delta - \epsilon) - \frac{(\beta + A) (\beta + \delta)}{\delta}.
\end{aligned}$$

Therefore,

$$x_{2N_0-3} > \frac{R_1 \left(A + x_{2N_0-2}\right) - \alpha - \beta x_{2N_0-2} - \delta x_{2N_0-4}}{\beta + \delta - x_{2N_0-2}},$$

where

$$R_1 = \frac{\beta + \delta + A - \epsilon}{\delta} R_0 - \frac{(\beta + A)(\beta + \delta)}{\delta}.$$

It follows by induction that for $k \geq 0$,

$$x_{2N_0-(2k+1)} > \frac{R_k\left(A + x_{2N_0-2k}\right) - \alpha - \beta x_{2N_0-2k} - \delta x_{2N_0-2(k+1)}}{\beta + \delta - x_{2N_0-2k}}$$

with

$$R_{k+1} = \frac{\beta + \delta + A - \epsilon}{\delta} R_k - \frac{(\beta + A)(\beta + \delta)}{\delta}.$$

Clearly, for N_0 and k sufficiently large, this leads to a contradiction and the proof of the lemma is complete. ∎

We are now ready to present the proof of Theorem 5.195.2.

PROOF

Clearly, every bounded solution of Eq.((5.195.1) converges to a (not necessarily prime) period-two solution. So assume for the sake of contradiction that Eq.(5.195.1) has an unbounded solution. We will assume that

$$\lim_{n\to\infty} x_{2n+1} = \infty \text{ and } \lim_{n\to\infty} x_{2n} = \beta + \delta,$$

with the subsequence of even terms of the solution being eventually decreasing and the subsequence of odd terms being eventually increasing. The case where the behavior of the even and odd subsequences is reversed is similar and will be omitted.

Then from (5.195.10) we obtain

$$x_{2n+3} - x_{2n+1} < \frac{\delta}{A + x_{2n+2}}\left(x_{2n+1} - x_{2n-1}\right)$$

$$< \frac{\delta}{\beta + \delta + A}\left(x_{2n+1} - x_{2n-1}\right), \quad \text{for } n \geq 0.$$

Therefore,

$$x_{2n+1} - x_{2n-1} < \frac{\delta}{(\beta + \delta + A)^n}\left(x_1 - x_{-1}\right)$$

and by summing up we find

$$x_{2n+1} - x_1 < \frac{\delta}{\beta + A}.$$

This contradicts the hypothesis that

$$\lim_{n\to\infty} x_{2n+1} = \infty$$

and completes the proof of the theorem. ∎

Conjecture 5.195.4 *Assume that*

$$\gamma > \beta + \delta + A.$$

Show that every positive and bounded solution of Eq.(5.195.1) converges to the positive equilibrium.

5.196 Equation #196 : $x_{n+1} = \dfrac{\alpha + \beta x_n + \gamma x_{n-1} + \delta x_{n-2}}{A + C x_{n-1}}$

The boundedness character of solutions of this equation was investigated in [49]. See also Theorem 2.7.1 where we established that every solution of the equation is bounded. Eq.(#196) can be written in the normalized form

$$x_{n+1} = \frac{\alpha + \beta x_n + \gamma x_{n-1} + x_{n-2}}{A + x_{n-1}}, \quad n = 0, 1, \ldots \tag{5.196.1}$$

with positive parameters α, β, γ, A and with arbitrary nonnegative initial conditions x_{-2}, x_{-1}, x_0.

By Theorems 5.23.2 and 5.23.3 it follows that when

$$\beta + \gamma + 1 \leq A,$$

the equilibrium of Eq.(5.196.1) is globally asymptotically stable.

Conjecture 5.196.1 *Show that for the equilibrium $\bar{x}$ of Eq.(5.196.1),*

Local Asymptotic Stability $\Rightarrow$ Global Attractivity.

Conjecture 5.196.2 *Assume that*

$$\beta + \gamma + 1 > A.$$

Show that Eq.(5.196.1) has solutions that do not converge to the equilibrium point or to a periodic solution.

5.197 Equation #197 : $x_{n+1} = \dfrac{\alpha + \beta x_n + \gamma x_{n-1} + \delta x_{n-2}}{A + D x_{n-2}}$

Eq.(#197) can be written in the normalized form

$$x_{n+1} = \frac{\alpha + \beta x_n + \gamma x_{n-1} + x_{n-2}}{A + x_{n-2}}, \quad n = 0, 1, \ldots \tag{5.197.1}$$

with positive parameters α, β, γ, A and with arbitrary nonnegative initial conditions x_{-2}, x_{-1}, x_0.

The boundedness character of solutions of this equation was investigated in [49]. See also Theorem 3.2.1 where we established that the equation has unbounded solutions when

$$\gamma > \beta + 1 + A.$$

By Theorems 5.23.2 and 5.23.3 it follows that when

$$\beta + \gamma + 1 \leq A,$$

the equilibrium of Eq.(5.197.1) is globally asymptotically stable.

Conjecture 5.197.1 *Assume that*

$$\gamma > \beta + 1 + A.$$

Show that every bounded solution of Eq.(5.197.1) converges to the equilibrium.

Conjecture 5.197.2 *Assume that*

$$\gamma < \beta + 1 + A.$$

Show that for the equilibrium $\bar{x}$ of Eq.(5.197.1),

Local Asymptotic Stability $\Rightarrow$ ***Global Attractivity.***

Conjecture 5.197.3 *Assume that*

$$\gamma = \beta + 1 + A.$$

Show that every solution of Eq.(5.197.1) converges to a (not necessarily prime) period-two solution.

Conjecture 5.197.4 *Assume that*

$$A - \beta - 1 < \gamma < \beta + 1 + A.$$

Show that Eq.(5.197.1) has bounded solutions that do not converge to the equilibrium point or to a periodic solution.

5.198 Equation #198 : $x_{n+1} = \dfrac{\alpha + \beta x_n + \gamma x_{n-1} + \delta x_{n-2}}{Bx_n + Cx_{n-1}}$

Eq.(#198) can be written in the normalized form

$$x_{n+1} = \frac{\alpha + x_n + \gamma x_{n-1} + \delta x_{n-2}}{x_n + Cx_{n-1}}, \quad n = 0, 1, \ldots \tag{5.198.1}$$

with positive parameters $\alpha, \gamma, \delta, C$ and with arbitrary positive initial conditions x_{-2}, x_{-1}, x_0.

The boundedness character of solutions of this equation was investigated in [69]. See also Theorem 3.4.1 where we established that the equation has unbounded solutions when

$$\delta > C + \frac{\gamma}{C}.$$

By Theorem 5.221.1 it follows that every solution of Eq.(#198) is bounded if

$$\delta < C + \frac{\gamma}{C}.$$

Conjecture 5.198.1 *Assume that*

$$0 < \delta \leq C + \frac{\gamma}{C}.$$

Show that for the equilibrium of Eq.(5.198.1),

Local Asymptotic Stabilty $\Longrightarrow$ ***Global Attractivity.***

Open Problem 5.198.1 *Assume that*

$$\delta > C + \frac{\gamma}{C}.$$

(i) *Determine the set of all initial conditions for which the solutions of Eq.(5.198.1) converge to the equilibrium.*

(ii) *Determine the set of all initial conditions for which the solutions of Eq.(5.198.1) are unbounded.*

(iii) *Determine all possible periodic solutions of Eq.(5.198.1).*

Conjecture 5.198.2 *It follows from the work in Section 4.2 that Eq.(5.198.1) has a unique prime period-two solution. Show that this period-two solution of Eq.(5.198.1) is locally asymptotically stable.*

Conjecture 5.198.3 *Show that Eq.(5.198.1) has bounded solutions that do not converge to the equilibrium point or to a periodic solution.*

5.199 Equation #199 : $x_{n+1} = \dfrac{\alpha + \beta x_n + \gamma x_{n-1} + \delta x_{n-2}}{Bx_n + Dx_{n-2}}$

Eq.(#199) can be written in the normalized form

$$x_{n+1} = \frac{\alpha + x_n + \gamma x_{n-1} + \delta x_{n-2}}{x_n + Dx_{n-2}}, \quad n = 0, 1, \ldots \tag{5.199.1}$$

with positive parameters $\alpha, \gamma, \delta, D$ and with arbitrary positive initial conditions x_{-2}, x_{-1}, x_0.

The boundedness character of solutions of this equation was investigated in [49]. See also Theorem 3.3.1 where we established that the equation has unbounded solutions when

$$\gamma > 1 + \delta.$$

Open Problem 5.199.1 *Investigate the periodic nature of solutions and the global stability of the equilibrium of Eq.(5.199.1).*

Conjecture 5.199.1 *Assume that*

$$\gamma > 1 + \delta.$$

Show that every bounded solution of Eq.(5.199.1) converges to the equilibrium.

Conjecture 5.199.2 *Assume that*

$$\gamma < 1 + \delta.$$

Show that for the equilibrium $\bar{x}$ of Eq.(5.199.1),

Local Asymptotic Stability* $\Rightarrow$ *Global Attractivity.

Conjecture 5.199.3 *Assume that*

$$\gamma = 1 + \delta.$$

Show that every solution of Eq.(5.199.1) converges to a (not necessarily prime) period-two solution.

Conjecture 5.199.4 *Assume that*

$$\gamma < 1 + \delta.$$

Show that Eq.(5.199.1) has bounded solutions that do not converge to the equilibrium point or to a periodic solution.

5.200 Equation #200 : $x_{n+1} = \dfrac{\alpha + \beta x_n + \gamma x_{n-1} + \delta x_{n-2}}{C x_{n-1} + D x_{n-2}}$

The boundedness character of this equation was investigated in [69]. See also Theorem 2.3.2 where we established that every solution of this equation is bounded. Eq.(#200) can be written in the normalized form

$$x_{n+1} = \frac{\alpha + \beta x_n + x_{n-1} + \delta x_{n-2}}{x_{n-1} + D x_{n-2}}, \quad n = 0, 1, \ldots \tag{5.200.1}$$

with positive parameters α, β, δ, D and with arbitrary positive initial conditions x_{-2}, x_{-1}, x_0.

Open Problem 5.200.1 *Investigate the periodic nature of solutions and the global stability of the equilibrium of Eq.(5.200.1).*

Conjecture 5.200.1 *Show that for the equilibrium $\bar{x}$ of Eq.(5.200.1),*

Local Asymptotic Stability* $\Rightarrow$ *Global Attractivity.

Conjecture 5.200.2 *Show that Eq.(5.200.1) has solutions that do not converge to the equilibrium point or to a periodic solution.*

Conjecture 5.200.3 *It follows from the work in Section 4.2 that Eq.(5.200.1) has a unique prime period-two solution. Show that this period-two solution of Eq.(5.200.1) is locally asymptotically stable.*

5.201 Equation #201 : $x_{n+1} = \dfrac{\alpha + \beta x_n + \gamma x_{n-1}}{A + B x_n + C x_{n-1}}$

Eq.(#201) can be written in the normalized form

$$x_{n+1} = \frac{\alpha + \beta x_n + x_{n-1}}{A + B x_n + x_{n-1}}, \quad n = 0, 1, \ldots \tag{5.201.1}$$

with positive parameters α, γ, A, C and with arbitrary nonnegative initial conditions x_{-2}, x_{-1}, x_0.

Conjecture 5.201.1 *Show that for the equilibrium $\bar{x}$ of Eq.(5.201.1),*

Local Asymptotic Stabilty* $\Rightarrow$ *Global Attractivity.

Conjecture 5.201.2 *It follows from the work in Section 4.2 that Eq.(5.201.1) has a unique prime period-two solution. Show that this period-two solution of Eq.(5.201.1) is locally asymptotically stable.*

For the remainder of this section we allow the parameters of the equation in the title to be nonnegative and the initial conditions to be arbitrary nonnegative real numbers such that the denominator is always positive. We summarize some of the highlights of the 49 special cases contained in the second-order rational difference equation

$$x_{n+1} = \frac{\alpha + \beta x_n + \gamma x_{n-1}}{A + Bx_n + Cx_{n-1}}, \quad n = 0, 1, \dots . \qquad (5.201.2)$$

Of the 49 special cases of Eq.(5.201.2), three cases are **trivial:**

#1, #6, and #11,

six are **linear:**

#5, #9, #41, #45, #53, and #117,

four cases are **reducible to linear:**

#2, #3, #7, and #10,

four are **Riccati equations:**

#17, #23, #42, and #65,

and four special cases are reducible to **Riccati equations:**

#18, #30, #47, and #72.

Therefore there remain only 28 special cases of second-order rational equations to be investigated. The character of solutions of these equations is summarized in the Table. As we see in the Table, in seven special cases it is known that every solution converges to an equilibrium:

#20, #24, #26, #55, #84, #101, and #105.

In seven special cases there is a **period-two trichotomy:**

#29, #46, #54, #71, #83, #118, and #165.

In five special cases, every solution is bounded and in each case there exist infinitely many prime period-two solutions, or there exists a unique prime period-two solution. In all of these five cases it is known that every solution converges to a (not necessarily prime) period-two solution:

#32, #74, #86, #109, and #145.

One of the 28 cases is the well-known **Lyness's Equation:** #43.

In five special cases **we conjecture** that every solution converges to the equilibrium:

#66, #68, #119, #141, and #166.

In three special cases **we conjecture** that the equilibrium is locally asymptotically stable in some region of the parameters and in the complement of the closure of this region there exists a unique prime period-two solution. These equations are:

#153, #168, and #201.

For each of these three equations **we conjecture** that the prime period-two solution is locally asymptotically stable and that every solution converges to a (not necessarily prime) period-two solution.

Summary of the Behavior of the 28 Nontrivial Second-Order Rational Difference Equations

#20 : $x_{n+1} = \frac{\alpha}{Bx_n + x_{n-1}}$	**ESC$\bar{x}$**			
#24 : $x_{n+1} = \frac{\beta x_n}{1 + x_{n-1}}$ **Pielou's Equation**	**ESC**			
#26 : $x_{n+1} = \frac{x_n}{Bx_n + x_{n-1}}$	**ESC$\bar{x}$**			
#29 : $x_{n+1} = \frac{x_{n-1}}{A + x_n}$			**P_2-Tricho**	
#32 : $x_{n+1} = \frac{x_{n-1}}{Bx_n + x_{n-1}}$		**ESCP_2**		
#43 : $x_{n+1} = \frac{\alpha + x_n}{x_{n-1}}$ **Lyness's Equation**				
#46 : $x_{n+1} = \frac{\alpha + x_{n-1}}{x_n}$			**Part of a P_2-Tricho**	
#54 : $x_{n+1} = \beta + \frac{x_{n-1}}{x_n}$			**First P_2-Tricho**	
#55 : $x_{n+1} = \gamma + \frac{x_n}{x_{n-1}}$	**ESC$\bar{x}$**			
#66 : $x_{n+1} = \frac{\alpha + x_n}{A + x_{n-1}}$				**Conjecture: ESC$\bar{x}$**
#68 : $x_{n+1} = \frac{\alpha + x_n}{x_n + Cx_{n-1}}$				**Conjecture: ESC$\bar{x}$**

$\#71 : x_{n+1} = \frac{\alpha+x_{n-1}}{A+x_n}$			**P_2-Tricho**		
$\#74 : x_{n+1} = \frac{\alpha+x_{n-1}}{Bx_n+x_{n-1}}$		**$ESCP_2$**			
$\#83 : x_{n+1} = \frac{x_n+\gamma x_{n-1}}{A+x_n}$			**P_2-Tricho**		
$\#84 : x_{n+1} = \frac{\beta x_n+x_{n-1}}{A+x_{n-1}}$	**ESC**				
$\#86 : x_{n+1} = \frac{\beta x_n+x_{n-1}}{Bx_n+x_{n-1}}$		**$ESCP_2$**			
$\#101 : x_{n+1} = \frac{1}{A+Bx_n+x_{n-1}}$	**ESC$\bar{x}$**				
$\#105 : x_{n+1} = \frac{\beta x_n}{1+Bx_n+x_{n-1}}$	**ESC**				
$\#109 : x_{n+1} = \frac{x_{n-1}}{A+Bx_n+x_{n-1}}$		**$ESCP_2$**			
$\#118 : x_{n+1} = \frac{\alpha+\beta x_n+\gamma x_{n-1}}{x_n}$			**P_2-Tricho**		
$\#119 : x_{n+1} = \frac{\alpha+\beta x_n+\gamma x_{n-1}}{x_{n-1}}$				**Conjecture: ESC$\bar{x}$**	
$\#141 : x_{n+1} = \frac{\alpha+x_n}{A+Bx_n+x_{n-1}}$				**Conjecture: ESC$\bar{x}$**	
$\#145 : x_{n+1} = \frac{\alpha+x_{n-1}}{A+Bx_n+x_{n-1}}$		**$ESCP_2$**			
$\#153 : x_{n+1} = \frac{\beta x_n+x_{n-1}}{A+Bx_n+x_{n-1}}$					**Conjecture: $ESCP_2$**

$\#165 : x_{n+1} = \frac{\alpha+\beta x_n+x_{n-1}}{A+x_n}$	**P_2-Tricho**		
$\#166 : x_{n+1} = \frac{\alpha+x_n+\gamma x_{n-1}}{A+x_{n-1}}$		**Conjecture: ESC$\bar{x}$**	
$\#168 : x_{n+1} = \frac{\alpha+x_n+\gamma x_{n-1}}{Bx_n+x_{n-1}}$			**Conjecture: ESCP_2**
$\#201 : x_{n+1} = \frac{\alpha+\beta x_n+x_{n-1}}{A+Bx_n+x_{n-1}}$			**Conjecture: ESCP_2**

Conjecture 5.201.3 *Assume that*

$$\gamma, A, B \in [0, \infty) \text{ and } \alpha, A + B, \gamma + A \in (0, \infty).$$

Show that the positive equilibrium of each of the following two equations,

$$x_{n+1} = \frac{\alpha + x_n}{A + Bx_n + x_{n-1}}, \quad n = 0, 1, \ldots$$

and

$$x_{n+1} = \frac{\alpha + x_n + \gamma x_{n-1}}{A + x_{n-1}}, \quad n = 0, 1, \ldots,$$

is globally asymptotically stable.

The two equations in Conjecture 5.201.3 include the five special cases:

#66, #68, #119, #141, and #166.

We believe that any claims in the literature, made prior to the submission date of this manuscript, that the conjecture has been confirmed for any of these five special cases are not correct.

Conjecture 5.201.4 *Assume that*

$$\alpha, A \in [0, \infty) \text{ and } \alpha + A, \beta, B \in (0, \infty).$$

(*i*) *Show that the unique two-cycle of the difference equation*

$$x_{n+1} = \frac{\alpha + \beta x_n + x_{n-1}}{A + Bx_n + x_{n-1}}, \quad n = 0, 1 \ldots, \tag{5.201.3}$$

which exists when

$$\beta + A < 1$$

and

$$4\alpha < (1 - \beta - A)\left[B(1 - \beta - A) - (1 + 3\beta - A)\right],$$

is locally asymptotically stable.

(ii) Show that when

$$\beta + A \geq 1$$

or

$$4\alpha \geq (1 - \beta - A)\left[B(1 - \beta - A) - (1 + 3\beta - A)\right],$$

the equilibrium of Eq.(5.201.3) is locally stable and a global attractor of every positive solution.

(iii) Show that every solution of Eq.(5.201.3) converges to a (not necessarily prime) period-two solution.

The equation in Conjecture 5.201.4 includes the three special cases:

#153, #168, and #201.

5.202 Equation #202 : $x_{n+1} = \dfrac{\alpha + \beta x_n + \gamma x_{n-1}}{A + Bx_n + Dx_{n-2}}$

Eq.(#202) can be written in the normalized form

$$x_{n+1} = \frac{\alpha + x_n + \gamma x_{n-1}}{A + x_n + Dx_{n-2}}, \quad n = 0, 1, \ldots \tag{5.202.1}$$

with positive parameters α, γ, A, D and with arbitrary nonnegative initial conditions x_{-2}, x_{-1}, x_0.

The boundedness character of solutions of this equation was investigated in [49]. See also Theorem 3.3.1 where we established that the equation has unbounded solutions when

$$\gamma > 1 + A.$$

By Theorems 5.23.2 and 5.23.3 it follows that when

$$\gamma + 1 \leq A,$$

the equilibrium of Eq.(5.202.1) is globally asymptotically stable.

Conjecture 5.202.1 *Assume that*

$$\gamma > 1 + A.$$

Show that every bounded solution of Eq.(5.202.1) converges to the equilibrium.

Conjecture 5.202.2 *Assume that*

$$\gamma < 1 + A.$$

Show that for the equilibrium $\bar{x}$ of Eq.(5.202.1),

Local Asymptotic Stability $\Rightarrow$ ***Global Attractivity.***

Conjecture 5.202.3 *Assume that*

$$\gamma = 1 + A.$$

Show that every solution of Eq.(5.202.1) converges to a (not necessarily prime) period-two solution.

Conjecture 5.202.4 *Assume that*

$$A - 1 < \gamma < 1 + A.$$

Show that Eq.(5.202.1) has bounded solutions that do not converge to the equilibrium point or to a periodic solution.

5.203 Equation #203 : $x_{n+1} = \dfrac{\alpha + \beta x_n + \gamma x_{n-1}}{A + Cx_{n-1} + Dx_{n-2}}$

The boundedness character of solutions of this equation was investigated in [69]. See also Theorem 2.3.1 where we established that every solution of the equation is bounded. Eq.(#203) can be written in the normalized form

$$x_{n+1} = \frac{\alpha + \beta x_n + x_{n-1}}{A + x_{n-1} + Dx_{n-2}}, \quad n = 0, 1, \ldots \tag{5.203.1}$$

with positive parameters α, β, A, D and with arbitrary nonnegative initial conditions x_{-2}, x_{-1}, x_0.

By Theorems 5.23.2 and 5.23.3 it follows that when

$$\beta + 1 \leq A$$

the equilibrium of Eq.(5.203.1) is globally asymptotically stable.

Conjecture 5.203.1 *It follows from the work in Section 4.2 that Eq.(5.203.1) has a unique prime period-two solution. Show that this period-two solution of Eq.(5.203.1) is locally asymptotically stable.*

Conjecture 5.203.2 *Show that for the equilibrium $\bar{x}$ of Eq.(5.203.1),*

$$\textbf{\textit{Local Asymptotic Stability}} \Rightarrow \textbf{\textit{Global Attractivity.}}$$

Conjecture 5.203.3 *Assume that*

$$\beta + 1 > A.$$

Show that Eq.(5.203.1) has solutions that do not converge to the equilibrium point or to a periodic solution.

5.204 Equation #204 : $x_{n+1} = \dfrac{\alpha + \beta x_n + \gamma x_{n-1}}{Bx_n + Cx_{n-1} + Dx_{n-2}}$

The boundedness character of solutions of this equation was investigated in [69]. See also Theorem 2.1.1 where we established that every solution of this equation is bounded. Eq.(#204) can be written in the normalized form

$$x_{n+1} = \frac{\alpha + x_n + \gamma x_{n-1}}{x_n + Cx_{n-1} + Dx_{n-2}}, \quad n = 0, 1, \ldots \tag{5.204.1}$$

with positive parameters α, γ, C, D and with arbitrary positive initial conditions x_{-2}, x_{-1}, x_0.

Open Problem 5.204.1 *Investigate the periodic nature of solutions and the global stability of the equilibrium of Eq.(5.204.1).*

Conjecture 5.204.1 *Show that for the equilibrium $\bar{x}$ of Eq.(5.204.1),*

$$\textbf{\textit{Local Asymptotic Stability}} \Rightarrow \textbf{\textit{Global Attractivity.}}$$

Conjecture 5.204.2 *It follows from the work in Section 4.2 that Eq.(5.204.1) has a unique prime period-two solution. Show that this period-two solution of Eq.(5.204.1) is locally asymptotically stable.*

Conjecture 5.204.3 *Show that Eq.(5.204.1) has solutions that do not converge to the equilibrium point or to a periodic solution.*

5.205 Equation #205 : $x_{n+1} = \dfrac{\alpha + \beta x_n + \delta x_{n-2}}{A + Bx_n + Cx_{n-1}}$

Eq.(#205) can be written in the normalized form

$$x_{n+1} = \frac{\alpha + x_n + \delta x_{n-2}}{A + x_n + Cx_{n-1}}, \quad n = 0, 1, \ldots \tag{5.205.1}$$

with positive parameters α, δ, A, C and with arbitrary nonnegative initial conditions x_{-2}, x_{-1}, x_0.

The boundedness character of solutions of this equation was investigated in [69]. See also Theorem 3.4.1 where we established that the equation has unbounded solutions when

$$\delta > A + C.$$

By Theorem 5.221.1 it follows that every solution of Eq.(5.205.1) is bounded when

$$\delta < A + C.$$

By Theorems 5.23.2 and 5.23.3 it follows that when

$$\delta + 1 \leq A,$$

the equilibrium of Eq.(5.205.1) is globally asymptotically stable.

Conjecture 5.205.1 *Assume that*

$$\delta \leq A + C.$$

Show that for the equilibrium $\bar{x}$ of Eq.(5.205.1),

Local Asymptotic Stabilty $\Longrightarrow$ Global Attractivity.

Open Problem 5.205.1 *Assume that*

$$\delta > A + C.$$

(*i*) *Determine the set of all initial conditions for which the solutions of Eq.(5.205.1) converge to the equilibrium.*

(*ii*) *Determine the set of all initial conditions for which the solutions of Eq.(5.205.1) are unbounded.*

(*iii*) *Determine all possible periodic solutions of Eq.(5.205.1).*

Conjecture 5.205.2 *Assume that*

$$\delta + 1 > A.$$

Show that Eq.(5.205.1) has bounded solutions that do not converge to the equilibrium point or to a periodic solution.

5.206 Equation #206 : $x_{n+1} = \dfrac{\alpha + \beta x_n + \delta x_{n-2}}{A + Bx_n + Dx_{n-2}}$

Eq.(#206) can be written in the normalized form

$$x_{n+1} = \frac{\alpha + x_n + \delta x_{n-2}}{A + x_n + Dx_{n-2}}, \quad n = 0, 1, \ldots \tag{5.206.1}$$

with positive parameters α, δ, A, D and with arbitrary nonnegative initial conditions x_{-2}, x_{-1}, x_0.

By Theorems 5.23.2 and 5.23.3 it follows that when

$$\delta + 1 \leq A,$$

the equilibrium of Eq.(5.206.1) is globally asymptotically stable.

Conjecture 5.206.1 *Show that for the equilibrium $\bar{x}$ of Eq.(5.206.1),*

Local Asymptotic Stability ⇒ Global Attractivity.

Conjecture 5.206.2 *Assume that*

$$\delta + 1 > A.$$

Show that Eq.(5.206.1) has solutions that do not converge to the equilibrium point or to a periodic solution.

5.207 Equation #207 : $x_{n+1} = \dfrac{\alpha + \beta x_n + \delta x_{n-2}}{A + Cx_{n-1} + Dx_{n-2}}$

The boundedness character of solutions of this equation was investigated in [69]. See also Theorem 2.3.1 where we established that every solution of the equation is bounded. Eq.(#207) can be written in the normalized form

$$x_{n+1} = \frac{\alpha + \beta x_n + x_{n-2}}{A + Cx_{n-1} + x_{n-2}}, \quad n = 0, 1, \ldots \tag{5.207.1}$$

with positive parameters α, β, A, C and with arbitrary nonnegative initial conditions x_{-2}, x_{-1}, x_0.

By Theorems 5.23.2 and 5.23.3 it follows that when

$$\beta + 1 \leq A$$

the equilibrium of Eq.(5.207.1) is globally asymptotically stable.

Conjecture 5.207.1 *Show that for the equilibrium $\bar{x}$ of Eq.(5.207.1),*

Local Asymptotic Stability* $\Rightarrow$ *Global Attractivity.

Conjecture 5.207.2 *Assume that*

$$\beta + 1 > A.$$

Show that Eq.(5.207.1) has solutions that do not converge to the equilibrium point or to a periodic solution.

5.208 Equation #208 : $x_{n+1} = \dfrac{\alpha + \beta x_n + \delta x_{n-2}}{Bx_n + Cx_{n-1} + Dx_{n-2}}$

The boundedness character of solutions of this equation was investigated in [69]. See also Theorem 2.1.1 where we established that every solution of this equation is bounded. Eq.(#208) can be written in the normalized form

$$x_{n+1} = \frac{\alpha + x_n + \delta x_{n-2}}{x_n + Cx_{n-1} + Dx_{n-2}}, \quad n = 0, 1, \ldots \tag{5.208.1}$$

with positive parameters α, δ, C, D and with arbitrary positive initial conditions x_{-2}, x_{-1}, x_0.

Open Problem 5.208.1 *Investigate the periodic nature of solutions and the global stability of the equilibrium of Eq.(5.208.1).*

Conjecture 5.208.1 *Show that for the equilibrium $\bar{x}$ of Eq.(5.208.1),*

Local Asymptotic Stability* $\Rightarrow$ *Global Attractivity.

Conjecture 5.208.2 *Show that Eq.(5.208.1) has solutions that do not converge to the equilibrium point or to a periodic solution.*

5.209 Equation #209 : $x_{n+1} = \dfrac{\alpha + \gamma x_{n-1} + \delta x_{n-2}}{A + Bx_n + Cx_{n-1}}$

Eq.(#209) can be written in the normalized form

$$x_{n+1} = \frac{\alpha + x_{n-1} + \delta x_{n-2}}{A + Bx_n + x_{n-1}}, \quad n = 0, 1, \ldots \tag{5.209.1}$$

with positive parameters α, δ, A, B and with arbitrary nonnegative initial conditions x_{-2}, x_{-1}, x_0.

The boundedness character of solutions of this equation was investigated in [69]. See also Theorem 3.4.1 where we established that the equation has unbounded solutions when

$$\delta > A + B.$$

By Theorem 5.221.1 it follows that every solution of Eq.(5.209.1) is bounded when

$$\delta < A + B.$$

By Theorems 5.23.2 and 5.23.3 it follows that when

$$\delta + 1 \leq A,$$

the equilibrium of Eq.(5.209.1) is globally asymptotically stable.

Conjecture 5.209.1 *Assume that*

$$\delta \leq A + B.$$

Show that for the equilibrium $\bar{x}$ of Eq.(5.209.1),

Local Asymptotic Stabilty $\Longrightarrow$ Global Attractivity.

Open Problem 5.209.1 *Assume that*

$$\delta > A + B.$$

(i) Determine the set of all initial conditions for which the solutions of Eq.(5.209.1) converge to the equilibrium.

(ii) Determine the set of all initial conditions for which the solutions of Eq.(5.209.1) are unbounded.

(iii) Determine all possible periodic solutions of Eq.(5.209.1).

Conjecture 5.209.2 *It follows from the work in Section 4.2 that Eq.(5.209.1) has a unique prime period-two solution. Show that this period-two solution of Eq.(5.209.1) is locally asymptotically stable.*

Conjecture 5.209.3 *Assume that*

$$\delta + 1 > A.$$

Show that Eq.(5.209.1) has bounded solutions that do not converge to the equilibrium point or to a periodic solution.

5.210 Equation #210 : $x_{n+1} = \dfrac{\alpha + \gamma x_{n-1} + \delta x_{n-2}}{A + Bx_n + Dx_{n-2}}$

Eq.(#210) can be written in the normalized form

$$x_{n+1} = \frac{\alpha + \gamma x_{n-1} + x_{n-2}}{A + Bx_n + x_{n-2}}, \quad n = 0, 1, \ldots \tag{5.210.1}$$

with positive parameters α, γ, A, B and with arbitrary nonnegative initial conditions x_{-2}, x_{-1}, x_0.

The boundedness character of solutions of this equation was investigated in [49]. See also Theorem 3.3.1 where we established that the equation has unbounded solutions when

$$\gamma > 1 + A.$$

By Theorems 5.23.2 and 5.23.3 it follows that when

$$\gamma + 1 \leq A,$$

the equilibrium of Eq.(5.210.1) is globally asymptotically stable.

Conjecture 5.210.1 *Assume that*

$$\gamma > 1 + A.$$

Show that every bounded solution of Eq.(5.210.1) converges to the equilibrium.

Conjecture 5.210.2 *Assume that*

$$\gamma < 1 + A.$$

Show that for the equilibrium $\bar{x}$ of Eq.(5.210.1),

Local Asymptotic Stability ⇒ Global Attractivity.

Conjecture 5.210.3 *Assume that*

$$\gamma = 1 + A.$$

Show that every solution of Eq.(5.210.1) converges to a (not necessarily prime) period-two solution.

Conjecture 5.210.4 *Assume that*

$$A - 1 < \gamma < 1 + A.$$

Show that Eq.(5.210.1) has bounded solutions that do not converge to the equilibrium point or to a periodic solution.

5.211 Equation #211 : $x_{n+1} = \dfrac{\alpha + \gamma x_{n-1} + \delta x_{n-2}}{A + Cx_{n-1} + Dx_{n-2}}$

Eq.(#211) can be written in the normalized form

$$x_{n+1} = \frac{\alpha + x_{n-1} + \delta x_{n-2}}{A + x_{n-1} + Dx_{n-2}}, \quad n = 0, 1, \ldots \tag{5.211.1}$$

with positive parameters α, δ, A, D and with arbitrary nonnegative initial conditions x_{-2}, x_{-1}, x_0.

By Theorems 5.23.2 and 5.23.3 it follows that when

$$\delta + 1 \leq A,$$

the equilibrium of Eq.(5.211.1) is globally asymptotically stable.

Conjecture 5.211.1 *It follows from the work in Section 4.2 that Eq.(5.211.1) has a unique prime period-two solution. Show that this period-two solution of Eq.(5.211.1) is locally asymptotically stable.*

Conjecture 5.211.2 *Show that for the equilibrium $\bar{x}$ of Eq.(5.211.1),*

Local Asymptotic Stability $\Rightarrow$ ***Global Attractivity.***

Conjecture 5.211.3 *Assume that*

$$\delta + 1 > A.$$

Show that Eq.(5.211.1) has solutions that do not converge to the equilibrium point or to a periodic solution.

5.212 Equation #212 : $x_{n+1} = \dfrac{\alpha + \gamma x_{n-1} + \delta x_{n-2}}{Bx_n + Cx_{n-1} + Dx_{n-2}}$

The boundedness character of solutions of this equation was investigated in [69]. See also Theorem 2.1.1 where we established that every solution of this equation is bounded. Eq.(#212) can be written in the normalized form

$$x_{n+1} = \frac{\alpha + x_{n-1} + \delta x_{n-2}}{Bx_n + x_{n-1} + Dx_{n-2}}, \quad n = 0, 1, \ldots \tag{5.212.1}$$

with positive parameters α, δ, B, D and with arbitrary nonnegative initial conditions x_{-2}, x_{-1}, x_0 such that the denominator is always positive.

Open Problem 5.212.1 *Investigate the periodic nature of solutions and the global stability of the equilibrium of Eq.(5.212.1).*

Conjecture 5.212.1 *Show that for the equilibrium $\bar{x}$ of Eq.(5.212.1),*

Local Asymptotic Stability* $\Rightarrow$ *Global Attractivity.

Conjecture 5.212.2 *It follows from the work in Section 4.2 that Eq.(5.212.1) has a unique prime period-two solution. Show that this period-two solution of Eq.(5.212.1) is locally asymptotically stable.*

Conjecture 5.212.3 *Show that Eq.(5.212.1) has solutions that do not converge to the equilibrium point or to a periodic solution.*

5.213 Equation #213 : $x_{n+1} = \dfrac{\beta x_n + \gamma x_{n-1} + \delta x_{n-2}}{A + Bx_n + Cx_{n-1}}$

Eq.(#213) can be written in the normalized form

$$x_{n+1} = \frac{x_n + \gamma x_{n-1} + \delta x_{n-2}}{A + x_n + Cx_{n-1}}, \quad n = 0, 1, \ldots \tag{5.213.1}$$

with positive parameters γ, δ, A, C and with arbitrary nonnegative initial conditions x_{-2}, x_{-1}, x_0.

The boundedness character of solutions of this equation was investigated in [69]. See also Theorem 3.4.1 where we established that the equation has unbounded solutions when

$$\delta > A + C + \frac{\gamma}{C}.$$

By Theorem 5.221.1 it follows that every solution of Eq.(5.213.1) is bounded when

$$\delta < A + C + \frac{\gamma}{C}.$$

By Theorems 5.23.2 and 5.23.3 it follows that when

$$A \geq 1 + \gamma + \delta,$$

the zero equilibrium of Eq.(5.213.1) is globally asymptotically stable.

When

$$A < 1 + \gamma + \delta,$$

Eq.(5.213.1) has the unique positive equilibrium point

$$\bar{x} = \frac{1 + \gamma + \delta - A}{1 + C}.$$

Conjecture 5.213.1 *Assume that*

$$0 < \delta \leq A + C + \frac{\gamma}{C}.$$

Show that for the positive equilibrium $\bar{x}$ of Eq.(5.213.1) and with positive initial conditions,

Local Asymptotic Stabilty $\Longrightarrow$ Global Attractivity.

Open Problem 5.213.1 *Assume that*

$$\delta > A + C + \frac{\gamma}{C}.$$

(*i*) *Determine the set of all positive initial conditions for which the solutions of Eq.(5.213.1) converge to the positive equilibrium $\bar{x}$.*

(*ii*) *Determine the set of all positive initial conditions for which the solutions of Eq.(5.213.1) are unbounded.*

(*iii*) *Determine all possible periodic solutions of Eq.(5.213.1).*

Conjecture 5.213.2 *It follows from the work in Section 4.2 that Eq.(5.213.1) has a unique prime period-two solution. Show that this period-two solution of Eq.(5.213.1) is locally asymptotically stable.*

Conjecture 5.213.3 *Show that Eq.(5.213.1) has bounded solutions that do not converge to an equilibrium point or to a periodic solution.*

5.214 Equation #214 : $x_{n+1} = \dfrac{\beta x_n + \gamma x_{n-1} + \delta x_{n-2}}{A + Bx_n + Dx_{n-2}}$

Eq.(#214) can be written in the normalized form

$$x_{n+1} = \frac{x_n + \gamma x_{n-1} + \delta x_{n-2}}{A + x_n + Dx_{n-2}}, \quad n = 0, 1, \ldots \tag{5.214.1}$$

with positive parameters γ, δ, A, D and with arbitrary nonnegative initial conditions x_{-2}, x_{-1}, x_0.

The boundedness character of solutions of this equation was investigated in [49]. See also Theorem 3.3.1 where we established that the equation has unbounded solutions when

$$\gamma > 1 + \delta + A.$$

By Theorems 5.23.2 and 5.23.3 it follows that when

$$A \geq 1 + \gamma + \delta,$$

the zero equilibrium of Eq.(5.214.1) is globally asymptotically stable.

When

$$A < 1 + \gamma + \delta,$$

Eq.(5.214.1) has the unique positive equilibrium point

$$\bar{x} = \frac{1 + \gamma + \delta - A}{1 + D}.$$

Conjecture 5.214.1 *Assume that*

$$\gamma > 1 + \delta + A.$$

Show that every positive and bounded solution of Eq.(5.214.1) converges to the positive equilibrium $\bar{x}$.

Conjecture 5.214.2 *Assume that*

$$\gamma < 1 + \delta + A.$$

Show that for the positive equilibrium $\bar{x}$ *of Eq.(5.214.1) and with positive initial conditions,*

$$\textit{Local Asymptotic Stability} \Rightarrow \textit{Global Attractivity}.$$

Conjecture 5.214.3 *Assume that*

$$\gamma = 1 + \delta + A.$$

Show that every solution of Eq.(5.214.1) converges to a (not necessarily prime) period-two solution.

Conjecture 5.214.4 *Assume that*

$$A - 1 - \delta < \gamma < 1 + \delta + A.$$

Show that Eq.(5.214.1) has bounded solutions that do not converge to an equilibrium point or to a periodic solution.

5.215 Equation #215 : $x_{n+1} = \frac{\beta x_n + \gamma x_{n-1} + \delta x_{n-2}}{A + Cx_{n-1} + Dx_{n-2}}$

The boundedness character of solutions of this equation was investigated in [69]. See also Theorem 2.3.1 where we established that every solution of this equation is bounded. Eq.(#215) can be written in the normalized form

$$x_{n+1} = \frac{\beta x_n + x_{n-1} + \delta x_{n-2}}{A + x_{n-1} + Dx_{n-2}}, \quad n = 0, 1, \ldots \tag{5.215.1}$$

with positive parameters β,δ,A,D and with arbitrary nonnegative initial conditions x_{-2}, x_{-1}, x_0.

By Theorems 5.23.2 and 5.23.4 it follows that when

$$A \geq 1 + \beta + \delta,$$

the zero equilibrium of Eq.(5.215.1) is globally asymptotically stable.

When

$$A < 1 + \beta + \delta,$$

Eq.(5.215.1) has the unique positive equilibrium point

$$\bar{x} = \frac{1 + \beta + \delta - A}{1 + D}.$$

Conjecture 5.215.1 *It follows from the work in Section 4.2 that Eq.(5.215.1) has a unique prime period-two solution. Show that this period-two solution of Eq.(5.215.1) is locally asymptotically stable.*

Conjecture 5.215.2 *Show that for the positive equilibrium $\bar{x}$ of Eq.(5.215.1) and with positive initial conditions,*

$$\textbf{\textit{Local Asymptotic Stability}} \Rightarrow \textbf{\textit{Global Attractivity.}}$$

Conjecture 5.215.3 *Assume that*

$$A < 1 + \beta + \delta.$$

Show that Eq.(5.215.1) has solutions that do not converge to an equilibrium point or to a periodic solution.

5.216 Equation #216 : $x_{n+1} = \dfrac{\beta x_n + \gamma x_{n-1} + \delta x_{n-2}}{Bx_n + Cx_{n-1} + Dx_{n-2}}$

Eq.(#216) can be written in the normalized form

$$x_{n+1} = \frac{x_n + \gamma x_{n-1} + \delta x_{n-2}}{x_n + Cx_{n-1} + Dx_{n-2}}, \quad n = 0, 1, \ldots \tag{5.216.1}$$

with positive parameters γ, δ, C, D and with arbitrary positive initial conditions x_{-2}, x_{-1}, x_0.

Open Problem 5.216.1 *Investigate the periodic nature of solutions and the global stability of the equilibrium of Eq.(5.216.1).*

Conjecture 5.216.1 *Show that for the equilibrium $\bar{x}$ of Eq.(5.216.1),*

$$\textbf{\textit{Local Asymptotic Stability}} \Rightarrow \textbf{\textit{Global Attractivity.}}$$

Conjecture 5.216.2 *It follows from the work in Section 4.2 that Eq.(5.216.1) has a unique prime period-two solution. Show that this period-two solution of Eq.(5.216.1) is locally asymptotically stable.*

Conjecture 5.216.3 *Show that Eq.(5.216.1) has solutions that do not converge to the equilibrium point or to a periodic solution.*

5.217 Equation #217 : $x_{n+1} = \dfrac{\alpha + \beta x_n + \gamma x_{n-1}}{A + Bx_n + Cx_{n-1} + Dx_{n-2}}$

Eq.(#217) can be written in the normalized form

$$x_{n+1} = \frac{\alpha + x_n + \gamma x_{n-1}}{A + x_n + Cx_{n-1} + Dx_{n-2}}, \quad n = 0, 1, \ldots \tag{5.217.1}$$

with positive parameters α, γ, A, C, D and with arbitrary nonnegative initial conditions x_{-2}, x_{-1}, x_0.

By Theorems 5.23.2 and 5.23.3 it follows that when

$$\gamma + 1 \leq A,$$

the equilibrium of Eq.(5.217.1)is globally asymptotically stable.

Conjecture 5.217.1 *It follows from the work in Section 4.2 that Eq.(5.217.1) has a unique prime period-two solution. Show that this period-two solution of Eq.(5.217.1) is locally asymptotically stable.*

Conjecture 5.217.2 *Show that for the equilibrium $\bar{x}$ of Eq.(5.217.1),*

Local Asymptotic Stability ⇒ Global Attractivity.

Conjecture 5.217.3 *Assume that*

$$\gamma + 1 > A.$$

Show that Eq.(5.217.1) has solutions that do not converge to the equilibrium point or to a periodic solution.

5.218 Equation #218 : $x_{n+1} = \dfrac{\alpha + \beta x_n + \delta x_{n-2}}{A + Bx_n + Cx_{n-1} + Dx_{n-2}}$

Eq.(#218) can be written in the normalized form

$$x_{n+1} = \frac{\alpha + x_n + \delta x_{n-2}}{A + x_n + Cx_{n-1} + Dx_{n-2}}, \quad n = 0, 1, \ldots \tag{5.218.1}$$

with positive parameters α, δ, A, C, D and with arbitrary nonnegative initial conditions x_{-2}, x_{-1}, x_0.

By Theorems 5.23.2 and 5.23.3 it follows that when

$$\delta + 1 \leq A,$$

the equilibrium of Eq.(5.218.1) is globally asymptotically stable.

Conjecture 5.218.1 *Show that for the equilibrium $\bar{x}$ of Eq.(5.218.1),*

Local Asymptotic Stability* $\Rightarrow$ *Global Attractivity.

Conjecture 5.218.2 *Assume that*

$$\delta + 1 > A.$$

Show that Eq.(5.218.1) has solutions that do not converge to the equilibrium point or to a periodic solution.

5.219 Equation #219 : $x_{n+1} = \dfrac{\alpha + \gamma x_{n-1} + \delta x_{n-2}}{A + Bx_n + Cx_{n-1} + Dx_{n-2}}$

Eq.(#219) can be written in the normalized form

$$x_{n+1} = \frac{\alpha + \gamma x_{n-1} + x_{n-2}}{A + Bx_n + Cx_{n-1} + x_{n-2}}, \quad n = 0, 1, \ldots \tag{5.219.1}$$

with positive parameters α, γ, A, B, C and with arbitrary nonnegative initial conditions x_{-2}, x_{-1}, x_0.

By Theorems 5.23.2 and 5.23.3 it follows that when

$$\gamma + 1 \leq A,$$

the equilibrium of Eq.(5.219.1) is globally asymptotically stable.

Conjecture 5.219.1 *It follows from the work in Section 4.2 that Eq.(5.219.1) has a unique prime period-two solution. Show that this period-two solution of Eq.(5.219.1) is locally asymptotically stable.*

Conjecture 5.219.2 *Show that for the equilibrium $\bar{x}$ of Eq.(5.219.1),*

Local Asymptotic Stability* $\Rightarrow$ *Global Attractivity.

Conjecture 5.219.3 *Assume that*

$$\gamma + 1 > A.$$

Show that Eq.(5.219.1) has solutions that do not converge to the equilibrium point or to a periodic solution.

5.220 Equation #220 : $x_{n+1} = \dfrac{\beta x_n + \gamma x_{n-1} + \delta x_{n-2}}{A + Bx_n + Cx_{n-1} + Dx_{n-2}}$

Eq.(#220) can be written in the normalized form

$$x_{n+1} = \frac{x_n + \gamma x_{n-1} + \delta x_{n-2}}{A + x_n + Cx_{n-1} + Dx_{n-2}}, \quad n = 0, 1, \ldots \tag{5.220.1}$$

with positive parameters γ, δ, A, C, D and with arbitrary nonnegative initial conditions x_{-2}, x_{-1}, x_0.

By Theorems 5.23.2 and 5.23.4 it follows that when

$$A \geq 1 + \gamma + \delta,$$

the zero equilibrium of Eq.(5.220.1) is globally asymptotically stable.

When

$$A < 1 + \gamma + \delta,$$

Eq.(5.220.1) has the unique positive equilibrium point

$$\bar{x} = \frac{1 + \gamma + \delta - A}{1 + C + D}.$$

Conjecture 5.220.1 *It follows from the work in Section 4.2 that Eq.(5.220.1) has a unique prime period-two solution. Show that this period-two solution of Eq.(5.220.1) is locally asymptotically stable.*

Conjecture 5.220.2 *Show that for the positive equilibrium $\bar{x}$ of Eq.(5.220.1) and with positive initial conditions,*

Local Asymptotic Stability $\Rightarrow$ Global Attractivity.

Conjecture 5.220.3 *Assume that*

$$A < 1 + \gamma + \delta.$$

Show that Eq.(5.220.1) has solutions that do not converge to an equilibrium point $\bar{x}$ or to a periodic solution.

5.221 Equation #221 : $x_{n+1} = \dfrac{\alpha + \beta x_n + \gamma x_{n-1} + \delta x_{n-2}}{A + Bx_n + Cx_{n-1}}$

In this section we investigate the global behavior of solutions of the equation

$$x_{n+1} = \frac{\alpha + \beta x_n + \gamma x_{n-1} + \delta x_{n-2}}{A + Bx_n + Cx_{n-1}}, \quad n = 0, 1, \dots \tag{5.221.1}$$

with nonnegative parameters α, β, γ, A, with positive parameters δ, B, C, and with arbitrary nonnegative initial conditions x_{-2}, x_{-1}, x_0.

The boundedness character of solutions of this equation was investigated in [69]. See also Theorem 3.4.1 where we established that the equation has unbounded solutions when

$$\delta > A + B\frac{\gamma}{C} + C\frac{\beta}{B}.$$

By Theorems 5.23.2 and 5.23.3 it follows that when

$$\beta + \gamma + \delta \leq A, \tag{5.221.2}$$

the equilibrium of Eq.(5.221.1) is globally asymptotically stable.

The following theorem is a new result about the boundedness of solutions of Eq.(5.221.1).

Theorem 5.221.1 *Every solution of Eq.(5.221.1) is bounded if*

(i)

$$\delta < A + B\frac{\gamma}{C} + C\frac{\beta}{B}, \tag{5.221.3}$$

or

(ii)

$$\delta = A + B\frac{\gamma}{C} + C\frac{\beta}{B}, \quad \text{and } \alpha = \beta\gamma = 0, \tag{5.221.4}$$

or

(iii)

$$\delta = A, \ \ \beta = \gamma = 0, \ \ \text{and } \alpha > 0\,. \tag{5.221.5}$$

PROOF Assume for the sake of contradiction that Eq.(5.221.1) has an unbounded solution $\{x_n\}$. Then there exists a subsequence $\{x_{n_i}\}$ such that

$$x_{n_i+1} \to \infty \tag{5.221.6}$$

and, for every i,

$$x_{n_i+1} > x_j, \quad \text{for all } j < n_i + 1. \tag{5.221.7}$$

From

$$x_{n_i+1} = \frac{\alpha + \beta x_{n_i} + \gamma x_{n_i-1} + \delta x_{n_i-2}}{A + Bx_{n_i} + Cx_{n_i-1}}$$

it follows that

$$x_{n_i-2} \to \infty \tag{5.221.8}$$

and the sequences

$$\{x_{n_i}\}, \{x_{n_i-1}\} \text{ are bounded.}$$

Similarly, it follows that

$$\{x_{n_i-3}\}, \{x_{n_i-4}\} \text{ are bounded.}$$

Clearly,

$$x_{n_i} = \frac{\alpha + \beta x_{n_i-1} + \gamma x_{n_i-2} + \delta x_{n_i-3}}{A + Bx_{n_i-1} + Cx_{n_i-2}} \to \frac{\gamma}{C}$$

and

$$x_{n_i-1} = \frac{\alpha + \beta x_{n_i-2} + \gamma x_{n_i-3} + \delta x_{n_i-4}}{A + Bx_{n_i-2} + Cx_{n_i-3}} \to \frac{\beta}{B}.$$

(i) Let ϵ be a positive number such that

$$(B + C)\epsilon < -\delta + A + B\frac{\gamma}{C} + C\frac{\beta}{B}.$$

From (5.221.7) and (5.221.8) it follows that, eventually,

$$x_{n_i} > \frac{\gamma}{C} - \epsilon \text{ and } x_{n_i-1} > \frac{\beta}{B} - \epsilon.$$

Therefore, eventually,

$$x_{n_i+1} < x_{n_i-2},$$

which contradicts (5.221.7) and completes the proof of (i).
(ii) Assume that either

$$\alpha = \beta = 0 \text{ and } \gamma > 0 \tag{5.221.9}$$

or

$$\alpha = \gamma = 0 \text{ and } \beta > 0. \tag{5.221.10}$$

We will give the proof when (5.221.9) holds. The proof when (5.221.10) holds is similar and will be omitted.

In this case we have that, eventually,

$$x_{n_i} > \frac{\gamma}{C}.$$

In view of (5.221.7), we have

$$x_{n_i+1} > x_{n_i-2}$$

and so

$$x_{n_i-2} < \frac{\gamma x_{n_i-1}}{A + Bx_{n_i} + Cx_{n_i-1} - A - B\frac{\gamma}{C}} < \frac{\gamma}{C},$$

which is a contradiction.

(*iii*) When

$$\alpha > 0 \text{ and } \beta = \gamma = 0,$$

the equation reduces to the special case #149 for which we established in Section 5.149 that, when $A \geq \delta$, every solution converges to the equilibrium and so is bounded. ▮

The following theorem is a new result about the global attractivity of the equilibrium of Eq.(5.221.1).

Theorem 5.221.2 *Assume that*

$$\frac{\beta}{B} = \frac{\gamma}{C}. \tag{5.221.11}$$

Then the equilibrium $\bar{x}$ of Eq.(5.221.1) is a global attractor of all solutions of Eq.(5.221.1) if and only if

$$\delta \leq A + B\frac{\gamma}{C} + C\frac{\beta}{B}. \tag{5.221.12}$$

PROOF Eq.(5.221.1) can be written in the normalized form

$$x_{n+1} = \frac{\alpha + Bx_n + x_{n-1} + \delta x_{n-2}}{A + Bx_n + x_{n-1}}, \quad n = 0, 1, \ldots . \tag{5.221.13}$$

It suffices to show that the equilibrium of Eq.(5.221.13) is a global attractor of all solutions when

$$\delta \leq A + B + 1.$$

We divide the proof into the following three cases:

Case 1:

$$\delta > A - \alpha.$$

We claim that there exists N, sufficiently large, such that

$$x_n > 1, \text{ for } n \geq N.$$

Otherwise, for some $N \geq 0$,

$$x_{N+1} = \frac{\alpha + Bx_N + x_{N-1} + \delta x_{N-2}}{A + Bx_N + x_{N-1}} \leq 1.$$

This implies that

$$x_{N-2} < \frac{A-\alpha}{\delta} < 1$$

and, similarly,

$$x_{N-5} < (\frac{A-\alpha}{\delta})^2,$$

which eventually leads to a contradiction. Hence, our claim is true and by using the change of variables

$$y_n = x_n - 1$$

Eq.(5.221.13) becomes

$$y_{n+1} = \frac{\alpha - A + \delta + \delta y_{n-2}}{A + B + 1 + By_n + y_{n-1}}.$$

Now the result follows from our work in Section 5.149 where we established that every solution of the equation above converges to the equilibrium when $A + B + 1 \geq \delta$.

Case 2:

$$\delta = A - \alpha.$$

Observe that

$$x_{n+1} - 1 = \frac{(A-\alpha)(x_{n-2} - 1)}{A + Bx_n + x_{n-1}}$$

and so

$$|x_{n+1} - 1| \leq \frac{A-\alpha}{A}|x_{n-2} - 1,|$$

from which the result follows.

Case 3:

$$\delta < A - \alpha.$$

Here we claim that there exists N, sufficiently large, such that

$$x_n \leq \frac{A-\alpha}{\delta}, \quad \text{for } n \geq N.$$

Suppose for the sake of contradiction that for some $N \geq 0$,

$$x_{N+1} = \frac{\alpha + Bx_N + x_{N-1} + \delta x_{N-2}}{A + Bx_N + x_{N-1}} > \frac{A-\alpha}{\delta}.$$

From this it follows that

$$x_{N-2} > \frac{A}{\delta} \cdot \frac{A-\alpha}{\delta}$$

and, similarly,

$$x_{N-5} > (\frac{A}{\delta})^2 \cdot \frac{A-\alpha}{\delta},$$

which eventually leads to a contradiction. Therefore, our claim is true. Set

$$S = \limsup_{n\to\infty} x_n \text{ and } I = \liminf_{n\to\infty} x_n.$$

Then, clearly,

$$S \le \frac{\alpha + (B+\delta+1)S}{A+(B+1)S} \text{ and } I \ge \frac{\alpha + (B+\delta+1)I}{A+(B+1)I}$$

from which it follows that,

$$S = I = \bar{x}$$

and the proof is complete. ∎

Conjecture 5.221.1 *It follows from the work in Section 4.2 that Eq.(5.221.1) has a unique prime period-two solution. Show that this period-two solution of Eq.(5.221.1) is locally asymptotically stable.*

Conjecture 5.221.2 *Show that for the equilibrium $\bar{x}$ of Eq.(5.221.1),*

Local Asymptotic Stability* ⇒ *Global Attractivity.

Conjecture 5.221.3 *Show that Eq.(5.221.1) has bounded solutions that do not converge to the equilibrium point $\bar{x}$ or to a periodic solution.*

5.222 Equation #222 : $x_{n+1} = \dfrac{\alpha + \beta x_n + \gamma x_{n-1} + \delta x_{n-2}}{A + Bx_n + Dx_{n-2}}$

Eq.(#222) can be written in the normalized form

$$x_{n+1} = \frac{\alpha + x_n + \gamma x_{n-1} + \delta x_{n-2}}{A + x_n + Dx_{n-2}}, \quad n = 0, 1, \ldots \tag{5.222.1}$$

with positive parameters $\alpha, \gamma, \delta, A, D$ and with arbitrary nonnegative initial conditions x_{-2}, x_{-1}, x_0.

The boundedness character of solutions of this equation was investigated in [49]. See also Theorem 3.3.1 where we established that the equation has unbounded solutions when

$$\gamma > 1 + \delta + A.$$

By Theorems 5.23.2 and 5.23.3 it follows that when

$$\gamma + \delta + 1 \leq A,$$

the equilibrium of Eq.(5.222.1) is globally asymptotically stable.

Conjecture 5.222.1 *Assume that*

$$\gamma > 1 + \delta + A.$$

Show that every bounded solution of Eq.(5.222.1) converges to the equilibrium.

Conjecture 5.222.2 *Assume that*

$$\gamma < 1 + \delta + A.$$

Show that for the equilibrium $\bar{x}$ of Eq.(5.222.1),

Local Asymptotic Stability* $\Rightarrow$ *Global Attractivity.

Conjecture 5.222.3 *Assume that*

$$\gamma = 1 + \delta + A.$$

Show that every solution of Eq.(5.222.1) converges to a (not necessarily prime) period-two solution.

Conjecture 5.222.4 *Assume that*

$$A - \delta - 1 < \gamma < 1 + \delta + A.$$

Show that Eq.(5.222.1) has bounded solutions that do not converge to the equilibrium point or to a periodic solution.

5.223 Equation #223 : $x_{n+1} = \dfrac{\alpha + \beta x_n + \gamma x_{n-1} + \delta x_{n-2}}{A + C x_{n-1} + D x_{n-2}}$

The boundedness character of solutions of this equation was investigated in [69]. See also Theorem 2.3.1 where we established that every solution of the equation is bounded. Eq.(#223) can be written in the normalized form

$$x_{n+1} = \frac{\alpha + \beta x_n + x_{n-1} + \delta x_{n-2}}{A + x_{n-1} + D x_{n-2}}, \quad n = 0, 1, \ldots \tag{5.223.1}$$

with positive parameters $\alpha, \beta, \delta, A, D$ and with arbitrary nonnegative initial conditions x_{-2}, x_{-1}, x_0.

By Theorems 5.23.2 and 5.23.3 it follows that when

$$\beta + \delta + 1 \leq A,$$

the equilibrium of Eq.(5.223.1) is globally asymptotically stable.

Conjecture 5.223.1 *It follows from the work in Section 4.2 that Eq.(5.223.1) has a unique prime period-two solution. Show that this period-two solution of Eq.(5.223.1) is locally asymptotically stable.*

Conjecture 5.223.2 *Show that for the equilibrium $\bar{x}$ of Eq.(5.223.1),*

$$\textbf{\textit{Local Asymptotic Stability}} \Rightarrow \textbf{\textit{Global Attractivity.}}$$

Conjecture 5.223.3 *Assume that*

$$\beta + \delta + 1 > A.$$

Show that Eq.(5.223.1) has solutions that do not converge to the equilibrium point or to a periodic solution.

5.224 Equation #224 : $x_{n+1} = \dfrac{\alpha + \beta x_n + \gamma x_{n-1} + \delta x_{n-2}}{Bx_n + Cx_{n-1} + Dx_{n-2}}$

Eq.(#224) can be written in the normalized form

$$x_{n+1} = \frac{\alpha + x_n + \gamma x_{n-1} + \delta x_{n-2}}{x_n + Cx_{n-1} + Dx_{n-2}}, \quad n = 0, 1, \ldots \qquad (5.224.1)$$

with positive parameters $\alpha, \gamma, \delta, C, D$ and with arbitrary positive initial conditions x_{-2}, x_{-1}, x_0.

Open Problem 5.224.1 *Investigate the periodic nature of solutions and the global stability of the equilibrium of Eq.(5.224.1).*

Conjecture 5.224.1 *It follows from the work in Section 4.2 that Eq.(5.224.1) has a unique prime period-two solution. Show that this period-two solution of Eq.(5.224.1) is locally asymptotically stable.*

Conjecture 5.224.2 *Show that for the equilibrium $\bar{x}$ of Eq.(5.224.1),*

$$\textbf{\textit{Local Asymptotic Stability}} \Rightarrow \textbf{\textit{Global Attractivity.}}$$

Conjecture 5.224.3 *Show that Eq.(5.224.1) has solutions that do not converge to the equilibrium point or to a periodic solution.*

5.225 Equation #225 : $x_{n+1} = \dfrac{\alpha + \beta x_n + \gamma x_{n-1} + \delta x_{n-2}}{A + Bx_n + Cx_{n-1} + Dx_{n-2}}$

The equation

$$x_{n+1} = \frac{\alpha + \beta x_n + \gamma x_{n-1} + \delta x_{n-2}}{A + Bx_n + Cx_{n-1} + Dx_{n-2}}, \quad n = 0, 1, \ldots . \tag{5.225.1}$$

and all its 225 special cases were the subject of investigation in this book. For related work see Sections 4.2 and 5.23. Throughout this section we allow the parameters of Eq.(5.225.1) to be nonnegative and the initial conditions to be arbitrary nonnegative real numbers such that the denominator is always positive. We summarize some of the highlights of the 225 special cases contained in Eq.(5.225.1).

Of the 225 special cases of Eq.(5.225.1), 39 special cases are trivial, linear, reducible to linear, Riccati equations, or reducible to Riccati equations. Another 28 special cases were the subject of investigation in the Kulenovic and Ladas book [175], which deals with the second-order rational difference equation

$$x_{n+1} = \frac{\alpha + \beta x_n + \gamma x_{n-1}}{A + Bx_n + Cx_{n-1}}, \quad n = 0, 1, \ldots . \tag{5.225.2}$$

See the Table in Section 5.201.

Therefore, there remain

$$225 - (39 + 28) = 158$$

special cases each of which is nonlinear third-order difference equation crying to be investigated. See Appendix A at the end of the book, which presents at a glance the boundedness character of each special case and gives some useful references and some highlights on the character of their solutions.

Concerning the boundedness character of solutions of the 225 special cases of Eq.(5.225.1), we have made the following remarkable conjecture. See [69].

Conjecture 5.225.1 *Show that in 135 special cases of Eq.(5.225.1), every solution of the equation is bounded and that in the remaining 90 special cases, the equation has unbounded solutions in some range of the parameters and for some initial conditions.*

As this book goes to print, we are proud to say that there remain **only five** special cases, out of a total number of 225, whose boundedness has not been established yet. In 135 special cases of Eq.(5.225.1) we have shown that every solution of the equation is bounded and in 85 cases we have shown that the equation has unbounded solutions in some range of the parameters and for some initial conditions.

For five special cases we have conjectured that they have unbounded solutions in some range of their parameters and for some initial conditions but we are not yet able to confirm it. They are the following:

#28, #44, #56, #70, #120.

See Conjecture 3.0.1.

It is interesting to note that the very first rational equation shown to have unbounded solutions is

$$\#54: \qquad x_{n+1} = \beta + \frac{x_{n-1}}{x_n}, \quad n = 0, 1, \ldots .$$

See [16]. As you can see from Appendix B we have made substantial progress in determining the boundedness character of a large number of rational difference equations. See also the results in Chapters 2 and 3. The following conjecture shows the importance of **boundedness** of solutions in rational difference equations.

Conjecture 5.225.2 *In each of the 135 special cases of Eq.(5.225.1), where every solution of the equation is bounded and with positive initial conditions, show that for the positive equilibrium*

Local Asymptotic Stability $\Rightarrow$ ***Global Asymptotic Stability.***

The very first rational equation discovered for which

Local Asymptotic Stability $\not\Rightarrow$ **Global Asymptotic Stability**

is

$$\#157: \qquad x_{n+1} = \frac{\beta x_n + \delta x_{n-2}}{A + Bx_n + Cx_{n-1}}, \quad n = 0, 1, \ldots .$$

See [48]. Actually, this surprising property is true **only** in the following 14 special cases of third-order rational difference equations:

#80, #92, #98, #149, #157, #161, #174,
#180, #186, #198, #205, #209, #213, #221.

Open Problem 5.225.1 *Determine all special cases in the rational difference equation*

$$x_n = \frac{\alpha + \sum_{i=1}^{k} \beta_i x_{n-i}}{A + \sum_{i=1}^{k} B_i x_{n-i}}, \quad n = 0, 1, \ldots \qquad (5.225.3)$$

with $k \geq 4$, *where*

Local Asymptotic Stability $\not\Rightarrow$ ***Global Asymptotic Stability.***

For the general rational equation (5.225.3) we offer the following powerful conjecture that, roughly speaking, states that

Local Asymptotic Stability and Boundedness ⇒ Global Attracticity.

Conjecture 5.225.3 *In a region S of the parameters of Eq.(5.225.3), assume that every positive solution of the equation is bounded and that Eq.(5.225.3) has an equilibrium point $\bar{x}$ that is* ***Locally Asymptotically Stable****. Then $\bar{x}$ is a* ***Global Attractor*** *of all positive solutions of Eq.(5.225.3).*

How do we establish that every solution of a rational difference equation is bounded?

How do we establish that a rational equation has unbounded solutions in some range of its parameters and for some initial conditions?

Theorems 2.1.1 and 2.3.4 from Chapter 2 provide the answer to the first question in a lot of special cases of Eq.(5.225.3). But how about the remaining cases? Is there a **recognizable pattern** that predicts the boundedness character of a rational difference equation?

The following Table presents at a glance the number of special cases contained in Eq.(5.225.3) for each value of the order k of the equation, the number of cases where **ESB**, the number of cases where **∃US**, and the number of cases established by Theorem 2.1.1 and Theorem 2.3.4.

Order	**Total Cases**	**ESB**	**∃US**	**Established by Thm 2.1.1**	**Established by Thm 2.3.4**
$k = 1$	9	7	2	7	0
$k = 2$	49	35	14	27	6
$k = 3$	225	135	90	91	22
$k = 4$	961	542	419	291	126
k	$(2^{k+1} - 1)^2$	?	?	$4 \cdot 3^k - 2 \cdot 2^k - 1$	?

Open Problem 5.225.2 *Determine the numbers in the blocks above where the three question marks appear in the Table.*

Conjecture 5.225.4 *The numbers under the columns* ***ESB*** *and* ***∃US*** *in the above Table are still conjectures for the values of $k \in \{3, 4\}$. Confirm these conjectures.*

See [14], [49], [66], and [69].

In contrast to second-order rational difference equations, a large number of third-order rational equations exhibit **chaotic behavior** in some range of the parameters. In this range, there is **sensitive dependence** on initial conditions, and there exist **dense orbits.** They are the following 124 special cases:

$$\begin{array}{lllll}
\#25, & \#27, & \#39-40, & \#56, & \#58, \\
\#60, & \#62-63, & \#67, & \#69, & \#77-78, \\
\#80-82, & \#85, & \#87-96, & \#98-100, & \#106-108, \\
\#114-116, & \#120, & \#122, & \#124, & \#126-127, \\
\#130-132, & \#134-136, & \#138-140, & \#142-144, & \#149-152, \\
\#154-164, & \#167, & \#169-178, & \#180-200, & \#202-225.
\end{array}$$

See Appendix A.

Among the 158 special cases mentioned before, there remain only 34 special cases of third-order rational equations without chaos, and they are as follows:

In five third-order special cases of Eq.(5.225.1), **ESC$\bar{x}$:**

$$\#22, \#102, \#103, \#104, \#133.$$

In one third-order special case **ESCP$_2$:**

$$\#21.$$

In 16 third-order special cases we have shown that there is a **P$_k$-Tricho:**

$$\begin{array}{l}
\#31, \#33, \#35, \#36, \#38, \#48, \\
\#50, \#64, \#73, \#75, \#97, \#110, \\
\#113, \#128, \#146, \#179.
\end{array}$$

In six third-order special cases we have conjectured that we have a **P$_k$-Tricho:**

$$\#28, \#44, \#51, \#59, \#70, \#123.$$

Finally, in six third-order special cases we conjecture that **ESCP$_2$:**

$$\#34, \#76, \#111, \#112, \#147, \#148.$$

The character of solutions of these 34 special cases is summarized in the Table at the end of the section.

As we saw in Section 4.2, Eq.(5.225.1) contains 49 special cases, each of which has a unique prime period-two solution in some range of the parameters. In one of these special cases,

$$\#30: \qquad x_{n+1} = \frac{x_{n-1}}{A + x_{n-1}}, \quad n = 0, 1, \ldots,$$

the unique prime period-two solution of the equation

$$\ldots, 0, 1-A, \ldots,$$

which exists when

$$A \in (0,1),$$

is **not** locally asymptotically stable.

For the remaining 48 special cases we offer the following conjecture:

Conjecture 5.225.5 *Show that the unique prime period-two solution of Eq.(5.225.1), which exists, if and only if,*

$$\beta + \delta + A < 1$$

and

$$4\alpha < (1-\beta-\delta-A)\left[(B+D)(1-\beta-\delta-A) - (1+3\beta+3\delta-A)\right],$$

is locally asymptotically stable provided that

$$\alpha + \beta + \delta + B + D > 0.$$

See also Open Problems 4.1.1 and 4.1.2.

Summary of the Behavior of Nontrivial Third-Order Rational Equations without Chaotic Behavior

$\#21 : x_{n+1} = \frac{\alpha}{Bx_n+Dx_{n-2}}$		**ESCP$_2$**	
$\#22 : x_{n+1} = \frac{\alpha}{Cx_{n-1}+x_{n-2}}$	**ESC$\bar{x}$**		
$\#28 : x_{n+1} = \frac{x_n}{Cx_{n-1}+x_{n-2}}$			**Part of a P$_6$-Tricho Conjecture**
$\#31 : x_{n+1} = \frac{x_{n-1}}{A+x_{n-2}}$			**P$_2$-Tricho**
$\#33 : x_{n+1} = \frac{x_{n-1}}{Bx_n+x_{n-2}}$			**P$_2$-Tricho**
$\#34 : x_{n+1} = \frac{x_{n-1}}{Cx_{n-1}+Dx_{n-2}}$		**Conjecture: ESCP$_2$**	
$\#35 : x_{n+1} = \frac{x_{n-2}}{A+x_n}$			**P$_3$-Tricho**
$\#36 : x_{n+1} = \frac{x_{n-2}}{A+x_{n-1}}$			**P$_3$-Tricho**
$\#38 : x_{n+1} = \frac{x_{n-2}}{Bx_n+x_{n-1}}$			**P$_3$-Tricho**
$\#44 : x_{n+1} = \frac{\alpha+x_n}{x_{n-2}}$			**Part of a P$_6$-Tricho Conjecture**
$\#48 : x_{n+1} = \frac{\alpha+x_{n-1}}{x_{n-2}}$			**Part of a P$_2$-Tricho**
$\#50 : x_{n+1} = \frac{\alpha+x_{n-2}}{x_n}$			**P$_5$-Tricho**

$\#51 : x_{n+1} = \frac{\alpha+x_{n-2}}{x_{n-1}}$			**Part of a P_4-Tricho Conjecture**
$\#59 : x_{n+1} = \frac{\beta x_n+x_{n-2}}{x_{n-1}}$			**Conjecture: P_4-Tricho**
$\#64 : x_{n+1} = \delta + \frac{x_{n-1}}{x_{n-2}}$			**P_2-Tricho**
$\#70 : x_{n+1} = \frac{\alpha+x_n}{Cx_{n-1}+x_{n-2}}$			**Conjecture: P_6-Tricho**
$\#73 : x_{n+1} = \frac{\alpha+x_{n-1}}{A+x_{n-2}}$			**P_2-Tricho**
$\#75 : x_{n+1} = \frac{\alpha+x_{n-1}}{Bx_n+x_{n-2}}$			**Part of a P_2-Tricho**
$\#76 : x_{n+1} = \frac{\alpha+x_{n-1}}{x_{n-1}+Dx_{n-2}}$		**Conjecture: $ESCP_2$**	
$\#97 : x_{n+1} = \frac{\gamma x_{n-1}+x_{n-2}}{A+x_{n-2}}$			**P_2-Tricho**
$\#102 : x_{n+1} = \frac{1}{A+Bx_n+x_{n-2}}$	**ESC$\bar{x}$**		
$\#103 : x_{n+1} = \frac{1}{A+Cx_{n-1}+x_{n-2}}$	**ESC$\bar{x}$**		
$\#104 : x_{n+1} = \frac{1}{x_n+Cx_{n-1}+Dx_{n-2}}$	**ESC$\bar{x}$**		
$\#110 : x_{n+1} = \frac{x_{n-1}}{A+Bx_n+x_{n-2}}$			**P_2-Tricho**
$\#111 : x_{n+1} = \frac{x_{n-1}}{A+x_{n-1}+Dx_{n-2}}$		**Conjecture: $ESCP_2$**	

$\#112 : x_{n+1} = \frac{x_{n-1}}{Bx_n+x_{n-1}+Dx_{n-2}}$		**Conjecture: ESCP$_2$**	
$\#113 : x_{n+1} = \frac{x_{n-2}}{A+Bx_n+x_{n-1}}$			**P$_3$-Tricho**
$\#123 : x_{n+1} = \frac{\alpha+\beta x_n+x_{n-2}}{x_{n-1}}$			**Conjecture: P$_4$-Tricho**
$\#128 : x_{n+1} = \frac{\alpha+\gamma x_{n-1}+x_{n-2}}{x_{n-2}}$			**P$_2$-Tricho**
$\#133 : x_{n+1} = \frac{1}{A+Bx_n+Cx_{n-1}+x_{n-2}}$	**ESC$\bar{x}$**		
$\#146 : x_{n+1} = \frac{\alpha+x_{n-1}}{A+Bx_n+x_{n-2}}$			**P$_2$-Tricho**
$\#147 : x_{n+1} = \frac{\alpha+x_{n-1}}{A+x_{n-1}+Dx_{n-2}}$		**Conjecture: ESCP$_2$**	
$\#148 : x_{n+1} = \frac{\alpha+x_{n-1}}{Bx_n+x_{n-1}+Dx_{n-2}}$		**Conjecture: ESCP$_2$**	
$\#179 : x_{n+1} = \frac{\alpha+\gamma x_{n-1}+x_{n-2}}{A+x_{n-2}}$			**P$_2$-Tricho**

Appendix A

Table on the Global Character of the 225 Special Cases of $x_{n+1} = \dfrac{\alpha + \beta x_n + \gamma x_{n-1} + \delta x_{n-2}}{A + Bx_n + Cx_{n-1} + Dx_{n-2}}$

A boldfaced **B** indicates that every solution of the equation in this special case is bounded and a boldfaced **U** indicates that the equation in this special case has unbounded solutions in some range of its parameters and for some initial conditions.

A boldfaced **B*** next to an equation indicates that we only conjecture that every solution of the equation is bounded and a boldfaced **U*** indicates that we only conjecture that the equation has unbounded solutions in some range of its parameters and for some initial conditions. Next to each case, we have also provided some relevant references and results on its global character.

In addition to **B**, **B***, **U**, and **U*** we will also use the following abbreviations:

iff stands for **"if and only if."**

ESB stands for **"every solution of the equation is bounded."**

∃US stands for **"there exist unbounded solutions."**

LAS stands for **"locally asymptotically stable"** or **"local asymptotic stability."**

GAS stands for **"globally asymptotically stable"** or **"global asymptotic stability."**

GA stands for **"global attractivity"** or **"global attractor."**
In this book by the abbreviation GA we mean that every positive solution of the equation has a finite limit.

ESC$\bar{x}$ stands for **"every solution of the equation converges to the unique equilibrium point of the equation."**

ESC stands for **"every solution of the equation converges to the zero equilibrium or to the positive equilibrium of the equation."**

EPSC$\bar{x}$ stands for **"every positive solution of the equation converges to the positive equilibrium point of the equation."**

EBSC$\bar{x}$ stands for **"every positive and bounded solution of the equation converges to the positive equilibrium point of the equation."**

$\exists!$ **P_2-solution** stands for **"the equation has a unique prime period-two cycle."**

ESP$_k$ stands for **"every solution of the equation is periodic with (not necessarily prime) period k."**

ESCP$_k$ stands for **"every solution of the equation converges to a (not necessarily prime) period-k solution."**

EBSCP$_k$ stands for **"every bounded solution of the equation converges to a (not necessarily prime) period-k solution."**

Has P$_k$-Tricho stands for **"the equation has period-k trichotomy."**

1	$x_{n+1} = \alpha$	**B**	Thm 2.1.1. This equation is trivial.
2	$x_{n+1} = \frac{\alpha}{x_n}$	**B**	Thm 2.1.1. **ESP$_2$.** **Periodicity destroys boundedness.**
3	$x_{n+1} = \frac{\alpha}{x_{n-1}}$	**B**	Thm 2.1.1. **ESP$_4$.** **Periodicity destroys boundedness.**
4	$x_{n+1} = \frac{\alpha}{x_{n-2}}$	**B**	Thm 2.1.1. **ESP$_6$.** **Periodicity destroys boundedness.**
5	$x_{n+1} = \beta x_n$	**U**	This is a linear equation.
6	$x_{n+1} = \beta$	**B**	Thm 2.1.1. This equation is trivial.
7	$x_{n+1} = \frac{x_n}{x_{n-1}}$	**B**	Thm 2.2.1. **ESP$_6$.** **Periodicity destroys boundedness.**
8	$x_{n+1} = \frac{x_n}{x_{n-2}}$	**U**	Reducible to linear equation.
9	$x_{n+1} = \gamma x_{n-1}$	**U**	**This is the only linear equation with P$_2$-Tricho.**
10	$x_{n+1} = \frac{x_{n-1}}{x_n}$	**U**	Reducible to linear equation and part of a **P$_2$-Tricho.**

11	$x_{n+1} = \gamma$	**B**	Thm 2.1.1. This equation is trivial.
12	$x_{n+1} = \frac{x_{n-1}}{x_{n-2}}$	**U**	Reducible to linear equation and part of a **$\mathbf{P_2}$-Tric**
13	$x_{n+1} = \delta x_{n-2}$	**U**	**This is the only linear equation with $\mathbf{P_3}$-Tric**
14	$x_{n+1} = \frac{x_{n-2}}{x_n}$	**U**	Reducible to linear equation and part of a **$\mathbf{P_5}$-Tric** See #50.
15	$x_{n+1} = \frac{x_{n-2}}{x_{n-1}}$	**U**	Reducible to linear equation.
16	$x_{n+1} = \delta$	**B**	Thm 2.1.1. This equation is trivial.
17	$x_{n+1} = \frac{1}{A+x_n}$	**B**	Thm 2.1.1. This is a **Riccati equation** ; **ESC$\bar{x}$**.
18	$x_{n+1} = \frac{1}{A+x_{n-1}}$	**B**	Thm 2.1.1. This is a **Riccati-type equation** ; **ESC$\bar{x}$**.
19	$x_{n+1} = \frac{1}{A+x_{n-2}}$	**B**	Thm 2.1.1. This is a **Riccati-type equation** ; **ESC$\bar{x}$**.
20	$x_{n+1} = \frac{1}{Bx_n+x_{n-1}}$	**B**	Thm 2.1.1. **ESC$\bar{x}$** by Thm 5.17.2. **Periodicity destroys boundedness.** See [208], [103], and [175, p. 55].

21	$x_{n+1} = \frac{1}{Bx_n + x_{n-2}}$	**B**	Thm 2.1.1. **ESCP$_2$** by Thm 5.17.2. See [91] and [103].
22	$x_{n+1} = \frac{1}{Cx_{n-1} + x_{n-2}}$	**B**	Thm 2.1.1. **ESC$\bar{x}$** by Thm 5.17.2. See [103] and [208].
23	$x_{n+1} = \frac{\beta x_n}{1 + x_n}$	**B**	Thm 2.1.1. This is the **Beverton-Holt Equation** ; **ESC.** See [83].
24	$x_{n+1} = \frac{\beta x_n}{1 + x_{n-1}}$	**B**	Thm 2.2.1 or Thm 2.3.4. **Pielou's Equation** ; **ESC.** See [63], [157], [175], [182], and [186].
25	$x_{n+1} = \frac{\beta x_n}{1 + x_{n-2}}$	**B**	Thm 2.3.3 or Thm 2.3.4. **Conjecture:** $1 < \beta < \frac{3+\sqrt{5}}{2} \Rightarrow$ **EPSC$\bar{x}$.** **Conjecture: There exist solutions that do not converge to the equilibrium or to a periodic solution.** See [157].
26	$x_{n+1} = \frac{x_n}{Bx_n + x_{n-1}}$	**B**	Thm 2.1.1. **ESC$\bar{x}$.** See [175, p. 58].
27	$x_{n+1} = \frac{x_n}{Bx_n + x_{n-2}}$	**B**	Thm 2.1.1. **Conjecture:** $-1 + \sqrt{2} < B < 1 \Rightarrow$ **ESC$\bar{x}$.** **Conjecture: There exist solutions that do not converge to an equilibrium or to a periodic solution.**

28	$x_{n+1} = \frac{x_n}{Cx_{n-1}+x_{n-2}}$	**U***	**This equation has not been investigated yet.** **This equation is part of a $\mathbf{P_6}$-Tricho Conjecture** ; See #70.
29	$x_{n+1} = \frac{x_{n-1}}{A+x_n}$	**U**	Thm 3.1.1. **Has $\mathbf{P_2}$-Tricho.** **When $A = 1$, a positive solution converges to zero iff $x_{n-1} > x_n$, for all $n \geq 0$.** See [110], [112], [133], [143], [146], [148], [149], [175], [226], [227], and [233].
30	$x_{n+1} = \frac{x_{n-1}}{A+x_{n-1}}$	**B**	Thm 2.1.1. This is a **Riccati-type equation.** **∃! $\mathbf{P_2}$-solution and it is not LAS** ; **ESC$\mathbf{P_2}$.**
31	$x_{n+1} = \frac{x_{n-1}}{A+x_{n-2}}$	**U**	Thm 3.2.1. **Has $\mathbf{P_2}$-Tricho.** **Conjecture: EBSC$\bar{x}$ when $A < 1$.** See [17] and [70].
32	$x_{n+1} = \frac{x_{n-1}}{Bx_n+x_{n-1}}$	**B**	Thm 2.1.1. **∃! $\mathbf{P_2}$-solution when $B \neq 1$ and it is LAS when $B > 1$.** **There exist infinitely many period-two solutions when $B = 1$.** **ESC$\mathbf{P_2}$** by Theorem 1.6.6. See [175, p. 60].
33	$x_{n+1} = \frac{x_{n-1}}{Bx_n+x_{n-2}}$	**U**	Thm 3.3.1. This equation is part of **a $\mathbf{P_2}$-Tricho** ; See #146. **Conjecture: EBSC$\bar{x}$.** See [72].

34	$x_{n+1} = \frac{x_{n-1}}{x_{n-1}+Dx_{n-2}}$	**B**	Thm 2.1.1. **∃! $\mathbf{P_2}$-solution when $D \neq 1$ and it is LAS when $D > 1$.** **There exist infinitely many period-two solutions when $D = 1$.** **Conjecture: $\mathbf{ESCP_2}$.**
35	$x_{n+1} = \frac{x_{n-2}}{A+x_n}$	**U**	Thm 3.5.1. **Has $\mathbf{P_3}$-Tricho.** **Conjecture: EBSC$\bar{x}$ when $A < 1$.** See [60] and [146].
36	$x_{n+1} = \frac{x_{n-2}}{A+x_{n-1}}$	**U**	Thm 3.5.1. **Has $\mathbf{P_3}$-Tricho.** **Conjecture: EBSC$\bar{x}$ when $A < 1$.** See [60] and [146].
37	$x_{n+1} = \frac{x_{n-2}}{A+x_{n-2}}$	**B**	Thm 2.1.1. This is a **Riccati-type equation.** **$\mathbf{ESCP_3}$.**
38	$x_{n+1} = \frac{x_{n-2}}{Bx_n+x_{n-1}}$	**U**	Thm 3.5.1. Part of a **$\mathbf{P_3}$-Tricho.** See [60] and [146].
39	$x_{n+1} = \frac{x_{n-2}}{Bx_n+x_{n-2}}$	**B**	Thm 2.1.1. ∃! **$\mathbf{P_3}$-solution** when $B < 1$. **Conjecture:** $B < 1+\sqrt{2} \Rightarrow$ **EPSC$\bar{x}$.** **Conjecture: $\mathbf{ESCP_{19}}$ when $B > 123$.** **Conjecture: There exist solutions that do not converge to the equilibrium or to a periodic solution.** See [60].

40	$x_{n+1} = \frac{x_{n-2}}{Cx_{n-1}+x_{n-2}}$	**B**	Thm 2.1.1. ∃! **P_3-solution** when $C < 1$. **Conjecture:** $C < \frac{1+\sqrt{5}}{2} \Rightarrow$ **EPSC$\bar{x}$.** **Conjecture: ESCP$_{13}$ when** $C > 8$. **Conjecture: There exist solutions that do not converge to the equilibrium or to a periodic solution.** See [60].
41	$x_{n+1} = \alpha + \beta x_n$	**U**	This is a linear equation.
42	$x_{n+1} = \beta + \frac{\alpha}{x_n}$	**B**	Thm 2.1.1. This is a **Riccati Equation** ; **ESC$\bar{x}$.**
43	$x_{n+1} = \frac{\alpha+x_n}{x_{n-1}}$	**B**	Thm 2.2.1. **Lyness's Equation.** **No nontrivial solution has a limit.** **It possesses the invariant:** $(\alpha + x_n + x_{n-1})(1 + \frac{1}{x_n})(1 + \frac{1}{x_{n-1}}) =$ **const.,** **for all** $n \geq 0$. **Periodicity destroys boundedness.** **Conjecture: There exist solutions that do not converge to the equilibrium or to a periodic solution.** See [20], [21], [22], [62], [116], [124], [157], [158], [174], [175, p. 70], [189], [193], [198], [199], [215], [216], [193], and [237].
44	$x_{n+1} = \frac{\alpha+x_n}{x_{n-2}}$	**U***	**This equation has not been investigated yet.** **This equation is part of a P$_6$-Tricho Conjecture** ; See #70.
45	$x_{n+1} = \alpha + \gamma x_{n-1}$	**U**	This equation is linear.

46	$x_{n+1} = \frac{\alpha+x_{n-1}}{x_n}$	**U**	Thm 3.1.1. This equation is part of a **P_2-Tricho** ; See #165. See [110] and [175].
47	$x_{n+1} = \gamma + \frac{\alpha}{x_{n-1}}$	**B**	Thm 2.1.1. This is a **Riccati-type equation** ; **ESC$\bar{x}$**.
48	$x_{n+1} = \frac{\alpha+x_{n-1}}{x_{n-2}}$	**U**	Thm 3.2.1. This equation is part of a **P_2-Tricho** ; See #146. **Conjecture: EBSC$\bar{x}$.** See [72].
49	$x_{n+1} = \alpha + \delta x_{n-2}$	**U**	This equation is linear.
50	$x_{n+1} = \frac{\alpha+x_{n-2}}{x_n}$	**U**	Thm 3.6.1. **Has P_5-Tricho.** See [54], [59], and [151].
51	$x_{n+1} = \frac{\alpha+x_{n-2}}{x_{n-1}}$	**U**	**This equation is part of a P_4-Tricho Conjecture.** See [47].
52	$x_{n+1} = \delta + \frac{\alpha}{x_{n-2}}$	**B**	Thm 2.1.1. This is a **Riccati-type equation** ; **ESC$\bar{x}$**.
53	$x_{n+1} = \beta x_n + \gamma x_{n-1}$	**U**	This is a linear equation.

54	$x_{n+1} = \beta + \frac{x_{n-1}}{x_n}$	U	Thm 3.1.1 ; **Has P_2-Tricho.** **The very first P_2-Tricho.** **EBSC$\bar{x}$ when $\beta < 1$ by Theorem 4.2.2.** See [16] and [175, p. 70].
55	$x_{n+1} = \gamma + \frac{x_n}{x_{n-1}}$	B	Thm 2.2.1 or Thm 2.3.4. **ESC$\bar{x}$.** See [175, p. 70].
56	$x_{n+1} = \frac{\beta x_n + x_{n-1}}{x_{n-2}}$	U*	**This equation has not been investigated yet for $\beta \neq 1$.** **For $\beta = 1$, ESB.** **For $\beta = 1$, it possesses the invariant:** $(x_n + x_{n-1})(1 + \frac{1}{x_n})(1 + \frac{1}{x_{n-1}}) =$ **const.,** **for all $n \geq 0$.** **Conjecture: $\exists$US iff $\beta \neq 1$.** **Conjecture: There exist bounded solutions that do not converge to the equilibrium or to a periodic solution.**
57	$x_{n+1} = \beta x_n + \delta x_{n-2}$	U	This is a linear equation.
58	$x_{n+1} = \beta + \frac{x_{n-2}}{x_n}$	B	Thm 2.4.1. **Conjecture:** $-1 + \sqrt{2} < \beta < 1 \Rightarrow$ **ESC$\bar{x}$.** **Conjecture: ESCP$_{19}$** when $\beta < \frac{1}{123}$. **Conjecture: There exist solutions that do not converge to the equilibrium or to a periodic solution.** **Periodicity destroys boundedness.** See [49], [65], and [87].
59	$x_{n+1} = \frac{\beta x_n + x_{n-2}}{x_{n-1}}$	U	**ESCP$_4$ when $\beta = 1$; $\exists$US when $\beta < 1$.** **Conjecture: Has P_4-Tricho.** **Conjecture: EBSC$\bar{x}$ when $\beta < 1$.** See [59], [150], and [222].

60	$x_{n+1} = \delta + \frac{x_n}{x_{n-2}}$	B	Thm 2.3.3 or Thm 2.3.4. **Conjecture:** $\delta > -1 + \sqrt{2} \Rightarrow$ **ESC**$\bar{x}$. **Conjecture: There exist solutions that do not converge to the equilibrium or to a periodic solution.** See [69].
61	$x_{n+1} = \gamma x_{n-1} + \delta x_{n-2}$	U	This equation is linear.
62	$x_{n+1} = \frac{\gamma x_{n-1} + x_{n-2}}{x_n}$	U	Thm 3.1.1. $\exists$**US** when $\gamma \geq 1$. **Conjecture:** $\frac{-1+\sqrt{3}}{2} < \gamma < 1 \Rightarrow$ **ESC**$\bar{x}$. **Conjecture: EBSC**$\bar{x}$ when $\gamma > 1$. **EBSCP**$_2$ when $\gamma = 1$. **Conjecture: There exist bounded solutions that do not converge to the equilibrium or to a periodic solution.** See [67] and [76].
63	$x_{n+1} = \gamma + \frac{x_{n-2}}{x_{n-1}}$	B	Thm 2.6.1. **Conjecture:** $\frac{-1+\sqrt{5}}{2} < \gamma < 1 \Rightarrow$ **ESC**$\bar{x}$. **Conjecture: ESCP**$_{13}$ **when** $\gamma < \frac{1}{8}$. **Conjecture: There exist solutions that do not converge to the equilibrium or to a periodic solution.** See [49] and [65].
64	$x_{n+1} = \delta + \frac{x_{n-1}}{x_{n-2}}$	U	Thm 3.2.1. **Has P**$_2$**-Tricho.** **Conjecture: EBSC**$\bar{x}$ **when** $\delta < 1$. See [70].

65	$x_{n+1} = \frac{\alpha + x_n}{A + x_n}$	**B**	Thm 2.1.1. This is the **Riccati Equation** with **Riccati number:** $\frac{\beta A - \alpha B}{(\beta + A)^2} \leq \frac{1}{4}$. **ESC**$\bar{x}$; See [45], [120], [124], [126], and [175, p. 17].
66	$x_{n+1} = \frac{\alpha + x_n}{A + x_{n-1}}$	**B**	Thm 2.2.1 or Thm 2.3.4. **Conjecture: ESC**$\bar{x}$. **Any claims prior to July 2007 that this conjecture has been confirmed are not correct.** See [104], [157], [158], and [175].
67	$x_{n+1} = \frac{\alpha + x_n}{A + x_{n-2}}$	**B**	Thm 2.3.3 or Thm 2.3.4. **Conjecture: For the equilibrium** $\bar{x}$, **LAS** $\Rightarrow$ **GA.** **Conjecture: There exist solutions that do not converge to an equilibrium or to a periodic solution.** See [157].
68	$x_{n+1} = \frac{\alpha + x_n}{Bx_n + x_{n-1}}$	**B**	Thm 2.1.1. **Conjecture: ESC**$\bar{x}$. **Any claims prior to July 2007 that this conjecture has been confirmed are not correct.** See [175, p. 82].
69	$x_{n+1} = \frac{\alpha + x_n}{x_n + Dx_{n-2}}$	**B**	Thm 2.1.1. **Conjecture: For the equilibrium** $\bar{x}$ **LAS** $\Rightarrow$ **GA.** **Conjecture: There exist solutions that do not converge to the equilibrium or to a periodic solution.**

70	$x_{n+1} = \frac{\alpha + x_n}{Cx_{n-1} + x_{n-2}}$	**U***	**Conjecture: Has $\mathbf{P_6}$-Tricho.** **This equation has not been investigated yet.** See [59].
71	$x_{n+1} = \frac{\alpha + x_{n-1}}{A + x_n}$	**U**	Thm 3.1.1. **Has $\mathbf{P_2}$-Tricho.** **EBSC$\bar{x}$ when $A < 1$ by Thm 4.2.2.** See [110] or [175, p. 89].
72	$x_{n+1} = \frac{\alpha + x_{n-1}}{A + x_{n-1}}$	**B**	Thm 2.1.1. This is a **Riccati-type equation** ; **ESC$\bar{x}$**.
73	$x_{n+1} = \frac{\alpha + x_{n-1}}{A + x_{n-2}}$	**U**	Thm 3.2.1. **Has $\mathbf{P_2}$-Tricho.** **Conjecture: EBSC$\bar{x}$** when $A < 1$. See [17] and [72].
74	$x_{n+1} = \frac{\alpha + x_{n-1}}{Bx_n + x_{n-1}}$	**B**	Thm 2.1.1. **∃! $\mathbf{P_2}$-solution when** $B > 1 + 4\alpha$ **and it is LAS.** **ESC$\mathbf{P_2}$.** See [15], [102], and [175, p. 92].
75	$x_{n+1} = \frac{\alpha + x_{n-1}}{Bx_n + x_{n-2}}$	**U**	Thm 3.3.1. This equation is part of a **$\mathbf{P_2}$-Tricho** ; See #146. **Conjecture: EBSC$\bar{x}$.** See [72].

76	$x_{n+1} = \frac{\alpha + x_{n-1}}{x_{n-1} + Dx_{n-2}}$	B	Thm 2.1.1. **For the equilibrium $\bar{x}$,** **$\mathbf{D} \leq \mathbf{1} + 4\alpha \Rightarrow$ GAS.** **$\exists!$ $\mathbf{P_2}$-solution when $\mathbf{D} > \mathbf{1} + 4\alpha$** **and it is LAS.** **Conjecture: $\mathbf{ESCP_2}$.** See [102], which extends and unifies #74 and #76.
77	$x_{n+1} = \frac{\alpha + x_{n-2}}{A + x_n}$	B	Thm 2.5.1. **Conjecture: For the equilibrium $\bar{x}$,** **LAS $\Rightarrow$ GA.** **Conjecture: There exist solutions that** **do not converge to the equilibrium** **or to a periodic solution.** **Periodicity destroys boundedness.** See [49] and [65].
78	$x_{n+1} = \frac{\alpha + x_{n-2}}{A + x_{n-1}}$	B	Thm 2.7.1. **Conjecture: For the equilibrium $\bar{x}$,** **LAS $\Rightarrow$ GA.** **Conjecture: There exist solutions that** **do not converge to the equilibrium** **or to a periodic solution.** **Periodicity destroys boundedness.** See [49] and [65].
79	$x_{n+1} = \frac{\alpha + x_{n-2}}{A + x_{n-2}}$	B	Thm 2.1.1. This is a **Riccati-type equation** ; **ESC$\bar{x}$.**
80	$x_{n+1} = \frac{\alpha + x_{n-2}}{Bx_n + x_{n-1}}$	U	Thm 3.4.1. **$\exists$US for all positive values** **of the parameters.** **For the equilibrium $\bar{x}$,** **LAS $\not\Rightarrow$ GA.** **Conjecture: There exist bounded** **solutions that do not converge** **to the equilibrium or** **to a periodic solution.** See [61] and [69].

81	$x_{n+1} = \frac{\alpha + x_{n-2}}{Bx_n + x_{n-2}}$	B	Thm 2.1.1. **Conjecture: For the equilibrium $\bar{x}$, LAS $\Rightarrow$ GA.** **Conjecture: There exist solutions that do not converge to the equilibrium or to a periodic solution.** See [90].
82	$x_{n+1} = \frac{\alpha + x_{n-2}}{Cx_{n-1} + x_{n-2}}$	B	Thm 2.1.1. **Conjecture: For the equilibrium $\bar{x}$, LAS $\Rightarrow$ GA.** **Conjecture: There exist solutions that do not converge to the equilibrium or to a periodic solution.** See [125].
83	$x_{n+1} = \frac{x_n + \gamma x_{n-1}}{A + x_n}$	U	Thm 3.1.1. **EBSC$\bar{x}$ when $\gamma > 1 + \mathbf{A}$ by Thm 4.2.2.** **Has $\mathbf{P_2}$-Tricho.** See [175] and [179].
84	$x_{n+1} = \frac{\beta x_n + x_{n-1}}{A + x_{n-1}}$	B	Thm 2.2.1 or Thm 2.3.4. **ESC.** See [175, p. 109] and [180].
85	$x_{n+1} = \frac{x_n + \gamma x_{n-1}}{A + x_{n-2}}$	U	Thm 3.2.1. **Conjecture: EBSC$\bar{x}$ when $\gamma > 1 + A$.** **Conjecture: For the positive equilibrium $\bar{x}$, LAS $\Rightarrow$ GA when $\gamma < 1 + A$.** **Conjecture: $\mathbf{ESCP_2}$ when $\gamma = 1 + A$.** **Conjecture: There exist bounded solutions that do not converge to an equilibrium or to a periodic solution.** See [49] and [69].

86	$x_{n+1} = \frac{\beta x_n + x_{n-1}}{B x_n + x_{n-1}}$	**B**	Thm 2.1.1. **∃! $\mathbf{P_2}$-solution and it is LAS.** **$\mathbf{ESCP_2}$.** See [15], [175, p. 113], [183], and [205].
87	$x_{n+1} = \frac{x_n + \gamma x_{n-1}}{x_n + D x_{n-2}}$	**U**	Thm 3.3.1. **Conjecture: EBSC$\bar{x}$ when $\gamma > 1$.** **Conjecture: For the equilibrium $\bar{x}$, LAS ⇒ GA when $\gamma < 1$.** **Conjecture: $\mathbf{ESCP_2}$ when $\gamma = 1$.** **Conjecture: There exist bounded solutions that do not converge to the equilibrium or to a periodic solution.**
88	$x_{n+1} = \frac{\beta x_n + x_{n-1}}{x_{n-1} + D x_{n-2}}$	**B**	Thm 2.8.1. **∃! $\mathbf{P_2}$-solution and we conjecture that it is LAS.** **Conjecture: For the equilibrium $\bar{x}$, LAS ⇒ GA.** **Conjecture: There exist solutions that do not converge to the equilibrium or to a periodic solution.** See [49] and [119].
89	$x_{n+1} = \frac{\beta x_n + \delta x_{n-2}}{1 + x_n}$	**B**	Thm 2.5.1. **Conjecture: For the positive equilibrium $\bar{x}$, LAS ⇒ GA.** **Conjecture: There exist solutions that do not converge to an equilibrium or to a periodic solution.**
90	$x_{n+1} = \frac{\beta x_n + \delta x_{n-2}}{1 + x_{n-1}}$	**B**	Thm 2.7.1. **Conjecture: For the positive equilibrium $\bar{x}$, LAS ⇒ GA.** **Conjecture: There exist solutions that do not converge to an equilibrium or to a periodic solution.**

91	$x_{n+1} = \frac{\beta x_n + x_{n-2}}{A + x_{n-2}}$	B	Thm 2.3.3 or Thm 2.3.4. **Conjecture: For the positive equilibrium $\bar{x}$, LAS $\Rightarrow$ GA.** **Conjecture: There exist solutions that do not converge to an equilibrium or to a periodic solution.** See [69].
92	$x_{n+1} = \frac{x_n + \delta x_{n-2}}{x_n + C x_{n-1}}$	U	Thm 3.4.1. **For the equilibrium $\bar{x}$, LAS $\not\Rightarrow$ GA.** **ESB iff $\delta \leq C$.** **Conjecture: For the equilibrium $\bar{x}$ in the region where ESB, LAS $\Rightarrow$ GA.** **Conjecture: There exist bounded solutions that do not converge to the equilibrium or to a periodic solution.** See [61] and [69].
93	$x_{n+1} = \frac{\beta x_n + x_{n-2}}{B x_n + x_{n-2}}$	B	Thm 2.1.1. **Conjecture: For the equilibrium $\bar{x}$, LAS $\Rightarrow$ GA.** **Conjecture: There exist solutions that do not converge to the equilibrium or to a periodic solution.**
94	$x_{n+1} = \frac{\beta x_n + x_{n-2}}{C x_{n-1} + x_{n-2}}$	B	Thm 2.9.1 **Conjecture: For the equilibrium $\bar{x}$, LAS $\Rightarrow$ GA.** **Conjecture: There exist solutions that do not converge to an equilibrium or to a periodic solution.** See [152].

95	$x_{n+1} = \frac{\gamma x_{n-1}+x_{n-2}}{A+x_n}$	**U**	Thm 3.1.1. **Conjecture: EBSC$\bar{x}$ when $\gamma > 1 + A$.** **Conjecture: For the positive equilibrium $\bar{x}$, LAS $\Rightarrow$ GAS when $\gamma < 1 + A$.** **ESCP$_2$ when $\gamma = 1 + A$.** **Conjecture: There exist bounded solutions that do not converge to an equilibrium or to a periodic solution.** See [67].
96	$x_{n+1} = \frac{\gamma x_{n-1}+x_{n-2}}{A+x_{n-1}}$	**B**	Thm 2.7.1. **Conjecture: For the positive equilibrium $\bar{x}$, LAS $\Rightarrow$ GA.** **Conjecture: There exist solutions that do not converge to an equilibrium or to a periodic solution.** See [49].
97	$x_{n+1} = \frac{\gamma x_{n-1}+x_{n-2}}{A+x_{n-2}}$	**U**	Thm 3.2.1. **Has P$_2$-Tricho.** **Conjecture: EBSC$\bar{x}$ when $\gamma > 1 + A$.** See [70] and [128].
98	$x_{n+1} = \frac{x_{n-1}+\delta x_{n-2}}{Bx_n+x_{n-1}}$	**U**	Thm 3.4.1. **For the equilibrium $\bar{x}$, LAS $\not\Rightarrow$ GA.** **ESB iff $\delta \leq B$.** **Conjecture: For the equilibrium $\bar{x}$ in the region where ESB, LAS $\Rightarrow$ GA.** **$\exists!$ P$_2$-solution and we conjecture that it is LAS.** **Conjecture: There exist solutions that do not converge to the equilibrium or to a periodic solution.** See [61].

99	$x_{n+1} = \frac{\gamma x_{n-1}+x_{n-2}}{Bx_n+x_{n-2}}$	**U**	Thm 3.3.1. **Conjecture: EBSC$\bar{x}$ when $\gamma > 1$.** **Conjecture: For the equilibrium $\bar{x}$, LAS $\Rightarrow$ GA when $\gamma < 1$.** **Conjecture: EBSCP$_2$ when $\gamma = 1$.** **Conjecture: There exist bounded solutions that do not converge to the equilibrium or to a periodic solution.** See [56].
100	$x_{n+1} = \frac{\gamma x_{n-1}+x_{n-2}}{Cx_{n-1}+x_{n-2}}$	**B**	Thm 2.1.1. **Conjecture: For the equilibrium $\bar{x}$, LAS $\Rightarrow$ GA.** **$\exists!$ P$_2$-solution and it is LAS.** **Conjecture: There exist solutions that do not converge to the equilibrium or to a periodic solution.** See [140] and [141].
101	$x_{n+1} = \frac{1}{A+Bx_n+x_{n-1}}$	**B**	Thm 2.1.1. **ESC$\bar{x}$.** See [175, p. 71] and [208].
102	$x_{n+1} = \frac{1}{A+Bx_n+x_{n-2}}$	**B**	Thm 2.1.1. **ESC$\bar{x}$.** See [208].
103	$x_{n+1} = \frac{1}{A+Cx_{n-1}+x_{n-2}}$	**B**	Thm 2.1.1. **ESC$\bar{x}$.** See [208].
104	$x_{n+1} = \frac{1}{Bx_n+Cx_{n-1}+x_{n-2}}$	**B**	Thm 2.1.1. **ESC$\bar{x}$.** See [103] and [208].

105	$x_{n+1} = \frac{\beta x_n}{1+Bx_n+x_{n-1}}$	B	Thm 2.1.1. **ESC by Thms 5.23.2, 5.23.4, and 1.6.3.**
106	$x_{n+1} = \frac{\beta x_n}{1+Bx_n+x_{n-2}}$	B	Thm 2.1.1. **Conjecture: For the positive equilibrium $\bar{x}$, LAS $\Rightarrow$ GA.** **Conjecture: There exist solutions that do not converge to an equilibrium or to a periodic solution.**
107	$x_{n+1} = \frac{\beta x_n}{1+Cx_{n-1}+x_{n-2}}$	B	Thm 2.3.1 or Thm 2.3.4. **Conjecture: For the positive equilibrium $\bar{x}$, LAS $\Rightarrow$ GA.** **Conjecture: There exist solutions that do not converge to an equilibrium or to a periodic solution.** See [157].
108	$x_{n+1} = \frac{x_n}{x_n+Cx_{n-1}+Dx_{n-2}}$	B	Thm 2.1.1. **Conjecture: For the equilibrium $\bar{x}$, LAS $\Rightarrow$ GA.** **Conjecture: There exist solutions that do not converge to the equilibrium or to a periodic solution.**
109	$x_{n+1} = \frac{x_{n-1}}{A+Bx_n+x_{n-1}}$	B	Thm 2.1.1. **$\exists!$ P_2-solution when $0 \leq A < 1$ and $B \neq 1$ and it is LAS when $0 \leq A < 1$ and $B > 1$.** **There exist infinitely many period-two solutions when $0 \leq A < 1$ and $B = 1$.** **$ESCP_2$ by Thm 1.6.6.** See [61] and [175, p. 133].

110	$x_{n+1} = \frac{x_{n-1}}{A+Bx_n+x_{n-2}}$	**U**	Thm 3.3.1. **Has P_2-Tricho.** **Conjecture: EBSC$\bar{x}$ when $A < 1$.** See [72].
111	$x_{n+1} = \frac{x_{n-1}}{A+x_{n-1}+Dx_{n-2}}$	**B**	Thm 2.1.1. **∃! P_2-solution** **when $0 \leq A < 1$ and $D \neq 1$** **and it is LAS** **when $0 \leq A < 1$ and $D > 1$.** **There exist infinitely many** **period-two solutions** **when $0 \leq A < 1$ and $D = 1$.** **Conjecture: ESCP_2.**
112	$x_{n+1} = \frac{x_{n-1}}{Bx_n+x_{n-1}+Dx_{n-2}}$	**B**	Thm 2.1.1. **∃! P_2-solution when $B+D \neq 1$** **and it is LAS when $B+D > 1$.** **Conjecture: ESCP_2.**
113	$x_{n+1} = \frac{x_{n-2}}{A+Bx_n+x_{n-1}}$	**U**	Thm 3.5.1. **Has P_3-Tricho.** See [60] and [146].
114	$x_{n+1} = \frac{x_{n-2}}{A+Bx_n+x_{n-2}}$	**B**	Thm 2.1.1. **Conjecture: For the positive** **equilibrium $\bar{x}$, LAS ⇒ GA.** **Conjecture: There exist solutions** **that do not converge** **to an equilibrium or** **to a periodic solution.**

115	$x_{n+1} = \frac{x_{n-2}}{A+Cx_{n-1}+x_{n-2}}$	B	Thm 2.1.1. **Conjecture: For the positive equilibrium $\bar{x}$, LAS $\Rightarrow$ GA.** **Conjecture: There exist solutions that do not converge to an equilibrium or to a periodic solution.**
116	$x_{n+1} = \frac{x_{n-2}}{Bx_n+Cx_{n-1}+x_{n-2}}$	B	Thm 2.1.1. **Conjecture: For the equilibrium $\bar{x}$, LAS $\Rightarrow$ GA.** **Conjecture: There exist solutions that do not converge to the equilibrium or to a periodic solution.**
117	$x_{n+1} = 1 + \beta x_n + \gamma x_{n-1}$	U	This is a linear equation.
118	$x_{n+1} = \frac{\alpha+x_n+\gamma x_{n-1}}{x_n}$	U	Thm 3.1.1. **Has P_2-Tricho.** See [175, p. 137].
119	$x_{n+1} = \frac{\alpha+\beta x_n+x_{n-1}}{x_{n-1}}$	B	Thm 2.2.1 or Thm 2.3.4. **Conjecture: ESC$\bar{x}$.** Can be transformed to #66 with $\alpha > A$, which remains a conjecture. See [175, p. 137].

120	$x_{n+1} = \frac{\alpha+\beta x_n+x_{n-1}}{x_{n-2}}$	**U***	**ESP_8 when $\alpha = \beta = 1$.** **Todd's Equation.** **Conjecture: ∃US iff $\beta \neq 1$.** **When $\beta = 1$, it possesses the invariant:** $(\alpha + \sum_{i=0}^{2} x_{n-i}) \prod_{j=0}^{2}(1 + \frac{1}{x_{n-j}})$ **=const., for all $n \geq 0$.** **Conjecture: There exist solutions that do not converge to the equilibrium or to a periodic solution.**
121	$x_{n+1} = 1 + \beta x_n + \delta x_{n-2}$	**U**	This equation is linear.
122	$x_{n+1} = \frac{\alpha+x_n+\delta x_{n-2}}{x_n}$	**B**	Thm 2.5.1. **Conjecture: For the equilibrium $\bar{x}$, LAS ⇒ GA.** **Conjecture: There exist solutions that do not converge to the equilibrium or to a periodic solution.**
123	$x_{n+1} = \frac{\alpha+\beta x_n+x_{n-2}}{x_{n-1}}$	**U**	**Conjecture: Has P_4-Tricho.** **Conjecture: ESC$\bar{x}$ when $\beta > 1$.** See [59], [150], and [222].
124	$x_{n+1} = \frac{\alpha+\beta x_n+x_{n-2}}{x_{n-2}}$	**B**	Thm 2.3.3 or Thm 2.3.4. **Conjecture: For the equilibrium $\bar{x}$, LAS ⇒ GA.** **Conjecture: There exist solutions that do not converge to the equilibrium or to a periodic solution.** See [69].

125	$x_{n+1} = 1 + \gamma x_{n-1} + \delta x_{n-2}$	U	This equation is linear.
126	$x_{n+1} = \frac{\alpha+\gamma x_{n-1}+x_{n-2}}{x_n}$	U	Thm 3.1.1. $\exists$**US** when $\gamma \geq 1$. **Conjecture: EBSC$\bar{x}$ when $\gamma > 1$.** **EBSCP$_2$ when $\gamma = 1$.** **Conjecture: For the equilibrium $\bar{x}$, LAS $\Rightarrow$ GA.** **Conjecture: There exist solutions that do not converge to the equilibrium or to a periodic solution.** See [46] and [67].
127	$x_{n+1} = \frac{\alpha+x_{n-1}+\delta x_{n-2}}{x_{n-1}}$	B	Thm 2.7.1. **Conjecture: For the equilibrium $\bar{x}$, LAS $\Rightarrow$ GA.** **Conjecture: There exist solutions that do not converge to the equilibrium or to a periodic solution.** See [49].
128	$x_{n+1} = \frac{\alpha+\gamma x_{n-1}+x_{n-2}}{x_{n-2}}$	U	Thm 3.2.1. **Has P$_2$-Tricho.** **Conjecture: EBSC$\bar{x}$ when $\gamma > 1$.** See [70].
129	$x_{n+1} = \beta x_n + \gamma x_{n-1} + \delta x_{n-2}$	U	This equation is linear.

130	$x_{n+1} = \frac{x_n+\gamma x_{n-1}+\delta x_{n-2}}{x_n}$	**U**	Thm 3.1.1. **Conjecture: EBSC$\bar{x}$ when $\gamma > 1+\delta$.** **Conjecture: For the equilibrium $\bar{x}$, LAS $\Rightarrow$ GA when $\gamma < 1+\delta$.** **Conjecture: ESCP$_2$ when $\gamma = 1+\delta$.** **Conjecture: There exist bounded solutions that do not converge to the equilibrium or to a periodic solution.** See [67].
131	$x_{n+1} = \frac{\beta x_n+x_{n-1}+\delta x_{n-2}}{x_{n-1}}$	**B**	Thm 2.7.1. **Conjecture: For the equilibrium $\bar{x}$, LAS $\Rightarrow$ GA.** **Conjecture: There exist solutions which do not converge to the equilibrium or to a periodic solution.** See [49].
132	$x_{n+1} = \frac{\beta x_n+\gamma x_{n-1}+x_{n-2}}{x_{n-2}}$	**U**	Thm 3.2.1. **Conjecture: EBSC$\bar{x}$ when $\gamma > 1+\beta$.** **Conjecture: For the equilibrium $\bar{x}$, LAS $\Rightarrow$ GA when $\gamma < 1+\beta$.** **Conjecture: ESCP$_2$ when $\gamma = 1+\beta$.** **Conjecture: There exist bounded solutions that do not converge to the equilibrium or to a periodic solution.** See [49].
133	$x_{n+1} = \frac{1}{A+Bx_n+Cx_{n-1}+x_{n-2}}$	**B**	Thm 2.1.1. **ESC$\bar{x}$.** See [208].

134	$x_{n+1} = \frac{x_n}{A+Bx_n+Cx_{n-1}+x_{n-2}}$	B	Thm 2.1.1. **Conjecture: For the positive equilibrium $\bar{x}$, LAS $\Rightarrow$ GA.** **Conjecture: There exist solutions that do not converge to an equilibrium or to a periodic solution.**
135	$x_{n+1} = \frac{x_{n-1}}{A+Bx_n+Cx_{n-1}+x_{n-2}}$	B	Thm 2.1.1. **Conjecture: For the positive equilibrium $\bar{x}$, LAS $\Rightarrow$ GA.** **$\exists!$ P_2-solution and it is LAS.** **Conjecture: There exist solutions that do not converge to an equilibrium or to a periodic solution.**
136	$x_{n+1} = \frac{x_{n-2}}{A+Bx_n+Cx_{n-1}+x_{n-2}}$	B	Thm 2.1.1. **Conjecture: For the positive equilibrium $\bar{x}$, LAS $\Rightarrow$ GA.** **Conjecture: There exist solutions that do not converge to an equilibrium or to a periodic solution.**
137	$x_{n+1} = 1 + \beta x_n + \gamma x_{n-1} + \delta x_{n-2}$	U	This equation is linear.
138	$x_{n+1} = \frac{\alpha+x_n+\gamma x_{n-1}+\delta x_{n-2}}{x_n}$	U	Thm 3.1.1. **Conjecture: EBSC$\bar{x}$ when $\gamma > 1+\delta$.** **Conjecture: For the equilibriu $\bar{x}$, LAS $\Rightarrow$ GA when $\gamma < 1+\delta$.** **ESCP_2 when $\gamma = 1+\delta$.** **Conjecture: There exist bounded solutions that do not converge to the equilibrium or to a periodic solution.** See [67].

139	$x_{n+1} = \frac{\alpha+\beta x_n+x_{n-1}+\delta x_{n-2}}{x_{n-1}}$	**B**	Thm 2.7.1. **Conjecture: For the equilibrium $\bar{x}$, LAS ⇒ GA.** **Conjecture: There exist solutions that do not converge to the equilibrium or to a periodic solution.** See [49].
140	$x_{n+1} = \frac{\alpha+\beta x_n+\gamma x_{n-1}+x_{n-2}}{x_{n-2}}$	**U**	Thm 3.2.1. **Conjecture: EBSC$\bar{x}$ when $\gamma > 1+\beta$.** **Conjecture: For the equilibrium $\bar{x}$, LAS ⇒ GA when $\gamma < 1+\beta$.** **Conjecture: ESCP$_2$ when $\gamma = 1+\beta$.** **Conjecture: There exist bounded solutions that do not converge to the equilibrium or to a periodic solution.** See [49].
141	$x_{n+1} = \frac{\alpha+x_n}{A+Bx_n+x_{n-1}}$	**B**	Thm 2.1.1. **Conjecture: ESC$\bar{x}$.** **Any claims prior to July 2007 that this conjecture has been confirmed are not correct.** See [15] and [175, p. 141].
142	$x_{n+1} = \frac{\alpha+x_n}{A+x_n+Dx_{n-2}}$	**B**	Thm 2.1.1. **Conjecture: For the equilibrium $\bar{x}$, LAS ⇒ GA.** **Conjecture: There exist solutions that do not converge to the equilibrium or to a periodic solution.**

143	$x_{n+1} = \frac{\alpha+x_n}{A+Cx_{n-1}+x_{n-2}}$	**B**	Thm 2.3.1 or Thm 2.3.4. **Conjecture: For the equilibrium $\bar{x}$, LAS $\Rightarrow$ GA.** **Conjecture: There exist solutions that do not converge to the equilibrium or to a periodic solution.** See [69].
144	$x_{n+1} = \frac{\alpha+x_n}{x_n+Cx_{n-1}+Dx_{n-2}}$	**B**	Thm 2.1.1. **Conjecture: For the equilibrium $\bar{x}$, LAS $\Rightarrow$ GA.** **Conjecture: There exist solutions that do not converge to the equilibrium or to a periodic solution.**
145	$x_{n+1} = \frac{\alpha+x_{n-1}}{A+Bx_n+x_{n-1}}$	**B**	Thm 2.1.1. **$\exists!$ P_2-solution and it is LAS.** **$ESCP_2$.** See [15] and [175, p. 149].
146	$x_{n+1} = \frac{\alpha+x_{n-1}}{A+Bx_n+x_{n-2}}$	**U**	Thm 3.3.1. **Has P_2-Tricho.** **Conjecture: EBSC$\bar{x}$ when $A < 1$.** See [49] and [72].
147	$x_{n+1} = \frac{\alpha+x_{n-1}}{A+x_{n-1}+Dx_{n-2}}$	**B**	Thm 2.1.1. **Conjecture: For the equilibrium $\bar{x}$, LAS $\Rightarrow$ GA.** **Conjecture: $\exists!$ P_2-solution and we conjecture that it is LAS.** **Conjecture: $ESCP_2$.**

148	$x_{n+1} = \frac{\alpha+x_{n-1}}{Bx_n+x_{n-1}+Dx_{n-2}}$	**B**	Thm 2.1.1. **Conjecture: For the equilibrium $\bar{x}$, LAS $\Rightarrow$ GA.** **Conjecture: $\exists!$ P_2-solution and we conjecture that it is LAS.** **Conjecture: $ESCP_2$.**
149	$x_{n+1} = \frac{\alpha+x_{n-2}}{A+Bx_n+x_{n-1}}$	**U**	Thm 3.4.1. **ESB iff** $A \geq 1$. **For the equilibrium** $\bar{x}$, **LAS $\not\Rightarrow$ GA.** **ESC$\bar{x}$ iff** $A \geq 1$. **Conjecture: There exist bounded solutions that do not converge to the equilibrium or to a periodic solution.** See [61] and [69].
150	$x_{n+1} = \frac{\alpha+x_{n-2}}{A+Bx_n+x_{n-2}}$	**B**	Thm 2.1.1. **Conjecture: For the equilibrium $\bar{x}$, LAS $\Rightarrow$ GA.** **Conjecture: There exist solutions that do not converge to the equilibrium or to a periodic solution.**
151	$x_{n+1} = \frac{\alpha+x_{n-2}}{A+Cx_{n-1}+x_{n-2}}$	**B**	Thm 2.1.1. **Conjecture: For the equilibrium $\bar{x}$, LAS $\Rightarrow$ GA.** **Conjecture: There exist solutions that do not converge to the equilibrium or to a periodic solution.**

152	$x_{n+1} = \frac{\alpha + x_{n-2}}{Bx_n + Cx_{n-1} + x_{n-2}}$	B	Thm 2.1.1. **Conjecture: For the equilibrium $\bar{x}$, LAS $\Rightarrow$ GA.** **Conjecture: There exist solutions that do not converge to the equilibrium or to a periodic solution.**
153	$x_{n+1} = \frac{\beta x_n + x_{n-1}}{A + Bx_n + x_{n-1}}$	B	Thm 2.1.1. **$\exists!$ P_2-solution and we conjecture that it is LAS.** **Conjecture: $ESCP_2$.** See [175, p. 158].
154	$x_{n+1} = \frac{x_n + \gamma x_{n-1}}{A + x_n + Dx_{n-2}}$	U	Thm 3.3.1. **Conjecture: For the positive equilibrium $\bar{x}$, LAS $\Rightarrow$ GA when $A - 1 < \gamma < 1 + A$.** **Conjecture: EBSC$\bar{x}$ when $\gamma > 1 + A$.** **Conjecture: $ESCP_2$ when $\gamma = 1 + A$.** **Conjecture: There exist bounded solutions that do not converge to an equilibrium or to a periodic solution.** See [49].
155	$x_{n+1} = \frac{\beta x_n + x_{n-1}}{A + x_{n-1} + Dx_{n-2}}$	B	Thm 2.3.1 or Thm 2.3.4. **Conjecture: For the positive equilibrium $\bar{x}$, LAS $\Rightarrow$ GA.** **$\exists!$ P_2-solution and we conjecture that it is LAS.** **Conjecture: There exist solutions that do not converge to an equilibrium or to a periodic solution.** See [69].

156	$x_{n+1} = \frac{x_n+\gamma x_{n-1}}{x_n+Cx_{n-1}+Dx_{n-2}}$	**B**	Thm 2.1.1. **Conjecture: For the equilibrium $\bar{x}$, LAS $\Rightarrow$ GA.** **$\exists!$ P$_2$-solution and we conjecture that it is LAS.** **Conjecture: There exist solutions that do not converge to the equilibrium or to a periodic solution.**
157	$x_{n+1} = \frac{\beta x_n+x_{n-2}}{A+Bx_n+x_{n-1}}$	**U**	Thm 3.4.1. **For the positive equilibrium $\bar{x}$, LAS $\not\Rightarrow$ GA.** **ESB iff** $A+\frac{\beta}{B}\geq 1$. **Conjecture: For the positive equilibrium $\bar{x}$ in the region where ESB, LAS $\Rightarrow$ GA.** **Conjecture: There exist bounded solutions that do not converge to an equilibrium or to a periodic solution.** See [48] and [61].
158	$x_{n+1} = \frac{x_n+\delta x_{n-2}}{A+x_n+Dx_{n-2}}$	**B**	Thm 2.1.1. **Conjecture: For the positive equilibrium $\bar{x}$, LAS $\Rightarrow$ GA.** **Conjecture: There exist solutions that do not converge to an equilibrium or to a periodic solution.**

159	$x_{n+1} = \frac{\beta x_n + x_{n-2}}{A + C x_{n-1} + x_{n-2}}$	**B**	Thm 2.3.1 or Thm 2.3.4. **Conjecture: For the positive equilibrium $\bar{x}$, LAS $\Rightarrow$ GA.** **Conjecture: There exist solutions that do not converge to an equilibrium or to a periodic solution.** See [69].
160	$x_{n+1} = \frac{\beta x_n + x_{n-2}}{B x_n + C x_{n-1} + x_{n-2}}$	**B**	Thm 2.1.1. **Conjecture: For the equilibrium $\bar{x}$, LAS $\Rightarrow$ GA.** **Conjecture: There exist solutions that do not converge to the equilibrium or to a periodic solution.**
161	$x_{n+1} = \frac{x_{n-1} + \delta x_{n-2}}{A + B x_n + x_{n-1}}$	**U**	Thm 3.4.1. **For the positive equilibrium $\bar{x}$, LAS $\not\Rightarrow$ GA.** **ESB iff $\delta \leq A + B$.** **Conjecture: For the positive equilibrium $\bar{x}$ in the region where ESB, LAS $\Rightarrow$ GA.** **$\exists$! P_2-solution and we conjecture that it is LAS.** **Conjecture: There exist bounded solutions that do not converge to an equilibrium or to a periodic solution.** See [61] and [69].

162	$x_{n+1} = \frac{\gamma x_{n-1}+x_{n-2}}{A+Bx_n+x_{n-2}}$	U	Thm 3.3.1. **Conjecture: EBSC$\bar{x}$ when $\gamma > 1 + A$.** **Conjecture: For the positive equilibrium $\bar{x}$, LAS $\Rightarrow$ GA when $\gamma < 1 + A$.** **Conjecture: EBSCP$_2$ when $\gamma = 1 + A$.** **Conjecture: There exist bounded solutions that do not converge to an equilibrium or to a periodic solution.**
163	$x_{n+1} = \frac{x_{n-1}+\delta x_{n-2}}{A+x_{n-1}+Dx_{n-2}}$	B	Thm 2.1.1. **Conjecture: For the positive equilibrium $\bar{x}$, LAS $\Rightarrow$ GA.** **$\exists!$ P$_2$-solution and we conjecture that it is LAS.** **Conjecture: There exist solutions that do not converge to an equilibrium or to a periodic solution.**
164	$x_{n+1} = \frac{x_{n-1}+\delta x_{n-2}}{Bx_n+x_{n-1}+Dx_{n-2}}$	B	Thm 2.1.1. **Conjecture: For the equilibrium $\bar{x}$, LAS $\Rightarrow$ GA.** **$\exists!$ P$_2$-solution and we conjecture that it is LAS.** **Conjecture: There exist solutions that do not converge to the equilibrium or to a periodic solution.**
165	$x_{n+1} = \frac{\alpha+x_n+\gamma x_{n-1}}{A+x_n}$	U	Thm 3.1.1. **Has P$_2$-Tricho.** **EBSC$\bar{x}$ when $\gamma > 1 + A$ by Theorem 4.2.2.** See [112] and [175, p. 167].

166	$x_{n+1} = \frac{\alpha+\beta x_n+x_{n-1}}{A+x_{n-1}}$	B	Thm 2.2.1 or Thm 2.3.4. **Conjecture: ESC$\bar{x}$.** See [69], [134], and [175, p. 172].
167	$x_{n+1} = \frac{\alpha+\beta x_n+x_{n-1}}{A+x_{n-2}}$	U	Thm 3.2.1. **Conjecture: EBSC$\bar{x}$ when $\beta+A<1$.** **Conjecture: For the equilibrium $\bar{x}$, LAS $\Rightarrow$ GA when $\beta+A>1$.** **Conjecture: ESCP$_2$ when $\beta+A=1$.** **Conjecture: There exist bounded solutions that do not converge to the equilibrium or to a periodic solution.**
168	$x_{n+1} = \frac{\alpha+x_n+\gamma x_{n-1}}{x_n+Cx_{n-1}}$	B	Thm 2.1.1. **Conjecture: For the equilibrium $\bar{x}$, LAS $\Rightarrow$ GA.** **$\exists$! P$_2$-solution and we conjecture that it is LAS.** **Conjecture: ESCP$_2$.** See [175, p. 175].
169	$x_{n+1} = \frac{\alpha+x_n+\gamma x_{n-1}}{x_n+Dx_{n-2}}$	U	Thm 3.3.1. **Conjecture: EBSC$\bar{x}$ when $\gamma>1$.** **Conjecture: For the equilibrium $\bar{x}$, LAS $\Rightarrow$ GA when $\gamma<1$.** **Conjecture: ESCP$_2$ when $\gamma=1$.** **Conjecture: There exist bounded solutions that do not converge to the equilibrium or to a periodic solution.** See [49] and [70].

170	$x_{n+1} = \frac{\alpha+\beta x_n+x_{n-1}}{x_{n-1}+Dx_{n-2}}$	**B**	Thm 2.8.1. **Conjecture: For the equilibrium $\bar{x}$, LAS $\Rightarrow$ GA.** **$\exists$! P_2-solution and we conjecture that it is LAS.** **Conjecture: There exist solutions that do not converge to the equilibrium or to a periodic solution.** See [49].
171	$x_{n+1} = \frac{\alpha+\beta x_n+x_{n-2}}{A+x_n}$	**B**	Thm 2.5.1. **Conjecture: For the equilibrium $\bar{x}$, LAS $\Rightarrow$ GA.** **Conjecture: There exist solutions that do not converge to the equilibrium or to a periodic solution.**
172	$x_{n+1} = \frac{\alpha+\beta x_n+\delta x_{n-2}}{1+x_{n-1}}$	**B**	Thm 2.7.1. **Conjecture: For the equilibrium $\bar{x}$, LAS $\Rightarrow$ GA.** **Conjecture: There exist solutions that do not converge to the equilibrium or to a periodic solution.**
173	$x_{n+1} = \frac{\alpha+\beta x_n+x_{n-2}}{A+x_{n-2}}$	**B**	Thm 2.3.3 or Thm 2.3.4. **Conjecture: For the equilibrium $\bar{x}$, LAS $\Rightarrow$ GA.** **Conjecture: There exist solutions that do not converge to the equilibrium or to a periodic solution.** See [69].

174	$x_{n+1} = \frac{\alpha+x_n+\delta x_{n-2}}{x_n+Cx_{n-1}}$	U	Thm 3.4.1. **For the equilibrium $\bar{x}$, LAS $\not\Rightarrow$GA.** **Conjecture: ESB iff $\delta \leq C$.** **Conjecture: For the equilibrium $\bar{x}$ in the region where ESB, LAS $\Rightarrow$ GA.** **Conjecture: There exist bounded solutions that do not converge to the equilibrium or to a periodic solution.** See [69].
175	$x_{n+1} = \frac{\alpha+\beta x_n+x_{n-2}}{Bx_n+x_{n-2}}$	B	Thm 2.1.1. **Conjecture: For the equilibrium $\bar{x}$, LAS $\Rightarrow$ GA.** **Conjecture: There exist solutions that do not converge to the equilibrium or to a periodic solution.**
176	$x_{n+1} = \frac{\alpha+\beta x_n+x_{n-2}}{Cx_{n-1}+x_{n-2}}$	B	Thm 2.9.1 **Conjecture: For the equilibrium $\bar{x}$, LAS $\Rightarrow$ GA.** **Conjecture: There exist solutions that do not converge to the equilibrium or to a periodic solution.** See [152].

177	$x_{n+1} = \frac{\alpha+\gamma x_{n-1}+x_{n-2}}{A+x_n}$	U	Thm 3.1.1. **Conjecture: EBSC$\bar{x}$ when $\gamma > 1+A$.** **Conjecture: For the equilibrium $\bar{x}$, LAS $\Rightarrow$ GA when $\gamma < 1+A$.** **Conjecture: ESCP$_2$ when $\gamma = 1+A$.** **Conjecture: There exist bounded solutions that do not converge to the equilibrium or to a periodic solution.** See [67].
178	$x_{n+1} = \frac{\alpha+\gamma x_{n-1}+\delta x_{n-2}}{1+x_{n-1}}$	B	Thm 2.7.1. **Conjecture: For the equilibrium $\bar{x}$, LAS $\Rightarrow$ GA.** **Conjecture: There exist solutions that do not converge to the equilibrium or to a periodic solution.**
179	$x_{n+1} = \frac{\alpha+\gamma x_{n-1}+x_{n-2}}{A+x_{n-2}}$	U	Thm 3.2.1. **Has P$_2$-Tricho.** **Conjecture: EBSC$\bar{x}$ when $\gamma > 1+A$.** See [70].

180	$x_{n+1} = \frac{\alpha + x_{n-1} + \delta x_{n-2}}{Bx_n + x_{n-1}}$	U	Thm 3.4.1. **Conjecture: ESB iff $\delta \leq B$.** **For the equilibrium $\bar{x}$, LAS $\not\Rightarrow$ GA.** **Conjecture: For the equilibrium $\bar{x}$ in the region where ESB, LAS $\Rightarrow$ GA.** **$\exists!$ P_2 solution and we conjecture that it is LAS.** **Conjecture: There exist bounded solutions that do not converge to the equilibrium or to a periodic solution.** See [61] and [69].
181	$x_{n+1} = \frac{\alpha + \gamma x_{n-1} + x_{n-2}}{Bx_n + x_{n-2}}$	U	Thm 3.3.1. **Conjecture: EBSC$\bar{x}$ when $\gamma > 1$.** **Conjecture: For the equilibrium $\bar{x}$, LAS $\Rightarrow$ GA when $\gamma < 1$.** **Conjecture: EBSCP_2 when $\gamma = 1$.** **Conjecture: There exist bounded solutions that do not converge to the equilibrium or to a periodic solution.** See [49].
182	$x_{n+1} = \frac{\alpha + \gamma x_{n-1} + x_{n-2}}{Cx_{n-1} + x_{n-2}}$	B	Thm 2.1.1. **Conjecture: For the equilibrium $\bar{x}$, LAS $\Rightarrow$ GA.** **$\exists!$ P_2-solution and we conjecture that it is LAS.** **Conjecture: There exist solutions that do not converge to the equilibrium or to a periodic solution.**

183	$x_{n+1} = \frac{x_n+\gamma x_{n-1}+\delta x_{n-2}}{A+x_n}$	**U**	Thm 3.1.1. **Conjecture: EBSC$\bar{x}$ when $\gamma > 1+\delta+A$.** **Conjecture: For the positive equilibrium $\bar{x}$, LAS $\Rightarrow$ GA when $\gamma < 1+\delta+A$.** **ESCP$_2$ when $\gamma = 1+\delta+A$.** **Conjecture: There exist bounded solutions that do not converge to an equilibrium or to a periodic solution.** See [67].
184	$x_{n+1} = \frac{\beta x_n+x_{n-1}+\delta x_{n-2}}{A+x_{n-1}}$	**B**	Thm 2.7.1. **Conjecture: For the positive equilibrium $\bar{x}$, LAS $\Rightarrow$ GA.** **Conjecture: There exist solutions that do not converge to an equilibrium or to a periodic solution.**
185	$x_{n+1} = \frac{\beta x_n+\gamma x_{n-1}+x_{n-2}}{A+x_{n-2}}$	**U**	Thm 3.2.1. **Conjecture: EBSC$\bar{x}$ when $\gamma > 1+\beta+A$.** **Conjecture: For the positive equilibrium $\bar{x}$, LAS $\Rightarrow$ GA when $\gamma < 1+\beta+A$.** **Conjecture: ESCP$_2$ when $\gamma = 1+\beta+A$.** **Conjecture: There exist bounded solutions that do not converge to an equilibrium or to a periodic solution.**

186	$x_{n+1} = \frac{x_n+\gamma x_{n-1}+\delta x_{n-2}}{x_n+Cx_{n-1}}$	**U**	Thm 3.4.1. **For the equilibrium $\bar{x}$, LAS $\not\Rightarrow$ GA. Conjecture: ESB iff $\delta \leq \frac{\gamma}{C}+C$. Conjecture: For the equilibrium $\bar{x}$ in the region where ESB, LAS $\Rightarrow$ GA. $\exists!$ P_2-solution and we conjecture that it is LAS. Conjecture: There exist bounded solutions that do not converge to the equilibrium or to a periodic solution.** See [69].
187	$x_{n+1} = \frac{x_n+\gamma x_{n-1}+\delta x_{n-2}}{x_n+Dx_{n-2}}$	**U**	Thm 3.3.1. **Conjecture: EBSC$\bar{x}$ when $\gamma > 1+\delta$. Conjecture: For the equilibrium $\bar{x}$, LAS $\Rightarrow$ GA when $\gamma < 1+\delta$. Conjecture: ESCP_2 when $\gamma = 1+\delta$. Conjecture: There exist bounded solutions that do not converge to the equilibrium or to a periodic solution.** See [49].

188	$x_{n+1} = \frac{\beta x_n + x_{n-1} + \delta x_{n-2}}{x_{n-1} + D x_{n-2}}$	**B**	Thm 2.3.2 or Thm 2.3.4. **Conjecture: For the equilibrium $\bar{x}$, LAS $\Rightarrow$ GA.** **$\exists!$ P_2-solution and we conjecture that it is LAS.** **Conjecture: There exist solutions that do not converge to the equilibrium or to a periodic solution.** See [69].
189	$x_{n+1} = \frac{\alpha + x_n}{A + x_n + C x_{n-1} + D x_{n-2}}$	**B**	Thm 2.1.1. **Conjecture: For the equilibrium $\bar{x}$, LAS $\Rightarrow$ GA.** **Conjecture: There exist solutions that do not converge to the equilibrium or to a periodic solution.**
190	$x_{n+1} = \frac{\alpha + x_{n-1}}{A + B x_n + x_{n-1} + D x_{n-2}}$	**B**	Thm 2.1.1. **Conjecture: For the equilibrium $\bar{x}$, LAS $\Rightarrow$ GA.** **$\exists!$ P_2-solution and we conjecture that it is LAS.** **Conjecture: There exist solutions that do not converge to the equilibrium or to a periodic solution.**
191	$x_{n+1} = \frac{\alpha + x_{n-2}}{A + B x_n + C x_{n-1} + x_{n-2}}$	**B**	Thm 2.1.1. **Conjecture: For the equilibrium $\bar{x}$, LAS $\Rightarrow$ GA.** **Conjecture: There exist solutions that do not converge to the equilibrium or to a periodic solution.**

192	$x_{n+1} = \frac{x_n + \gamma x_{n-1}}{A + x_n + C x_{n-1} + D x_{n-2}}$	B	Thm 2.1.1. **Conjecture: For the positive equilibrium $\bar{x}$, LAS $\Rightarrow$ GA. $\exists!$ P_2-solution and we conjecture that it is LAS. Conjecture: There exist solutions that do not converge to an equilibrium or to a periodic solution.**
193	$x_{n+1} = \frac{x_n + \delta x_{n-2}}{A + x_n + C x_{n-1} + D x_{n-2}}$	B	Thm 2.1.1. **Conjecture: For the positive equilibrium $\bar{x}$, LAS $\Rightarrow$ GA. Conjecture: There exist solutions that do not converge to an equilibrium or to a periodic solution.**
194	$x_{n+1} = \frac{x_{n-1} + \delta x_{n-2}}{A + B x_n + x_{n-1} + D x_{n-2}}$	B	Thm 2.1.1. **Conjecture: For the positive equilibrium $\bar{x}$, LAS $\Rightarrow$ GA. $\exists!$ P_2-solution and we conjecture that it is LAS. Conjecture: There exist solutions that do not converge to an equilibrium or to a periodic solution.**

195	$x_{n+1} = \frac{\alpha+\beta x_n+\gamma x_{n-1}+\delta x_{n-2}}{A+x_n}$	**U**	Thm 3.1.1. **ESCP$_2$ when** $\gamma = \beta + \delta + A$. **Conjecture: EBSC**$\bar{x}$ **when** $\gamma > \beta + \delta + A$. **Conjecture: For the equilibrium** $\bar{x}$, **LAS** $\Rightarrow$ **GA** **when** $\gamma < \beta + \delta + A$. **Conjecture: There exist bounded solutions that do not converge to the equilibrium or to a periodic solution.** See [67].
196	$x_{n+1} = \frac{\alpha+\beta x_n+\gamma x_{n-1}+x_{n-2}}{A+x_{n-1}}$	**B**	Thm 2.7.1. **Conjecture: For the equilibrium** $\bar{x}$, **LAS** $\Rightarrow$ **GA.** **Conjecture: There exist solutions that do not converge to the equilibrium or to a periodic solution.**
197	$x_{n+1} = \frac{\alpha+\beta x_n+\gamma x_{n-1}+x_{n-2}}{A+x_{n-2}}$	**U**	Thm 3.2.1. **Conjecture: EBSC**$\bar{x}$ **when** $\gamma > 1 + \beta + A$. **Conjecture: For the equilibrium** $\bar{x}$, **LAS** $\Rightarrow$ **GA** **when** $\gamma < 1 + \beta + A$. **Conjecture: ESCP$_2$** **when** $\gamma = 1 + \beta + A$. **Conjecture: There exist bounded solutions that do not converge to the equilibrium or to a periodic solution.**

198	$x_{n+1} = \frac{\alpha+x_n+\gamma x_{n-1}+\delta x_{n-2}}{x_n+Cx_{n-1}}$	**U**	Thm 3.4.1. **Conjecture: ESB iff $\delta \leq C + \frac{\gamma}{C}$.** **Conjecture: For the equilibrium $\bar{x}$, LAS $\not\Rightarrow$ GA.** **However in the region where ESB, LAS $\Rightarrow$ GA.** **$\exists!$ P$_2$-solution and we conjecture that it is LAS.** **Conjecture: There exist bounded solutions that do not converge to the equilibrium or to a periodic solution.** See [69].
199	$x_{n+1} = \frac{\alpha+x_n+\gamma x_{n-1}+\delta x_{n-2}}{x_n+Dx_{n-2}}$	**U**	Thm 3.3.1. **Conjecture: EBSC$\bar{x}$ when $\gamma > 1+\delta$.** **Conjecture: For the equilibrium $\bar{x}$, LAS $\Rightarrow$ GA when $\gamma < 1+\delta$.** **Conjecture: ESCP$_2$ when $\gamma = 1+\delta$.** **Conjecture: There exist bounded solutions that do not converge to the equilibrium or to a periodic solution.** See [49].
200	$x_{n+1} = \frac{\alpha+\beta x_n+x_{n-1}+\delta x_{n-2}}{x_{n-1}+Dx_{n-2}}$	**B**	Thm 2.3.2. **Conjecture: For the equilibrium $\bar{x}$, LAS $\Rightarrow$ GA.** **$\exists!$ P$_2$ solution and we conjecture that it is LAS.** **Conjecture: There exist solutions that do not converge to the equilibrium or to a periodic solution.** See [69].

201	$x_{n+1} = \frac{\alpha+\beta x_n+x_{n-1}}{A+Bx_n+x_{n-1}}$	B	Thm 2.1.1. **Conjecture: For the equilibrium** $\bar{x}$, **LAS ⇒ GA.** **∃! P_2 solution iff** $\beta + A < 1$ and 4α $< (1-\beta-A)\,[B(1-\beta-A)-(1+3\beta-A)]$ **and we conjecture that it is LAS.** **Conjecture:** $\bar{x}$ **is GAS when either** $\beta + A \geq 1$ **or** 4α $\geq (1-\beta-A)\,[B(1-\beta-A)-(1+3\beta-A)]$. **Conjecture: ESCP$_2$.**
202	$x_{n+1} = \frac{\alpha+x_n+\gamma x_{n-1}}{A+x_n+Dx_{n-2}}$	U	Thm 3.3.1. **Conjecture: EBSC$\bar{x}$ when** $\gamma > 1 + A$. **Conjecture: For the equilibrium** $\bar{x}$, **LAS ⇒ GA when** $\gamma < 1 + A$. **Conjecture: ESCP$_2$ when** $\gamma = 1 + A$. **Conjecture: There exist bounded solutions that do not converge to the equilibrium or to a periodic solution.** See [69].
203	$x_{n+1} = \frac{\alpha+\beta x_n+x_{n-1}}{A+x_{n-1}+Dx_{n-2}}$	B	Thm 2.3.1 or Thm 2.3.4. **Conjecture: For the equilibrium** $\bar{x}$, **LAS ⇒ GA.** **∃! P_2-solution and we conjecture that it is LAS.** **Conjecture: There exist solutions that do not converge to the equilibrium or to a periodic solution.** See [69].

204	$x_{n+1} = \frac{\alpha+x_n+\gamma x_{n-1}}{x_n+Cx_{n-1}+Dx_{n-2}}$	B	Thm 2.1.1. **Conjecture: For the equilibrium $\bar{x}$, LAS $\Rightarrow$ GA.** **$\exists!$ P_2-solution and we conjecture that it is LAS.** **Conjecture: There exist solutions that do not converge to the equilibrium or to a periodic solution.**
205	$x_{n+1} = \frac{\alpha+x_n+\delta x_{n-2}}{A+x_n+Cx_{n-1}}$	U	Thm 3.4.1. **Conjecture: ESB iff $\delta \leq A+C$.** **Conjecture: For the equilibrium $\bar{x}$, LAS $\not\Rightarrow$ GA.** **However in the region where ESB, LAS $\Rightarrow$ GA.** **Conjecture: There exist bounded solutions that do not converge to the equilibrium or to a periodic solution.** See [49].
206	$x_{n+1} = \frac{\alpha+x_n+\delta x_{n-2}}{A+x_n+Dx_{n-2}}$	B	Thm 2.1.1. **Conjecture: For the equilibrium $\bar{x}$, LAS $\Rightarrow$ GA.** **Conjecture: There exist solutions that do not converge to the equilibrium or to a periodic solution.**
207	$x_{n+1} = \frac{\alpha+\beta x_n+x_{n-2}}{A+Cx_{n-1}+x_{n-2}}$	B	Thm 2.3.1 or Thm 2.3.4. **Conjecture: For the equilibrium $\bar{x}$, LAS $\Rightarrow$ GA.** **Conjecture: There exist solutions that do not converge to the equilibrium or to a periodic solution.** See [69].

208	$x_{n+1} = \frac{\alpha+x_n+\delta x_{n-2}}{x_n+Cx_{n-1}+Dx_{n-2}}$	B	Thm 2.1.1. **Conjecture: For the equilibrium $\bar{x}$, LAS $\Rightarrow$ GA.** **Conjecture: There exist solutions that do not converge to the equilibrium or to a periodic solution.**
209	$x_{n+1} = \frac{\alpha+x_{n-1}+\delta x_{n-2}}{A+Bx_n+x_{n-1}}$	U	Thm 3.4.1. **Conjecture: ESB iff** $\delta \leq A+B$. **Conjecture: For the equilibrium $\bar{x}$, LAS $\not\Rightarrow$ GA.** **However in the region where ESB, LAS $\Rightarrow$ GA.** **$\exists!$ P_2-solution and we conjecture that it is LAS.** **Conjecture: There exist bounded solutions that do not converge to the equilibrium or to a periodic solution.** See [49].
210	$x_{n+1} = \frac{\alpha+\gamma x_{n-1}+x_{n-2}}{A+Bx_n+x_{n-2}}$	U	Thm 3.3.1. **Conjecture: EBSC$\bar{x}$ when $\gamma > 1+A$.** **Conjecture: For the equilibrium $\bar{x}$, LAS $\Rightarrow$ GA when $\gamma < 1+A$.** **Conjecture: ESCP_2 when $\gamma = 1+A$.** **Conjecture: There exist bounded solutions that do not converge to the equilibrium or to a periodic solution.** See [49].

211	$x_{n+1} = \frac{\alpha+x_{n-1}+\delta x_{n-2}}{A+x_{n-1}+Dx_{n-2}}$	**B**	Thm 2.1.1. **Conjecture: For the equilibrium $\bar{x}$, LAS $\Rightarrow$ GA.** **$\exists$! P_2-solution and we conjecture that it is LAS.** **Conjecture: There exist solutions that do not converge to the equilibrium or to a periodic solution.**
212	$x_{n+1} = \frac{\alpha+x_{n-1}+\delta x_{n-2}}{Bx_n+x_{n-1}+Dx_{n-2}}$	**B**	Thm 2.1.1. **Conjecture: For the equilibrium $\bar{x}$, LAS $\Rightarrow$ GA.** **$\exists$! P_2-solution and we conjecture that it is LAS.** **Conjecture: There exist solutions that do not converge to the equilibrium or to a periodic solution.**
213	$x_{n+1} = \frac{x_n+\gamma x_{n-1}+\delta x_{n-2}}{A+x_n+Cx_{n-1}}$	**U**	Thm 3.4.1. **Conjecture: ESB iff $\delta \leq A + \frac{\gamma}{C} + C$.** **Conjecture: For the positive equilibrium $\bar{x}$, LAS $\not\Rightarrow$ GA.** **However in the region where ESB, LAS $\Rightarrow$ GA.** **$\exists$! P_2-solution and we conjecture that it is LAS.** **Conjecture: There exist bounded solutions that do not converge to an equilibrium or to a periodic solution.** See [69].

214	$x_{n+1} = \frac{x_n+\gamma x_{n-1}+\delta x_{n-2}}{A+x_n+Dx_{n-2}}$	U	Thm 3.3.1. **Conjecture: EBSC$\bar{x}$ when $\gamma > 1+\delta+A$.** **Conjecture: For the positive equilibrium $\bar{x}$, LAS $\Rightarrow$ GA when $\gamma < 1+\delta+A$.** **Conjecture: ESCP$_2$ when $\gamma = 1+\delta+A$.** **Conjecture: There exist bounded solutions that do not converge to an equilibrium or to a periodic solution.** See [49].
215	$x_{n+1} = \frac{\beta x_n+x_{n-1}+\delta x_{n-2}}{A+x_{n-1}+Dx_{n-2}}$	B	Thm 2.3.1 or Thm 2.3.4. **Conjecture: For the positive equilibrium $\bar{x}$, LAS $\Rightarrow$ GA.** **$\exists!$ P$_2$-solution and we conjecture that it is LAS.** **Conjecture: There exist solutions that do not converge to an equilibrium or to a periodic solution.** See [69].
216	$x_{n+1} = \frac{x_n+\gamma x_{n-1}+\delta x_{n-2}}{x_n+Cx_{n-1}+Dx_{n-2}}$	B	Thm 2.1.1. **Conjecture: For the equilibrium $\bar{x}$, LAS $\Rightarrow$ GA.** **$\exists!$ P$_2$-solution and we conjecture that it is LAS.** **Conjecture: There exist solutions that do not converge to the equilibrium or to a periodic solution.**

217	$x_{n+1} = \frac{\alpha+x_n+\gamma x_{n-1}}{A+x_n+Cx_{n-1}+Dx_{n-2}}$	B	Thm 2.1.1. **Conjecture: For the equilibrium $\bar{x}$, LAS $\Rightarrow$ GA.** **$\exists!$ P_2-solution and we conjecture that it is LAS.** **Conjecture: There exist solutions that do not converge to the equilibrium or to a periodic solution.**
218	$x_{n+1} = \frac{\alpha+x_n+\delta x_{n-2}}{A+x_n+Cx_{n-1}+Dx_{n-2}}$	B	Thm 2.1.1. **Conjecture: For the equilibrium $\bar{x}$, LAS $\Rightarrow$ GA.** **Conjecture: There exist solutions that do not converge to the equilibrium or to a periodic solution.**
219	$x_{n+1} = \frac{\alpha+\gamma x_{n-1}+x_{n-2}}{A+Bx_n+Cx_{n-1}+x_{n-2}}$	B	Thm 2.1.1. **Conjecture: For the equilibrium $\bar{x}$, LAS $\Rightarrow$ GA.** **$\exists!$ P_2-solution and we conjecture that it is LAS.** **Conjecture: There exist solutions that do not converge to the equilibrium or to a periodic solution.**
220	$x_{n+1} = \frac{x_n+\gamma x_{n-1}+\delta x_{n-2}}{A+x_n+Cx_{n-1}+Dx_{n-2}}$	B	Thm 2.1.1. **Conjecture: For the positive equilibrium $\bar{x}$, LAS $\Rightarrow$ GA.** **$\exists!$ P_2-solution and we conjecture that it is LAS.** **Conjecture: There exist solutions that do not converge to an equilibrium or to a periodic solution.**

221	$x_{n+1} = \frac{\alpha+\beta x_n+\gamma x_{n-1}+\delta x_{n-2}}{A+Bx_n+Cx_{n-1}}$	**U**	Thm 3.4.1. **Conjecture: ESB iff** $\delta \le A + B\frac{\gamma}{C} + C\frac{\beta}{B}$. **Conjecture: For the equilibrium** $\bar{x}$, **LAS** $\not\Rightarrow$ **GA.** **However in the region where ESB, LAS** $\Rightarrow$ **GA.** $\exists!$ **P_2 solution and we conjecture that it is LAS.** **Conjecture: There exist bounded solutions that do not converge to the equilibrium or to a periodic solution.** See [61] and [69].
222	$x_{n+1} = \frac{\alpha+x_n+\gamma x_{n-1}+\delta x_{n-2}}{A+x_n+Dx_{n-2}}$	**U**	Thm 3.3.1. **Conjecture: EBSC**$\bar{x}$ **when** $\gamma > 1+\delta+A$. **ESB iff** $\gamma \le 1+\delta+A$. **Conjecture: For the equilibrium** $\bar{x}$, **LAS** $\Rightarrow$ **GA when** $\gamma < 1+\delta+A$. **Conjecture: ESCP_2 when** $\gamma = 1+\delta+A$. **Conjecture: There exist bounded solutions that do not converge to the equilibrium or to a periodic solution.**
223	$x_{n+1} = \frac{\alpha+\beta x_n+x_{n-1}+\delta x_{n-2}}{A+x_{n-1}+Dx_{n-2}}$	**B**	Thm 2.3.1 or Thm 2.3.4. **Conjecture: For the equilibrium** $\bar{x}$, **LAS** $\Rightarrow$ **GA.** $\exists!$ **P_2-solution and we conjecture that it is LAS.** **Conjecture: There exist solutions that do not converge to the equilibrium or to a periodic solution.** See [69].

224	$x_{n+1} = \frac{\alpha+x_n+\gamma x_{n-1}+\delta x_{n-2}}{x_n+Cx_{n-1}+Dx_{n-2}}$	**B**	Thm 2.1.1. **Conjecture: For the equilibrium $\bar{x}$, LAS $\Rightarrow$ GA.** **$\exists!$ P_2-solution and we conjecture that it is LAS.** **Conjecture: There exist solutions that do not converge to the equilibrium or to a periodic solution.**
225	$x_{n+1} = \frac{\alpha+\beta x_n+x_{n-1}+\delta x_{n-2}}{A+Bx_n+x_{n-1}+Dx_{n-2}}$	**B**	Thm 2.1.1. **Conjecture: For the equilibrium $\bar{x}$, LAS $\Rightarrow$ GA.** **$\exists!$ P_2 solution iff $\beta+\delta+A<1$ and $4\alpha<(B+D)(1-\beta-\delta-A)^2$ $-(1-\beta-\delta-A)(1+3\beta+3\delta-A)$ and we conjecture that it is LAS.** **Conjecture: There exist solutions that do not converge to the equilibrium or to a periodic solution.**

Appendix B

Table of the Boundedness Character of the 736 Special Cases of Order Four of

$$x_{n+1} = \frac{\alpha + \beta x_n + \gamma x_{n-1} + \delta x_{n-2} + \epsilon x_{n-3}}{A + B x_n + C x_{n-1} + D x_{n-2} + E x_{n-3}}, \quad n = 0, 1, \ldots .$$

The above equation contains $(2^5-1)\times(2^5-1) = 961$ special cases, of which 225 are of order less than or equal to three (see Appendix A) and the remaining 736 are of order four and they are presented in this appendix.

For the definition of the number assigned to a special case see Section 2.3.

A boldfaced **B** indicates that every solution of the equation in this special case is bounded and a boldfaced **U** indicates that the equation in this special case has unbounded solutions in some range of its parameters and for some initial conditions.

A boldfaced **B*** next to an equation indicates that we only conjecture that every solution of the equation is bounded and a boldfaced **U*** indicates that we only conjecture that the equation has unbounded solutions in some range of its parameters and for some initial conditions.

258	$x_{n+1} = \frac{\epsilon x_{n-3}}{A}$	**U**	Linear
259	$x_{n+1} = \frac{\alpha+\epsilon x_{n-3}}{A}$	**U**	Linear
262	$x_{n+1} = \frac{\beta x_n+\epsilon x_{n-3}}{A}$	**U**	Linear
263	$x_{n+1} = \frac{\alpha+\beta x_n+\epsilon x_{n-3}}{A}$	**U**	Linear
264	$x_{n+1} = \frac{\epsilon x_{n-3}}{Bx_n}$	**U**	Reducible to linear
265	$x_{n+1} = \frac{\alpha+\epsilon x_{n-3}}{Bx_n}$	**U**	Thm 4.2.4
266	$x_{n+1} = \frac{\epsilon x_{n-3}}{A+Bx_n}$	**U**	Thm 4.2.4
267	$x_{n+1} = \frac{\alpha+\epsilon x_{n-3}}{A+Bx_n}$	**U**	Thm 4.2.4
268	$x_{n+1} = \frac{\beta x_n+\epsilon x_{n-3}}{Bx_n}$	**U**	Thm 4.2.4
269	$x_{n+1} = \frac{\alpha+\beta x_n+\epsilon x_{n-3}}{Bx_n}$	**U**	Thm 4.2.4
270	$x_{n+1} = \frac{\beta x_n+\epsilon x_{n-3}}{A+Bx_n}$	**U**	Thm 4.2.4
271	$x_{n+1} = \frac{\alpha+\beta x_n+\epsilon x_{n-3}}{A+Bx_n}$	**U**	Thm 4.2.4
274	$x_{n+1} = \frac{\gamma x_{n-1}+\epsilon x_{n-3}}{A}$	**U**	Linear
275	$x_{n+1} = \frac{\alpha+\gamma x_{n-1}+\epsilon x_{n-3}}{A}$	**U**	Linear
278	$x_{n+1} = \frac{\beta x_n+\gamma x_{n-1}+\epsilon x_{n-3}}{A}$	**U**	Linear
279	$x_{n+1} = \frac{\alpha+\beta x_n+\gamma x_{n-1}+\epsilon x_{n-3}}{A}$	**U**	Linear
280	$x_{n+1} = \frac{\gamma x_{n-1}+\epsilon x_{n-3}}{Bx_n}$	**U**	Thm 3.1.2. See also [58].
281	$x_{n+1} = \frac{\alpha+\gamma x_{n-1}+\epsilon x_{n-3}}{Bx_n}$	**U**	Thm 3.1.2. See also [58].
282	$x_{n+1} = \frac{\gamma x_{n-1}+\epsilon x_{n-3}}{A+Bx_n}$	**U**	Thm 3.1.2. See also [58].
283	$x_{n+1} = \frac{\alpha+\gamma x_{n-1}+\epsilon x_{n-3}}{A+Bx_n}$	**U**	Thm 3.1.2. See also [58].
284	$x_{n+1} = \frac{\beta x_n+\gamma x_{n-1}+\epsilon x_{n-3}}{Bx_n}$	**U**	Thm 3.1.2. See also [58].
285	$x_{n+1} = \frac{\alpha+\beta x_n+\gamma x_{n-1}+\epsilon x_{n-3}}{Bx_n}$	**U**	Thm 3.1.2. See also [58].
286	$x_{n+1} = \frac{\beta x_n+\gamma x_{n-1}+\epsilon x_{n-3}}{A+Bx_n}$	**U**	Thm 3.1.2. See also [58].
287	$x_{n+1} = \frac{\alpha+\beta x_n+\gamma x_{n-1}+\epsilon x_{n-3}}{A+Bx_n}$	**U**	Thm 3.1.2. See also [58].
288	$x_{n+1} = \frac{\epsilon x_{n-3}}{Cx_{n-1}}$	**U**	Reducible to linear
289	$x_{n+1} = \frac{\alpha+\epsilon x_{n-3}}{Cx_{n-1}}$	**U**	Reducible to case #46
290	$x_{n+1} = \frac{\epsilon x_{n-3}}{A+Cx_{n-1}}$	**U**	Reducible to case #29
291	$x_{n+1} = \frac{\alpha+\epsilon x_{n-3}}{A+Cx_{n-1}}$	**U**	Reducible to case #71
292	$x_{n+1} = \frac{\beta x_n+\epsilon x_{n-3}}{Cx_{n-1}}$	**U***	
293	$x_{n+1} = \frac{\alpha+\beta x_n+\epsilon x_{n-3}}{Cx_{n-1}}$	**U***	
294	$x_{n+1} = \frac{\beta x_n+\epsilon x_{n-3}}{A+Cx_{n-1}}$	**U***	
295	$x_{n+1} = \frac{\alpha+\beta x_n+\epsilon x_{n-3}}{A+Cx_{n-1}}$	**U***	
296	$x_{n+1} = \frac{\epsilon x_{n-3}}{Bx_n+Cx_{n-1}}$	**U***	
297	$x_{n+1} = \frac{\alpha+\epsilon x_{n-3}}{Bx_n+Cx_{n-1}}$	**U***	

298	$x_{n+1} = \frac{\epsilon x_{n-3}}{A+Bx_n+Cx_{n-1}}$	**U**	Thm 4.4.2
299	$x_{n+1} = \frac{\alpha+\epsilon x_{n-3}}{A+Bx_n+Cx_{n-1}}$	**B**	See [14]
300	$x_{n+1} = \frac{\beta x_n+\epsilon x_{n-3}}{Bx_n+Cx_{n-1}}$	**U**	See [65]
301	$x_{n+1} = \frac{\alpha+\beta x_n+\epsilon x_{n-3}}{Bx_n+Cx_{n-1}}$	**U**	See [65]
302	$x_{n+1} = \frac{\beta x_n+\epsilon x_{n-3}}{A+Bx_n+Cx_{n-1}}$	**U**	See [65]
303	$x_{n+1} = \frac{\alpha+\beta x_n+\epsilon x_{n-3}}{A+Bx_n+Cx_{n-1}}$	**U**	See [65]
304	$x_{n+1} = \frac{\gamma x_{n-1}+\epsilon x_{n-3}}{Cx_{n-1}}$	**U**	Reducible to case #54
305	$x_{n+1} = \frac{\alpha+\gamma x_{n-1}+\epsilon x_{n-3}}{Cx_{n-1}}$	**U**	Reducible to case #118
306	$x_{n+1} = \frac{\gamma x_{n-1}+\epsilon x_{n-3}}{A+Cx_{n-1}}$	**U**	Reducible to case #83
307	$x_{n+1} = \frac{\alpha+\gamma x_{n-1}+\epsilon x_{n-3}}{A+Cx_{n-1}}$	**U**	Reducible to case #165
308	$x_{n+1} = \frac{\beta x_n+\gamma x_{n-1}+\epsilon x_{n-3}}{Cx_{n-1}}$	**U***	
309	$x_{n+1} = \frac{\alpha+\beta x_n+\gamma x_{n-1}+\epsilon x_{n-3}}{Cx_{n-1}}$	**U***	
310	$x_{n+1} = \frac{\beta x_n+\gamma x_{n-1}+\epsilon x_{n-3}}{A+Cx_{n-1}}$	**U***	
311	$x_{n+1} = \frac{\alpha+\beta x_n+\gamma x_{n-1}+\epsilon x_{n-3}}{A+Cx_{n-1}}$	**U***	
312	$x_{n+1} = \frac{\gamma x_{n-1}+\epsilon x_{n-3}}{Bx_n+Cx_{n-1}}$	**U**	See [65]
313	$x_{n+1} = \frac{\alpha+\gamma x_{n-1}+\epsilon x_{n-3}}{Bx_n+Cx_{n-1}}$	**U**	See [65]
314	$x_{n+1} = \frac{\gamma x_{n-1}+\epsilon x_{n-3}}{A+Bx_n+Cx_{n-1}}$	**U**	See [65]
315	$x_{n+1} = \frac{\alpha+\gamma x_{n-1}+\epsilon x_{n-3}}{A+Bx_n+Cx_{n-1}}$	**U**	See [65]
316	$x_{n+1} = \frac{\beta x_n+\gamma x_{n-1}+\epsilon x_{n-3}}{Bx_n+Cx_{n-1}}$	**U**	See [65]
317	$x_{n+1} = \frac{\alpha+\beta x_n+\gamma x_{n-1}+\epsilon x_{n-3}}{Bx_n+Cx_{n-1}}$	**U**	See [65]
318	$x_{n+1} = \frac{\beta x_n+\gamma x_{n-1}+\epsilon x_{n-3}}{A+Bx_n+Cx_{n-1}}$	**U**	See [65]
319	$x_{n+1} = \frac{\alpha+\beta x_n+\gamma x_{n-1}+\epsilon x_{n-3}}{A+Bx_n+Cx_{n-1}}$	**U**	See [65]
322	$x_{n+1} = \frac{\delta x_{n-2}+\epsilon x_{n-3}}{A}$	**U**	Linear
323	$x_{n+1} = \frac{\alpha+\delta x_{n-2}+\epsilon x_{n-3}}{A}$	**U**	Linear
326	$x_{n+1} = \frac{\beta x_n+\delta x_{n-2}+\epsilon x_{n-3}}{A}$	**U**	Linear
327	$x_{n+1} = \frac{\alpha+\beta x_n+\delta x_{n-2}+\epsilon x_{n-3}}{A}$	**U**	Linear
328	$x_{n+1} = \frac{\delta x_{n-2}+\epsilon x_{n-3}}{Bx_n}$	**U***	
329	$x_{n+1} = \frac{\alpha+\delta x_{n-2}+\epsilon x_{n-3}}{Bx_n}$	**U***	
330	$x_{n+1} = \frac{\delta x_{n-2}+\epsilon x_{n-3}}{A+Bx_n}$	**U***	
331	$x_{n+1} = \frac{\alpha+\delta x_{n-2}+\epsilon x_{n-3}}{A+Bx_n}$	**U***	
332	$x_{n+1} = \frac{\beta x_n+\delta x_{n-2}+\epsilon x_{n-3}}{Bx_n}$	**U***	
333	$x_{n+1} = \frac{\alpha+\beta x_n+\delta x_{n-2}+\epsilon x_{n-3}}{Bx_n}$	**U***	
334	$x_{n+1} = \frac{\beta x_n+\delta x_{n-2}+\epsilon x_{n-3}}{A+Bx_n}$	**U***	
335	$x_{n+1} = \frac{\alpha+\beta x_n+\delta x_{n-2}+\epsilon x_{n-3}}{A+Bx_n}$	**U***	
338	$x_{n+1} = \frac{\gamma x_{n-1}+\delta x_{n-2}+\epsilon x_{n-3}}{A}$	**U**	Linear
339	$x_{n+1} = \frac{\alpha+\gamma x_{n-1}+\delta x_{n-2}+\epsilon x_{n-3}}{A}$	**U**	Linear
342	$x_{n+1} = \frac{\beta x_n+\gamma x_{n-1}+\delta x_{n-2}+\epsilon x_{n-3}}{A}$	**U**	Linear

343	$x_{n+1} = \frac{\alpha+\beta x_n+\gamma x_{n-1}+\delta x_{n-2}+\epsilon x_{n-3}}{A}$	**U**	Linear
344	$x_{n+1} = \frac{\gamma x_{n-1}+\delta x_{n-2}+\epsilon x_{n-3}}{Bx_n}$	**U**	Thm 3.1.2
345	$x_{n+1} = \frac{\alpha+\gamma x_{n-1}+\delta x_{n-2}+\epsilon x_{n-3}}{Bx_n}$	**U**	Thm 3.1.2
346	$x_{n+1} = \frac{\gamma x_{n-1}+\delta x_{n-2}+\epsilon x_{n-3}}{A+Bx_n}$	**U**	Thm 3.1.2
347	$x_{n+1} = \frac{\alpha+\gamma x_{n-1}+\delta x_{n-2}+\epsilon x_{n-3}}{A+Bx_n}$	**U**	Thm 3.1.2
348	$x_{n+1} = \frac{\beta x_n+\gamma x_{n-1}+\delta x_{n-2}+\epsilon x_{n-3}}{Bx_n}$	**U**	Thm 3.1.2
349	$x_{n+1} = \frac{\alpha+\beta x_n+\gamma x_{n-1}+\delta x_{n-2}+\epsilon x_{n-3}}{Bx_n}$	**U**	Thm 3.1.2
350	$x_{n+1} = \frac{\beta x_n+\gamma x_{n-1}+\delta x_{n-2}+\epsilon x_{n-3}}{A+Bx_n}$	**U**	Thm 3.1.2
351	$x_{n+1} = \frac{\alpha+\beta x_n+\gamma x_{n-1}+\delta x_{n-2}+\epsilon x_{n-3}}{A+Bx_n}$	**U**	Thm 3.1.2
352	$x_{n+1} = \frac{\delta x_{n-2}+\epsilon x_{n-3}}{Cx_{n-1}}$	**U***	
353	$x_{n+1} = \frac{\alpha+\delta x_{n-2}+\epsilon x_{n-3}}{Cx_{n-1}}$	**U***	
354	$x_{n+1} = \frac{\delta x_{n-2}+\epsilon x_{n-3}}{A+Cx_{n-1}}$	**U***	
355	$x_{n+1} = \frac{\alpha+\delta x_{n-2}+\epsilon x_{n-3}}{A+Cx_{n-1}}$	**U***	
356	$x_{n+1} = \frac{\beta x_n+\delta x_{n-2}+\epsilon x_{n-3}}{Cx_{n-1}}$	**U***	
357	$x_{n+1} = \frac{\alpha+\beta x_n+\delta x_{n-2}+\epsilon x_{n-3}}{Cx_{n-1}}$	**U***	
358	$x_{n+1} = \frac{\beta x_n+\delta x_{n-2}+\epsilon x_{n-3}}{A+Cx_{n-1}}$	**U***	
359	$x_{n+1} = \frac{\alpha+\beta x_n+\delta x_{n-2}+\epsilon x_{n-3}}{A+Cx_{n-1}}$	**U***	
360	$x_{n+1} = \frac{\delta x_{n-2}+\epsilon x_{n-3}}{Bx_n+Cx_{n-1}}$	**U**	See [65]
361	$x_{n+1} = \frac{\alpha+\delta x_{n-2}+\epsilon x_{n-3}}{Bx_n+Cx_{n-1}}$	**U**	See [65]
362	$x_{n+1} = \frac{\delta x_{n-2}+\epsilon x_{n-3}}{A+Bx_n+Cx_{n-1}}$	**U**	See [65]
363	$x_{n+1} = \frac{\alpha+\delta x_{n-2}+\epsilon x_{n-3}}{A+Bx_n+Cx_{n-1}}$	**U**	See [65]
364	$x_{n+1} = \frac{\beta x_n+\delta x_{n-2}+\epsilon x_{n-3}}{Bx_n+Cx_{n-1}}$	**U**	See [65]
365	$x_{n+1} = \frac{\alpha+\beta x_n+\delta x_{n-2}+\epsilon x_{n-3}}{Bx_n+Cx_{n-1}}$	**U**	See [65]
366	$x_{n+1} = \frac{\beta x_n+\delta x_{n-2}+\epsilon x_{n-3}}{A+Bx_n+Cx_{n-1}}$	**U**	See [65]
367	$x_{n+1} = \frac{\alpha+\beta x_n+\delta x_{n-2}+\epsilon x_{n-3}}{A+Bx_n+Cx_{n-1}}$	**U**	See [65]
368	$x_{n+1} = \frac{\gamma x_{n-1}+\delta x_{n-2}+\epsilon x_{n-3}}{Cx_{n-1}}$	**U***	
369	$x_{n+1} = \frac{\alpha+\gamma x_{n-1}+\delta x_{n-2}+\epsilon x_{n-3}}{Cx_{n-1}}$	**U***	
370	$x_{n+1} = \frac{\gamma x_{n-1}+\delta x_{n-2}+\epsilon x_{n-3}}{A+Cx_{n-1}}$	**U***	
371	$x_{n+1} = \frac{\alpha+\gamma x_{n-1}+\delta x_{n-2}+\epsilon x_{n-3}}{A+Cx_{n-1}}$	**U***	
372	$x_{n+1} = \frac{\beta x_n+\gamma x_{n-1}+\delta x_{n-2}+\epsilon x_{n-3}}{Cx_{n-1}}$	**U***	
373	$x_{n+1} = \frac{\alpha+\beta x_n+\gamma x_{n-1}+\delta x_{n-2}+\epsilon x_{n-3}}{Cx_{n-1}}$	**U***	
374	$x_{n+1} = \frac{\beta x_n+\gamma x_{n-1}+\delta x_{n-2}+\epsilon x_{n-3}}{A+Cx_{n-1}}$	**U***	
375	$x_{n+1} = \frac{\alpha+\beta x_n+\gamma x_{n-1}+\delta x_{n-2}+\epsilon x_{n-3}}{A+Cx_{n-1}}$	**U***	
376	$x_{n+1} = \frac{\gamma x_{n-1}+\delta x_{n-2}+\epsilon x_{n-3}}{Bx_n+Cx_{n-1}}$	**U**	See [65]
377	$x_{n+1} = \frac{\alpha+\gamma x_{n-1}+\delta x_{n-2}+\epsilon x_{n-3}}{Bx_n+Cx_{n-1}}$	**U**	See [65]
378	$x_{n+1} = \frac{\gamma x_{n-1}+\delta x_{n-2}+\epsilon x_{n-3}}{A+Bx_n+Cx_{n-1}}$	**U**	See [65]
379	$x_{n+1} = \frac{\alpha+\gamma x_{n-1}+\delta x_{n-2}+\epsilon x_{n-3}}{A+Bx_n+Cx_{n-1}}$	**U**	See [65]
380	$x_{n+1} = \frac{\beta x_n+\gamma x_{n-1}+\delta x_{n-2}+\epsilon x_{n-3}}{Bx_n+Cx_{n-1}}$	**U**	See [65]
381	$x_{n+1} = \frac{\alpha+\beta x_n+\gamma x_{n-1}+\delta x_{n-2}+\epsilon x_{n-3}}{Bx_n+Cx_{n-1}}$	**U**	See [65]

382	$x_{n+1} = \frac{\beta x_n+\gamma x_{n-1}+\delta x_{n-2}+\epsilon x_{n-3}}{A+Bx_n+Cx_{n-1}}$	**U**	See [65]
383	$x_{n+1} = \frac{\alpha+\beta x_n+\gamma x_{n-1}+\delta x_{n-2}+\epsilon x_{n-3}}{A+Bx_n+Cx_{n-1}}$	**U**	See [65]
384	$x_{n+1} = \frac{\epsilon x_{n-3}}{Dx_{n-2}}$	**U**	Reducible to linear
385	$x_{n+1} = \frac{\alpha+\epsilon x_{n-3}}{Dx_{n-2}}$	**U**	Thm 4.2.4
386	$x_{n+1} = \frac{\epsilon x_{n-3}}{A+Dx_{n-2}}$	**U**	Thm 4.2.4
387	$x_{n+1} = \frac{\alpha+\epsilon x_{n-3}}{A+Dx_{n-2}}$	**U**	Thm 4.2.4
388	$x_{n+1} = \frac{\beta x_n+\epsilon x_{n-3}}{Dx_{n-2}}$	**U**	See [65]
389	$x_{n+1} = \frac{\alpha+\beta x_n+\epsilon x_{n-3}}{Dx_{n-2}}$	**U**	See [65]
390	$x_{n+1} = \frac{\beta x_n+\epsilon x_{n-3}}{A+Dx_{n-2}}$	**U**	See [65]
391	$x_{n+1} = \frac{\alpha+\beta x_n+\epsilon x_{n-3}}{A+Dx_{n-2}}$	**U**	See [65]
392	$x_{n+1} = \frac{\epsilon x_{n-3}}{Bx_n+Dx_{n-2}}$	**U**	Thm 3.1.2. See also [236].
393	$x_{n+1} = \frac{\alpha+\epsilon x_{n-3}}{Bx_n+Dx_{n-2}}$	**U**	Thm 3.1.2. See also [236].
394	$x_{n+1} = \frac{\epsilon x_{n-3}}{A+Bx_n+Dx_{n-2}}$	**U**	Thm 3.1.2. See also [236].
395	$x_{n+1} = \frac{\alpha+\epsilon x_{n-3}}{A+Bx_n+Dx_{n-2}}$	**U**	Thm 3.1.2. See also [236].
396	$x_{n+1} = \frac{\beta x_n+\epsilon x_{n-3}}{Bx_n+Dx_{n-2}}$	**U**	See [65]
397	$x_{n+1} = \frac{\alpha+\beta x_n+\epsilon x_{n-3}}{Bx_n+Dx_{n-2}}$	**U**	See [65]
398	$x_{n+1} = \frac{\beta x_n+\epsilon x_{n-3}}{A+Bx_n+Dx_{n-2}}$	**U**	See [65]
399	$x_{n+1} = \frac{\alpha+\beta x_n+\epsilon x_{n-3}}{A+Bx_n+Dx_{n-2}}$	**U**	See [65]
400	$x_{n+1} = \frac{\gamma x_{n-1}+\epsilon x_{n-3}}{Dx_{n-2}}$	**U**	Thm 3.1.2
401	$x_{n+1} = \frac{\alpha+\gamma x_{n-1}+\epsilon x_{n-3}}{Dx_{n-2}}$	**U**	Thm 3.1.2
402	$x_{n+1} = \frac{\gamma x_{n-1}+\epsilon x_{n-3}}{A+Dx_{n-2}}$	**U**	Thm 3.1.2
403	$x_{n+1} = \frac{\alpha+\gamma x_{n-1}+\epsilon x_{n-3}}{A+Dx_{n-2}}$	**U**	Thm 3.1.2
404	$x_{n+1} = \frac{\beta x_n+\gamma x_{n-1}+\epsilon x_{n-3}}{Dx_{n-2}}$	**U***	
405	$x_{n+1} = \frac{\alpha+\beta x_n+\gamma x_{n-1}+\epsilon x_{n-3}}{Dx_{n-2}}$	**U***	
406	$x_{n+1} = \frac{\beta x_n+\gamma x_{n-1}+\epsilon x_{n-3}}{A+Dx_{n-2}}$	**U***	
407	$x_{n+1} = \frac{\alpha+\beta x_n+\gamma x_{n-1}+\epsilon x_{n-3}}{A+Dx_{n-2}}$	**U***	
408	$x_{n+1} = \frac{\gamma x_{n-1}+\epsilon x_{n-3}}{Bx_n+Dx_{n-2}}$	**U**	Thm 3.1.2. See also [58].
409	$x_{n+1} = \frac{\alpha+\gamma x_{n-1}+\epsilon x_{n-3}}{Bx_n+Dx_{n-2}}$	**U**	Thm 3.1.2. See also [58].
410	$x_{n+1} = \frac{\gamma x_{n-1}+\epsilon x_{n-3}}{A+Bx_n+Dx_{n-2}}$	**U**	Thm 3.1.2. See also [58].
411	$x_{n+1} = \frac{\alpha+\gamma x_{n-1}+\epsilon x_{n-3}}{A+Bx_n+Dx_{n-2}}$	**U**	Thm 3.1.2. See also [58].
412	$x_{n+1} = \frac{\beta x_n+\gamma x_{n-1}+\epsilon x_{n-3}}{Bx_n+Dx_{n-2}}$	**U**	Thm 3.1.2. See also [58].
413	$x_{n+1} = \frac{\alpha+\beta x_n+\gamma x_{n-1}+\epsilon x_{n-3}}{Bx_n+Dx_{n-2}}$	**U**	Thm 3.1.2. See also [58].
414	$x_{n+1} = \frac{\beta x_n+\gamma x_{n-1}+\epsilon x_{n-3}}{A+Bx_n+Dx_{n-2}}$	**U**	Thm 3.1.2. See also [58].
415	$x_{n+1} = \frac{\alpha+\beta x_n+\gamma x_{n-1}+\epsilon x_{n-3}}{A+Bx_n+Dx_{n-2}}$	**U**	Thm 3.1.2. See also [58].
416	$x_{n+1} = \frac{\epsilon x_{n-3}}{Cx_{n-1}+Dx_{n-2}}$	**U**	Thm 4.4.2
417	$x_{n+1} = \frac{\alpha+\epsilon x_{n-3}}{Cx_{n-1}+Dx_{n-2}}$	**U***	
418	$x_{n+1} = \frac{\epsilon x_{n-3}}{A+Cx_{n-1}+Dx_{n-2}}$	**U**	Thm 4.4.2
419	$x_{n+1} = \frac{\alpha+\epsilon x_{n-3}}{A+Cx_{n-1}+Dx_{n-2}}$	**B***	
420	$x_{n+1} = \frac{\beta x_n+\epsilon x_{n-3}}{Cx_{n-1}+Dx_{n-2}}$	**U***	

421	$x_{n+1} = \frac{\alpha+\beta x_n+\epsilon x_{n-3}}{Cx_{n-1}+Dx_{n-2}}$	**U***	
422	$x_{n+1} = \frac{\beta x_n+\epsilon x_{n-3}}{A+Cx_{n-1}+Dx_{n-2}}$	**B***	
423	$x_{n+1} = \frac{\alpha+\beta x_n+\epsilon x_{n-3}}{A+Cx_{n-1}+Dx_{n-2}}$	**B***	
424	$x_{n+1} = \frac{\epsilon x_{n-3}}{Bx_n+Cx_{n-1}+Dx_{n-2}}$	**U**	Thm 3.4.2
425	$x_{n+1} = \frac{\alpha+\epsilon x_{n-3}}{Bx_n+Cx_{n-1}+Dx_{n-2}}$	**U**	Thm 3.4.2
426	$x_{n+1} = \frac{\epsilon x_{n-3}}{A+Bx_n+Cx_{n-1}+Dx_{n-2}}$	**U**	Thm 3.4.2
427	$x_{n+1} = \frac{\alpha+\epsilon x_{n-3}}{A+Bx_n+Cx_{n-1}+Dx_{n-2}}$	**U**	Thm 3.4.2
428	$x_{n+1} = \frac{\beta x_n+\epsilon x_{n-3}}{Bx_n+Cx_{n-1}+Dx_{n-2}}$	**U**	Thm 3.4.2
429	$x_{n+1} = \frac{\alpha+\beta x_n+\epsilon x_{n-2}}{Bx_n+Cx_{n-1}+Dx_{n-2}}$	**U**	Thm 3.4.2
430	$x_{n+1} = \frac{\beta x_n+\epsilon x_{n-3}}{A+Bx_n+Cx_{n-1}+Dx_{n-2}}$	**U**	Thm 3.4.2
431	$x_{n+1} = \frac{\alpha+\beta x_n+\epsilon x_{n-3}}{A+Bx_n+Cx_{n-1}+Dx_{n-2}}$	**U**	Thm 3.4.2
432	$x_{n+1} = \frac{\gamma x_{n-1}+\epsilon x_{n-3}}{Cx_{n-1}+Dx_{n-2}}$	**U**	See [65]
433	$x_{n+1} = \frac{\alpha+\gamma x_{n-1}+\epsilon x_{n-3}}{Cx_{n-1}+Dx_{n-2}}$	**U**	See [65]
434	$x_{n+1} = \frac{\gamma x_{n-1}+\epsilon x_{n-3}}{A+Cx_{n-1}+Dx_{n-2}}$	**U**	See [65]
435	$x_{n+1} = \frac{\alpha+\gamma x_{n-1}+\delta x_{n-3}}{A+Cx_{n-1}+Dx_{n-2}}$	**U**	See [65]
436	$x_{n+1} = \frac{\beta x_n+\gamma x_{n-1}+\epsilon x_{n-3}}{Cx_{n-1}+Dx_{n-2}}$	**B***	
437	$x_{n+1} = \frac{\alpha+\beta x_n+\gamma x_{n-1}+\epsilon x_{n-3}}{Cx_{n-1}+Dx_{n-2}}$	**B***	
438	$x_{n+1} = \frac{\beta x_n+\gamma x_{n-1}+\epsilon x_{n-3}}{A+Cx_{n-1}+Dx_{n-2}}$	**B***	
439	$x_{n+1} = \frac{\alpha+\beta x_n+\gamma x_{n-1}+\epsilon x_{n-3}}{A+Cx_{n-1}+Dx_{n-2}}$	**B***	
440	$x_{n+1} = \frac{\gamma x_{n-1}+\epsilon x_{n-3}}{Bx_n+Cx_{n-1}+Dx_{n-2}}$	**U**	Thm 3.4.2
441	$x_{n+1} = \frac{\alpha+\gamma x_{n-1}+\epsilon x_{n-3}}{Bx_n+Cx_{n-1}+Dx_{n-2}}$	**U**	Thm 3.4.2
442	$x_{n+1} = \frac{\gamma x_{n-1}+\epsilon x_{n-3}}{A+Bx_n+Cx_{n-1}+Dx_{n-2}}$	**U**	Thm 3.4.2
443	$x_{n+1} = \frac{\alpha+\gamma x_{n-1}+\epsilon x_{n-3}}{A+Bx_n+Cx_{n-1}+Dx_{n-2}}$	**U**	Thm 3.4.2
444	$x_{n+1} = \frac{\beta x_n+\gamma x_{n-1}+\epsilon x_{n-3}}{Bx_n+Cx_{n-1}+Dx_{n-2}}$	**U**	Thm 3.4.2
445	$x_{n+1} = \frac{\alpha+\beta x_n+\gamma x_{n-1}+\epsilon x_{n-3}}{Bx_n+Cx_{n-1}+Dx_{n-2}}$	**U**	Thm 3.4.2
446	$x_{n+1} = \frac{\beta x_n+\gamma x_{n-1}+\epsilon x_{n-3}}{A+Bx_n+Cx_{n-1}+Dx_{n-2}}$	**U**	Thm 3.4.2
447	$x_{n+1} = \frac{\alpha+\beta x_n+\gamma x_{n-1}+\epsilon x_{n-3}}{A+Bx_n+Cx_{n-1}+Dx_{n-2}}$	**U**	Thm 3.4.2
448	$x_{n+1} = \frac{\delta x_{n-2}+\epsilon x_{n-3}}{Dx_{n-2}}$	**U**	Thm 4.2.4
449	$x_{n+1} = \frac{\alpha+\delta x_{n-2}+\epsilon x_{n-3}}{Dx_{n-2}}$	**U**	Thm 4.2.4
450	$x_{n+1} = \frac{\delta x_{n-2}+\epsilon x_{n-3}}{A+Dx_{n-2}}$	**U**	Thm 4.2.4
451	$x_{n+1} = \frac{\alpha+\delta x_{n-2}+\epsilon x_{n-3}}{A+Dx_{n-2}}$	**U**	Thm 4.2.4
452	$x_{n+1} = \frac{\beta x_n+\delta x_{n-2}+\epsilon x_{n-3}}{Dx_{n-2}}$	**U**	See [65]
453	$x_{n+1} = \frac{\alpha+\beta x_n+\delta x_{n-2}+\epsilon x_{n-3}}{Dx_{n-2}}$	**U**	See [65]
454	$x_{n+1} = \frac{\beta x_n+\delta x_{n-2}+\epsilon x_{n-3}}{A+Dx_{n-2}}$	**U**	See [65]
455	$x_{n+1} = \frac{\alpha+\beta x_n+\delta x_{n-2}+\epsilon x_{n-3}}{A+Dx_{n-2}}$	**U**	See [65]
456	$x_{n+1} = \frac{\delta x_{n-2}+\epsilon x_{n-3}}{Bx_n+Dx_{n-2}}$	**U**	See [65]
457	$x_{n+1} = \frac{\alpha+\delta x_{n-2}+\epsilon x_{n-3}}{Bx_n+Dx_{n-2}}$	**U**	See [65]
458	$x_{n+1} = \frac{\delta x_{n-2}+\epsilon x_{n-3}}{A+Bx_n+Dx_{n-2}}$	**U**	See [65]
459	$x_{n+1} = \frac{\alpha+\delta x_{n-2}+\epsilon x_{n-3}}{A+Bx_n+Dx_{n-2}}$	**U**	See [65]

460	$x_{n+1} = \frac{\beta x_n+\delta x_{n-2}+\epsilon x_{n-3}}{Bx_n+Dx_{n-2}}$	**U**	See [65]
461	$x_{n+1} = \frac{\alpha+\beta x_n+\delta x_{n-2}+\epsilon x_{n-3}}{Bx_n+Dx_{n-2}}$	**U**	See [65]
462	$x_{n+1} = \frac{\beta x_n+\delta x_{n-2}+\epsilon x_{n-3}}{A+Bx_n+Dx_{n-2}}$	**U**	See [65]
463	$x_{n+1} = \frac{\alpha+\beta x_n+\delta x_{n-2}+\epsilon x_{n-3}}{A+Bx_n+Dx_{n-2}}$	**U**	See [65]
464	$x_{n+1} = \frac{\gamma x_{n-1}+\delta x_{n-2}+\epsilon x_{n-3}}{Dx_{n-2}}$	**U**	Thm 3.1.2
465	$x_{n+1} = \frac{\alpha+\gamma x_{n-1}+\delta x_{n-2}+\epsilon x_{n-3}}{Dx_{n-2}}$	**U**	Thm 3.1.2
466	$x_{n+1} = \frac{\gamma x_{n-1}+\delta x_{n-2}+\epsilon x_{n-3}}{A+Dx_{n-2}}$	**U**	Thm 3.1.2
467	$x_{n+1} = \frac{\alpha+\gamma x_{n-1}+\delta x_{n-2}+\epsilon x_{n-3}}{A+Dx_{n-2}}$	**U**	Thm 3.1.2
468	$x_{n+1} = \frac{\beta x_n+\gamma x_{n-1}+\delta x_{n-2}+\epsilon x_{n-3}}{Dx_{n-2}}$	**U**	Thm 3.1.2
469	$x_{n+1} = \frac{\alpha+\beta x_n+\gamma x_{n-1}+\delta x_{n-2}+\epsilon x_{n-3}}{Dx_{n-2}}$	**U**	Thm 3.1.2
470	$x_{n+1} = \frac{\beta x_n+\gamma x_{n-1}+\delta x_{n-2}+\epsilon x_{n-3}}{A+Dx_{n-2}}$	**U**	Thm 3.1.2
471	$x_{n+1} = \frac{\alpha+\beta x_n+\gamma x_{n-1}+\delta x_{n-2}+\epsilon x_{n-3}}{A+Dx_{n-2}}$	**U**	Thm 3.1.2
472	$x_{n+1} = \frac{\gamma x_{n-1}+\delta x_{n-2}+\epsilon x_{n-3}}{Bx_n+Dx_{n-2}}$	**U**	Thm 3.1.2. See also [58].
473	$x_{n+1} = \frac{\alpha+\gamma x_{n-1}+\delta x_{n-2}+\epsilon x_{n-3}}{Bx_n+Dx_{n-2}}$	**U**	Thm 3.1.2. See also [58].
474	$x_{n+1} = \frac{\gamma x_{n-1}+\delta x_{n-2}+\epsilon x_{n-3}}{A+Bx_n+Dx_{n-2}}$	**U**	Thm 3.1.2. See also [58].
475	$x_{n+1} = \frac{\alpha+\gamma x_{n-1}+\delta x_{n-2}+\epsilon x_{n-3}}{A+Bx_n+Dx_{n-2}}$	**U**	Thm 3.1.2. See also [58].
476	$x_{n+1} = \frac{\beta x_n+\gamma x_{n-1}+\delta x_{n-2}+\epsilon x_{n-3}}{Bx_n+Dx_{n-2}}$	**U**	Thm 3.1.2. See also [58].
477	$x_{n+1} = \frac{\alpha+\beta x_n+\gamma x_{n-1}+\delta x_{n-2}+\epsilon x_{n-3}}{Bx_n+Dx_{n-2}}$	**U**	Thm 3.1.2. See also [58].
478	$x_{n+1} = \frac{\beta x_n+\gamma x_{n-1}+\delta x_{n-2}+\epsilon x_{n-3}}{A+Bx_n+Dx_{n-2}}$	**U**	Thm 3.1.2. See also [58].
479	$x_{n+1} = \frac{\alpha+\beta x_n+\gamma x_{n-1}+\delta x_{n-2}+\epsilon x_{n-3}}{A+Bx_n+Dx_{n-2}}$	**U**	Thm 3.1.2. See also [58].
480	$x_{n+1} = \frac{\delta x_{n-2}+\epsilon x_{n-3}}{Cx_{n-1}+Dx_{n-2}}$	**U**	See [65]
481	$x_{n+1} = \frac{\alpha+\delta x_{n-2}+\epsilon x_{n-3}}{Cx_{n-1}+Dx_{n-2}}$	**U**	See [65]
482	$x_{n+1} = \frac{\delta x_{n-2}+\epsilon x_{n-3}}{A+Cx_{n-1}+Dx_{n-2}}$	**U**	See [65]
483	$x_{n+1} = \frac{\alpha+\delta x_{n-2}+\epsilon x_{n-3}}{A+Cx_{n-1}+Dx_{n-2}}$	**U**	See [65]
484	$x_{n+1} = \frac{\beta x_n+\delta x_{n-2}+\epsilon x_{n-3}}{Cx_{n-1}+Dx_{n-2}}$	**B***	
485	$x_{n+1} = \frac{\alpha+\beta x_n+\delta x_{n-2}+\epsilon x_{n-3}}{Cx_{n-1}+Dx_{n-2}}$	**B***	
486	$x_{n+1} = \frac{\beta x_n+\delta x_{n-2}+\epsilon x_{n-3}}{A+Cx_{n-1}+Dx_{n-2}}$	**B***	
487	$x_{n+1} = \frac{\alpha+\beta x_n+\delta x_{n-2}+\epsilon x_{n-3}}{A+Cx_{n-1}+Dx_{n-2}}$	**B***	
488	$x_{n+1} = \frac{\delta x_{n-2}+\epsilon x_{n-3}}{Bx_n+Cx_{n-1}+Dx_{n-2}}$	**U**	Thm 3.4.2
489	$x_{n+1} = \frac{\alpha+\delta x_{n-2}+\epsilon x_{n-3}}{Bx_n+Cx_{n-1}+Dx_{n-2}}$	**U**	Thm 3.4.2
490	$x_{n+1} = \frac{\delta x_{n-2}+\epsilon x_{n-3}}{A+Bx_n+Cx_{n-1}+Dx_{n-2}}$	**U**	Thm 3.4.2
491	$x_{n+1} = \frac{\alpha+\delta x_{n-2}+\epsilon x_{n-3}}{A+Bx_n+Cx_{n-1}+Dx_{n-2}}$	**U**	Thm 3.4.2
492	$x_{n+1} = \frac{\beta x_n+\delta x_{n-2}+\epsilon x_{n-3}}{Bx_n+Cx_{n-1}+Dx_{n-2}}$	**U**	Thm 3.4.2
493	$x_{n+1} = \frac{\alpha+\beta x_n+\delta x_{n-2}+\epsilon x_{n-3}}{Bx_n+Cx_{n-1}+Dx_{n-2}}$	**U**	Thm 3.4.2
494	$x_{n+1} = \frac{\beta x_n+\delta x_{n-2}+\epsilon x_{n-3}}{A+Bx_n+Cx_{n-1}+Dx_{n-2}}$	**U**	Thm 3.4.2
495	$x_{n+1} = \frac{\alpha+\beta x_n+\delta x_{n-2}+\epsilon x_{n-3}}{A+Bx_n+Cx_{n-1}+Dx_{n-2}}$	**U**	Thm 3.4.2
496	$x_{n+1} = \frac{\gamma x_{n-1}+\delta x_{n-2}+\epsilon x_{n-3}}{Cx_{n-1}+Dx_{n-2}}$	**U**	See [65]
497	$x_{n+1} = \frac{\alpha+\gamma x_{n-1}+\delta x_{n-2}+\epsilon x_{n-3}}{Cx_{n-1}+Dx_{n-2}}$	**U**	See [65]

498	$x_{n+1} = \frac{\gamma x_{n-1}+\delta x_{n-2}+\epsilon x_{n-3}}{A+Cx_{n-1}+Dx_{n-2}}$	**U**	See [65]
499	$x_{n+1} = \frac{\alpha+\gamma x_{n-1}+\delta x_{n-2}+\epsilon x_{n-3}}{A+Cx_{n-1}+Dx_{n-2}}$	**U**	See [65]
500	$x_{n+1} = \frac{\beta x_n+\gamma x_{n-1}+\delta x_{n-2}+\epsilon x_{n-3}}{Cx_{n-1}+Dx_{n-2}}$	**B***	
501	$x_{n+1} = \frac{\alpha+\beta x_n+\gamma x_{n-1}+\delta x_{n-2}+\epsilon x_{n-3}}{Cx_{n-1}+Dx_{n-2}}$	**B***	
502	$x_{n+1} = \frac{\beta x_n+\gamma x_{n-1}+\delta x_{n-2}+\epsilon x_{n-3}}{A+Cx_{n-1}+Dx_{n-2}}$	**B***	
503	$x_{n+1} = \frac{\alpha+\beta x_n+\gamma x_{n-1}+\delta x_{n-2}+\epsilon x_{n-3}}{A+Cx_{n-1}+Dx_{n-2}}$	**B***	
504	$x_{n+1} = \frac{\gamma x_{n-1}+\delta x_{n-2}+\epsilon x_{n-3}}{Bx_n+Cx_{n-1}+Dx_{n-2}}$	**U**	Thm 3.4.2
505	$x_{n+1} = \frac{\alpha+\gamma x_{n-1}+\delta x_{n-2}+\epsilon x_{n-3}}{Bx_n+Cx_{n-1}+Dx_{n-2}}$	**U**	Thm 3.4.2
506	$x_{n+1} = \frac{\gamma x_{n-1}+\delta x_{n-2}+\epsilon x_{n-3}}{A+Bx_n+Cx_{n-1}+Dx_{n-2}}$	**U**	Thm 3.4.2
507	$x_{n+1} = \frac{\alpha+\gamma x_{n-1}+\delta x_{n-2}+\epsilon x_{n-3}}{A+Bx_n+Cx_{n-1}+Dx_{n-2}}$	**U**	Thm 3.4.2
508	$x_{n+1} = \frac{\beta x_n+\gamma x_{n-1}+\delta x_{n-2}+\epsilon x_{n-3}}{Bx_n+Cx_{n-1}+Dx_{n-2}}$	**U**	Thm 3.4.2
509	$x_{n+1} = \frac{\alpha+\beta x_n+\gamma x_{n-1}+\delta x_{n-2}+\epsilon x_{n-3}}{Bx_n+Cx_{n-1}+Dx_{n-2}}$	**U**	Thm 3.4.2
510	$x_{n+1} = \frac{\beta x_n+\gamma x_{n-1}+\delta x_{n-2}+\epsilon x_{n-3}}{A+Bx_n+Cx_{n-1}+Dx_{n-2}}$	**U**	Thm 3.4.2
511	$x_{n+1} = \frac{\alpha+\beta x_n+\gamma x_{n-1}+\delta x_{n-2}+\epsilon x_{n-3}}{A+Bx_n+Cx_{n-1}+Dx_{n-2}}$	**U**	Thm 3.4.2
513	$x_{n+1} = \frac{\alpha}{Ex_{n-3}}$	**B**	Thm 2.1.1
515	$x_{n+1} = \frac{\alpha}{A+Ex_{n-3}}$	**B**	Thm 2.1.1
516	$x_{n+1} = \frac{\beta x_n}{Ex_{n-3}}$	**U**	Reducible to linear
517	$x_{n+1} = \frac{\alpha+\beta x_n}{Ex_{n-3}}$	**U***	
518	$x_{n+1} = \frac{\beta x_n}{A+Ex_{n-3}}$	**B**	Thm 2.3.4
519	$x_{n+1} = \frac{\alpha+\beta x_n}{A+Ex_{n-3}}$	**B**	Thm 2.3.4
521	$x_{n+1} = \frac{\alpha}{Bx_n+Ex_{n-3}}$	**B**	Thm 2.1.1
523	$x_{n+1} = \frac{\alpha}{A+Bx_n+Ex_{n-3}}$	**B**	Thm 2.1.1
524	$x_{n+1} = \frac{\beta x_n}{Bx_n+Ex_{n-3}}$	**B**	Thm 2.1.1
525	$x_{n+1} = \frac{\alpha+\beta x_n}{Bx_n+Ex_{n-3}}$	**B**	Thm 2.1.1
526	$x_{n+1} = \frac{\beta x_n}{A+Bx_n+Ex_{n-3}}$	**B**	Thm 2.1.1
527	$x_{n+1} = \frac{\alpha+\beta x_n}{A+Bx_n+Ex_{n-3}}$	**B**	Thm 2.1.1
528	$x_{n+1} = \frac{\gamma x_{n-1}}{Ex_{n-3}}$	**B**	Reducible to # 7
529	$x_{n+1} = \frac{\alpha+\gamma x_{n-1}}{Ex_{n-3}}$	**B**	Reducible to # 43
530	$x_{n+1} = \frac{\gamma x_{n-1}}{A+Ex_{n-3}}$	**B**	Thm 2.3.4 or reducible to #24
531	$x_{n+1} = \frac{\alpha+\gamma x_{n-1}}{A+Ex_{n-3}}$	**B**	Thm 2.3.4 or reducible to # 66
532	$x_{n+1} = \frac{\beta x_n+\gamma x_{n-1}}{Ex_{n-3}}$	**U***	
533	$x_{n+1} = \frac{\alpha+\beta x_n+\gamma x_{n-1}}{Ex_{n-3}}$	**U***	
534	$x_{n+1} = \frac{\beta x_n+\gamma x_{n-1}}{A+Ex_{n-3}}$	**B**	Thm 2.3.4
535	$x_{n+1} = \frac{\alpha+\beta x_n+\gamma x_{n-1}}{A+Ex_{n-3}}$	**B**	Thm 2.3.4
536	$x_{n+1} = \frac{\gamma x_{n-1}}{Bx_n+Ex_{n-3}}$	**B***	
537	$x_{n+1} = \frac{\alpha+\gamma x_{n-1}}{Bx_n+Ex_{n-3}}$	**U***	
538	$x_{n+1} = \frac{\gamma x_{n-1}}{A+Bx_n+Ex_{n-3}}$	**B**	Thm 2.3.4
539	$x_{n+1} = \frac{\alpha+\gamma x_{n-1}}{A+Bx_n+Ex_{n-3}}$	**B**	Thm 2.3.4
540	$x_{n+1} = \frac{\beta x_n+\gamma x_{n-1}}{Bx_n+Ex_{n-3}}$	**B***	

541	$x_{n+1} = \frac{\alpha+\beta x_n+\gamma x_{n-1}}{Bx_n+Ex_{n-3}}$	**B***	
542	$x_{n+1} = \frac{\beta x_n+\gamma x_{n-1}}{A+Bx_n+Ex_{n-3}}$	**B**	Thm 2.3.4
543	$x_{n+1} = \frac{\alpha+\beta x_n+\gamma x_{n-1}}{A+Bx_n+Ex_{n-3}}$	**B**	Thm 2.3.4
545	$x_{n+1} = \frac{\alpha}{Cx_{n-1}+Ex_{n-3}}$	**B**	Thm 2.1.1
547	$x_{n+1} = \frac{\alpha}{A+Cx_{n-1}+Ex_{n-3}}$	**B**	Thm 2.1.1
548	$x_{n+1} = \frac{\beta x_n}{Cx_{n-1}+Ex_{n-3}}$	**U***	
549	$x_{n+1} = \frac{\alpha+\beta x_n}{Cx_{n-1}+Ex_{n-3}}$	**U***	
550	$x_{n+1} = \frac{\beta x_n}{A+Cx_{n-1}+Ex_{n-3}}$	**B**	Thm 2.3.4
551	$x_{n+1} = \frac{\alpha+\beta x_n}{A+Cx_{n-1}+Ex_{n-3}}$	**B**	Thm 2.3.4
553	$x_{n+1} = \frac{\alpha}{Bx_n+Cx_{n-1}+Ex_{n-3}}$	**B**	Thm 2.1.1
555	$x_{n+1} = \frac{\alpha}{A+Bx_n+Cx_{n-1}+Ex_{n-3}}$	**B**	Thm 2.1.1
556	$x_{n+1} = \frac{\beta x_n}{Bx_n+Cx_{n-1}+Ex_{n-3}}$	**B**	Thm 2.1.1
557	$x_{n+1} = \frac{\alpha+\beta x_n}{Bx_n+Cx_{n-1}+Ex_{n-3}}$	**B**	Thm 2.1.1
558	$x_{n+1} = \frac{\beta x_n}{A+Bx_n+Cx_{n-1}+Ex_{n-3}}$	**B**	Thm 2.1.1
559	$x_{n+1} = \frac{\alpha+\beta x_n}{A+Bx_n+Cx_{n-1}+Ex_{n-3}}$	**B**	Thm 2.1.1
560	$x_{n+1} = \frac{\gamma x_{n-1}}{Cx_{n-1}+Ex_{n-3}}$	**B**	Thm 2.1.1
561	$x_{n+1} = \frac{\alpha+\gamma x_{n-1}}{Cx_{n-1}+Ex_{n-3}}$	**B**	Thm 2.1.1
562	$x_{n+1} = \frac{\gamma x_{n-1}}{A+Cx_{n-1}+Ex_{n-3}}$	**B**	Thm 2.1.1
563	$x_{n+1} = \frac{\alpha+\gamma x_{n-1}}{A+Cx_{n-1}+Ex_{n-3}}$	**B**	Thm 2.1.1
564	$x_{n+1} = \frac{\beta x_n+\gamma x_{n-1}}{Cx_{n-1}+Ex_{n-3}}$	**U***	
565	$x_{n+1} = \frac{\alpha+\beta x_n+\gamma x_{n-1}}{Cx_{n-1}+Ex_{n-3}}$	**B***	
566	$x_{n+1} = \frac{\beta x_n+\gamma x_{n-1}}{A+Cx_{n-1}+Ex_{n-3}}$	**B**	Thm 2.3.4
567	$x_{n+1} = \frac{\alpha+\beta x_n+\gamma x_{n-1}}{A+Cx_{n-1}+Ex_{n-3}}$	**B**	Thm 2.3.4
568	$x_{n+1} = \frac{\gamma x_{n-1}}{Bx_n+Cx_{n-1}+Ex_{n-3}}$	**B**	Thm 2.1.1
569	$x_{n+1} = \frac{\alpha+\gamma x_{n-1}}{Bx_n+Cx_{n-1}+Ex_{n-3}}$	**B**	Thm 2.1.1
570	$x_{n+1} = \frac{\gamma x_{n-1}}{A+Bx_n+Cx_{n-1}+Ex_{n-3}}$	**B**	Thm 2.1.1
571	$x_{n+1} = \frac{\alpha+\gamma x_{n-1}}{A+Bx_n+Cx_{n-1}+Ex_{n-3}}$	**B**	Thm 2.1.1
572	$x_{n+1} = \frac{\beta x_n+\gamma x_{n-1}}{Bx_n+Cx_{n-1}+Ex_{n-3}}$	**B**	Thm 2.1.1
573	$x_{n+1} = \frac{\alpha+\beta x_n+\gamma x_{n-1}}{Bx_n+Cx_{n-1}+Ex_{n-3}}$	**B**	Thm 2.1.1
574	$x_{n+1} = \frac{\beta x_n+\gamma x_{n-1}}{A+Bx_n+Cx_{n-1}+Ex_{n-3}}$	**B**	Thm 2.1.1
575	$x_{n+1} = \frac{\alpha+\beta x_n+\gamma x_{n-1}}{A+Bx_n+Cx_{n-1}+Ex_{n-3}}$	**B**	Thm 2.1.1
576	$x_{n+1} = \frac{\delta x_{n-2}}{Ex_{n-3}}$	**U**	Reducible to linear
577	$x_{n+1} = \frac{\alpha+\delta x_{n-2}}{Ex_{n-3}}$	**U***	
578	$x_{n+1} = \frac{\delta x_{n-2}}{A+Ex_{n-3}}$	**U***	
579	$x_{n+1} = \frac{\alpha+\delta x_{n-2}}{A+Ex_{n-3}}$	**B**	See [14]
580	$x_{n+1} = \frac{\beta x_n+\delta x_{n-2}}{Ex_{n-3}}$	**U***	
581	$x_{n+1} = \frac{\alpha+\beta x_n+\delta x_{n-2}}{Ex_{n-3}}$	**U***	
582	$x_{n+1} = \frac{\beta x_n+\delta x_{n-2}}{A+Ex_{n-3}}$	**B***	
583	$x_{n+1} = \frac{\alpha+\beta x_n+\delta x_{n-2}}{A+Ex_{n-3}}$	**B***	

584	$x_{n+1} = \frac{\delta x_{n-2}}{Bx_n+Ex_{n-3}}$	U*
585	$x_{n+1} = \frac{\alpha+\delta x_{n-2}}{Bx_n+Ex_{n-3}}$	U*
586	$x_{n+1} = \frac{\delta x_{n-2}}{A+Bx_n+Ex_{n-3}}$	U*
587	$x_{n+1} = \frac{\alpha+\delta x_{n-2}}{A+Bx_n+Ex_{n-3}}$	B*
588	$x_{n+1} = \frac{\beta x_n+\delta x_{n-2}}{Bx_n+Ex_{n-3}}$	B*
589	$x_{n+1} = \frac{\alpha+\beta x_n+\delta x_{n-2}}{Bx_n+Ex_{n-3}}$	B*
590	$x_{n+1} = \frac{\beta x_n+\delta x_{n-2}}{A+Bx_n+Ex_{n-3}}$	B*
591	$x_{n+1} = \frac{\alpha+\beta x_n+\delta x_{n-2}}{A+Bx_n+Ex_{n-3}}$	B*
592	$x_{n+1} = \frac{\gamma x_{n-1}+\delta x_{n-2}}{Ex_{n-3}}$	U*
593	$x_{n+1} = \frac{\alpha+\gamma x_{n-1}+\delta x_{n-2}}{Ex_{n-3}}$	U*
594	$x_{n+1} = \frac{\gamma x_{n-1}+\delta x_{n-2}}{A+Ex_{n-3}}$	U*
595	$x_{n+1} = \frac{\alpha+\gamma x_{n-1}+\delta x_{n-2}}{A+Ex_{n-3}}$	U*
596	$x_{n+1} = \frac{\beta x_n+\gamma x_{n-1}+\delta x_{n-2}}{Ex_{n-3}}$	U*
597	$x_{n+1} = \frac{\alpha+\beta x_n+\gamma x_{n-1}+\delta x_{n-2}}{Ex_{n-3}}$	U*
598	$x_{n+1} = \frac{\beta x_n+\gamma x_{n-1}+\delta x_{n-2}}{A+Ex_{n-3}}$	U*
599	$x_{n+1} = \frac{\alpha+\beta x_n+\gamma x_{n-1}+\delta x_{n-2}}{A+Ex_{n-3}}$	U*
600	$x_{n+1} = \frac{\gamma x_{n-1}+\delta x_{n-2}}{Bx_n+Ex_{n-3}}$	U*
601	$x_{n+1} = \frac{\alpha+\gamma x_{n-1}+\delta x_{n-2}}{Bx_n+Ex_{n-3}}$	U*
602	$x_{n+1} = \frac{\gamma x_{n-1}+\delta x_{n-2}}{A+Bx_n+Ex_{n-3}}$	B*
603	$x_{n+1} = \frac{\alpha+\gamma x_{n-1}+\delta x_{n-2}}{A+Bx_n+Ex_{n-3}}$	B*
604	$x_{n+1} = \frac{\beta x_n+\gamma x_{n-1}+\delta x_{n-2}}{Bx_n+Ex_{n-3}}$	B*
605	$x_{n+1} = \frac{\alpha+\beta x_n+\gamma x_{n-1}+\delta x_{n-2}}{Bx_n+Ex_{n-3}}$	B*
606	$x_{n+1} = \frac{\beta x_n+\gamma x_{n-1}+\delta x_{n-2}}{A+Bx_n+Ex_{n-3}}$	B*
607	$x_{n+1} = \frac{\alpha+\beta x_n+\gamma x_{n-1}+\delta x_{n-2}}{A+Bx_n+Ex_{n-3}}$	B*
608	$x_{n+1} = \frac{\delta x_{n-2}}{Cx_{n-1}+Ex_{n-3}}$	U*
609	$x_{n+1} = \frac{\alpha+\delta x_{n-2}}{Cx_{n-1}+Ex_{n-3}}$	U*
610	$x_{n+1} = \frac{\delta x_{n-2}}{A+Cx_{n-1}+Ex_{n-3}}$	U*
611	$x_{n+1} = \frac{\alpha+\delta x_{n-2}}{A+Cx_{n-1}+Ex_{n-3}}$	U*
612	$x_{n+1} = \frac{\beta x_n+\delta x_{n-2}}{Cx_{n-1}+Ex_{n-3}}$	U*
613	$x_{n+1} = \frac{\alpha+\beta x_n+\delta x_{n-2}}{Cx_{n-1}+Ex_{n-3}}$	U*
614	$x_{n+1} = \frac{\beta x_n+\delta x_{n-2}}{A+Cx_{n-1}+Ex_{n-3}}$	U*
615	$x_{n+1} = \frac{\alpha+\beta x_n+\delta x_{n-2}}{A+Cx_{n-1}+Ex_{n-3}}$	U*
616	$x_{n+1} = \frac{\delta x_{n-2}}{Bx_n+Cx_{n-1}+Ex_{n-3}}$	U*
617	$x_{n+1} = \frac{\alpha+\delta x_{n-2}}{Bx_n+Cx_{n-1}+Ex_{n-3}}$	U*
618	$x_{n+1} = \frac{\delta x_{n-2}}{A+Bx_n+Cx_{n-1}+Ex_{n-3}}$	U*
619	$x_{n+1} = \frac{\alpha+\delta x_{n-2}}{A+Bx_n+Cx_{n-1}+Ex_{n-3}}$	U*
620	$x_{n+1} = \frac{\beta x_n+\delta x_{n-2}}{Bx_n+Cx_{n-1}+Ex_{n-3}}$	U*
621	$x_{n+1} = \frac{\alpha+\beta x_n+\delta x_{n-2}}{Bx_n+Cx_{n-1}+Ex_{n-3}}$	U*

622	$x_{n+1} = \frac{\beta x_n+\delta x_{n-2}}{A+Bx_n+Cx_{n-1}+Ex_{n-3}}$	**U***	
623	$x_{n+1} = \frac{\alpha+\beta x_n+\delta x_{n-2}}{A+Bx_n+Cx_{n-1}+Ex_{n-3}}$	**U***	
624	$x_{n+1} = \frac{\gamma x_{n-1}+\delta x_{n-2}}{Cx_{n-1}+Ex_{n-3}}$	**U***	
625	$x_{n+1} = \frac{\alpha+\gamma x_{n-1}+\delta x_{n-2}}{Cx_{n-1}+Ex_{n-3}}$	**U***	
626	$x_{n+1} = \frac{\gamma x_{n-1}+\delta x_{n-2}}{A+Cx_{n-1}+Ex_{n-3}}$	**U***	
627	$x_{n+1} = \frac{\alpha+\gamma x_{n-1}+\delta x_{n-2}}{A+Cx_{n-1}+Ex_{n-3}}$	**U***	
628	$x_{n+1} = \frac{\beta x_n+\gamma x_{n-1}+\delta x_{n-2}}{Cx_{n-1}+Ex_{n-3}}$	**U***	
629	$x_{n+1} = \frac{\alpha+\beta x_n+\gamma x_{n-1}+\delta x_{n-2}}{Cx_{n-1}+Ex_{n-3}}$	**U***	
630	$x_{n+1} = \frac{\beta x_n+\gamma x_{n-1}+\delta x_{n-2}}{A+Cx_{n-1}+Ex_{n-3}}$	**U***	
631	$x_{n+1} = \frac{\alpha+\beta x_n+\gamma x_{n-1}+\delta x_{n-2}}{A+Cx_{n-1}+Ex_{n-3}}$	**U***	
632	$x_{n+1} = \frac{\gamma x_{n-1}+\delta x_{n-2}}{Bx_n+Cx_{n-1}+Ex_{n-3}}$	**U***	
633	$x_{n+1} = \frac{\alpha+\gamma x_{n-1}+\delta x_{n-2}}{Bx_n+Cx_{n-1}+Ex_{n-3}}$	**U***	
634	$x_{n+1} = \frac{\gamma x_{n-1}+\delta x_{n-2}}{A+Bx_n+Cx_{n-1}+Ex_{n-3}}$	**U***	
635	$x_{n+1} = \frac{\alpha+\gamma x_{n-1}+\delta x_{n-2}}{A+Bx_n+Cx_{n-1}+Ex_{n-3}}$	**U***	
636	$x_{n+1} = \frac{\beta x_n+\gamma x_{n-1}+\delta x_{n-2}}{Bx_n+Cx_{n-1}+Ex_{n-3}}$	**U***	
637	$x_{n+1} = \frac{\alpha+\beta x_n+\gamma x_{n-1}+\delta x_{n-2}}{Bx_n+Cx_{n-1}+Ex_{n-3}}$	**U***	
638	$x_{n+1} = \frac{\beta x_n+\gamma x_{n-1}+\delta x_{n-2}}{A+Bx_n+Cx_{n-1}+Ex_{n-3}}$	**U***	
639	$x_{n+1} = \frac{\alpha+\beta x_n+\gamma x_{n-1}+\delta x_{n-2}}{A+Bx_n+Cx_{n-1}+Ex_{n-3}}$	**U***	
641	$x_{n+1} = \frac{\alpha}{Dx_{n-2}+Ex_{n-3}}$	**B**	Thm 2.1.1
643	$x_{n+1} = \frac{\alpha}{A+Dx_{n-2}+Ex_{n-3}}$	**B**	Thm 2.1.1
644	$x_{n+1} = \frac{\beta x_n}{Dx_{n-2}+Ex_{n-3}}$	**U***	
645	$x_{n+1} = \frac{\alpha+\beta x_n}{Dx_{n-2}+Ex_{n-3}}$	**U***	
646	$x_{n+1} = \frac{\beta x_n}{A+Dx_{n-2}+Ex_{n-3}}$	**B**	Thm 2.3.4
647	$x_{n+1} = \frac{\alpha+\beta x_n}{A+Dx_{n-2}+Ex_{n-3}}$	**B**	Thm 2.3.4
649	$x_{n+1} = \frac{\alpha}{Bx_n+Dx_{n-2}+Ex_{n-3}}$	**B**	Thm 2.1.1
651	$x_{n+1} = \frac{\alpha}{A+Bx_n+Dx_{n-2}+Ex_{n-3}}$	**B**	Thm 2.1.1
652	$x_{n+1} = \frac{\beta x_n}{Bx_n+Dx_{n-2}+Ex_{n-3}}$	**B**	Thm 2.1.1
653	$x_{n+1} = \frac{\alpha+\beta x_n}{Bx_n+Dx_{n-2}+Ex_{n-3}}$	**B**	Thm 2.1.1
654	$x_{n+1} = \frac{\beta x_n}{A+Bx_n+Dx_{n-2}+Ex_{n-3}}$	**B**	Thm 2.1.1
655	$x_{n+1} = \frac{\alpha+\beta x_n}{A+Bx_n+Dx_{n-2}+Ex_{n-3}}$	**B**	Thm 2.1.1
656	$x_{n+1} = \frac{\gamma x_{n-1}}{Dx_{n-2}+Ex_{n-3}}$	**B***	
657	$x_{n+1} = \frac{\alpha+\gamma x_{n-1}}{Dx_{n-2}+Ex_{n-3}}$	**B***	
658	$x_{n+1} = \frac{\gamma x_{n-1}}{A+Dx_{n-2}+Ex_{n-3}}$	**B**	Thm 2.3.4
659	$x_{n+1} = \frac{\alpha+\gamma x_{n-1}}{A+Dx_{n-2}+Ex_{n-3}}$	**B**	Thm 2.3.4
660	$x_{n+1} = \frac{\beta x_n+\gamma x_{n-1}}{Dx_{n-2}+Ex_{n-3}}$	**U***	
661	$x_{n+1} = \frac{\alpha+\beta x_n+\gamma x_{n-1}}{Dx_{n-2}+Ex_{n-3}}$	**U***	
662	$x_{n+1} = \frac{\beta x_n+\gamma x_{n-1}}{A+Dx_{n-2}+Ex_{n-3}}$	**B**	Thm 2.3.4
663	$x_{n+1} = \frac{\alpha+\beta x_n+\gamma x_{n-1}}{A+Dx_{n-2}+Ex_{n-3}}$	**B**	Thm 2.3.4
664	$x_{n+1} = \frac{\gamma x_{n-1}}{Bx_n+Dx_{n-2}+Ex_{n-3}}$	**B***	

665	$x_{n+1} = \frac{\alpha+\gamma x_{n-1}}{Bx_n+Dx_{n-2}+Ex_{n-3}}$	**B***	
666	$x_{n+1} = \frac{\gamma x_{n-1}}{A+Bx_n+Dx_{n-2}+Ex_{n-3}}$	**B**	Thm 2.3.4
667	$x_{n+1} = \frac{\alpha+\gamma x_{n-1}}{A+Bx_n+Dx_{n-2}+Ex_{n-3}}$	**B**	Thm 2.3.4
668	$x_{n+1} = \frac{\beta x_n+\gamma x_{n-1}}{Bx_n+Dx_{n-2}+Ex_{n-3}}$	**B***	
669	$x_{n+1} = \frac{\alpha+\beta x_n+\gamma x_{n-1}}{Bx_n+Dx_{n-2}+Ex_{n-3}}$	**B***	
670	$x_{n+1} = \frac{\beta x_n+\gamma x_{n-1}}{A+Bx_n+Dx_{n-2}+Ex_{n-3}}$	**B**	Thm 2.3.4
671	$x_{n+1} = \frac{\alpha+\beta x_n+\gamma x_{n-1}}{A+Bx_n+Dx_{n-2}+Ex_{n-3}}$	**B**	Thm 2.3.4
673	$x_{n+1} = \frac{\alpha}{Cx_{n-1}+Dx_{n-2}+Ex_{n-3}}$	**B**	Thm 2.1.1
675	$x_{n+1} = \frac{\alpha}{A+Cx_{n-1}+Dx_{n-2}+Ex_{n-3}}$	**B**	Thm 2.1.1
676	$x_{n+1} = \frac{\beta x_n}{Cx_{n-1}+Dx_{n-2}+Ex_{n-3}}$	**U***	
677	$x_{n+1} = \frac{\alpha+\beta x_n}{Cx_{n-1}+Dx_{n-2}+Ex_{n-3}}$	**U***	
678	$x_{n+1} = \frac{\beta x_n}{A+Cx_{n-1}+Dx_{n-2}+Ex_{n-3}}$	**B**	Thm 2.3.4
679	$x_{n+1} = \frac{\alpha+\beta x_n}{A+Cx_{n-1}+Dx_{n-2}+Ex_{n-3}}$	**B**	Thm 2.3.4
681	$x_{n+1} = \frac{\alpha}{Bx_n+Cx_{n-1}+Dx_{n-2}+Ex_{n-3}}$	**B**	Thm 2.1.1
683	$x_{n+1} = \frac{\alpha}{A+Bx_n+Cx_{n-1}+Dx_{n-2}+Ex_{n-3}}$	**B**	Thm 2.1.1
684	$x_{n+1} = \frac{\beta x_n}{Bx_n+Cx_{n-1}+Dx_{n-2}+Ex_{n-3}}$	**B**	Thm 2.1.1
685	$x_{n+1} = \frac{\alpha+\beta x_n}{Bx_n+Cx_{n-1}+Dx_{n-2}+Ex_{n-3}}$	**B**	Thm 2.1.1
686	$x_{n+1} = \frac{\beta x_n}{A+Bx_n+Cx_{n-1}+Dx_{n-2}+Ex_{n-3}}$	**B**	Thm 2.1.1
687	$x_{n+1} = \frac{\alpha+\beta x_n}{A+Bx_n+Cx_{n-1}+Dx_{n-2}+Ex_{n-3}}$	**B**	Thm 2.1.1
688	$x_{n+1} = \frac{\gamma x_{n-1}}{Cx_{n-1}+Dx_{n-2}+Ex_{n-3}}$	**B**	Thm 2.1.1
689	$x_{n+1} = \frac{\alpha+\gamma x_{n-1}}{Cx_{n-1}+Dx_{n-2}+Ex_{n-3}}$	**B**	Thm 2.1.1
690	$x_{n+1} = \frac{\gamma x_{n-1}}{A+Cx_{n-1}+Dx_{n-2}+Ex_{n-3}}$	**B**	Thm 2.1.1
691	$x_{n+1} = \frac{\alpha+\gamma x_{n-1}}{A+Cx_{n-1}+Dx_{n-2}+Ex_{n-3}}$	**B**	Thm 2.1.1
692	$x_{n+1} = \frac{\beta x_n+\gamma x_{n-1}}{Cx_{n-1}+Dx_{n-2}+Ex_{n-3}}$	**U***	
693	$x_{n+1} = \frac{\alpha+\beta x_n+\gamma x_{n-1}}{Cx_{n-1}+Dx_{n-2}+Ex_{n-3}}$	**B***	
694	$x_{n+1} = \frac{\beta x_n+\gamma x_{n-1}}{A+Cx_{n-1}+Dx_{n-2}+Ex_{n-3}}$	**B**	Thm 2.3.4
695	$x_{n+1} = \frac{\alpha+\beta x_n+\gamma x_{n-1}}{A+Cx_{n-1}+Dx_{n-2}+Ex_{n-3}}$	**B**	Thm 2.3.4
696	$x_{n+1} = \frac{\gamma x_{n-1}}{Bx_n+Cx_{n-1}+Dx_{n-2}+Ex_{n-3}}$	**B**	Thm 2.1.1
697	$x_{n+1} = \frac{\alpha+\gamma x_{n-1}}{Bx_n+Cx_{n-1}+Dx_{n-2}+Ex_{n-3}}$	**B**	Thm 2.1.1
698	$x_{n+1} = \frac{\gamma x_{n-1}}{A+Bx_n+Cx_{n-1}+Dx_{n-2}+Ex_{n-3}}$	**B**	Thm 2.1.1
699	$x_{n+1} = \frac{\alpha+\gamma x_{n-1}}{A+Bx_n+Cx_{n-1}+Dx_{n-2}+Ex_{n-3}}$	**B**	Thm 2.1.1
700	$x_{n+1} = \frac{\beta x_n+\gamma x_{n-1}}{Bx_n+Cx_{n-1}+Dx_{n-2}+Ex_{n-3}}$	**B**	Thm 2.1.1
701	$x_{n+1} = \frac{\alpha+\beta x_n+\gamma x_{n-1}}{Bx_n+Cx_{n-1}+Dx_{n-2}+Ex_{n-3}}$	**B**	Thm 2.1.1
702	$x_{n+1} = \frac{\beta x_n+\gamma x_{n-1}}{A+Bx_n+Cx_{n-1}+Dx_{n-2}+Ex_{n-3}}$	**B**	Thm 2.1.1
703	$x_{n+1} = \frac{\alpha+\beta x_n+\gamma x_{n-1}}{A+Bx_n+Cx_{n-1}+Dx_{n-2}+Ex_{n-3}}$	**B**	Thm 2.1.1
704	$x_{n+1} = \frac{\delta x_{n-2}}{Dx_{n-2}+Ex_{n-3}}$	**B**	Thm 2.1.1
705	$x_{n+1} = \frac{\alpha+\delta x_{n-2}}{Dx_{n-2}+Ex_{n-3}}$	**B**	Thm 2.1.1
706	$x_{n+1} = \frac{\delta x_{n-2}}{A+Dx_{n-2}+Ex_{n-3}}$	**B**	Thm 2.1.1
707	$x_{n+1} = \frac{\alpha+\delta x_{n-2}}{A+Dx_{n-2}+Ex_{n-3}}$	**B**	Thm 2.1.1
708	$x_{n+1} = \frac{\beta x_n+\delta x_{n-2}}{Dx_{n-2}+Ex_{n-3}}$	**B***	

709	$x_{n+1} = \frac{\alpha+\beta x_n+\delta x_{n-2}}{Dx_{n-2}+Ex_{n-3}}$	**B***	
710	$x_{n+1} = \frac{\beta x_n+\delta x_{n-2}}{A+Dx_{n-2}+Ex_{n-3}}$	**B**	Thm 2.3.4
711	$x_{n+1} = \frac{\alpha+\beta x_n+\delta x_{n-2}}{A+Dx_{n-2}+Ex_{n-3}}$	**B**	Thm 2.3.4
712	$x_{n+1} = \frac{\delta x_{n-2}}{Bx_n+Dx_{n-2}+Ex_{n-3}}$	**B**	Thm 2.1.1
713	$x_{n+1} = \frac{\alpha+\delta x_{n-2}}{Bx_n+Dx_{n-2}+Ex_{n-3}}$	**B**	Thm 2.1.1
714	$x_{n+1} = \frac{\delta x_{n-2}}{A+Bx_n+Dx_{n-2}+Ex_{n-3}}$	**B**	Thm 2.1.1
715	$x_{n+1} = \frac{\alpha+\delta x_{n-2}}{A+Bx_n+Dx_{n-2}+Ex_{n-3}}$	**B**	Thm 2.1.1
716	$x_{n+1} = \frac{\beta x_n+\delta x_{n-2}}{Bx_n+Dx_{n-2}+Ex_{n-3}}$	**B**	Thm 2.1.1
717	$x_{n+1} = \frac{\alpha+\beta x_n+\delta x_{n-2}}{Bx_n+Dx_{n-2}+Ex_{n-3}}$	**B**	Thm 2.1.1
718	$x_{n+1} = \frac{\beta x_n+\delta x_{n-2}}{A+Bx_n+Dx_{n-2}+Ex_{n-3}}$	**B**	Thm 2.1.1
719	$x_{n+1} = \frac{\alpha+\beta x_n+\delta x_{n-2}}{A+Bx_n+Dx_{n-2}+Ex_{n-3}}$	**B**	Thm 2.1.1
720	$x_{n+1} = \frac{\gamma x_{n-1}+\delta x_{n-2}}{Dx_{n-2}+Ex_{n-3}}$	**B***	
721	$x_{n+1} = \frac{\alpha+\gamma x_{n-1}+\delta x_{n-2}}{Dx_{n-2}+Ex_{n-3}}$	**B***	
722	$x_{n+1} = \frac{\gamma x_{n-1}+\delta x_{n-2}}{A+Dx_{n-2}+Ex_{n-3}}$	**B**	Thm 2.3.4
723	$x_{n+1} = \frac{\alpha+\gamma x_{n-1}+\delta x_{n-2}}{A+Dx_{n-2}+Ex_{n-3}}$	**B**	Thm 2.3.4
724	$x_{n+1} = \frac{\beta x_n+\gamma x_{n-1}+\delta x_{n-2}}{Dx_{n-2}+Ex_{n-3}}$	**B***	
725	$x_{n+1} = \frac{\alpha+\beta x_n+\gamma x_{n-1}+\delta x_{n-2}}{Dx_{n-2}+Ex_{n-3}}$	**B***	
726	$x_{n+1} = \frac{\beta x_n+\gamma x_{n-1}+\delta x_{n-2}}{A+Dx_{n-2}+Ex_{n-3}}$	**B**	Thm 2.3.4
727	$x_{n+1} = \frac{\alpha+\beta x_n+\gamma x_{n-1}+\delta x_{n-2}}{A+Dx_{n-2}+Ex_{n-3}}$	**B**	Thm 2.3.4
728	$x_{n+1} = \frac{\gamma x_{n-1}+\delta x_{n-2}}{Bx_n+Dx_{n-2}+Ex_{n-3}}$	**B***	
729	$x_{n+1} = \frac{\alpha+\gamma x_{n-1}+\delta x_{n-2}}{Bx_n+Dx_{n-2}+Ex_{n-3}}$	**B***	
730	$x_{n+1} = \frac{\gamma x_{n-1}+\delta x_{n-2}}{A+Bx_n+Dx_{n-2}+Ex_{n-3}}$	**B**	Thm 2.3.4
731	$x_{n+1} = \frac{\alpha+\gamma x_{n-1}+\delta x_{n-2}}{A+Bx_n+Dx_{n-2}+Ex_{n-3}}$	**B**	Thm 2.3.4
732	$x_{n+1} = \frac{\beta x_n+\gamma x_{n-1}+\delta x_{n-2}}{Bx_n+Dx_{n-2}+Ex_{n-3}}$	**B***	
733	$x_{n+1} = \frac{\alpha+\beta x_n+\gamma x_{n-1}+\delta x_{n-2}}{Bx_n+Dx_{n-2}+Ex_{n-3}}$	**B***	
734	$x_{n+1} = \frac{\beta x_n+\gamma x_{n-1}+\delta x_{n-2}}{A+Bx_n+Dx_{n-2}+Ex_{n-3}}$	**B**	Thm 2.3.4
735	$x_{n+1} = \frac{\alpha+\beta x_n+\gamma x_{n-1}+\delta x_{n-2}}{A+Bx_n+Dx_{n-2}+Ex_{n-3}}$	**B**	Thm 2.3.4
736	$x_{n+1} = \frac{\delta x_{n-2}}{Cx_{n-1}+Dx_{n-2}+Ex_{n-3}}$	**B**	Thm 2.1.1
737	$x_{n+1} = \frac{\alpha+\delta x_{n-2}}{Cx_{n-1}+Dx_{n-2}+Ex_{n-3}}$	**B**	Thm 2.1.1
738	$x_{n+1} = \frac{\delta x_{n-2}}{A+Cx_{n-1}+Dx_{n-2}+Ex_{n-3}}$	**B**	Thm 2.1.1
739	$x_{n+1} = \frac{\alpha+\delta x_{n-2}}{A+Cx_{n-1}+Dx_{n-2}+Ex_{n-3}}$	**B**	Thm 2.1.1
740	$x_{n+1} = \frac{\beta x_n+\delta x_{n-2}}{Cx_{n-1}+Dx_{n-2}+Ex_{n-3}}$	**B***	
741	$x_{n+1} = \frac{\alpha+\beta x_n+\delta x_{n-2}}{Cx_{n-1}+Dx_{n-2}+Ex_{n-3}}$	**B***	
742	$x_{n+1} = \frac{\beta x_n+\delta x_{n-2}}{A+Cx_{n-1}+Dx_{n-2}+Ex_{n-3}}$	**B**	Thm 2.3.4
743	$x_{n+1} = \frac{\alpha+\beta x_n+\delta x_{n-2}}{A+Cx_{n-1}+Dx_{n-2}+Ex_{n-3}}$	**B**	Thm 2.3.4
744	$x_{n+1} = \frac{\delta x_{n-2}}{Bx_n+Cx_{n-1}+Dx_{n-2}+Ex_{n-3}}$	**B**	Thm 2.1.1
745	$x_{n+1} = \frac{\alpha+\delta x_{n-2}}{Bx_n+Cx_{n-1}+Dx_{n-2}+Ex_{n-3}}$	**B**	Thm 2.1.1
746	$x_{n+1} = \frac{\delta x_{n-2}}{A+Bx_n+Cx_{n-1}+Dx_{n-2}+Ex_{n-3}}$	**B**	Thm 2.1.1
747	$x_{n+1} = \frac{\alpha+\delta x_{n-2}}{A+Bx_n+Cx_{n-1}+Dx_{n-2}+Ex_{n-3}}$	**B**	Thm 2.1.1

748	$x_{n+1}=\frac{\beta x_n+\delta x_{n-2}}{Bx_n+Cx_{n-1}+Dx_{n-2}+Ex_{n-3}}$	**B**	Thm 2.1.1
749	$x_{n+1}=\frac{\alpha+\beta x_n+\delta x_{n-2}}{Bx_n+Cx_{n-1}+Dx_{n-2}+Ex_{n-3}}$	**B**	Thm 2.1.1
750	$x_{n+1}=\frac{\beta x_n+\delta x_{n-2}}{A+Bx_n+Cx_{n-1}+Dx_{n-2}+Ex_{n-3}}$	**B**	Thm 2.1.1
751	$x_{n+1}=\frac{\alpha+\beta x_n+\delta x_{n-2}}{A+Bx_n+Cx_{n-1}+Dx_{n-2}+Ex_{n-3}}$	**B**	Thm 2.1.1
752	$x_{n+1}=\frac{\gamma x_{n-1}+\delta x_{n-2}}{Cx_{n-1}+Dx_{n-2}+Ex_{n-3}}$	**B**	Thm 2.1.1
753	$x_{n+1}=\frac{\alpha+\gamma x_{n-1}+\delta x_{n-2}}{Cx_{n-1}+Dx_{n-2}+Ex_{n-3}}$	**B**	Thm 2.1.1
754	$x_{n+1}=\frac{\gamma x_{n-1}+\delta x_{n-2}}{A+Cx_{n-1}+Dx_{n-2}+Ex_{n-3}}$	**B**	Thm 2.1.1
755	$x_{n+1}=\frac{\alpha+\gamma x_{n-1}+\delta x_{n-2}}{A+Cx_{n-1}+Dx_{n-2}+Ex_{n-3}}$	**B**	Thm 2.1.1
756	$x_{n+1}=\frac{\beta x_n+\gamma x_{n-1}+\delta x_{n-2}}{Cx_{n-1}+Dx_{n-2}+Ex_{n-3}}$	**B***	
757	$x_{n+1}=\frac{\alpha+\beta x_n+\gamma x_{n-1}+\delta x_{n-2}}{Cx_{n-1}+Dx_{n-2}+Ex_{n-3}}$	**B***	
758	$x_{n+1}=\frac{\beta x_n+\gamma x_{n-1}+\delta x_{n-2}}{A+Cx_{n-1}+Dx_{n-2}+Ex_{n-3}}$	**B**	Thm 2.3.4
759	$x_{n+1}=\frac{\alpha+\beta x_n+\gamma x_{n-1}+\delta x_{n-2}}{A+Cx_{n-1}+Dx_{n-2}+Ex_{n-3}}$	**B**	Thm 2.3.4
760	$x_{n+1}=\frac{\gamma x_{n-1}+\delta x_{n-2}}{Bx_n+Cx_{n-1}+Dx_{n-2}+Ex_{n-3}}$	**B**	Thm 2.1.1
761	$x_{n+1}=\frac{\alpha+\gamma x_{n-1}+\delta x_{n-2}}{Bx_n+Cx_{n-1}+Dx_{n-2}+Ex_{n-3}}$	**B**	Thm 2.1.1
762	$x_{n+1}=\frac{\gamma x_{n-1}+\delta x_{n-2}}{A+Bx_n+Cx_{n-1}+Dx_{n-2}+Ex_{n-3}}$	**B**	Thm 2.1.1
763	$x_{n+1}=\frac{\alpha+\gamma x_{n-1}+\delta x_{n-2}}{A+Bx_n+Cx_{n-1}+Dx_{n-2}+Ex_{n-3}}$	**B**	Thm 2.1.1
764	$x_{n+1}=\frac{\beta x_n+\gamma x_{n-1}+\delta x_{n-2}}{Bx_n+Cx_{n-1}+Dx_{n-2}+Ex_{n-3}}$	**B**	Thm 2.1.1
765	$x_{n+1}=\frac{\alpha+\beta x_n+\gamma x_{n-1}+\delta x_{n-2}}{Bx_n+Cx_{n-1}+Dx_{n-2}+Ex_{n-3}}$	**B**	Thm 2.1.1
766	$x_{n+1}=\frac{\beta x_n+\gamma x_{n-1}+\delta x_{n-2}}{A+Bx_n+Cx_{n-1}+Dx_{n-2}+Ex_{n-3}}$	**B**	Thm 2.1.1
767	$x_{n+1}=\frac{\alpha+\beta x_n+\gamma x_{n-1}+\delta x_{n-2}}{A+Bx_n+Cx_{n-1}+Dx_{n-2}+Ex_{n-3}}$	**B**	Thm 2.1.1
768	$x_{n+1}=\frac{\epsilon x_{n-3}}{Ex_{n-3}}$	**B**	Thm 2.1.1
769	$x_{n+1}=\frac{\alpha+\epsilon x_{n-3}}{Ex_{n-3}}$	**B**	Thm 2.1.1
770	$x_{n+1}=\frac{\epsilon x_{n-3}}{A+Ex_{n-3}}$	**B**	Thm 2.1.1
771	$x_{n+1}=\frac{\alpha+\epsilon x_{n-3}}{A+Ex_{n-3}}$	**B**	Thm 2.1.1
772	$x_{n+1}=\frac{\beta x_n+\epsilon x_{n-3}}{Ex_{n-3}}$	**B**	Thm 2.3.4
773	$x_{n+1}=\frac{\alpha+\beta x_n+\epsilon x_{n-3}}{Ex_{n-3}}$	**B**	Thm 2.3.4
774	$x_{n+1}=\frac{\beta x_n+\epsilon x_{n-3}}{A+Ex_{n-3}}$	**B**	Thm 2.3.4
775	$x_{n+1}=\frac{\alpha+\beta x_n+\epsilon x_{n-3}}{A+Ex_{n-3}}$	**B**	Thm 2.3.4
776	$x_{n+1}=\frac{\epsilon x_{n-3}}{Bx_n+Ex_{n-3}}$	**B**	Thm 2.1.1
777	$x_{n+1}=\frac{\alpha+\epsilon x_{n-3}}{Bx_n+Ex_{n-3}}$	**B**	Thm 2.1.1
778	$x_{n+1}=\frac{\epsilon x_{n-3}}{A+Bx_n+Ex_{n-3}}$	**B**	Thm 2.1.1
779	$x_{n+1}=\frac{\alpha+\epsilon x_{n-3}}{A+Bx_n+Ex_{n-3}}$	**B**	Thm 2.1.1
780	$x_{n+1}=\frac{\beta x_n+\epsilon x_{n-3}}{Bx_n+Ex_{n-3}}$	**B**	Thm 2.1.1
781	$x_{n+1}=\frac{\alpha+\beta x_n+\epsilon x_{n-3}}{Bx_n+Ex_{n-3}}$	**B**	Thm 2.1.1
782	$x_{n+1}=\frac{\beta x_n+\epsilon x_{n-3}}{A+Bx_n+Ex_{n-3}}$	**B**	Thm 2.1.1
783	$x_{n+1}=\frac{\alpha+\beta x_n+\epsilon x_{n-3}}{A+Bx_n+Ex_{n-3}}$	**B**	Thm 2.1.1
784	$x_{n+1}=\frac{\gamma x_{n-1}+\epsilon x_{n-3}}{Ex_{n-3}}$	**B**	Thm 2.3.4 or reducible to #55
785	$x_{n+1}=\frac{\alpha+\gamma x_{n-1}+\epsilon x_{n-3}}{Ex_{n-3}}$	**B**	Thm 2.3.4 or reducible to #119
786	$x_{n+1}=\frac{\gamma x_{n-1}+\epsilon x_{n-3}}{A+Ex_{n-3}}$	**B**	Thm 2.3.4 or reducible to # 84

787	$x_{n+1} = \frac{\alpha+\gamma x_{n-1}+\epsilon x_{n-3}}{A+Ex_{n-3}}$	**B**	Thm 2.3.4 or reducible to # 166
788	$x_{n+1} = \frac{\beta x_n+\gamma x_{n-1}+\epsilon x_{n-3}}{Ex_{n-3}}$	**B**	Thm 2.3.4
789	$x_{n+1} = \frac{\alpha+\beta x_n+\gamma x_{n-1}+\epsilon x_{n-3}}{Ex_{n-3}}$	**B**	Thm 2.3.4
790	$x_{n+1} = \frac{\beta x_n+\gamma x_{n-1}+\epsilon x_{n-3}}{A+Ex_{n-3}}$	**B**	Thm 2.3.4
791	$x_{n+1} = \frac{\alpha+\beta x_n+\gamma x_{n-1}+\epsilon x_{n-3}}{A+Ex_{n-3}}$	**B**	Thm 2.3.4
792	$x_{n+1} = \frac{\gamma x_{n-1}+\epsilon x_{n-3}}{Bx_n+Ex_{n-3}}$	**B***	
793	$x_{n+1} = \frac{\alpha+\gamma x_{n-1}+\epsilon x_{n-3}}{Bx_n+Ex_{n-3}}$	**B***	
794	$x_{n+1} = \frac{\gamma x_{n-1}+\epsilon x_{n-3}}{A+Bx_n+Ex_{n-3}}$	**B**	Thm 2.3.4
795	$x_{n+1} = \frac{\alpha+\gamma x_{n-1}+\epsilon x_{n-3}}{A+Bx_n+Ex_{n-3}}$	**B**	Thm 2.3.4
796	$x_{n+1} = \frac{\beta x_n+\gamma x_{n-1}+\epsilon x_{n-3}}{Bx_n+Ex_{n-3}}$	**B**	Thm 2.3.4
797	$x_{n+1} = \frac{\alpha+\beta x_n+\gamma x_{n-1}+\epsilon x_{n-3}}{Bx_n+Ex_{n-3}}$	**B**	Thm 2.3.4
798	$x_{n+1} = \frac{\beta x_n+\gamma x_{n-1}+\epsilon x_{n-3}}{A+Bx_n+Ex_{n-3}}$	**B**	Thm 2.3.4
799	$x_{n+1} = \frac{\alpha+\beta x_n+\gamma x_{n-1}+\epsilon x_{n-3}}{A+Bx_n+Ex_{n-3}}$	**B**	Thm 2.3.4
800	$x_{n+1} = \frac{\epsilon x_{n-3}}{Cx_{n-1}+Ex_{n-3}}$	**B**	Thm 2.1.1
801	$x_{n+1} = \frac{\alpha+\epsilon x_{n-3}}{Cx_{n-1}+Ex_{n-3}}$	**B**	Thm 2.1.1
802	$x_{n+1} = \frac{\epsilon x_{n-3}}{A+Cx_{n-1}+Ex_{n-3}}$	**B**	Thm 2.1.1
803	$x_{n+1} = \frac{\alpha+\epsilon x_{n-3}}{A+Cx_{n-1}+Ex_{n-3}}$	**B**	Thm 2.1.1
804	$x_{n+1} = \frac{\beta x_n+\epsilon x_{n-3}}{Cx_{n-1}+Ex_{n-3}}$	**B***	
805	$x_{n+1} = \frac{\alpha+\beta x_n+\epsilon x_{n-3}}{Cx_{n-1}+Ex_{n-3}}$	**B***	
806	$x_{n+1} = \frac{\beta x_n+\epsilon x_{n-3}}{A+Cx_{n-1}+Ex_{n-3}}$	**B**	Thm 2.3.4
807	$x_{n+1} = \frac{\alpha+\beta x_n+\epsilon x_{n-3}}{A+Cx_{n-1}+Ex_{n-3}}$	**B**	Thm 2.3.4
808	$x_{n+1} = \frac{\epsilon x_{n-3}}{Bx_n+Cx_{n-1}+Ex_{n-3}}$	**B**	Thm 2.1.1
809	$x_{n+1} = \frac{\alpha+\epsilon x_{n-3}}{Bx_n+Cx_{n-1}+Ex_{n-3}}$	**B**	Thm 2.1.1
810	$x_{n+1} = \frac{\epsilon x_{n-3}}{A+Bx_n+Cx_{n-1}+Ex_{n-3}}$	**B**	Thm 2.1.1
811	$x_{n+1} = \frac{\alpha+\epsilon x_{n-3}}{A+Bx_n+Cx_{n-1}+Ex_{n-3}}$	**B**	Thm 2.1.1
812	$x_{n+1} = \frac{\beta x_n+\epsilon x_{n-3}}{Bx_n+Cx_{n-1}+Ex_{n-3}}$	**B**	Thm 2.1.1
813	$x_{n+1} = \frac{\alpha+\beta x_n+\epsilon x_{n-3}}{Bx_n+Cx_{n-1}+Ex_{n-3}}$	**B**	Thm 2.1.1
814	$x_{n+1} = \frac{\beta x_n+\epsilon x_{n-3}}{A+Bx_n+Cx_{n-1}+Ex_{n-3}}$	**B**	Thm 2.1.1
815	$x_{n+1} = \frac{\alpha+\beta x_n+\epsilon x_{n-3}}{A+Bx_n+Cx_{n-1}+Ex_{n-3}}$	**B**	Thm 2.1.1
816	$x_{n+1} = \frac{\gamma x_{n-1}+\epsilon x_{n-3}}{Cx_{n-1}+Ex_{n-3}}$	**B**	Thm 2.1.1
817	$x_{n+1} = \frac{\alpha+\gamma x_{n-1}+\epsilon x_{n-3}}{Cx_{n-1}+Ex_{n-3}}$	**B**	Thm 2.1.1
818	$x_{n+1} = \frac{\gamma x_{n-1}+\epsilon x_{n-3}}{A+Cx_{n-1}+Ex_{n-3}}$	**B**	Thm 2.1.1
819	$x_{n+1} = \frac{\alpha+\gamma x_{n-1}+\epsilon x_{n-3}}{A+Cx_{n-1}+Ex_{n-3}}$	**B**	Thm 2.1.1
820	$x_{n+1} = \frac{\beta x_n+\gamma x_{n-1}+\epsilon x_{n-3}}{Cx_{n-1}+Ex_{n-3}}$	**B**	Thm 2.3.4
821	$x_{n+1} = \frac{\alpha+\beta x_n+\gamma x_{n-1}+\epsilon x_{n-3}}{Cx_{n-1}+Ex_{n-3}}$	**B**	Thm 2.3.4
822	$x_{n+1} = \frac{\beta x_n+\gamma x_{n-1}+\epsilon x_{n-3}}{A+Cx_{n-1}+Ex_{n-3}}$	**B**	Thm 2.3.4
823	$x_{n+1} = \frac{\alpha+\beta x_n+\gamma x_{n-1}+\epsilon x_{n-3}}{A+Cx_{n-1}+Ex_{n-3}}$	**B**	Thm 2.3.4
824	$x_{n+1} = \frac{\gamma x_{n-1}+\epsilon x_{n-3}}{Bx_n+Cx_{n-1}+Ex_{n-3}}$	**B**	Thm 2.1.1
825	$x_{n+1} = \frac{\alpha+\gamma x_{n-1}+\epsilon x_{n-3}}{Bx_n+Cx_{n-1}+Ex_{n-3}}$	**B**	Thm 2.1.1
826	$x_{n+1} = \frac{\gamma x_{n-1}+\epsilon x_{n-3}}{A+Bx_n+Cx_{n-1}+Ex_{n-3}}$	**B**	Thm 2.1.1

827	$x_{n+1} = \frac{\alpha+\gamma x_{n-1}+\epsilon x_{n-3}}{A+Bx_n+Cx_{n-1}+Ex_{n-3}}$	**B**	Thm 2.1.1
828	$x_{n+1} = \frac{\beta x_n+\gamma x_{n-1}+\epsilon x_{n-3}}{Bx_n+Cx_{n-1}+Ex_{n-3}}$	**B**	Thm 2.1.1
829	$x_{n+1} = \frac{\alpha+\beta x_n+\gamma x_{n-1}+\epsilon x_{n-3}}{Bx_n+Cx_{n-1}+Ex_{n-3}}$	**B**	Thm 2.1.1
830	$x_{n+1} = \frac{\beta x_n+\gamma x_{n-1}+\epsilon x_{n-3}}{A+Bx_n+Cx_{n-1}+Ex_{n-3}}$	**B**	Thm 2.1.1
831	$x_{n+1} = \frac{\alpha+\beta x_n+\gamma x_{n-1}+\epsilon x_{n-3}}{A+Bx_n+Cx_{n-1}+Ex_{n-3}}$	**B**	Thm 2.1.1
832	$x_{n+1} = \frac{\delta x_{n-2}+\epsilon x_{n-3}}{Ex_{n-3}}$	**B**	See [14]
833	$x_{n+1} = \frac{\alpha+\delta x_{n-2}+\epsilon x_{n-3}}{Ex_{n-3}}$	**B**	See [14]
834	$x_{n+1} = \frac{\delta x_{n-2}+\epsilon x_{n-3}}{A+Ex_{n-3}}$	**B**	See [14] and [50]
835	$x_{n+1} = \frac{\alpha+\delta x_{n-2}+\epsilon x_{n-3}}{A+Ex_{n-3}}$	**B**	See [14]
836	$x_{n+1} = \frac{\beta x_n+\delta x_{n-2}+\epsilon x_{n-3}}{Ex_{n-3}}$	**B***	
837	$x_{n+1} = \frac{\alpha+\beta x_n+\delta x_{n-2}+\epsilon x_{n-3}}{Ex_{n-3}}$	**B***	
838	$x_{n+1} = \frac{\beta x_n+\delta x_{n-2}+\epsilon x_{n-3}}{A+Ex_{n-3}}$	**B***	
839	$x_{n+1} = \frac{\alpha+\beta x_n+\delta x_{n-2}+\epsilon x_{n-3}}{A+Ex_{n-3}}$	**B***	
840	$x_{n+1} = \frac{\delta x_{n-2}+\epsilon x_{n-3}}{Bx_n+Ex_{n-3}}$	**B***	
841	$x_{n+1} = \frac{\alpha+\delta x_{n-2}+\epsilon x_{n-3}}{Bx_n+Ex_{n-3}}$	**B***	
842	$x_{n+1} = \frac{\delta x_{n-2}+\epsilon x_{n-3}}{A+Bx_n+Ex_{n-3}}$	**B***	
843	$x_{n+1} = \frac{\alpha+\delta x_{n-2}+\epsilon x_{n-3}}{A+Bx_n+Ex_{n-3}}$	**B***	
844	$x_{n+1} = \frac{\beta x_n+\delta x_{n-2}+\epsilon x_{n-3}}{Bx_n+Ex_{n-3}}$	**B***	
845	$x_{n+1} = \frac{\alpha+\beta x_n+\delta x_{n-2}+\epsilon x_{n-3}}{Bx_n+Ex_{n-3}}$	**B***	
846	$x_{n+1} = \frac{\beta x_n+\delta x_{n-2}+\epsilon x_{n-3}}{A+Bx_n+Ex_{n-3}}$	**B***	
847	$x_{n+1} = \frac{\alpha+\beta x_n+\delta x_{n-2}+\epsilon x_{n-3}}{A+Bx_n+Ex_{n-3}}$	**B***	
848	$x_{n+1} = \frac{\gamma x_{n-1}+\delta x_{n-2}+\epsilon x_{n-3}}{Ex_{n-3}}$	**U***	
849	$x_{n+1} = \frac{\alpha+\gamma x_{n-1}+\delta x_{n-2}+\epsilon x_{n-3}}{Ex_{n-3}}$	**U***	
850	$x_{n+1} = \frac{\gamma x_{n-1}+\delta x_{n-2}+\epsilon x_{n-3}}{A+Ex_{n-3}}$	**U***	
851	$x_{n+1} = \frac{\alpha+\gamma x_{n-1}+\delta x_{n-2}+\epsilon x_{n-3}}{A+Ex_{n-3}}$	**U***	
852	$x_{n+1} = \frac{\beta x_n+\gamma x_{n-1}+\delta x_{n-2}+\epsilon x_{n-3}}{Ex_{n-3}}$	**U***	
853	$x_{n+1} = \frac{\alpha+\beta x_n+\gamma x_{n-1}+\delta x_{n-2}+\epsilon x_{n-3}}{Ex_{n-3}}$	**U***	
854	$x_{n+1} = \frac{\beta x_n+\gamma x_{n-1}+\delta x_{n-2}+\epsilon x_{n-3}}{A+Ex_{n-3}}$	**U***	
855	$x_{n+1} = \frac{\alpha+\beta x_n+\gamma x_{n-1}+\delta x_{n-2}+\epsilon x_{n-3}}{A+Ex_{n-3}}$	**U***	
856	$x_{n+1} = \frac{\gamma x_{n-1}+\delta x_{n-2}+\epsilon x_{n-3}}{Bx_n+Ex_{n-3}}$	**B***	
857	$x_{n+1} = \frac{\alpha+\gamma x_{n-1}+\delta x_{n-2}+\epsilon x_{n-3}}{Bx_n+Ex_{n-3}}$	**B***	
858	$x_{n+1} = \frac{\gamma x_{n-1}+\delta x_{n-2}+\epsilon x_{n-3}}{A+Bx_n+Ex_{n-3}}$	**B***	
859	$x_{n+1} = \frac{\alpha+\gamma x_{n-1}+\delta x_{n-2}+\epsilon x_{n-3}}{A+Bx_n+Ex_{n-3}}$	**B***	
860	$x_{n+1} = \frac{\beta x_n+\gamma x_{n-1}+\delta x_{n-2}+\epsilon x_{n-3}}{Bx_n+Ex_{n-3}}$	**B***	
861	$x_{n+1} = \frac{\alpha+\beta x_n+\gamma x_{n-1}+\delta x_{n-2}+\epsilon x_{n-3}}{Bx_n+Ex_{n-3}}$	**B***	
862	$x_{n+1} = \frac{\beta x_n+\gamma x_{n-1}+\delta x_{n-2}+\epsilon x_{n-3}}{A+Bx_n+Ex_{n-3}}$	**B***	
863	$x_{n+1} = \frac{\alpha+\beta x_n+\gamma x_{n-1}+\delta x_{n-2}+\epsilon x_{n-3}}{A+Bx_n+Ex_{n-3}}$	**B***	
864	$x_{n+1} = \frac{\delta x_{n-2}+\epsilon x_{n-3}}{Cx_{n-1}+Ex_{n-3}}$	**U***	

865	$x_{n+1} = \frac{\alpha+\delta x_{n-2}+\epsilon x_{n-3}}{Cx_{n-1}+Ex_{n-3}}$	**U***	
866	$x_{n+1} = \frac{\delta x_{n-2}+\epsilon x_{n-3}}{A+Cx_{n-1}+Ex_{n-3}}$	**U***	
867	$x_{n+1} = \frac{\alpha+\delta x_{n-2}+\epsilon x_{n-3}}{A+Cx_{n-1}+Ex_{n-3}}$	**U***	
868	$x_{n+1} = \frac{\beta x_n+\delta x_{n-2}+\epsilon x_{n-3}}{Cx_{n-1}+Ex_{n-3}}$	**U***	
869	$x_{n+1} = \frac{\alpha+\beta x_n+\delta x_{n-2}+\epsilon x_{n-3}}{Cx_{n-1}+Ex_{n-3}}$	**U***	
870	$x_{n+1} = \frac{\beta x_n+\delta x_{n-2}+\epsilon x_{n-3}}{A+Cx_{n-1}+Ex_{n-3}}$	**U***	
871	$x_{n+1} = \frac{\alpha+\beta x_n+\delta x_{n-2}+\epsilon x_{n-3}}{A+Cx_{n-1}+Ex_{n-3}}$	**U***	
872	$x_{n+1} = \frac{\delta x_{n-2}+\epsilon x_{n-3}}{Bx_n+Cx_{n-1}+Ex_{n-3}}$	**U***	
873	$x_{n+1} = \frac{\alpha+\delta x_{n-2}+\epsilon x_{n-3}}{Bx_n+Cx_{n-1}+Ex_{n-3}}$	**U***	
874	$x_{n+1} = \frac{\delta x_{n-2}+\epsilon x_{n-3}}{A+Bx_n+Cx_{n-1}+Ex_{n-3}}$	**U***	
875	$x_{n+1} = \frac{\alpha+\delta x_{n-2}+\epsilon x_{n-3}}{A+Bx_n+Cx_{n-1}+Ex_{n-3}}$	**U***	
876	$x_{n+1} = \frac{\beta x_n+\delta x_{n-2}+\epsilon x_{n-3}}{Bx_n+Cx_{n-1}+Ex_{n-3}}$	**U***	
877	$x_{n+1} = \frac{\alpha+\beta x_n+\delta x_{n-2}+\epsilon x_{n-3}}{Bx_n+Cx_{n-1}+Ex_{n-3}}$	**U***	
878	$x_{n+1} = \frac{\beta x_n+\delta x_{n-2}+\epsilon x_{n-3}}{A+Bx_n+Cx_{n-1}+Ex_{n-3}}$	**U***	
879	$x_{n+1} = \frac{\alpha+\beta x_n+\delta x_{n-2}+\epsilon x_{n-3}}{A+Bx_n+Cx_{n-1}+Ex_{n-3}}$	**U***	
880	$x_{n+1} = \frac{\gamma x_{n-1}+\delta x_{n-2}+\epsilon x_{n-3}}{Cx_{n-1}+Ex_{n-3}}$	**U***	
881	$x_{n+1} = \frac{\alpha+\gamma x_{n-1}+\delta x_{n-2}+\epsilon x_{n-3}}{Cx_{n-1}+Ex_{n-3}}$	**U***	
882	$x_{n+1} = \frac{\gamma x_{n-1}+\delta x_{n-2}+\epsilon x_{n-3}}{A+Cx_{n-1}+Ex_{n-3}}$	**U***	
883	$x_{n+1} = \frac{\alpha+\gamma x_{n-1}+\delta x_{n-2}+\epsilon x_{n-3}}{A+Cx_{n-1}+Ex_{n-3}}$	**U***	
884	$x_{n+1} = \frac{\beta x_n+\gamma x_{n-1}+\delta x_{n-2}+\epsilon x_{n-3}}{Cx_{n-1}+Ex_{n-3}}$	**U***	
885	$x_{n+1} = \frac{\alpha+\beta x_n+\gamma x_{n-1}+\delta x_{n-2}+\epsilon x_{n-3}}{Cx_{n-1}+Ex_{n-3}}$	**U***	
886	$x_{n+1} = \frac{\beta x_n+\gamma x_{n-1}+\delta x_{n-2}+\epsilon x_{n-3}}{A+Cx_{n-1}+Ex_{n-3}}$	**U***	
887	$x_{n+1} = \frac{\alpha+\beta x_n+\gamma x_{n-1}+\delta x_{n-2}+\epsilon x_{n-3}}{A+Cx_{n-1}+Ex_{n-3}}$	**U***	
888	$x_{n+1} = \frac{\gamma x_{n-1}+\delta x_{n-2}+\epsilon x_{n-3}}{Bx_n+Cx_{n-1}+Ex_{n-3}}$	**U***	
889	$x_{n+1} = \frac{\alpha+\gamma x_{n-1}+\delta x_{n-2}+\epsilon x_{n-3}}{Bx_n+Cx_{n-1}+Ex_{n-3}}$	**U***	
890	$x_{n+1} = \frac{\gamma x_{n-1}+\delta x_{n-2}+\epsilon x_{n-3}}{A+Bx_n+Cx_{n-1}+Ex_{n-3}}$	**U***	
891	$x_{n+1} = \frac{\alpha+\gamma x_{n-1}+\delta x_{n-2}+\epsilon x_{n-3}}{A+Bx_n+Cx_{n-1}+Ex_{n-3}}$	**U***	
892	$x_{n+1} = \frac{\beta x_n+\gamma x_{n-1}+\delta x_{n-2}+\epsilon x_{n-3}}{Bx_n+Cx_{n-1}+Ex_{n-3}}$	**U***	
893	$x_{n+1} = \frac{\alpha+\beta x_n+\gamma x_{n-1}+\delta x_{n-2}+\epsilon x_{n-3}}{Bx_n+Cx_{n-1}+Ex_{n-3}}$	**U***	
894	$x_{n+1} = \frac{\beta x_n+\gamma x_{n-1}+\delta x_{n-2}+\epsilon x_{n-3}}{A+Bx_n+Cx_{n-1}+Ex_{n-3}}$	**U***	
895	$x_{n+1} = \frac{\alpha+\beta x_n+\gamma x_{n-1}+\delta x_{n-2}+\epsilon x_{n-3}}{A+Bx_n+Cx_{n-1}+Ex_{n-3}}$	**U***	
896	$x_{n+1} = \frac{\epsilon x_{n-3}}{Dx_{n-2}+Ex_{n-3}}$	**B**	Thm 2.1.1
897	$x_{n+1} = \frac{\alpha+\epsilon x_{n-3}}{Dx_{n-2}+Ex_{n-3}}$	**B**	Thm 2.1.1
898	$x_{n+1} = \frac{\epsilon x_{n-3}}{A+Dx_{n-2}+Ex_{n-3}}$	**B**	Thm 2.1.1
899	$x_{n+1} = \frac{\alpha+\epsilon x_{n-3}}{A+Dx_{n-2}+Ex_{n-3}}$	**B**	Thm 2.1.1
900	$x_{n+1} = \frac{\beta x_n+\epsilon x_{n-3}}{Dx_{n-2}+Ex_{n-3}}$	**B***	
901	$x_{n+1} = \frac{\alpha+\beta x_n+\epsilon x_{n-3}}{Dx_{n-2}+Ex_{n-3}}$	**B***	
902	$x_{n+1} = \frac{\beta x_n+\epsilon x_{n-3}}{A+Dx_{n-2}+Ex_{n-3}}$	**B**	Thm 2.3.4
903	$x_{n+1} = \frac{\alpha+\beta x_n+\epsilon x_{n-3}}{A+Dx_{n-2}+Ex_{n-3}}$	**B**	Thm 2.3.4

904	$x_{n+1} = \frac{\epsilon x_{n-3}}{Bx_n+Dx_{n-2}+Ex_{n-3}}$	**B**	Thm 2.1.1
905	$x_{n+1} = \frac{\alpha+\epsilon x_{n-3}}{Bx_n+Dx_{n-2}+Ex_{n-3}}$	**B**	Thm 2.1.1
906	$x_{n+1} = \frac{\epsilon x_{n-3}}{A+Bx_n+Dx_{n-2}+Ex_{n-3}}$	**B**	Thm 2.1.1
907	$x_{n+1} = \frac{\alpha+\epsilon x_{n-3}}{A+Bx_n+Dx_{n-2}+Ex_{n-3}}$	**B**	Thm 2.1.1
908	$x_{n+1} = \frac{\beta x_n+\epsilon x_{n-3}}{Bx_n+Dx_{n-2}+Ex_{n-3}}$	**B**	Thm 2.1.1
909	$x_{n+1} = \frac{\alpha+\beta x_n+\epsilon x_{n-3}}{Bx_n+Dx_{n-2}+Ex_{n-3}}$	**B**	Thm 2.1.1
910	$x_{n+1} = \frac{\beta x_n+\epsilon x_{n-3}}{A+Bx_n+Dx_{n-2}+Ex_{n-3}}$	**B**	Thm 2.1.1
911	$x_{n+1} = \frac{\alpha+\beta x_n+\epsilon x_{n-3}}{A+Bx_n+Dx_{n-2}+Ex_{n-3}}$	**B**	Thm 2.1.1
912	$x_{n+1} = \frac{\gamma x_{n-1}+\epsilon x_{n-3}}{Dx_{n-2}+Ex_{n-3}}$	**B***	
913	$x_{n+1} = \frac{\alpha+\gamma x_{n-1}+\epsilon x_{n-3}}{Dx_{n-2}+Ex_{n-3}}$	**B***	
914	$x_{n+1} = \frac{\gamma x_{n-1}+\epsilon x_{n-3}}{A+Dx_{n-2}+Ex_{n-3}}$	**B**	Thm 2.3.4
915	$x_{n+1} = \frac{\alpha+\gamma x_{n-1}+\epsilon x_{n-3}}{A+Dx_{n-2}+Ex_{n-3}}$	**B**	Thm 2.3.4
916	$x_{n+1} = \frac{\beta x_n+\gamma x_{n-1}+\epsilon x_{n-3}}{Dx_{n-2}+Ex_{n-3}}$	**B***	
917	$x_{n+1} = \frac{\alpha+\beta x_n+\gamma x_{n-1}+\epsilon x_{n-3}}{Dx_{n-2}+Ex_{n-3}}$	**B***	
918	$x_{n+1} = \frac{\beta x_n+\gamma x_{n-1}+\epsilon x_{n-3}}{A+Dx_{n-2}+Ex_{n-3}}$	**B**	Thm 2.3.4
919	$x_{n+1} = \frac{\alpha+\beta x_n+\gamma x_{n-1}+\epsilon x_{n-3}}{A+Dx_{n-2}+Ex_{n-3}}$	**B**	Thm 2.3.4
920	$x_{n+1} = \frac{\gamma x_{n-1}+\epsilon x_{n-3}}{Bx_n+Dx_{n-2}+Ex_{n-3}}$	**B***	
921	$x_{n+1} = \frac{\alpha+\gamma x_{n-1}+\epsilon x_{n-3}}{Bx_n+Dx_{n-2}+Ex_{n-3}}$	**B***	
922	$x_{n+1} = \frac{\gamma x_{n-1}+\epsilon x_{n-3}}{A+Bx_n+Dx_{n-2}+Ex_{n-3}}$	**B**	Thm 2.3.4
923	$x_{n+1} = \frac{\alpha+\gamma x_{n-1}+\epsilon x_{n-3}}{A+Bx_n+Dx_{n-2}+Ex_{n-3}}$	**B**	Thm 2.3.4
924	$x_{n+1} = \frac{\beta x_n+\gamma x_{n-1}+\epsilon x_{n-3}}{Bx_n+Dx_{n-2}+Ex_{n-3}}$	**B***	
925	$x_{n+1} = \frac{\alpha+\beta x_n+\gamma x_{n-1}+\epsilon x_{n-3}}{Bx_n+Dx_{n-2}+Ex_{n-3}}$	**B***	
926	$x_{n+1} = \frac{\beta x_n+\gamma x_{n-1}+\epsilon x_{n-3}}{A+Bx_n+Dx_{n-2}+Ex_{n-3}}$	**B**	Thm 2.3.4
927	$x_{n+1} = \frac{\alpha+\beta x_n+\gamma x_{n-1}+\epsilon x_{n-3}}{A+Bx_n+Dx_{n-2}+Ex_{n-3}}$	**B**	Thm 2.3.4
928	$x_{n+1} = \frac{\epsilon x_{n-3}}{Cx_{n-1}+Dx_{n-2}+Ex_{n-3}}$	**B**	Thm 2.1.1
929	$x_{n+1} = \frac{\alpha+\epsilon x_{n-3}}{Cx_{n-1}+Dx_{n-2}+Ex_{n-3}}$	**B**	Thm 2.1.1
930	$x_{n+1} = \frac{\epsilon x_{n-3}}{A+Cx_{n-1}+Dx_{n-2}+Ex_{n-3}}$	**B**	Thm 2.1.1
931	$x_{n+1} = \frac{\alpha+\epsilon x_{n-3}}{A+Cx_{n-1}+Dx_{n-2}+Ex_{n-3}}$	**B**	Thm 2.1.1
932	$x_{n+1} = \frac{\beta x_n+\epsilon x_{n-3}}{Cx_{n-1}+Dx_{n-2}+Ex_{n-3}}$	**B***	
933	$x_{n+1} = \frac{\alpha+\beta x_n+\epsilon x_{n-3}}{Cx_{n-1}+Dx_{n-2}+Ex_{n-3}}$	**B***	
934	$x_{n+1} = \frac{\beta x_n+\epsilon x_{n-3}}{A+Cx_{n-1}+Dx_{n-2}+Ex_{n-3}}$	**B**	Thm 2.3.4
935	$x_{n+1} = \frac{\alpha+\beta x_n+\epsilon x_{n-3}}{A+Cx_{n-1}+Dx_{n-2}+Ex_{n-3}}$	**B**	Thm 2.3.4
936	$x_{n+1} = \frac{\epsilon x_{n-3}}{Bx_n+Cx_{n-1}+Dx_{n-2}+Ex_{n-3}}$	**B**	Thm 2.1.1
937	$x_{n+1} = \frac{\alpha+\epsilon x_{n-3}}{Bx_n+Cx_{n-1}+Dx_{n-2}+Ex_{n-3}}$	**B**	Thm 2.1.1
938	$x_{n+1} = \frac{\epsilon x_{n-3}}{A+Bx_n+Cx_{n-1}+Dx_{n-2}+Ex_{n-3}}$	**B**	Thm 2.1.1
939	$x_{n+1} = \frac{\alpha+\epsilon x_{n-3}}{A+Bx_n+Cx_{n-1}+Dx_{n-2}+Ex_{n-3}}$	**B**	Thm 2.1.1
940	$x_{n+1} = \frac{\beta x_n+\epsilon x_{n-3}}{Bx_n+Cx_{n-1}+Dx_{n-2}+Ex_{n-3}}$	**B**	Thm 2.1.1
941	$x_{n+1} = \frac{\alpha+\beta x_n+\epsilon x_{n-3}}{Bx_n+Cx_{n-1}+Dx_{n-2}+Ex_{n-3}}$	**B**	Thm 2.1.1
942	$x_{n+1} = \frac{\beta x_n+\epsilon x_{n-3}}{A+Bx_n+Cx_{n-1}+Dx_{n-2}+Ex_{n-3}}$	**B**	Thm 2.1.1

943	$x_{n+1} = \frac{\alpha+\beta x_n+\epsilon x_{n-3}}{A+Bx_n+Cx_{n-1}+Dx_{n-2}+Ex_{n-3}}$	**B**	Thm 2.1.1
944	$x_{n+1} = \frac{\gamma x_{n-1}+\epsilon x_{n-3}}{Cx_{n-1}+Dx_{n-2}+Ex_{n-3}}$	**B**	Thm 2.1.1
945	$x_{n+1} = \frac{\alpha+\gamma x_{n-1}+\epsilon x_{n-3}}{Cx_{n-1}+Dx_{n-2}+Ex_{n-3}}$	**B**	Thm 2.1.1
946	$x_{n+1} = \frac{\gamma x_{n-1}+\epsilon x_{n-3}}{A+Cx_{n-1}+Dx_{n-2}+Ex_{n-3}}$	**B**	Thm 2.1.1
947	$x_{n+1} = \frac{\alpha+\gamma x_{n-1}+\epsilon x_{n-3}}{A+Cx_{n-1}+Dx_{n-2}+Ex_{n-3}}$	**B**	Thm 2.1.1
948	$x_{n+1} = \frac{\beta x_n+\gamma x_{n-1}+\epsilon x_{n-3}}{Cx_{n-1}+Dx_{n-2}+Ex_{n-3}}$	**B***	
949	$x_{n+1} = \frac{\alpha+\beta x_n+\gamma x_{n-1}+\epsilon x_{n-3}}{Cx_{n-1}+Dx_{n-2}+Ex_{n-3}}$	**B***	
950	$x_{n+1} = \frac{\beta x_n+\gamma x_{n-1}+\epsilon x_{n-3}}{A+Cx_{n-1}+Dx_{n-2}+Ex_{n-3}}$	**B**	Thm 2.3.4
951	$x_{n+1} = \frac{\alpha+\beta x_n+\gamma x_{n-1}+\epsilon x_{n-3}}{A+Cx_{n-1}+Dx_{n-2}+Ex_{n-3}}$	**B**	Thm 2.3.4
952	$x_{n+1} = \frac{\gamma x_{n-1}+\epsilon x_{n-3}}{Bx_n+Cx_{n-1}+Dx_{n-2}+Ex_{n-3}}$	**B**	Thm 2.1.1
953	$x_{n+1} = \frac{\alpha+\gamma x_{n-1}+\epsilon x_{n-3}}{Bx_n+Cx_{n-1}+Dx_{n-2}+Ex_{n-3}}$	**B**	Thm 2.1.1
954	$x_{n+1} = \frac{\gamma x_{n-1}+\epsilon x_{n-3}}{A+Bx_n+Cx_{n-1}+Dx_{n-2}+Ex_{n-3}}$	**B**	Thm 2.1.1
955	$x_{n+1} = \frac{\alpha+\gamma x_{n-1}+\epsilon x_{n-3}}{A+Bx_n+Cx_{n-1}+Dx_{n-2}+Ex_{n-3}}$	**B**	Thm 2.1.1
956	$x_{n+1} = \frac{\beta x_n+\gamma x_{n-1}+\epsilon x_{n-3}}{Bx_n+Cx_{n-1}+Dx_{n-2}+Ex_{n-3}}$	**B**	Thm 2.1.1
957	$x_{n+1} = \frac{\alpha+\beta x_n+\gamma x_{n-1}+\epsilon x_{n-3}}{Bx_n+Cx_{n-1}+Dx_{n-2}+Ex_{n-3}}$	**B**	Thm 2.1.1
958	$x_{n+1} = \frac{\beta x_n+\gamma x_{n-1}+\epsilon x_{n-3}}{A+Bx_n+Cx_{n-1}+Dx_{n-2}+Ex_{n-3}}$	**B**	Thm 2.1.1
959	$x_{n+1} = \frac{\alpha+\beta x_n+\gamma x_{n-1}+\epsilon x_{n-3}}{A+Bx_n+Cx_{n-1}+Dx_{n-2}+Ex_{n-3}}$	**B**	Thm 2.1.1
960	$x_{n+1} = \frac{\delta x_{n-2}+\epsilon x_{n-3}}{Dx_{n-2}+Ex_{n-3}}$	**B**	Thm 2.1.1
961	$x_{n+1} = \frac{\alpha+\delta x_{n-2}+\epsilon x_{n-3}}{Dx_{n-2}+Ex_{n-3}}$	**B**	Thm 2.1.1
962	$x_{n+1} = \frac{\delta x_{n-2}+\epsilon x_{n-3}}{A+Dx_{n-2}+Ex_{n-3}}$	**B**	Thm 2.1.1
963	$x_{n+1} = \frac{\alpha+\delta x_{n-2}+\epsilon x_{n-3}}{A+Dx_{n-2}+Ex_{n-3}}$	**B**	Thm 2.1.1
964	$x_{n+1} = \frac{\beta x_n+\delta x_{n-2}+\epsilon x_{n-3}}{Dx_{n-2}+Ex_{n-3}}$	**B**	Thm 2.3.4
965	$x_{n+1} = \frac{\alpha+\beta x_n+\delta x_{n-2}+\epsilon x_{n-3}}{Dx_{n-2}+Ex_{n-3}}$	**B**	Thm 2.3.4
966	$x_{n+1} = \frac{\beta x_n+\delta x_{n-2}+\epsilon x_{n-3}}{A+Dx_{n-2}+Ex_{n-3}}$	**B**	Thm 2.3.4
967	$x_{n+1} = \frac{\alpha+\beta x_n+\delta x_{n-2}+\epsilon x_{n-3}}{A+Dx_{n-2}+Ex_{n-3}}$	**B**	Thm 2.3.4
968	$x_{n+1} = \frac{\delta x_{n-2}+\epsilon x_{n-3}}{Bx_n+Dx_{n-2}+Ex_{n-3}}$	**B**	Thm 2.1.1
969	$x_{n+1} = \frac{\alpha+\delta x_{n-2}+\epsilon x_{n-3}}{Bx_n+Dx_{n-2}+Ex_{n-3}}$	**B**	Thm 2.1.1
970	$x_{n+1} = \frac{\delta x_{n-2}+\epsilon x_{n-3}}{A+Bx_n+Dx_{n-2}+Ex_{n-3}}$	**B**	Thm 2.1.1
971	$x_{n+1} = \frac{\alpha+\delta x_{n-2}+\epsilon x_{n-3}}{A+Bx_n+Dx_{n-2}+Ex_{n-3}}$	**B**	Thm 2.1.1
972	$x_{n+1} = \frac{\beta x_n+\delta x_{n-2}+\epsilon x_{n-3}}{Bx_n+Dx_{n-2}+Ex_{n-3}}$	**B**	Thm 2.1.1
973	$x_{n+1} = \frac{\alpha+\beta x_n+\delta x_{n-2}+\epsilon x_{n-3}}{Bx_n+Dx_{n-2}+Ex_{n-3}}$	**B**	Thm 2.1.1
974	$x_{n+1} = \frac{\beta x_n+\delta x_{n-2}+\epsilon x_{n-3}}{A+Bx_n+Dx_{n-2}+Ex_{n-3}}$	**B**	Thm 2.1.1
975	$x_{n+1} = \frac{\alpha+\beta x_n+\delta x_{n-2}+\epsilon x_{n-3}}{A+Bx_n+Dx_{n-2}+Ex_{n-3}}$	**B**	Thm 2.1.1
976	$x_{n+1} = \frac{\gamma x_{n-1}+\delta x_{n-2}+\epsilon x_{n-3}}{Dx_{n-2}+Ex_{n-3}}$	**B**	Thm 2.3.4
977	$x_{n+1} = \frac{\alpha+\gamma x_{n-1}+\delta x_{n-2}+\epsilon x_{n-3}}{Dx_{n-2}+Ex_{n-3}}$	**B**	Thm 2.3.4
978	$x_{n+1} = \frac{\gamma x_{n-1}+\delta x_{n-2}+\epsilon x_{n-3}}{A+Dx_{n-2}+Ex_{n-3}}$	**B**	Thm 2.3.4
979	$x_{n+1} = \frac{\alpha+\gamma x_{n-1}+\delta x_{n-2}+\epsilon x_{n-3}}{A+Dx_{n-2}+Ex_{n-3}}$	**B**	Thm 2.3.4
980	$x_{n+1} = \frac{\beta x_n+\gamma x_{n-1}+\delta x_{n-2}+\epsilon x_{n-3}}{Dx_{n-2}+Ex_{n-3}}$	**B**	Thm 2.3.4
981	$x_{n+1} = \frac{\alpha+\beta x_n+\gamma x_{n-1}+\delta x_{n-2}+\epsilon x_{n-3}}{Dx_{n-2}+Ex_{n-3}}$	**B**	Thm 2.3.4

982	$x_{n+1} = \frac{\beta x_n+\gamma x_{n-1}+\delta x_{n-2}+\epsilon x_{n-3}}{A+Dx_{n-2}+Ex_{n-3}}$	B	Thm 2.3.4
983	$x_{n+1} = \frac{\alpha+\beta x_n+\gamma x_{n-1}+\delta x_{n-2}+\epsilon x_{n-3}}{A+Dx_{n-2}+Ex_{n-3}}$	B	Thm 2.3.4
984	$x_{n+1} = \frac{\gamma x_{n-1}+\delta x_{n-2}+\epsilon x_{n-3}}{Bx_n+Dx_{n-2}+Ex_{n-3}}$	B*	
985	$x_{n+1} = \frac{\alpha+\gamma x_{n-1}+\delta x_{n-2}+\epsilon x_{n-3}}{Bx_n+Dx_{n-2}+Ex_{n-3}}$	B*	
986	$x_{n+1} = \frac{\gamma x_{n-1}+\delta x_{n-2}+\epsilon x_{n-3}}{A+Bx_n+Dx_{n-2}+Ex_{n-3}}$	B	Thm 2.3.4
987	$x_{n+1} = \frac{\alpha+\gamma x_{n-1}+\delta x_{n-2}+\epsilon x_{n-3}}{A+Bx_n+Dx_{n-2}+Ex_{n-3}}$	B	Thm 2.3.4
988	$x_{n+1} = \frac{\beta x_n+\gamma x_{n-1}+\delta x_{n-2}+\epsilon x_{n-3}}{Bx_n+Dx_{n-2}+Ex_{n-3}}$	B	Thm 2.3.4
989	$x_{n+1} = \frac{\alpha+\beta x_n+\gamma x_{n-1}+\delta x_{n-2}+\epsilon x_{n-3}}{Bx_n+Dx_{n-2}+Ex_{n-3}}$	B	Thm 2.3.4
990	$x_{n+1} = \frac{\beta x_n+\gamma x_{n-1}+\delta x_{n-2}+\epsilon x_{n-3}}{A+Bx_n+Dx_{n-2}+Ex_{n-3}}$	B	Thm 2.3.4
991	$x_{n+1} = \frac{\alpha+\beta x_n+\gamma x_{n-1}+\delta x_{n-2}+\epsilon x_{n-3}}{A+Bx_n+Dx_{n-2}+Ex_{n-3}}$	B	Thm 2.3.4
992	$x_{n+1} = \frac{\delta x_{n-2}+\epsilon x_{n-3}}{Cx_{n-1}+Dx_{n-2}+Ex_{n-3}}$	B	Thm 2.1.1
993	$x_{n+1} = \frac{\alpha+\delta x_{n-2}+\epsilon x_{n-3}}{Cx_{n-1}+Dx_{n-2}+Ex_{n-3}}$	B	Thm 2.1.1
994	$x_{n+1} = \frac{\delta x_{n-2}+\epsilon x_{n-3}}{A+Cx_{n-1}+Dx_{n-2}+Ex_{n-3}}$	B	Thm 2.1.1
995	$x_{n+1} = \frac{\alpha+\delta x_{n-2}+\epsilon x_{n-3}}{A+Cx_{n-1}+Dx_{n-2}+Ex_{n-3}}$	B	Thm 2.1.1
996	$x_{n+1} = \frac{\beta x_n+\delta x_{n-2}+\epsilon x_{n-3}}{Cx_{n-1}+Dx_{n-2}+Ex_{n-3}}$	B*	
997	$x_{n+1} = \frac{\alpha+\beta x_n+\delta x_{n-2}+\epsilon x_{n-3}}{Cx_{n-1}+Dx_{n-2}+Ex_{n-3}}$	B*	
998	$x_{n+1} = \frac{\beta x_n+\delta x_{n-2}+\epsilon x_{n-3}}{A+Cx_{n-1}+Dx_{n-2}+Ex_{n-3}}$	B	Thm 2.3.4
999	$x_{n+1} = \frac{\alpha+\beta x_n+\delta x_{n-2}+\epsilon x_{n-3}}{A+Cx_{n-1}+Dx_{n-2}+Ex_{n-3}}$	B	Thm 2.3.4
1000	$x_{n+1} = \frac{\delta x_{n-2}+\epsilon x_{n-3}}{Bx_n+Cx_{n-1}+Dx_{n-2}+Ex_{n-3}}$	B	Thm 2.1.1
1001	$x_{n+1} = \frac{\alpha+\delta x_{n-2}+\epsilon x_{n-3}}{Bx_n+Cx_{n-1}+Dx_{n-2}+Ex_{n-3}}$	B	Thm 2.1.1
1002	$x_{n+1} = \frac{\delta x_{n-2}+\epsilon x_{n-3}}{A+Bx_n+Cx_{n-1}+Dx_{n-2}+Ex_{n-3}}$	B	Thm 2.1.1
1003	$x_{n+1} = \frac{\alpha+\delta x_{n-2}+\epsilon x_{n-3}}{A+Bx_n+Cx_{n-1}+Dx_{n-2}+Ex_{n-3}}$	B	Thm 2.1.1
1004	$x_{n+1} = \frac{\beta x_n+\delta x_{n-2}+\epsilon x_{n-3}}{Bx_n+Cx_{n-1}+Dx_{n-2}+Ex_{n-3}}$	B	Thm 2.1.1
1005	$x_{n+1} = \frac{\alpha+\beta x_n+\delta x_{n-2}+\epsilon x_{n-3}}{Bx_n+Cx_{n-1}+Dx_{n-2}+Ex_{n-3}}$	B	Thm 2.1.1
1006	$x_{n+1} = \frac{\beta x_n+\delta x_{n-2}+\epsilon x_{n-3}}{A+Bx_n+Cx_{n-1}+Dx_{n-2}+Ex_{n-3}}$	B	Thm 2.1.1
1007	$x_{n+1} = \frac{\alpha+\beta x_n+\delta x_{n-2}+\epsilon x_{n-3}}{A+Bx_n+Cx_{n-1}+Dx_{n-2}+Ex_{n-3}}$	B	Thm 2.1.1
1008	$x_{n+1} = \frac{\gamma x_{n-1}+\delta x_{n-2}+\epsilon x_{n-3}}{Cx_{n-1}+Dx_{n-2}+Ex_{n-3}}$	B	Thm 2.1.1
1009	$x_{n+1} = \frac{\alpha+\gamma x_{n-1}+\delta x_{n-2}+\epsilon x_{n-3}}{Cx_{n-1}+Dx_{n-2}+Ex_{n-3}}$	B	Thm 2.1.1
1010	$x_{n+1} = \frac{\gamma x_{n-1}+\delta x_{n-2}+\epsilon x_{n-3}}{A+Cx_{n-1}+Dx_{n-2}+Ex_{n-3}}$	B	Thm 2.1.1
1011	$x_{n+1} = \frac{\alpha+\gamma x_{n-1}+\delta x_{n-2}+\epsilon x_{n-3}}{A+Cx_{n-1}+Dx_{n-2}+Ex_{n-3}}$	B	Thm 2.1.1
1012	$x_{n+1} = \frac{\beta x_n+\gamma x_{n-1}+\delta x_{n-2}+\epsilon x_{n-3}}{Cx_{n-1}+Dx_{n-2}+Ex_{n-3}}$	B	Thm 2.3.4
1013	$x_{n+1} = \frac{\alpha+\beta x_n+\gamma x_{n-1}+\delta x_{n-2}+\epsilon x_{n-3}}{Cx_{n-1}+Dx_{n-2}+Ex_{n-3}}$	B	Thm 2.3.4
1014	$x_{n+1} = \frac{\beta x_n+\gamma x_{n-1}+\delta x_{n-2}+\epsilon x_{n-3}}{A+Cx_{n-1}+Dx_{n-2}+Ex_{n-3}}$	B	Thm 2.3.4
1015	$x_{n+1} = \frac{\alpha+\beta x_n+\gamma x_{n-1}+\delta x_{n-2}+\epsilon x_{n-3}}{A+Cx_{n-1}+Dx_{n-2}+Ex_{n-3}}$	B	Thm 2.3.4
1016	$x_{n+1} = \frac{\gamma x_{n-1}+\delta x_{n-2}+\epsilon x_{n-3}}{Bx_n+Cx_{n-1}+Dx_{n-2}+Ex_{n-3}}$	B	Thm 2.1.1
1017	$x_{n+1} = \frac{\alpha+\gamma x_{n-1}+\delta x_{n-2}+\epsilon x_{n-3}}{Bx_n+Cx_{n-1}+Dx_{n-2}+Ex_{n-3}}$	B	Thm 2.1.1
1018	$x_{n+1} = \frac{\gamma x_{n-1}+\delta x_{n-2}+\epsilon x_{n-3}}{A+Bx_n+Cx_{n-1}+Dx_{n-2}+Ex_{n-3}}$	B	Thm 2.1.1
1019	$x_{n+1} = \frac{\alpha+\gamma x_{n-1}+\delta x_{n-2}+\epsilon x_{n-3}}{A+Bx_n+Cx_{n-1}+Dx_{n-2}+Ex_{n-3}}$	B	Thm 2.1.1

1020	$x_{n+1} = \frac{\beta x_n+\gamma x_{n-1}+\delta x_{n-2}+\epsilon x_{n-3}}{Bx_n+Cx_{n-1}+Dx_{n-2}+Ex_{n-3}}$	**B**	Thm 2.1.1
1021	$x_{n+1} = \frac{\alpha+\beta x_n+\gamma x_{n-1}+\delta x_{n-2}+\epsilon x_{n-3}}{Bx_n+Cx_{n-1}+Dx_{n-2}+Ex_{n-3}}$	**B**	Thm 2.1.1
1022	$x_{n+1} = \frac{\beta x_n+\gamma x_{n-1}+\delta x_{n-2}+\epsilon x_{n-3}}{A+Bx_n+Cx_{n-1}+Dx_{n-2}+Ex_{n-3}}$	**B**	Thm 2.1.1
1023	$x_{n+1} = \frac{\alpha+\beta x_n+\gamma x_{n-1}+\delta x_{n-2}+\epsilon x_{n-3}}{A+Bx_n+Cx_{n-1}+Dx_{n-2}+Ex_{n-3}}$	**B**	Thm 2.1.1

Bibliography

[1] R. Abu-Saris, Characterization of rational periodic sequences, *J. Difference Equ. Appl.,* **6**(2000), 233-242.

[2] R. Abu-Saris, Global behavior of rational sequences involving piecewise power function, *J. Comp. Anal. Appl.*, **2**(2000), 103-109.

[3] R. Abu-Saris, A note of the attractivity of period-four solutions of a third order rational difference equations, *J. Difference Equ. Appl.,* (to appear).

[4] R. Abu-Saris, Globally attracting periodic cycles: necessary and sufficient conditions with applications, *J. Difference Equ. Appl.,* (to appear).

[5] R. Abu-Saris and Kh. Al-Dosary, Invariants: A functional-equation approach, *J. Difference Equ. Appl.,* **10**(2004), 1037-1040.

[6] R. Abu-Saris, K. Al-Dosary, and Q. Al-Hassan, Characterization and transformation of invariants, *J. Difference Equ. Appl.,* **10**(2003), 869-877.

[7] R. Abu-Saris and Q. Al-Hassan, On global periodicity of difference equations, *J. Math. Anal. Appl.,* **283**(2003), 468-477.

[8] R. Abu-Saris and Q. Al-Hassan, On globally attracting two-cycles of second order difference equations, *J. Difference Equ. Appl.*, (to appear).

[9] R. Abu-Saris and R. Devault, On the recursive sequence $y_{n+1} = A + \dfrac{y_n}{y_{n-k}}$, *Appl. Math. Lett.*, **6** (2003), 173-178.

[10] R. Abu-Saris and R. DeVault, On the asymptotic behavior of $x_{n+1} = a_n x_n / x_{n-1}$, *J. Math. Anal. Appl.*, **280**(2003), 148-154.

[11] R. Abu-Saris and N. Al-Jubouri, Characterization of rational periodic sequences II, *J. Difference Equ. Appl.,* (to appear).

[12] R. Agarwal, *Difference Equations and Inequalities, Theory, Methods and Applications*, Marcel Dekker Inc., New York, 1992.

[13] K.T. Alligood, T. Sauer, and J.A. Yorke, *CHAOS An Introduction to Dynamical Systems*, Springer-Verlag, New York, Berlin, Heidelberg, Tokyo, 1997.

[14] A.M. Amleh, E. Camouzis, and G. Ladas, On the boundedness character of rational equations, Part 2, *J. Difference Equ. Appl.*, **12** (2006), 637-650.

[15] A.M. Amleh, E. Camouzis, and G. Ladas, On second-order rational difference equations, Part 1 and Part 2, *J. Difference Equ. Appl.*, (2007).

[16] A.M. Amleh, D.A. Georgiou, E.A. Grove, and G. Ladas, On the recursive sequence $x_{n+1} = \alpha + \frac{x_{n-1}}{x_n}$, *J. Math. Anal. Appl.*, **233**(1999), 790-798.

[17] A.M. Amleh, V. Kirk, and G. Ladas, On the dynamics of $x_{n+1} = \frac{a + bx_{n-1}}{A + Bx_{n-2}}$, *Math. Sci. Res. Hot-Line*, **5**(2001), 1-15.

[18] F. Balibrea and A. Linero, On the periodic structure of delayed difference equations of the form $x_n = f(x_{n-k})$ on $\mathcal{I}$ and $\mathcal{S}^1$, *J. Difference Equ. Appl.*, **9**(2003), 359-371.

[19] F. Balibrea, A. Linero, G.S. Lopez, and S. Stevic, Global periodicity of $x_{n+k+1} = f_k(x_{n+k}) \cdots f_2(x_{n+2}) f_1(x_{n+1})$, *J. Difference Equ. Appl.*, (2007).

[20] E. Barbeau, B. Gelford, and S. Tanny, Periodicities of solutions of the generalized Lyness equation, *J. Difference Equ. Appl.*, **1**(1995), 291-306.

[21] G. Bastien and M. Rogalski, Global behavior of the solutions of Lyness' difference equation $u_{n+2}u_n = u_{n+1} + a$, *J. Difference Equ. Appl.*, **11** (2004), 997-1003.

[22] G. Bastien and M. Rogalski, Global behavior of the solutions of the k-lacunary order $2k$ Lyness' difference equation $u_n = \frac{u_{n-k} + a}{u_{n-2k}}$ in R_*^+ and of other more general equations, *J. Difference Equ. Appl.*, **13**(2007), 79-88.

[23] M.R. Bellavia and M.A. Radin, Long term behavior of the nonautonomous rational difference equation $x_{n+1} = \frac{A_n x_{n-1}}{B_n + x_n}$, *Math. Sci. Res. J.*, (to appear).

[24] M. Benedicks and L. Carleson, The dynamics of the Hénon map, *Ann. Math.*, **133**(1991), 73-169.

[25] K.S. Berenhaut, K.S. Dice, J.E. Foley, J.D. Iricanin, and S. Stevic, Periodic solutions of the rational difference equation $y_n = \frac{y_{n-3} + y_{n-4}}{y_{n-1}}$, *J. Difference Equ. Appl.*, **12** (2006), 183-189.

[26] K.S. Berenhaut and J.D. Foley, Product difference equation approximating rational equations, *Proceedings of the 2005 Conference on Differential and Difference Equations*, 2006.

[27] K.S. Berenhaut, J.D. Foley, and S. Stevic, The global attractivity of the rational difference equation $y_{n+1} = \frac{y_{n-k} + y_{n-m}}{1 + y_{n-k}y_{n-m}}$, *Appl. Math. Lett.*, **20**(2007), 54-58.

[28] K.S. Berenhaut, J.D. Foley, and S. Stevic, The global attractivity of the rational difference equation $y_{n+1} = 1 + \frac{y_{n-k}}{y_{n-m}}$, *Proc. Am. Math. Soc.*, **135**(2007), 1133-1140.

[29] K.S. Berenhaut, J.D. Foley, and S. Stevic, Quantitative bounds for the recursive sequence $y_n = A + \frac{y_{n-l}}{y_{n-k}}$, *Appl. Math. Lett.*, **19**(2006), 983-989.

[30] K.S. Berenhaut, J.D. Foley, and S. Stevic, The global attractivity of the rational difference equation $y_n = A + \left(\frac{y_{n-k}}{y_{n-m}}\right)^p$, *Proc. Am. Math. Soc.*, (2006).

[31] K.S. Berenhaut and S. Stevic, The global attractivity of a higher order rational difference equation, *J. Math. Anal. Appl.*, **326**(2007), 940-944.

[32] K.S. Berenhaut and S. Stevic, A note on the difference equation $x_{n+1} = \frac{1}{x_n x_{n-1}} + \frac{1}{x_{n-3}x_{n-4}}$, *J. Difference Equ. Appl.*, **11**(2005), 1225-1228.

[33] K.S. Berenhaut and S. Stevic, The difference equation $x_{n+1} = \alpha + \frac{x_{n-k}}{\sum_{i=0}^{k-1} b_i x_{n-i}}$, *J. Math. Anal. Appl.*, **326**(2007), 1466-1471.

[34] K.S. Berenhaut and S. Stevic, On positive nonoscillatory solutions of the difference equation $x_{n+1} = \alpha + \frac{x_{n-k}^p}{x_n^p}$, *J. Difference Equ. Appl.*, **12**(2006), 495-499.

[35] K.S. Berenhaut and S. Stevic, The behavior of the positive solutions of the difference equation $x_{n+1} = A + (\frac{x_{n-2}}{x_{n-1}})^p$, *J. Difference Equ. Appl.*, **12**(2006), 909-918.

[36] L. Berg, Asymptotische auflsung von differential-functionalgleichungen, *Math. Nachr.*, **17**(1959), 195-210.

[37] L. Berg, Asymptotische lsungen fr operatorgleichungen, *Z. Ang. Math. Mech.*, **45**(1965), 333-352.

[38] L. Berg, Asymptotische darstellungen und entwicklungen, Dt. Verlag Wiss., Berlin (1968).

[39] L. Berg, On the asymptotics of nonlinear difference equations, *Z. Anal. Anwendugen.*, **21**(2002), 1061-1074.

[40] L. Berg, Oscillating solutions of rational difference equations, *Rostock. Math. Colloq.*, **58**(2004), 31-35.

[41] L. Berg, Inclusion theorems for nonlinear difference equations with applications, *J. Difference Equ. Appl.*, **10**(2004), 399-408. Corrections: **11**(2005), 181-182.

[42] L. Berg, On the differnce equation $x_{n+1} = \dfrac{\beta x_n + \gamma x_{n-1}}{\gamma x_n + B x_{n-1}}$, *Rostock. Math. Colloq.*, **51**(2005), 3-11.

[43] L. Berg, Nonlinear difference equations with periodic solutions, *Rostock. Math. Colloq.*, **61**(2006), 12-19.

[44] L. Berg and L. V. Wolfersdorf, On a class of generalized autoconvolution equations of the third kind, *J. Anal. Appl.*, **24**(2005), 217-250.

[45] L. Brand, A sequence defined by a difference equation, *Am. Math. Monthly*, **62**(1955), 489-492.

[46] E. Camouzis, On rational third order difference equations, *Proceedings of the Eighth International Conference on Difference Equations and Applications*, July 28-Aug 2, 2003, Brno, Czech Republic.

[47] E. Camouzis, On the dynamics of $x_{n+1} = \frac{\alpha + x_{n-2}}{x_{n-1}}$, *Int. J. Appl. Math. Sci.*, **1**(2004), 133-149.

[48] E. Camouzis, Global analysis of solutions of $x_{n+1} = \frac{\beta x_n + \delta x_{n-2}}{A + B x_n + C x_{n-1}}$, *J. Math. Anal. Appl.*, **316**(2006), 616-627.

[49] E. Camouzis, On the boundedness of some rational difference equations, *J. Difference Equ. Appl.*, **12**(2006), 69-94.

[50] E. Camouzis, E. Chatterjee, and G. Ladas, On the dynamics of $x_{n+1} = \frac{\delta x_{n-2} + \epsilon x_{n-3}}{A + x_{n-3}}$, *J. Math. Anal. Appl.*, (2006).

[51] E. Camouzis, E. Chatterjee, G. Ladas, and E.P. Quinn, On third-order rational difference equations, Part 3, *J. Difference Equ. Appl.*, **10**(2004), 1119-1127.

[52] E. Camouzis, E. Chatterjee, G. Ladas, and E.P. Quinn, Progress report on the boundedness character of third-order rational equations, *J. Difference. Equ. Appl.*, **11**(2005), 1029-1035.

[53] E. Camouzis and R. DeVault, The forbidden set of $x_{n+1} = p + \dfrac{x_{n-1}}{x_n}$, *J. Difference Equ. Appl.*, **9**(2003), 739-750.

[54] E. Camouzis, R. DeVault, and W. Kosmala, On the period-five trichotomy of $x_{n+1} = \dfrac{p + x_{n-2}}{x_n}$, *J. Math. Anal. Appl.*, **291**(2004), 40-49.

[55] E. Camouzis, R. DeVault, and G. Ladas, On the recursive sequence $x_{n+1} = -1 + \dfrac{x_{n-1}}{x_n}$, *J. Difference Equ. Appl.*, **7**(2001), 477-482.

[56] E. Camouzis, R. Devault, and G. Papaschinopoulos, On the recursive sequence $x_{n+1} = \dfrac{\gamma x_{n-1} + x_{n-2}}{x_n + x_{n-2}}$, *Adv. Difference Equations*, **1**(2005), 31-40.

[57] E. Camouzis, C.H. Gibbons, and G. Ladas, On period-two convergence in rational equations, *J. Difference Equ. Appl.*, **9**(2003), 535-540.

[58] E. Camouzis, E.A. Grove, Y. Kostrov, and G. Ladas, Existence of unbounded solutions in rational equations, (to appear).

[59] E. Camouzis and G. Ladas, Three trichotomy conjectures, *J. Difference Equ. Appl.*, **8**(2002), 495-500.

[60] E. Camouzis and G. Ladas, On third order rational difference equations, Part 5, *J. Difference. Equ. Appl.*, **11**(2005), 553-562.

[61] E. Camouzis and G. Ladas, When does local stability imply global attractivity in rational equations?, *J. Difference. Equ. Appl.*, **12**(2006), 863-885.

[62] E. Camouzis and G. Ladas, When does periodicity destroys boundedness in rational difference equations?, *J. Difference. Equ. Appl.*, **12**(2006), 961-979.

[63] E. Camouzis and G. Ladas, Periodically forced Pielou's Equation, *J. Math. Anal. Appl.*, (2007).

[64] E. Camouzis and G. Ladas, Global convergence in difference equations, *Com. Appl. Nonlinear Anal.*, (2007).

[65] E. Camouzis and G. Ladas, On third-order rational difference equations, Part 1 and Part 2, *J. Difference Equ. Appl.*, (2007).

[66] E. Camouzis, G. Ladas, F. Palladino, and E.P. Quinn, On the boundedness character of rational equations, Part 1, *J. Difference Equ. Appl.*, **12**(2006), 503-523.

[67] E. Camouzis, G. Ladas, and E.P. Quinn, On the dynamics of $x_{n+1} = \dfrac{\alpha + \beta x_n + \gamma x_{n-1} + \delta x_{n-2}}{A + x_n}$, *J. Difference Equ. Appl.*, **10**(2004), 963-976.

[68] E. Camouzis, G. Ladas, and E.P. Quinn, On third-order rational difference equations, Part 2, *J. Difference Equ. Appl.*, **10**(2004), 1041-1047.

[69] E. Camouzis, G. Ladas, and E.P. Quinn, On third order rational difference equations, Part 6, *J. Difference Equ. Appl.*, **11**(2005), 759-777.

[70] E. Camouzis, G. Ladas, and H.D. Voulov, On the dynamics of $x_{n+1} = \dfrac{\alpha + \gamma x_{n-1} + \delta x_{n-2}}{A + x_{n-2}}$, *J. Difference Equ. Appl.*, **9**(2003), 731-738.

[71] D.M. Chan, E.R. Chang, M. Dehghan, C.M. Kent, R. Mazrooei-Sebdani, and H. Sedaghat, Asymptotic stability for difference equations with decreasing arguments, *J. Difference Equ. Appl.*, (to appear).

[72] E. Chatterjee, E.A. Grove, Y. Kostrov, and G. Ladas, On the trichotomy character of $x_{n+1} = \dfrac{\alpha + \gamma x_{n-1}}{A + Bx_n + x_{n-2}}$, *J. Difference Equ. Appl.*, **9**(2003), 1113-1128.

[73] A. Cima, A. Gasull, and V. Manosa, Dynamics of the third order Lyness' difference equation, *J. Difference Equ. Appl.*, (2007).

[74] C.W. Clark, A delayed recruitment model of population dynamics with an application to baleen whale populations, *J. Math. Biol.*, **3**(1976), 381-391.

[75] M.E. Clark and L.J. Gross, Periodic solutions to nonautonomous difference equations, *Math. Biosc.*, **102**(1990), 105-119.

[76] C.A. Clark, M.R.S. Kulenovic, and S. Valicenti On the dynamics of $x_{n+1} = \frac{\alpha x_{n-1} + \beta x_{n-2}}{A + x_n}$, *Math. Sci. Res. J.*, (to appear).

[77] D. Clark, M.R.S. Kulenović, and J.F. Selgrade, On a system of rational difference equations, *J. Difference Equ. Appl.*, **11**(2005), 565-580.

[78] A. Clark, E.S. Thomas, and D.R. Wilken, Continuous invariants for a class of difference equations, *Proceedings of the Tenth International Conference in Difference Equations and Applications*, Munich, Germany, 2005 (to appear).

[79] A. Clark, E.S. Thomas, and D.R. Wilken, A proof of the no rational invariant theorem, *J. Difference Equ. Appl.*, (to appear).

[80] J.H. Conway and H.S.M. Coxeter, Triangulated polygons and frieze patterns, *Math. Gaz.*, **57**(1973) n^0. 400, 87-94 and n^0. 401, 175-183.

[81] M. Csörnyei and M. Laczkovich, Some periodic and nonperiodic Recursions, *Monatsh. Math.*, **132**(2001), 215-236.

[82] J.M. Cushing, Periodically forced nonlinear systems of difference equations, *J. Difference Equ. Appl.*, **3**(1998), 547-561.

[83] J.M. Cushing and S.M. Henson, A periodically forced Beverton-Holt equation, *J. Difference Equ. Appl.*, **8**(2002), 1119-1120.

[84] R.L. Devaney, *A First Course in Chaotic Dynamical Systems: Theory and Experiment*, Addison-Wesley, Reading, MA, 1992.

[85] R. Devault and L. Galminas, Global stability of $x_{n+1} = \frac{A}{x_n^p} + \frac{B}{x_{n-1}^{\frac{1}{p}}}$, *J. Math. Anal. Appl.*, **231**(1999), 459-466.

[86] R. Devault, L. Galminas, E.J. Janowski, and G. Ladas, On the recursive sequence $x_{n+1} = \frac{A}{x_n} + \frac{B}{x_{n-1}}$, *J. Difference Equ. Appl.*, **6**(2000), 121-125.

[87] R. Devault, C.M. Kent, and W. Kosmala, On the recursive sequence $x_{n+1} = p + \frac{x_{n-k}}{x_n}$, *J. Difference Equ. Appl.*, **9**(2003), 721-730.

[88] R. Devault, V.L. Kocic, and G. Ladas, Global stability of a recursive sequence, *Dynamic Syst. Applications*, **1**(1992), 13-21.

[89] R. Devault, V.L. Kocic, and D. Stutson, Global behavior of solutions of the nonlinear difference equation $x_{n+1} = p_n + \frac{x_{n-1}}{x_n}$, *J. Difference Equ. Appl.*, **11**(2005), 707-719.

[90] R. Devault, W. Kosmala, G. Ladas, and S. W. Schultz, Global behavior of $x_{n+1} = \frac{p + y_{n-k}}{qy_n + y_{n-k}}$, *Nonlinear Anal.*, **47**(2001), 4743-4751.

[91] R. DeVault, G. Ladas, and S.W. Schultz, On the recursive sequence $x_{n+1} = \frac{A}{x_n} + \frac{1}{x_{n-2}}$, *Proc. Am. Math. Soc.*, **126**(1998), 3257-3261.

[92] R. DeVault, G. Ladas, and S.W. Schultz, On the recursive sequence $x_{n+1} = \frac{A}{x_n^p} + \frac{B}{x_{n-1}^q}$, *Proceedings of the Second International Conference on Difference Equations and Applications*, August 7-11, 1995, Veprém, Hungary, Gordon and Breach Science Publishers, 1997, 125-136.

[93] R. DeVault, G. Ladas, and S.W. Schultz, Necessary and sufficient conditions for the boundedness of solutions of $x_{n+1} = \frac{A}{x_n^p} + \frac{B}{x_{n-1}^q}$, *J. Difference Equ. Appl.*, **3**(1998), 259-266.

[94] R. DeVault, G. Ladas, and S.W. Schultz, On the recursive sequence $x_{n+1} = \frac{A}{x_n x_{n-1}} + \frac{1}{x_{n-3}x_{n-4}}$, *J. Difference Equ. Appl.*, **6**(2000), 481-483.

[95] S. Elaydi, *An Introduction to Difference Equations*, 3rd ed., Springer-Verlag, New York, 2005.

[96] S. Elaydi, *Discrete Chaos*, CRC Press, Boca Raton, FL, 2000.

[97] S. Elaydi and A.A. Yakubu, Global stability of cycles: Lotka-Volterra competition model with stocking, *J. Difference Equ. Appl.*, **8**(2002), 537-549.

[98] S. Elaydi and A.A. Yakubu, Open problems on the basins of attraction of stable cycles, *J. Difference Equ. Appl.*, **8**(2002), 755-760.

[99] S. Elaydi and R.J. Sacker, Global stability of periodic orbits of nonautonomous difference equations in population biology and the Cushing-Henson conjectures, *Proc. 8th Int. Conf. Difference Equat. Appl.*, Brno, Czech Republic, 2003.

[100] S. Elaydi and R.J. Sacker, Global stability of periodic orbits of nonautonomous difference equations and population biology, *J. Differential Equ.*, **208**(2004), 258-273.

[101] H.A. El-Metwally, E.A. Grove, and G. Ladas, A global convergence result with applications to periodic solutions, *J. Math. Anal. Appl.*, **245**(2000), 161-170.

[102] H.A. El-Metwally, E.A. Grove, G. Ladas, and L.C. McGrath, On the difference equation $y_{n+1} = \dfrac{y_{n-(2k+1)} + p}{y_{n-(2k+1)} + qy_{n-2l}}$, *Proceedings of the* 6th *International Conference on Difference Equations and Applications*, Augsburg, Germany, Chapman and Hall/CRC Press, 2004, 433-452.

[103] H.A. El-Metwally, E.A. Grove, G. Ladas, and H.D. Voulov, On the global attractivity and the periodic character of some difference equations, *J. Difference Equ. Appl.*, **7**(2001), 837-850.

[104] H.A. El-Morshedy, The global attractivity of difference equations of nonincreasing nonlinearities with applications, *Comp. Math. Appl.*, **45**(2003), 749-758.

[105] Problem # E3437 [1991,366]. *Am. Math. Monthly*, November, 1992.

[106] J.E. Franke, J.T. Hoag, and G. Ladas, Global attractivity and convergence to a two-cycle in a difference equation, *J. Difference Equ. Appl.*, **5**(1999), 203-210.

[107] Y. Fan, L. Wang, and W. Li, Global behavior of a higher order nonlinear difference equation, *J. Math. Anal. Appl.*, **299**(2004), 113-126.

[108] C.H. Gibbons, M.R.S. Kulenović, and G. Ladas, On the recursive sequence $x_{n+1} = \dfrac{\alpha + \beta x_{n-1}}{\gamma + x_n}$, *Math. Sci. Res. Hot-Line*, **4**(2000), 1-11.

[109] C.H. Gibbons, S. Kalabusic, and C.B. Overdeep, The dichotomy character of $x_{n+1} = \dfrac{\beta_n x_n + \gamma_n x_{n-1}}{A_n + B_n x_n}$ with period-two coefficients, (to appear).

[110] C.H. Gibbons, M.R.S. Kulenovic, and G. Ladas, On the recursive sequence $x_{n+1} = \frac{\alpha+\beta x_{n-1}}{\gamma+x_n}$, *Math. Sci. Res. Hot-Line*, **4**(2002), 1-11.

[111] C.H. Gibbons, M.R.S. Kulenović, and G. Ladas, On the dynamics of $x_{n+1} = \dfrac{\alpha + \beta x_n + \gamma x_{n-1}}{A + Bx_n}$, *Proceedings of the Fifth International Conference on Difference Equations and Applications*, Temuco, Chile, Taylor & Francis, London 2002, 141-158.

[112] C.H. Gibbons, M.R.S. Kulenović, G. Ladas, and H.D. Voulov, On the trichotomy character of $x_{n+1} = \dfrac{\alpha + \beta x_n + \gamma x_{n-1}}{A + x_n}$, *J. Difference Equ. Appl.*, **8**(2002), 75-92.

[113] C.H. Gibbons and C.B. Overdeep, On the trichotomy character of $x_{n+1} = \dfrac{\alpha_n + \gamma_n x_{n-1}}{A_n + B_n x_n}$ with period-two coefficients, (to appear).

[114] R.L. Graham, Problem #1343, *Math. Mag.*, **63**(1990), 125.

[115] R.L. Graham, D.E. Knuth, and O. Patashnic, *Concrete Mathematics. A foundation for computer science, Addison-Wesley Publishing Company*, Advanced Book Program, Reading MA, 1989.

[116] E.A. Grove, E.J. Janowski, C.M. Kent, and G. Ladas, On the rational recursive sequence $x_{n+1} = \dfrac{\alpha x_n + \beta}{(\gamma x_n + \delta)x_{n-1}}$, *Commun. Appl. Nonlinear Anal.*, **1**(1994), 61-72.

[117] E.A. Grove, C.M. Kent, and G. Ladas, Lyness equations with variable coefficients, *Proceedings of the Second International Conference on Difference Equations and Applications*, August 7-11, 1995, Vesprém, Hungary, Gordon and Breach Science Publishers, 1997, 281-288.

[118] E.A. Grove, C.M. Kent, and G. Ladas, Boundedness and persistence of the nonautonomous Lyness and max equations, *J. Difference Equ. Appl.*, **3**(1998), 241-258.

[119] E.A. Grove, Y. Kostrov, G. Ladas, and M. Predescu, On third-order rational difference equations, Part 4, *J. Difference Equ. Appl.*, **11**(2005), 261-269.

[120] E.A. Grove, Y. Kostrov, G. Ladas, and S. W. Schultz , Riccati difference equations with real period-2 coefficients, *Commun. Appl. Nonlinear Anal.*, **14**(2007), 33-56.

[121] E.A. Grove, M.R.S. Kulenović, and G. Ladas, Progress report on rational difference equations, *J. Difference Equ. Appl.*, **10**(2004), 1313-1327.

[122] E.A. Grove and G. Ladas, Periodicity in nonlinear difference equations, *Revista Cubo*, May(2002), 195-230.

[123] E.A. Grove and G. Ladas, On period-two solutions of $x_{n+1} = \frac{\alpha + \beta x_n + \gamma x_{n-1}}{A + Bx_n + Cx_{n-1}}$, *Proceedings of the* 7th *International Conference on Difference Equations and Applications*, Beijing, China, Fields Institute Communications, 2003.

[124] E.A. Grove and G. Ladas, *Periodicities in Nonlinear Difference Equations*, Chapman & Hall/CRC Press, 2005.

[125] E.A. Grove, G. Ladas, and L.C. McGrath, On the dynamics of $y_{n+1} = \frac{p+y_{n-2}}{qy_{n-1}+y_{n-2}}$, *Proceedings of the Sixth International Conference on Difference Equations, Augsburg, Germany, 2001 (Edited by B. Aulbach, S. Elaydi, and G. Ladas)*, Chapman and Hall/CRC, 2004, 425-431.

[126] E.A. Grove, G. Ladas, L.C. McGrath, and C.T. Teixeira, Existence and behavior of solutions of a rational system, *Commun. Appl. Nonlinear Anal.*, **8**(2001), 1-25.

[127] E.A. Grove, G. Ladas, and M. Predescu, On the periodic character of $x_{n+1} = \frac{px_{n-2l} + x_{n-(2k+1)}}{1 + x_{n-2l}}$, *Math. Sci. Res. J.*, 2002.

[128] E.A. Grove, G. Ladas, M. Predescu, and M. Radin, On the global character of $x_{n+1} = \frac{px_{n-1} + x_{n-2}}{q + x_{n-2}}$, *Math. Sci. Res. Hot-Line*, **5**(2001), 25-39.

[129] E.A. Grove, G. Ladas, M. Predescu, and M. Radin, On the global character of the difference equation $x_{n+1} = \frac{\alpha + \gamma x_{n-(2k+1)} + \delta x_{n-2l}}{A + x_{n-2l}}$, *J. Difference Equ. Appl.*, **9**(2003), 171-200.

[130] J. Guckenheimer and P. Holmes, *Nonlinear Oscillations, Dynamical Systems, and Bifurcations of Vector Fields*, Springer-Verlag, New York, Berlin, Heidelberg, Toyko, 1983.

[131] J. Hale and H. Kocak, *Dynamics and Bifurcations*, Springer-Verlag, New York, Berlin, Heidelberg, Toyko, 1991.

[132] M.L.J. Hautus and T.S. Bolis, Solution to problem E2721, *Am. Math. Monthly*, **86**(1979), 865-866.

[133] J. T. Hoag, Monotonicity of solutions converging to a saddle point equilibrium, *J. Math. Anal. Appl.*, **295**(2004), 10-14.

[134] Y.S. Huang and P.M. Knopf, Boundedness of positive solutions of second order rational difference equations, *J. Difference Equ. Appl.*, **10**(2004), 935-940.

[135] L.X. Hu, W.T. Li, and S. Stevic, Global asymptotic stability of a second-order rational difference equation, *J. Difference Equ. Appl.*, (to appear).

[136] Y.H. Su, W.T. Li, and S. Stevic, Dynamics of a higher order nonlinear rational difference equation, *J. Difference Equ. Appl.*, **11**(2005), 133-150.

[137] V. Hutson and K. Schmidtt, Persistence and the dynamics of biological systems, *Math. Biosc.*, **11**(1992), 1-71.

[138] E.J. Janowski, G. Ladas, and S. Valicenti, Lyness-type equations with period-two coefficients, *Proceedings of the Second International Conference on Difference Equations and Applications*, August 7-11, 1995, Vesprém, Hungary, Gordon and Breach Science Publishers, 1997, 327-334.

[139] A.J.E.M. Janssen and D.L.A. Tjaden, Solution to problem 86-2, *Math. Intelligencer*, **9**(1987), 40-43.

[140] S. Kalabucić and M.R.S. Kulenović, On the recursive sequence $x_{n+1} = \dfrac{\gamma x_{n-1} + \delta x_{n-2}}{Bx_{n-1} + Dx_{n-2}}$, *J. Difference Equ. Appl.*, **9**(2003), 701-720.

[141] S. Kalabucić, M.R.S. Kulenović, and C.B. Overdeep, Dynamics of the recursive sequence $x_{n+1} = \dfrac{\beta x_{n-l} + \delta x_{n-k}}{Bx_{n-l} + Dx_{n-k}}$, *J. Difference Equ. Appl.*, **10**(2004), 915-928.

[142] S. Kalabucić, M.R.S. Kulenović, and C.B. Overdeep, On the dynamics of $x_{n+1} = p_n + \dfrac{x_{n-l}}{x_n}$, *J. Difference Equ. Appl.*, **9**(2003), 1053-1056.

[143] S. Kalikow, P.M. Knopf, Y.S. Huang, and G. Nyerges, Convergence properties in the nonhyperbolic case $x_{n+1} = \dfrac{x_{n-l}}{1 + f(x_n)}$, *J. Math. Anal. Appl.*, **326**(2007), 456467.

[144] G. Karakostas, Asymptotic 2-periodic difference equations with diagonally self-invertible responses, *J. Difference Equ. Appl.*, **6**(2000), 329-335.

[145] G. Karakostas, Convergence of a difference equation via the full limiting sequences method, *Differential Equ. Dynamical Syst.*, **1**(1993), 289-294

[146] G.L. Karakostas and S. Stevic, On the recursive sequence $x_{n+1} = B + \dfrac{x_{n-k}}{a_0 x_n + \ldots + a_{k-1}x_{n-k+1} + \gamma}$, *J. Difference Equ. Appl.*, **10**(2004), 809-815.

[147] W.G. Kelley and A.C. Peterson, *Difference Equations*, Academic Press, New York, 1991.

[148] C.M. Kent, Convergence of solutions in a nonhyperbolic case, *Proceedings of the Third World Congress of Nonlinear Analysts*, July 19-16, 2000, Catania, Sicily, Italy, Elsevier Science Ltd., **47**(2001), 4651-4665.

[149] C.M. Kent, Convergence of solutions in a nonhyperbolic case with positive equilibrium, *Proceedings of the Sixth International Conference on Difference Equations and Applications: New Progress in Difference Equations*, August 2001, Augburg, Germany, Edited by B. Aulbach, S. Elaydi, and G. Ladas, Chapman & Hall/CRC (2004), 485-492.

[150] P.M. Knopf, Boundededness properties of the difference equation $x_{n+1} = \dfrac{\alpha + \beta x_n + \delta x_{n-2}}{x_{n-1}}$, *J. Difference Equ. Appl.*, (2007)

[151] P.M. Knopf and Y.S. Huang, On the period-five trichotomy of the rational equation $x_{n+1} = \dfrac{p + x_{n-2}}{x_n}$, *J. Difference Equ. Appl.*, (2007)

[152] P.M. Knopf and Y.S. Huang, On the boundedness and local stability of $x_{n+1} = \dfrac{\alpha + \beta x_n + x_{n-2}}{Cx_{n-1} + x_{n-2}}$, *J. Difference Equ. Appl.*, (2007)

[153] V.L. Kocic, A note on the nonautonomous Beverton-Holt model, *J. Difference Equ. Appl.*, (to appear).

[154] V.L. Kocic and G. Ladas, Attractivity in a second-order nonlinear difference equation, *J. Math. Anal. Appl.,* **180**(1993), 144-150.

[155] V.L. Kocic and G. Ladas, Permanence and global attractivity in nonlinear difference equations, *Proceedings of the First World Congresss of Nonlinear Analysis* (Tampa, Florida, Auigust 19-26, 1992, Walter de Gruyter, Berlin, New York, 1996.

[156] V.L. Kocic and G. Ladas, Global attractivity in a nonlinear second-order difference equations, *Commun. Pure Apll. Math.,* **XLVIII**, 1115-1122.

[157] V.L. Kocic and G. Ladas, *Global Behavior of Nonlinear Difference Equations of Higher Order with Applications*, Kluwer Academic Publishers, Dordrecht, 1993.

[158] V.L. Kocic, G. Ladas, and I.W. Rodrigues, On rational recursive sequences, *J. Math. Anal. Appl.,* **173**(1993), 127-157.

[159] V.L. Kocic, G. Ladas, E. Thomas, and G. Tzanetopoulos, On the stability of Lyness' equation, *Dynamics of Continuous, Discrete Impulsive Syst.,* **1**(1995), 245-254.

[160] V.L. Kocic, D. Stutson, and G. Arora, Global behavior of solutions of a nonautonomous delay logistic difference equation *J. Difference Equ. Appl.,* **10**(2004), 1267-1279.

[161] R. Kon, A note on attenuant cycles of population models with periodic carrying capacity, *J. Difference Equ. Appl.*, **10**(2004), 791-793.

[162] W.A. Kosmala, M.R.S. Kulenović, G. Ladas, and C.T. Teixeira, On the recursive sequence $y_{n+1} = \frac{p + y_{n-1}}{qy_n + y_{n-1}}$, *J. Math. Anal. Appl.*, **251**(2000), 571-586.

[163] U. Krause, Stability of positive solutions of nonlinear difference equations, *Proceedings of the First International Conference in Difference Equations*, (San Antonio, TX, 1994), Gordon and Breach Publ., 1995, 311-325.

[164] U. Krause, Perron's stability theorem for nonlinear mappings, *J. Math. Economics*, **15**(1986), 275-282.

[165] U. Krause, A theorem of Poincare's type for non-autonomous nonlinear difference equations, *Adv. Difference Equ.*, Gordon and Breach Publ., 1997, 363-369.

[166] U. Krause, Concave Perron-Frobenius theory and applications, *Nonlinear Anal.*, (TMA) **47**(2001), 1457-1466.

[167] U. Krause, A discrete nonlinear and non-autonomous model of consensus formation, *Commun. Difference Equ.*, Gordon and Breach Science Publ., 2000, 227-236.

[168] U. Krause, The asymptotic behavior of monotone difference equations of higher order, *Comp. Math. Appl.*, **42**(2001), 647-654.

[169] U. Krause, A local-global principle for difference equations, *Proceedings of the Sixth International Conference in Difference Equations*, Augsburg, Germany (2001), Chapman & Hall/CRC, 2004, 167-180.

[170] U. Krause, Time variant consensus formation in higher order dimensions, *Proceedings of the Eighth International Conference in Difference Equations*, Chapman & Hall/CRC, 2005, 185-191.

[171] U. Krause and T. Nesemann, Differenzengleichungen und discrete dynamische Systeme *Ene Einfhrung in Theorie and Anvendungen*, B.G. Teubner, Stuttgart/Leipsig, 1999.

[172] U. Krause and M. Pituk, Boundedness and stability for higher order difference equations, *J. Difference Equ. Appl.*, **10**(2004), 343-356.

[173] Y. Kuang and J.M. Cushing, Global stability in a nonlinear difference-delay equation model of flour beetle population growth, *J. Difference Equ. Appl.*, **2**(1996), 31-37.

[174] M.R.S. Kulenović, Invariants and related Liapunov functions for difference equations, *Appl. Math. Lett.*, **13**(2000), 1-8.

[175] M.R.S. Kulenović and G. Ladas, *Dynamics of Second Order Rational Difference Equations; with Open Problems and Conjectures*, Chapman & Hall/CRC Press, 2001.

[176] M.R.S. Kulenović, G. Ladas, I.F. Martins, and I.W. Rodrigues, The dynamics of $x_{n+1} = \dfrac{\alpha + \beta x_{n-1}}{A + Bx_n + Cx_{n-1}}$, facts and conjectures, *Comp. Math. Appl.*, **45**(2003), 1087-1099.

[177] M.R.S. Kulenović, G. Ladas, and C.B. Overdeep, On the dynamics of $x_{n+1} = p_n + \dfrac{x_{n-1}}{x_n}$ with a period-two coefficient, *J. Difference Equ. Appl.*, **10**(2004), 905-914.

[178] M.R.S. Kulenović, G. Ladas, and C.B. Overdeep, Open problems and conjectures on the dynamics of $x_{n+1} = p_n + \dfrac{x_{n-1}}{x_n}$ with a period-two Coefficient, *J. Difference Equ. Appl.*, **9**(2003), 1053-1056.

[179] M.R.S. Kulenović, G. Ladas, and N.R. Prokup, On the recursive sequence $x_{n+1} = \dfrac{\alpha x_n + \beta x_{n-1}}{1 + x_n}$, *J. Difference Equ. Appl.*, **5**(2000), 563-576.

[180] M.R.S. Kulenović, G. Ladas, and N.R. Prokup, A rational difference equation, *Comput. Math. Appl.*, **41**(2001), 671-678.

[181] M.R.S. Kulenović, G. Ladas, and W.S. Sizer, On the recursive sequence $x_{n+1} = \dfrac{\alpha x_n + \beta x_{n-1}}{\gamma x_n + Cx_{n-1}}$, *Math. Sci. Res. Hot-Line,* **2**(1998), no. 5, 1-16.

[182] M.R.S. Kulenović and O. Merino, Stability analysis of Pielou's equation with period-two coefficient, *J. Difference Equ. Appl.*, (to appear).

[183] M.R.S. Kulenović and O. Merino, Convergence to a period-two solution of a class of second order rational difference equations, (to appear).

[184] M.R.S. Kulenović and O. Merino, Global attractivity of the equation $x_{n+1} = \dfrac{px_n + x_{n-1}}{qx_n + x_{n-1}}$ for $q < p$, *J. Difference Equ. Appl.*, **12**(2006), 101-108.

[185] S.A. Kuruklis, The asymptotic stability of $x_{n+1} - ax_n + bx_{n-k} = 0$, *J. Math. Anal. Appl.*, **18**(1994), 8719-8731.

[186] S.A. Kuruklis and G. Ladas, Oscillation and global attractivity in a discrete delay logistic model, *Quart. Appl. Math.*, **L**(1992), 227-233.

[187] G. Ladas, Open problems and conjectures, *Proceedings of the First International Conference on Difference Equations*, (San Antonio, 1994), Gordon and Breach Science Publishers, Basel, 1995, 337-348.

[188] G. Ladas, Invariants for generalized Lyness equations, *J. Difference Equ. Appl.*, **1**(1995), 209-214.

[189] G. Ladas, On the recursive sequence $x_{n+1} = \dfrac{\alpha + \beta x_n + \gamma x_{n-1}}{A + Bx_n + Cx_{n-1}}$, *J. Difference Equ. Appl.*, **1**(1995), 317-321.

[190] G. Ladas, Progress report on $x_{n+1} = \frac{\alpha + \beta x_n + \gamma x_{n-1}}{A + Bx_n + Cx_{n-1}}$, *J. Difference Equ. Appl.,* **5**(1999), 211-215.

[191] G. Ladas, Open problems and conjectures, *J. Difference Equ. Appl.,* **4**(1998), 93-94.

[192] G. Ladas, On third-order rational difference equations, Part 1, *J. Difference Equ. Appl.,* **10**(2004), 869-879.

[193] G. Ladas, G. Tzanetopoulos, and E. Thomas, On the stability of Lyness's Equation, *Dynamics of Continuous, Discrete Impulsive Syst.,* **1**(1995), 245-254.

[194] G. Ladas, G. Tzanetopoulos, and A. Tovbis, On May's host parasitoid model, *J. Difference Equ. Appl.,* **2**(1996), 195-204.

[195] W.T. Li and H.R Sun, Dynamics of a rational difference equation, *Appl. Math. Comp.,* **163**(2005), 577-591.

[196] W.T. Li and H.R Sun, Global attractivity in a rational recursive sequence, *Dynamic Syst. Applications,* **11**(2002), 339-346.

[197] T.Y. Li and J.A. Yorke, Period three implies chaos, *Am. Math. Monthly,* **82**(1975), 985-992.

[198] R.C. Lyness, Note 1581, *Math. Gaz.,* **26**(1942), 62.

[199] R.C. Lyness, Note 1847, *Math. Gaz.,* **29**(1945), 231.

[200] E. Magnucka-Blandzi and J. Popenda, On the asymptotic behavior of a rational system of difference equations, *J. Difference Equ. Appl.,* **5**(1999), 271-286.

[201] E. Magnucka-Blandzi, Trajectories for the case of a rational system of difference equations, *Proceedings of the Fourth International Conference on Difference Equations,* August 27-31, 1998, Poznan, Poland, Gordon and Breach Science Publishers, 2000, 237-246.

[202] M. Martelli, *Introduction to Discrete Dynamical Systems and Chaos,* John Wiley & Sons, New York, 1999.

[203] L.F. Martins, A nonlinear recursion in the positive orthant of $\mathbf{R}^m$ (to appear).

[204] B.D. Mestel, On globally periodic solutions of the difference equation $x_{n+1} = \frac{f(x_n)}{x_{n-1}}$, *J. Difference Equ. Appl.,* **9**(2003), 201-209.

[205] R. Nussbaum, Global stability, two conjectures and maple, *Nonlinear Anal.,* **66**(2007), 1064-1090.

[206] W.T. Patula and H.D. Voulov, On the oscillation and periodic character of a third order rational difference equation, *Proc. Am. Math. Soc.* (to appear).

[207] H.D. Peitgen and D. Sanpe, *The Science of Fractal Images, Springer-Verlag*, New York, 1988.

[208] Ch.G. Philos, I.K. Purnaras, and Y.G. Sficas, Global attractivity in a nonlinear difference equation, *Appl. Math. Comput.*, **62**(1994), 249-258.

[209] E.C. Pielou, *An Introduction to Mathematical Ecology*, Wiley-Interscience, New York, 1969.

[210] E.C. Pielou, *Population and Community Ecology*, Gordon & Breach, New York, 1974.

[211] C. Robinson, *Dynamical Systems, Stability, Symbolic Dynamics, and Chaos*, CRC Press, Boca Raton, FL, 1995.

[212] D. Ruelle and F. Takens, On the nature of turbulence, *Commun. Math. Phys.*, **20**(1971), 167-192.

[213] H. Sedaghat, *Nonlinear Difference Equations, Theory and Applications to Social Science Models*, Kluwer Academic Publishers, Dordrecht, 2003.

[214] A.G. Sivak, On the periodicity of recursive sequences, *Proceedings of the Second International Conference on Difference Equations and Applications*, Gordon and Breach Science Publishers, 1997, 559-566.

[215] W.S. Sizer, Periodicity in the Lyness equation, *Math. Sci. Res. J.*, **7**(2003), 366-372.

[216] W.S. Sizer, Some periodic solutions of the Lyness equation, *Proceedings of the Fifth International Conference on Difference Equations and Applications*, January 3-7, 2000, Temuca, Chile, Gordon and Breach Science Publishers.

[217] H.L. Smith, Monotone dynamical systems. An introduction to the theory of competitive and cooperative systems, *Math. Surveys Monogr.*, Vol. **41** , Amer. Math. Soc., Providence, RI, 1995.

[218] H.L. Smith, Planar competitive and cooperative difference equations, *J. Difference Equ. Appl.*, **3**(1998), 335-357.

[219] H.L. Smith and H. Thieme, Monotone semiflows in scalar non-quasi-monotone functional differential equations, *J. Math. Anal. Appl.*, **150**(1990), 289-306.

[220] Y.H. Su, W.T. Li, and S. Stevic, Dynamics of a higher order rational difference equation, *J. Difference Equ. Appl.*, **11**(2005), 133-150.

[221] S. Stevic, On the recursive sequence$x_{n+1} = -\frac{1}{x_n} + \frac{A}{x_{n-1}}$, *Int. J. Math. Sci.*, **27**(2001), 1-6.

[222] S. Stevic, Periodic character of a class of difference equations, *J. Difference Equ. Appl.*, **10**(2004), 615-619.

[223] S. Stevic, A note on periodic character of a difference equation, *J. Difference Equ. Appl.*, **10**(2004), 929-932.

[224] S. Stevic, On the recursive sequence $x_{n+1} = \frac{\alpha + \sum_{i=1}^{k} \alpha_i x_{n-p_i}}{1 + \sum_{j=1}^{m} \beta_j x_{n-q_j}}$, *J. Difference Equ. Appl.*, **13**(2007), 41-46.

[225] P. Tacic, Convergence to equilibria on invariant d-hypersurfaces for strongly increasing discrete-type semigroups, *J. Math. Anal. Appl.* **141** (1990), 223-244.

[226] S. Taixiang, On nonoscillatory solution of the recursive sequence $x_{n+1} = p + \frac{x_{n-k}}{x_n}$, *J. Difference Equ. Appl.*, **11**(2005), 483-485.

[227] S. Taixiang and X. Hongjian, On the solutions of a class of difference equations, *J. Math. Anal. Appl.*, **311**(2005), 766-770.

[228] S. Taixiang and X. Hongjian, Global attractivity for a family of difference equations, *Appl. Math. Lett.*, (2006).

[229] S. Taixiang and X. Hongjian, On the system of rational difference equations $x_{n+1} = f(x_n, y_{n-k}), y_{n+1} = f(y_n, x_{n-k})$, *Adv. Difference Equ.*, (2006), 1-7.

[230] S. Taixiang and X. Hongjian, The periodic character of positive solutions of the difference equation $x_{n+1} = f(x_n, x_{n-k})$, *Comp. Math. Appl.*, **51**(2006), 1431-1436.

[231] S. Taixiang and X. Hongjian, On the global behavior of the nonlinear difference equation $x_{n+1} = f(p_n, x_{n-m}, x_{n-t(k+1)+1})$, *Discrete Dynamics in Nature and Science*, (2006).

[232] S. Taixiang and X. Hongjian, On the boundedness of the solutions of the difference equation $x_{n+1} = \frac{x_{n-1}}{p + x_n}$, *Discrete Dynamics Nat. Sci.*, (2006).

[233] S. Taixiang and X. Hongjian, On convergence of the solutions of the difference equation $x_{n+1} = 1 + \frac{x_{n-1}}{x_n}$, *J. Math. Anal. Appl.*, **325**(2007), 1491-1494.

[234] X.X. Yan and W.T. Li, Global attractivity for a class of nonlinear difference equations, *Soochow J. Math.*, **29**(2003), 327-338.

[235] X.X. Yan and W.T. Li, Global attractivity in a rational recursive sequence, *Appl. Math. Comp.,* **145**(2003), 1-12.

[236] Q. Wang, F. Zeng, G. Zang, and X. Liu, Dynamics of the difference equation $x_{n+1} = \frac{a + B_1 x_{n-1} + B_3 x_{n-3} + ... + B_{2k+1} x_{n-2k-1}}{A + B_0 x_n + B_2 x_{n-2} + ... + B_{2k} x_{n-2k}}$, *J. Difference Equ. Appl.,* **10**(2006), 399-417.

[237] E.C. Zeeman, Geometric unfolding of a difference equation, *http://www.math.utsa.edu/ecz/gu.html.*

[238] E. Zeidler, *Nonlinear Functional Analysis with its Applications, I. Fixed Point Theorems*, Springer, New York, 1986.

Index

Milton Keynes UK
Ingram Content Group UK Ltd.
UKHW021929071024
449327UK00022B/1741

9 780367 388294